Preface

We hope this book will be of value to both environmental managers and geomorphologists. Amongst the former, we have in mind especially engineers, including those concerned with planning, design, and construction; planners, be they urban, rural, regional, national, or environmental; and many landowners and developers. Our experience in working with such environmental managers suggests that they have recently become much more aware of the nature, relevance, and cost-effectiveness of geomorphological advice in the context of their own distinctive responsibilities. Our main concern is that the awareness should be further enhanced and that the results of applied geomorphology should be readily available to them.

We also write for geomorphology students who want to be guided towards those aspects of the subject that have found a practical application, and who want to share with us a geomorphologist's concern for the wise use of land. We do not argue either that all geomorphology can be applied or that the geomorphologist can usefully contribute to all problems of environmental management: that would be absurd. But we do believe that geomorphology can contribute towards the solution of many environmental problems and that the proposed solutions for such problems are often inadequate if they lack a geomorphological ingredient.

Much has changed in both geomorphology and environmental management since the first edition of this book was published in 1974. The environmental consciousness of that time has, if anything, grown in response to yet more environmental catastrophes and degradation, such as the ravages of the Sahelian drought, the destruction by post-volcanic mudflows at Armero in Colombia, and the devastation of flooding in Bangladesh. Less spectacularly, there is an increased realization that rates of soil erosion in places as far apart as the UK, Kenya, Nepal, the Caribbean, and China have risen as a result of mismanagement of agricultural land, often as a consequence of national policies as well as of local attitudes, pressures, and practices. In several countries, a statutory requirement to assess the 'environmental impact' of proposed developments has also enhanced environmental consciousness generally and the need for geomorphological assessments in particular.

The dangers and costs of inadequately assessing the geomorphological impact of development on terrain have become increasingly apparent. In the context of large engineering projects, the impact of the Aswan High Dam on coastal erosion of the Nile Delta is but one example that illustrates why it is often so important to assess the whole of the geomorphological system affected by engineering works prior to development. At the smaller scale of the development site it is also now widely recognized that many problems could have been avoided by adequate appraisal of terrain and materials prior to construction. For example, some of the prestigious buildings constructed in the Gulf States of the Middle East during the oil-boom years of the 1970s are now suffering salt attack arising from the ingress of saline groundwater into foundations, a problem that a prior appraisal of the sites could have prevented.

Since 1974, more geomorphologists have been employed in universities, government agencies, private companies, and as private individuals to undertake applied work. Their contributions, many of them unpublished or in relatively inaccessible conference proceedings, have increased greatly in number, making our task of synthesis more difficult yet even more necessary, and demanding changes to the content and balance of this new edition. At the same time, we and our collaborators have gained much more experience of working with clients who have responsibilities for environmental problems (e.g. Doornkamp *et al.*, 1980; Cooke *et al.*, 1982; Cooke, 1984). That experience has confirmed our belief in the importance of applying geomorphological research to the solution of environmental problems, and substantially modified our views on the ways it can best be achieved. In this revision, we therefore try to integrate both the wealth of new material and our own greatly extended practical experience.

Several texts on aspects of applied geomorphology have appeared since 1974, including *Applied Geomorphology* (1977*a*) edited by Hails, Verstappen's *Applied Geomorphology* (1983), and *Developments and Applications of Geomorphology* (1984) edited by Costa and Fleisher. In addition, several texts and conference proceedings have also appeared that deal with single themes (e.g. Morgan's (1986) *Soil Erosion and Conservation* and Brunsden and Prior's (1984) *Slope Instability*, and with field and laboratory techniques (e.g. Goudie, 1981). They allow us to refer to much material in a way that was not possible in the early 1970s, and also enable us to omit some of the material in the first edition which is now effectively covered elsewhere. However, our text remains different from others especially in so far as it tries to provide an integrated, systematic overview and a detailed commentary. Furthermore, it is designed to complement basic textbooks in geomorphology, because most of them deal very inadequately with applied research.

Our first edition was well received. It clearly served the needs of many courses in geology, geography, and environmental science and also provided a reference source for non-geomorphologists and teachers. We have retained the overall concept, but the weight of new material obliges us to make substantial changes: very little of the original text remains. Part of the problem in selecting material lies in one of geomorphology's great strengths—most of it is potentially applicable because it relates to the surface of the earth, the human domain. Our emphasis in this edition is more on application than on applicability. Some chapters have been deleted or radically reduced, such as those on land subsidence and landscape evaluation; others, like that on weathering, have been fundamentally altered; and completely new chapters or sections have been added to cover such important matters as approaches to the subject, applications of remote sensing techniques, monitoring change, sediment yield, glacial hazards, limestone terrain, neotectonics and related phenomena, and aeolian problems in drylands. In addition, new techniques have been developed, and there are signs that standards of practice have begun to emerge. As far as possible, we have incorporated these changes. Even so, our coverage is not comprehensive, and we have only been able to discuss briefly some important themes.

Within most chapters we have retained the original structure—discussing for each theme the applied context, the nature of the geomorphological system and ways of studying it, and selected major issues relating the system to environmental management. Within this structure we have downgraded the former emphasis on land systems, increased the number of case-studies, and incorporated photographs. We have also extended the geographical coverage of the case-studies, especially in North America and Europe. In doing so, we have referred mainly to accessible sources and, as far as possible, our use of European sources is referenced by English-language translations. We have used metric scales throughout, except where rescaling is impracticable.

Several colleagues have generously reviewed parts of the text: Professor D. Walling (Chapter 7), Professor K. J. Gregory and Dr G. E. Hollis (Chapter 6) and Professor C. Vita-Finzi (Chapter 13); Professor C. A. M. King (Chapter 10) kindly commented on our metamorphosis of her contribution to the first edition. Their comments have helped to eliminate errors of fact, emphasis, and omission. We are alone responsible for any remaining inadequacies. Denys Brunsden and David Jones have shared many of our applied projects in recent years, and we happily acknowledge our great debt to them. Geomorphological Services Ltd. has provided us with many opportunities for applied research. Our thanks are also due to Alick Newman and Lauren McClue of UCL Geography Department's drawing-office, who drew all of the maps and diagrams; and to Claudette John and Alison Wright (UCL), and Karen Korzeniewski (Nottingham) for word-processing most of the text.

GEOMORPHOLOGY
IN
ENVIRONMENTAL
MANAGEMENT

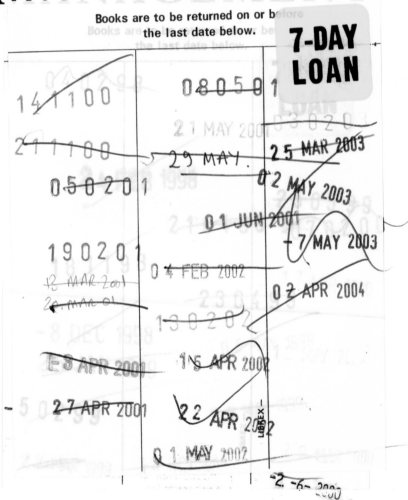
CLARENDON PRESS · OXFORD

1990

Oxford University Press, Great Clarendon Street, Oxford OX2 6DP

Oxford New York
Athens Auckland Bangkok Bogota Bombay
Buenos Aires Calcutta Cape Town Dar es Salaam
Delhi Florence Hong Kong Istanbul Karachi
Kuala Lumpur Madras Madrid Melbourne
Mexico City Nairobi Paris Singapore
Taipei Tokyo Toronto Warsaw
and associated companies in
Berlin Ibadan

Oxford is a trade mark of Oxford University Press

Published in the United States by
Oxford University Press Inc., New York

© R. U. Cooke and J. C. Doornkamp, 1990

Paperback reprinted 1993, 1994, 1997

British Library Cataloguing in Publication Data
Cooke, R. U. (Ronald Urwick)
Geomorphology in environmental management.—
2nd ed.
1. Applied geomorphology
I. Title   II. Doornkamp, John C.
(John Charles)
ISBN 0-19-874510-2
ISBN 0-19-874151-0 (Pbk)

Library of Congress Cataloging in Publication Data
Cooke, Ronald U.
Geomorphology in environmental management: a new introduction/
R. U. Cooke and J. C. Doornkamp.—2nd ed.
Bibliography: p. Includes index.
1. Geomorphology. 2. Environmental protection. I. Doornkamp,
John Charles. II. Title.
GB406.C64 1990 551.4—dc19 89-3027
ISBN 0-19-874510-2
ISBN 0-19-874151-0 (Pbk)

Printed in Great Britain
by The Alden Press
Oxford

Together with the publishers, we are grateful for permission to reproduce tables and figures from many copyright works; details are given in the underlines and references. We have been unable to contact a very small number of authors for permission to reproduce material from their work; we apologize to them, and hope that our acknow-ledgements and use of their material meets with their approval.

R.U.C.
J.C.D.

University College London
University of Nottingham
1 *January* 1988

Contents

x *Contents*

List of Figures

List of Tables

List of Plates

1 Geomorphology and Environmental Management

1.1 Perspectives

(a) *Some definitions*

Environmental management comprises a broad range of activities concerned with the human use of land, air, plants, and water. It has come to involve many specialist groups responsible for environmental planning and development within public authorities and private enterprises, including planners, engineers, landscape architects, politicians, lawyers, and professional administrators. It also includes many individuals or organizations with particular interests in some aspects of the natural environment, such as farmers, recreationalists, conservation groups, and property-owners. So pervasive and important have the problems of environmental management become, especially in the increasingly crucial context of public policy, that there are some who feel environmental management should now be recognized as a profession in its own right. Within the broad range of activities there is often a compelling need for information about the physical environment, so that those responsible for management commonly seek advice from several environmental sciences including meteorology, hydrology, ecology, pedology, geology, and geomorphology.

Geomorphology is the study of landforms, and in particular their nature, origin, processes of development, and material composition. This book focuses on those aspects of geomorphology that relate to human use of the natural environment and on the methods by which they can be used successfully in environmental management.

Most geomorphological research is *potentially applicable* to some problem of environmental management, although its applicability may not be apparent when it is first carried out. For example, numerous studies of deglacial history, which are academic exercises, contain within them information of considerable potential value to the aggregate industry; similarly the micro-erosion meter was designed to monitor rates of limestone weathering for academic research purposes, yet it has proved to be of value in the analysis of the rates of building-stone decay. But there is a fundamental difference between potential applicability and *application*: the latter involves either taking geomorphological data and injecting it successfully into the management process or, and increasingly, undertaking geomorphological studies within the management context (e.g. Brunsden *et al.*, 1978; Thornes, 1979; Jones, 1980).

Geomorphological contributions to environmental management are not exclusively carried out by professional geomorphologists. Many who use land in their work may become involved in assessing geomorphological problems, including especially design, construction, and agricultural engineers, and foresters, landscape architects, and planners.

(b) *Applying geomorphology*

The use of geomorphology in environmental management can be introduced most effectively by a few brief examples.

London was rapidly running short of suitable aggregates for the building industry during the great phase of development and reconstruction following the Second World War. The prime aggregate resources of the river Thames's terraces were depleted, or held hostage by construction on top of them. Where were new supplies to come from? S. W. Wooldridge and other geomorphologists had shown that in the Vale of St Albans there was a major gravel-train associated with an early Quaternary course of the Thames. This research, carried out largely because the history of the

Thames and the evolution of the landscapè of south-east England is one of great complexity and intrinsic interest, provided the key to future resources of aggregates around north London (e.g. Ministry of Town and Country Planning, 1948; below, Chapter 11). The same gravel-train is still being worked, for example to provide aggregates for the M25 motorway.

For many years, geomorphologists have studied the nature and rates of weathering in limestone terrain. They have examined the evolution of such small weathering features as clints, grykes, weathering pits, small rills, and blisters, and they have developed several methods of monitoring change. In a quite different context, many important buildings are built of limestone and, especially in polluted urban areas, they are suffering from the effects of weathering, thus creating a management problem of considerable public concern. Can the methods of studying limestone terrain successfully be applied to the study of building-stone decay and conservation? As Chapter 12 shows, the answer is clearly 'yes'.

A final example illustrates a more widespread justification for applying geomorphology to environmental management. Plate 1.1 shows a relatively unpretentious desert plain in South America: a fairly flat stony surface underlain by a primitive soil and a thick accumulation of calcium carbonate, fringing escarpments, and numerous small isolated dunes. This view, like so many, can inspire in the geomorphologist a plethora of research questions. Take the small *barchan* dunes,

for example: how are they distributed, what are they composed of, and what are their geometrical properties? Are they moving, and if so, in which directions, when, and how quickly? Why does their colour differ from that of the prevailing surface material? And how do all the dune characteristics relate to wind patterns and sand sources at present and in the past? All such questions are scientifically respectable and they can be answered by well-established methods of geomorphological analysis; the answers can illuminate an understanding of this piece of landscape and desert landscapes in general, and they may contribute to the growing body of geomorphological theory.

But this same view can be seen in a different light. To a European consultancy, part of this plain appeared to be suitable for the development of irrigation agriculture. As a result of the consultants' report, open canals were built to supply the area with water and the encumbrance of a stony surface was removed to expose a cultivable soil. A good crop of vegetables was grown in the first year of production. But problems soon emerged. The sand-dunes happen to be moving quickly across the stony surface—when they encounter a canal, it traps them and the supply of water is seriously disrupted. Removal of the stony surface meant that the underlying silt was prey to quick and easy erosion by the wind and, in places, the calcium carbonate on the soil profile (wrongly identified by the consultants as a marine deposit) became exposed. Productivity, and enthusiasm for

PLATE 1.1 *Barchan* dunes on a stony desert surface: ground for improvement?

the development, rapidly declined. The developers began to ask questions, questions identical to those posed by the geomorphologist, but they asked them too late. And, in any case, no geomorphologist was on hand to provide answers.

These very different anecdotes provide a single message: geomorphological studies may be of academic interest in their own right; but they may also be useful in environmental management.

1.2 The Rise of Geomorphology in Environmental Management

It is as well to remember that not only have there been applied geomorphological studies in the past, but also some very substantial and theoretically fundamental studies have been stimulated by the demands for solutions to environmental problems. Leonardo da Vinci's observations on fluvial processes were certainly provoked, in part, by his appreciation of the ravages of floods and alluviation (Alexander, 1982). G. K. Gilbert's (1914, 1917; Yochelson, 1980) seminal studies of debris transportation in running water were undertaken in the context of his efforts to understand the effects of hydraulic mining on the Sacramento River. More recently, the fundamental work of Robert E. Horton (1933, 1945) on the drainage basin and soil erosion by water, and of W. S. Chepil and his collaborators (e.g. Chepil and Woodruff, 1963) on soil erosion by wind was certainly sustained by the perceived seriousness of the soil-erosion problem in North America from the 1930s onwards. And there have been several major examinations since the early nineteenth century of the part played by human activity in influencing the nature of geomorphological change, including G. P. Marsh's (1864) *Man and Nature; or, Physical Geography as modified by Human Action*, R. L. Sherlock's (1922) *Man as a Geological Agent*, and K. Bryan's (1925) extended analyses of erosion and sedimentation in the Papago country of Arizona.

But such contributions form only a tiny proportion of geomorphological research. Even during the philosophical turmoil within geomorphology in the early 1960s, applied geomorphological work did not figure prominently. Although some of the inspiration for the changes at that time undoubtedly came from the work of Robert Horton, Luna Leopold, Walther Langbein, and others, all of whom had conspicuously succeeded in applying their geomorphological research to environmental problems, of the widely used texts available only Thornbury's (1954) now fading volume on the *Principles of Geomorphology* included an 'applied geomorphology' chapter. Since that time, the use of geomorphology in environmental management has begun to flourish for several reasons.

First, the study of process, which had been relatively unimportant during the early decades of the century and had been marked by only a few fundamental studies (e.g. Johnson (1919) on shore processes and Bagnold (1941) on aeolian processes), became a major focus of interest. The revitalized enthusiasm for the study of contemporary processes and changes at the earth's surface placed part of geomorphology's research frontier firmly in those areas of the subject of greatest potential value to resource and hazard appraisal, planning, and civil engineering. In the United Kingdom, this change was provoked, in part, by developments within geography in the 1960s that stimulated the wish to be more scientific and quantitative, promoted the interest in dynamics, systems, and methods of evaluating complex spatial and temporal interrelationships, and encouraged the objectives of defining laws and gaining respectability in other cognate disciplines. But it was a change forced mainly by academic dissatisfaction with the results of earlier geomorphological work, in which long-term landform evolution was the centure of attention, rather than by any stimulus external to the academic world.

Secondly, the trend towards process studies within geomorphology was accompanied by the development of more precise techniques for mapping the landscape and for monitoring change; and by the adoption of methods, such as those arising from systems analysis, capable of handling very complex, dynamic situations. As a result, geomorphological advice could be offered in a form that was more rigorous, and more intelligible and acceptable to those outside the subject who were seeking it.

Thirdly, the changes within the subject coincided with a resurgence of international interest in environmental problems and terrain resource evaluation. As the colonial era closed, newly independent countries also sought details of their natural environment as a basis for planning and development, in much the same way as countries such as Australia and the USSR had begun to appraise land in previous decades. Land was the major focus of interest in these appraisals and geomorphology was perceived to have an important contribution to make. Growing international concern with the environment was also generated by pressure groups, perhaps by a growing political awareness of the finite limits to some resources, and especially by a growing number of crises and catastrophes. In the desert realm, for example, desertification became an issue of alarming proportions and of vital concern to many poor countries; and it is one that has generated discussions and research in which geomorphologists have contributed substantially (e.g. Warren and Maizels, 1977). In addition, some major geomorphological hazards became apparent in the rapid, oil-boom development of desert coasts, especially in the Middle East. Here there emerged, for example, problems of inadequate aggregates, of alluvial-fan flooding, of the vulnerability of stone pavements to traffic, of the weathering effectiveness of salts, and of sand and dust movement (e.g. Cooke *et al.*, 1982). Elsewhere, specific disasters enhanced public awareness of geomorphological problems: the Aberfan disaster in 1966, when the failure of a coal waste-tip killed 115 people; the permafrost problems associated with the trans-Alaska oil pipelines since the late 1960s; the Yungay debris avalanche in Peru that killed 25 000 in 1970; the eruption of Mount St Helens, USA, in 1980, that caused *c.*\$3 billion of damage; and the mudflows from the eruption of the Nevado del Ruiz in Colombia which eliminated the town of Armero and killed *c.*21 000 in 1985.

The response to these concerns was profound, and it was often accompanied by legislation. For instance, after the passing of the National Environmental Policy Act by the US Congress in 1969, the need for 'environmental impact assessments' prior to development became widely accepted in the United States and other countries (e.g. Leopold *et al.*, 1971); and such assessments commonly required a geomorphological contribution. In the UK, the Department of the Environment frequently seeks geomorphological advice, for example in the context of highway planning, aggregate appraisal, and slope stability problems. There were also significant legislative changes taking place within local authorities. For example, efforts to control slope failure in Los Angeles are extensively recorded in planning and building

ordinances (Cooke, 1984). And litigation has frequently served to sharpen the edges of public policy control of environmental management (e.g. Coates, 1984).

Finally, and perhaps most significantly, the number of geomorphologists has grown substantially since the early 1960s, and an increasing number has begun more deliberately to try and serve the needs of environmental managers. Such efforts enhanced the contacts between geomorphologists and environmental managers and scientists in cognate fields in the context of common problem-solving, and thus provided a very valuable stimulus to, and gave a higher public profile to, geomorphological research. As a result, a wide range of planners and engineers—most of whom are hard taskmasters—have been successfully persuaded that geomorphologists can contribute effectively to the solutions of their problems, and often cheaply and at relatively short notice. Although applied geomorphological work has become technically much more sophisticated and more expensive in recent years, it is still often relatively cheap by comparison with, for example, some types of geotechnical surveys; also, low-technology research based on low-cost field-mapping and the analysis of available data is seen as being efficient, quick, and adequate in many developing countries with limited funds for environmental appraisals.

Not all the evidence of recent success is obvious. Certainly there are now several books that draw heavily on the experience (e.g. Coates, 1976; Doornkamp *et al.*, 1980; Cooke *et al.*, 1982; Craig and Craft, 1982; Verstappen, 1983; Costa and Fleisher, 1984; Doornkamp, 1985; Fookes and Vaughan, 1986). Certainly, too, one company and several consultants in the UK now provide a professional service; and a small but rising number of geomorphologists are finding their way into engineering and planning companies, consultancies, and public agencies. But many of the recent results of this success are contained within unpublished reports and in consultants' files. And the main, yet largely invisible success is in the number of organizations concerned with environmental management that have actually sought and used geomorphological advice in recent years. Of these groups, those concerned with planning, regional development, and engineering are the most important.

Amongst institutions in the UK that currently

employ geomorphologists are: the Institute of Hydrology, several regional water authorities, the Nature Conservancy Council, the Institute of Oceanographic Sciences, the Transport and Road Research Laboratory, the Land Resources Division of the Ministry of Overseas Development, and the Soil Survey of England and Wales. Companies that have used the advice of geomorphologists include: Sir William Halcrow and Partners; G. Wimpey Ltd.; Rendell Palmer and Tritton (consulting engineers); Engineering Geology Ltd.; Ove Arup and Partners; Sceptre Resources; Dames and Moore; Fugro Inc.; British Steel; the National Coal Board; Mobil North Sea; Howard Humphreys; Binnie and Partners; Scott, Wilson, Kirkpatrick; and Dar Al Handasah (Shair) and Partners. In addition, geomorphologists have worked for several government departments and international agencies including the D.o.E. (UK); the British Insurance Association; FAO; UNESCO; the World Bank; and the governments of Bahrain, Egypt, Saudi Arabia, the UAE, Oman, Sri Lanka, Hong Kong, Honduras, Bangladesh, and Colombia. The British experience is matched elsewhere. In North America, a wide range of public agencies employ geographers and geologists who have training in geomorphology, and there is a growing body of engineering geologists who consult on geomorphological problems. In both western and eastern Europe, and in India, China, and Australia, the same developments can be discerned. At the international level too, especially in UNESCO the potential value of geomorphology has begun to be recognized.

1.3 Geomorphology for Whom?

(a) *General perspective*

Environmental management comprises so many different problems, responsibilities, and agencies that broad generalizations about it are difficult. In general, however, environmental management normally involves four major phases: problem *identification*, policy *formulation* and project *planning*; policy/project *implementation*; and policy/project *evaluation*. Broadly speaking, there are two principal types of policy—those arising as a response to a crisis, and those relating to development initiatives.

The policies, projects, or problems may apply at very different scales of space and time, and may involve many different agencies and interests. This spectrum is exemplified in Fig. 1.1, which shows for North America a sample range of land uses and the relative contributions of a range of professional groups to study it. In this diagram, geomorphology is included within the earth sciences.

Management responsibilities depend very much on the time and space scales involved. Commonly the responsibilities are disseminated through a complex hierarchy of agencies. For example, Fig. 1.2 summarizes the hierarchy of agencies in Los Angeles County, California, and the agencies' responsibilities for the three major aspects of managing slope and flooding problems; survey, assessment, planning (which related to the phases of identification, planning, and evaluation); repair, maintenance, and construction (mainly arising in the implementation phase); and emergency action (which involves aspects of all four phases outlined above). Most of these agencies have a need for geomorphological advice: some of them employ engineering geomorphologists; others obtain their information through contracted research. Within Fig. 1.2 each relationship described by a blob-in-a-box provides opportunities for research into the relationship itself and the geomorphological contribution to it. For example, the blob linking *city* 'Building and Safety department' with 'survey, assessment, planning, and insurance', invites an investigation of the ways in which cities have legislated for slope failure problems through building codes and of the success they have achieved. These codes are complex documents that vary between cities and are changed from time to time, usually following storms (Cooke, 1984).

For any given problem the management process may have a temporal cyclicity, especially if the problem and its contingent policies and projects are related to ephemeral hazards, such as flooding, slope failure, or earthquakes. Fig. 1.3 shows a general model of management response to a hazard event, and Fig. 1.4 shows a more detailed and specific illustration for storms in Los Angeles that are usually accompanied by geomorphological problems.

Figs 1.1–1.4 raise several issues that require brief discussion. Firstly, the contribution of geomorphology varies according to the nature and formal responsibilities of the agencies, groups, or individuals involved. For example, the geomorphology required to predict the pattern of slope failure

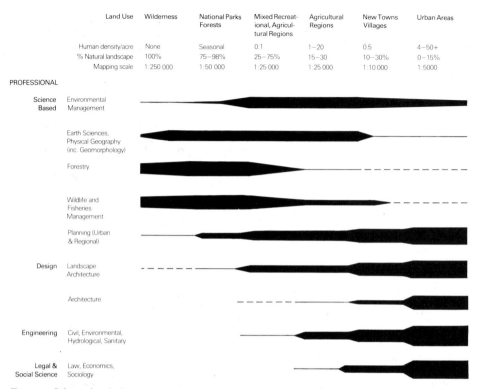

Land Use	Wilderness	National Parks Forests	Mixed Recreat-ional, Agricul-tural Regions	Agricultural Regions	New Towns Villages	Urban Areas
Human density/acre	None	Seasonal	0.1	1–20	0.5	4–50+
% Natural landscape	100%	75–98%	25–75%	15–30	10–30%	0–15%
Mapping scale	1:250 000	1:50 000	1:25 000	1:25 000	1:10 000	1:5000

PROFESSIONAL

Science Based	Environmental Management
	Earth Sciences, Physical Geography (inc. Geomorphology)
	Forestry
	Wildlife and Fisheries Management
	Planning (Urban & Regional)
Design	Landscape Architecture
	Architecture
Engineering	Civil, Environmental, Hydrological, Sanitary
Legal & Social Science	Law, Economics, Sociology

FIG. 1.1 Schematic relations between professional environmental interests, scale, and land use in North America (R. S. Dorney, pers. comm.)

for a national survey agency is likely to be quite different from that required by a city council facing an inquest into slope damage after a storm. Secondly, a corollary of this is that it is essential for the geomorphologist to be sensitive to clients' needs and the contexts in which he is working—to the views of politicians, to the nature of vested interests, financial limitations, time constraints, and technical resources, and to jurisdictional boundaries (e.g. Coates, 1984).

Thirdly, most geomorphological work done for environmental management serves the needs of clients who are almost certainly not trained geomorphologists: as indicated above, they may be from backgrounds as diverse as politics, architecture, the law, and farming. The research problems are posed by the clients, and the answers must therefore be provided in a form that the clients can understand. As a result, most applied geomorphological research—while it may be innovative in terms of methods and may ultimately feed back into geomorphological theory—must be cast in a form that is imposed from outside. Thus, while the work is undertaken using geomorphological

methods, it normally has to be *translated* for clients and critically *tested* by them. This process of translation and testing is fundamental, and one of the great attractions of applied work. It is well illustrated in professionally produced guides for decision-makers, such as those prepared by the US Geological Survey (Robinson and Spieker, 1978; Brown and Kockelman, 1983; see also Fig. 1.7 and Chapter 2).

Fourthly, the question arises, 'how do geomorphologists become involved in the management process?' The fundamental need here is for effective communication, in particular by bringing the results of research and the potential of research methodology to the attention of environmental managers through publications and personal discussions with the managers involved. There is a strong need to proselytize, and to point out the potential of geomorphology to minimize environmental damage, to maintain public safety, health, and welfare, and to identify, regulate, and monitor hazardous locations (e.g. Coates, 1984).

Fifthly, involvement must be timely to be effective. Fig. 1.5 suggests the degree of influence

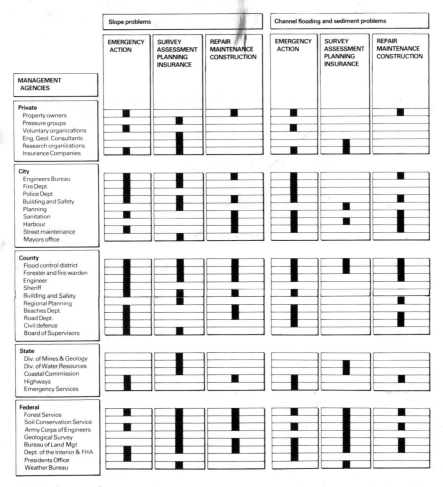

FIG. 1.2 Some relations between the hierarchy of management agencies and slope, channel-flooding, and sediment problems in Los Angeles County, California, USA (after Cooke, 1984)

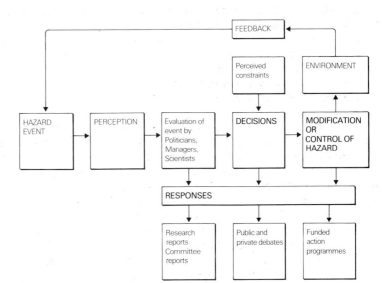

FIG. 1.3 General model of response to a hazardous event (after Kasperson, 1969)

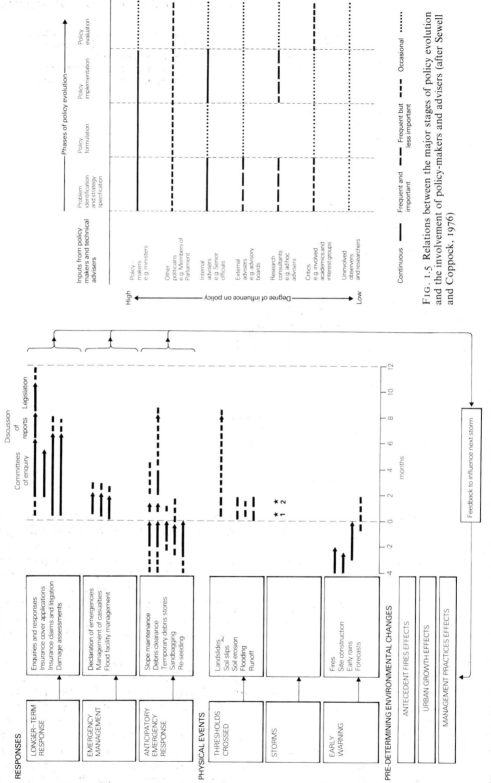

Fig. 1.5 Relations between the major stages of policy evolution and the involvement of policy-makers and advisers (after Sewell and Coppock, 1976)

Fig. 1.4 Model of storm-hazard responses in Los Angeles County, California, USA (after Cooke, 1984)

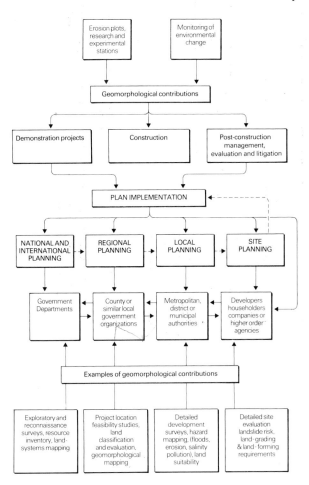

FIG. 1.6 Relations between planning scales, the hierarchy of management agencies, and potential geomorphological contributions to decision-making (after Cooke, 1982)

different groups may generally exert on the main phases of environmental management. In this diagram, geomorphologists may be involved as internal or external advisers, as research consultants, or as critics. Their involvement may be substantial in each of the phases, but it is particularly important in the planning phase and in the implementation phase (where relations with engineering are fundamental).

(b) *Geomorphology and planning*

In the context of the planning phase of environmental management, there may be a need for advice at international, national, regional, local, and site scales, and before, during, and after planning procedures (Fig. 1.6). Each scale has its own range of management agencies, and each tends to pose distinctive questions arising from its particular perspectives and responsibilities. At international and national scales, governments and supranational bodies commonly seek exploratory or reconnaissance surveys, resource inventories, and small-scale systematic mapping. FAO's (1978) use of LANDSAT imagery to assess the global pattern of soil degradation, a study that involved geomorphologists, is a good example. Fig. 1.7 shows an example of a national soil-erosion hazard map based on the translation of geomorphological data into hazard zones, a map of relevance to agricultural land-use planning in Zimbabwe (Stocking and Elwell, 1973). At a regional scale, data are often required to help select sites for particular purposes within a range of possibilities, and to evaluate relative terrain qualities. The appraisal of route corridors prior to road construction provides one illustration amongst many at this scale where geomorphologists have contributed successfully in recent years

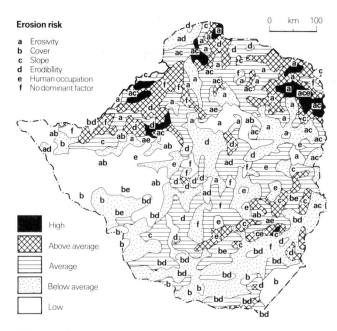

Erosion risk

a Erosivity
b Cover
c Slope
d Erodibility
e Human occupation
f No dominant factor

FIG. 1.7 Classification and pattern of
erosion risk in Zimbabwe (after
Stocking and Elwell, 1973)

High

Above average

Average

Below average

Low

Categories of erosion

	Erosivity J/mm/m²/h	Cover (mm of rainfall) Basal cover est. (%)	Slope (degrees)
■	above 11 000	below 500 0–2	above 8
▨	9000–11 000	400–600 1–4	6–8
▤	7000–9000	600–800 3–6	4–6
▦	5000–7000	800–1000 5–8	2–4
□	below 5000	above 1000 7~10	0–2

(e.g. Brunsden *et al.*, 1975; Jones *et al.*, 1983). More locally, the appraisal of hazards and resources is commonly required in rural areas (e.g. land evaluation for agriculture, irrigation, or forestry) and especially in the contexts of urban renewal and growth, and new-town planning.

Finally, site planning requires evaluation of such specific problems as slope failure and controls on land engineering. The work of numerous engineering geomorphologists on landslides affecting property in Los Angeles provides one example. Geomorphological appraisal of sediment yield to help predict reservoir longevity provides another. In such studies, both site conditions *and* regional situation of sites tend to be important, and it is the latter in which geomorphological assistance is often most effective.

Within the planning phase, geomorphological contributions will vary in type and importance. A broad general description of the geomorphological contributions is shown in Fig. 1.8 in which it can be seen that the most substantial work is usually involved in assembling existing data, acquiring new data, and analysing the data; as mentioned above, communication between the planner and the geomorphologist is fundamental throughout the process.

(c) *Geomorphology and engineering*

During the phase of policy or project implementation—the phase of development, construction, and continuing management of a development—geomorphological contributions are usually made

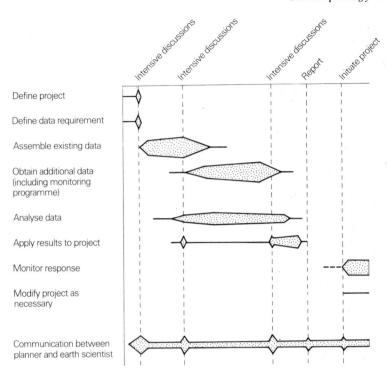

FIG. 1.8 Stages in the involvement of geomorphological investigations in the planning phase of a project (after Doornkamp, 1985)

in the broad context of civil engineering through collaboration with cognate specialists such as soil mechanic engineers and engineering geologists (Fig. 1.9). As Hutchinson (1979, p. 1) said:

The purpose of engineering geomorphology is to achieve an understanding of the nature of landscapes sufficient for engineering works to be carried out safely, predictably and economically within them. It starts with the premise that the earth's surface is, in general, a sensitive indicator of the more recent geological events, and that its morphology thus constitutes a most valuable source of information.

It will be clear that in most circumstances to be effective the geomorphological advice is best provided in the project planning phase *prior* to implementation. Nevertheless, continuing advice is often essential, especially when unexpected problems arise (Plate 1.2). Such problems are common. In the UK one infamous example is the case of the Sevenoaks bypass in Kent, in which excavation for the new road led to the reactivation of a 'fossil' solifluction lobe (Weeks, 1969). Internationally, slope failure is probably the single most common cause of unexpected problems (see Chapter 5).

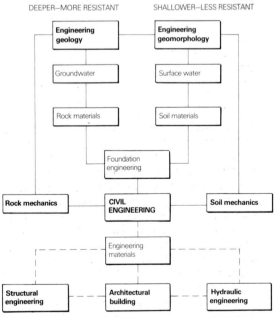

FIG. 1.9 Some relationships between geomorphology, geology, and engineering (after P. G. Fookes and M. Gray, 1987; © 1987 and redrawn by permission, J. Wiley and Sons Ltd.)

PLATE 1.2 Destruction of a paved road by unexpected flooding of a channel in Saudi Arabia: forewarned might have been forearmed

The requirements of engineers, although generally related to understanding terrain conditions, do vary with the kind of engineering work. Examples throughout the following chapters suggest that the most common engineering works requiring geomorphological assistance include construction of buildings and urban development, transportation facilities (especially roads, airports, and canals), utilities (e.g. pipelines), flood-control facilities, recreational facilities (e.g. Leighton, 1971), and coastal-protection and harbour works. Table 1.1 summarizes the time spent on a range of geomorphological studies at different stages in engineering projects. In all of them, the ability to interpret the ground evidence of site conditions—either from remote sensing imagery or in the field—is a central and most sought-after contribution by geomorphologists to engineering projects.

After construction is complete, the day-to-day maintenance and monitoring of facilities, the evaluation of policies, and the need to assess damage caused by hazardous events such as storms, may all require a geomorphological contribution. And increasingly, consequent litigation involves geomorphological advice being used in an adversarial role.

1.4 The Geomorphological Contribution

(a) *The kinds of geomorphological work*

Environmental management, and the geomorphological contributions to it, are not confined to any one environment: all environments require management, and most receive it. But the nature of geomorphological problems and potentials varies greatly, for instance between rural and urban areas, between climatic regions, and between different nations and styles of environmental management. Fig. 1.10 provides a general assessment of the main problems and potentials in a range of environments, and Fig. 1.11 shows the distribution of death-causing geomorphologically related hazards in the Third World.

Although geomorphological advice may be offered independently of other environmental information in any management study, this is not

TABLE 1.1 Example of geomorphological studies for engineering projects

A. Working practice

Stage	Liaison with engineers	Desk studies (Home based)	Field studies (Based on site)
I	Brief received from client—discussions with senior engineers and engineering geologist involved	Familiarization with project Examination of available literature and maps Air-photo interpretation	
	Brief re-examined		
II	Continuing discussions with engineer's field staff		Field mapping —investigation of landforms, materials, and processes —review of trial pit and borehole information (if available) Geomorphological map compilation
III		Derivative maps compiled Data additional to initial brief compiled Site investigation suggestions defined	
	Report with maps passed to client		

B. Summary of some investigations, their output and time involved

Project	Output	Time involved (Man-days)	
Highway engineering design. Nepal Reconnaissance survey. Dharan to Dhankuta mountain road Route length: 65 km	Maps of landslides, potential instability, and flood hazard, likely to affect route. Proformas and maps of conditions (c.f. soils, rocks, stability, drainage) along route	I II III	20 90 80
Airport site (proposed). Dubai Site investigation Ground area: 15 km²	Maps of surface materials foundation conditions, aggressive soils, migrating dunes, depth to watertable, site drainage Cross-sections between TP and BH	I II III	20 60 80
Industrial area. Dubai Investigation of aggressive soils Ground area: 10 km²	Maps of depth to watertable, aggressive (saline) soils, groundwater conductivity Data on salts present	I II III	5 20 80
Airport site (alternative). Dubai Reconnaissance survey Ground area: 15 km²	Maps of soil and bedrock materials, proposed location of boreholes and trial pits	I II III	1 8 8
Regional development. Khor Khwai Ra's al Khaimah Reconnaissance survey Ground area: 60 km²	Maps of geomorphology, surface materials, saline ground Trial pit descriptions	I II III	3 10 25
Aggregate Resources. Dubai Air-photo interpretation Ground area: 250 km²	Maps of air-photo tonal boundaries with geomorphological interpretation	Total of 50	
Irrigation scheme. S. Yemen Air-photo interpretation Ground area: 180 km²	Maps of air-photo tonal boundaries with geomorphological interpretation, drainage and selected cultural features	Total of 50	
Road design. South Wales in support of site investigation Route length: 4 km	Maps of slope form, drainage, stability, major material boundaries, and man-made structures	I II III	10 12 20
Urban expansion and regional development. Suez Reconnaissance survey and site investigation Ground area: 100 km²	Maps of geomorphology, surface soils and bedrock, depth to watertable, salinity of groundwater, flood hazards. Interpretation of exposures	I II III	25 100 80

Notes: TP = Trial Pits.   BH = Boreholes.

Source: Doornkamp *et al.* (1979).

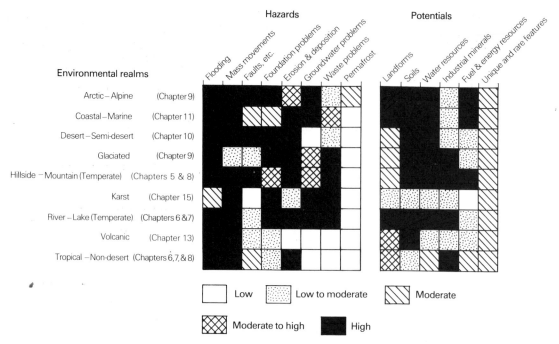

FIG. 1.10 Geomorphological hazards and potentials in different environments (based, in part, on Spangle *et al.*, 1974)

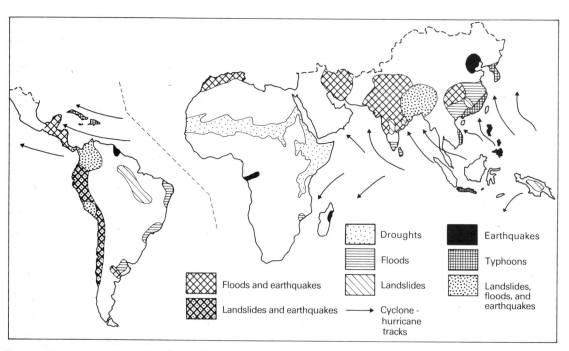

FIG. 1.11 Approximate distribution of natural hazards in the Third World which have involved the loss of more than 100 lives in a single incident during the past 50 years (after Doornkamp, 1985)

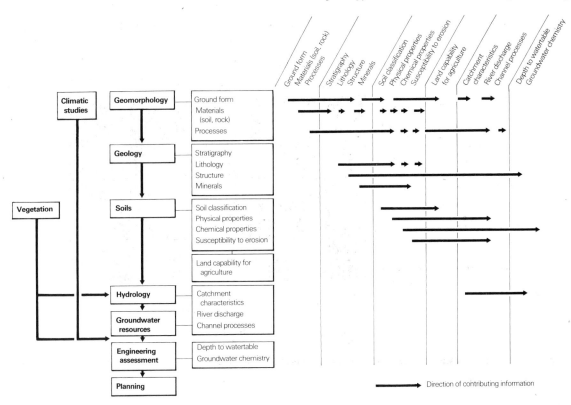

FIG. 1.12 Sequence of environmental data collection based on an initial geomorphological survey

normally the case. More commonly, geomorphological contributions form part of the *earth-science information* required, and it is provided in collaboration with other scientists. The theoretical links between the different aspects of earth-science are well known, but the ways in which the diverse material is integrated into a single report is not well established. Often, geomorphology may form the *basis* of terrain data collection (Fig. 1.12) because the data are relatively easy to recover; landforms cover the terrain comprehensively (unlike, say, vegetation); landforms are capable of hierarchical classification so that they can be studied at almost any scale of inquiry, from the site to the continent, that is appropriate to the problem in hand; and they can often be related predictively to other environmental variables, such as soils. Common ways of integrating earth-science information are through folios of environmental maps and reports, through sieve mapping, and through land-systems analysis (e.g. Chapter 2).

In general, geomorphologists contribute towards the study of natural hazards, environmental audits, resource evaluations, and impact assess-

ments in the planning phase, to resource evaluations and environmental impact assessments in the implementation phase, and to retrospective evaluations ('*ex-post* audits') in the evaluation phase (e.g. Jones, 1983). Each of these contributions is examined in subsequent chapters.

A modern trend amongst engineers and planners is to create a conceptual understanding of the environment in which a project is to be carried out before embarking upon the collection of data. Geomorphological considerations form a natural part of this work. For example, discussions concerning a proposed construction site in the mountains of Papua New Guinea quickly identified the fact that forest clearances on steep mountain slopes carrying a thick soil mantle in an area of very heavy rainfall would lead to landsliding. No field visit was needed to reach this conclusion. Existing geomorphological awareness and experience was enough to indicate at the outset, in the project planning phase, that landslides would be a hazard to the project, and would require close investigation and remedial engineering measures.

There are many similar situations where a geomorphological awareness can guide project planning during the planning phase. When this happens, data collection takes place for well-defined reasons in relation to the overall purposes of the project. This increases the efficiency and hence the cost-effectiveness of the project.

(b) *The skills used*

Requests for geomorphological assistance in environmental management usually arise from a need to appraise the nature—and especially the distribution and changes over time—of hazards and resources. The most common requirements are to map a landscape, or selected attributes of it, and/or to monitor the nature and causes of change. It is valuable to emphasize some of the analytical approaches the geomorphologist brings to these critical tasks of mapping and monitoring, because they arise directly from geomorphological training.

Most importantly, the skill of landscape interpretation in the field undoubtedly underlies the success of much recent geomorphological work in environmental management. It is not a new skill; indeed it is a traditional skill that may have been weakened within the subject by the fact that so many geomorphologists became preoccupied in the 1960s and 1970s with the minutiae of contemporary processes to the exclusion of a broader view of specific landscapes and their unique historical antecedents. It is a highly prized skill in environmental management—with its requirements of four-dimensional thinking (i.e. the dimensions of space, and the fourth dimension of time) at different scales and a great deal of often strenuous field-tramping—especially because it is practised by so few. At one level, it allows the accurate recording of geomorphological data in the form of maps that can often serve as the basis for a whole range of environmental surveys (see Chapter 2).

The skill of landform interpretation is also fundamental to the analysis, monitoring, and prediction of environmental change. Because environmental managers often require knowledge of past and present changes and prediction of future trends, and because they need to know quickly without giving any opportunity for long-term monitoring, it is frequently necessary to interpret change and process through the field interpretation of the surrogate evidence of landform and surface materials (see Chapter 3).

This skill, which is heavily based on experience, is also relevant to the complementary skill of recording change and mapping landscape from aerial photographs and satellite imagery. Here, geomorphologists have been in the forefront of those seeking to exploit the potential of remote sensing technology for environmental and other applied purposes (e.g. Townshend, 1981) because landforms are often clearly displayed on remote sensing imagery, and successive images of the same locations can reveal changes not easily monitored on the ground (see Chapter 2).

The applied geomorphologist also calls upon other skills. Techniques for the direct measurement and monitoring of processes in the field have been developed substantially in recent decades, especially in drainage basins, along coasts, and in aeolian environments. Closely allied to these skills are those related to field measurements of variables of predictive value in studying process (such as salinity in the context of salt weathering or soil properties in the context of soil erosion). Indirect monitoring of change by the analysis of successive remote sensing images finds an important counterpart in the growing efforts to monitor and explain environmental change through the use of historical archives (e.g. Hooke and Kain, 1982).

The ability to use hardware models in the analysis of processes and complex environmental situations under controlled laboratory conditions requires another set of skills of great applicability, as shown by recent studies in flumes, wind-tunnels, and climatic cabinets. Finally, the use of computers is clearly fundamental, for instance in creating digital simulation models, in analysing digital types of satellite data, and in preparing maps and other derivatives from field information.

(c) *Some inherent problems*

Geomorphological advice, contrary to the opinion of some scientists, is never value-free. Each scientist views an environmental problem from the perspective of his own experience, prejudices, and ability within the constraints placed upon him. Thus, the geomorphologist may well find himself in conflict with other scientists and with his clients. The conflicts can arise unpredictably and cause great difficulty. For example, it may be the geomorphologist's view that a hazardous location or a scarce area of marsh is best avoided; his client may prefer to design structures in the hazardous zone or across the marsh. Who is right? Equally, decisions in environmental management may not satisfactorily reflect the strongly held views and competent advice of the geomorphologist because other factors—usually financial or political—are found to be more compelling. What does the geomorphologist do? Again, the client commonly needs speedy, low-cost advice that can only be provided by prejudicing the highest standards of academic research: what does the geomorphologist do? Another frequent cause of difficulty arises when the client changes the nature of the demands placed on the geomorphologist during the course of an investigation. How flexible can the geomorphologist be? These, and other cognate problems, often figure prominently in controversial work in environmental management.

Because geomorphological contributions to environmental management have only become numerous recently, and because the contributions are so varied, relatively little progress has yet been made in codifying professional practice. A start has been made, however, for example through the Geological Society of London's working party report on mapping (Anon., 1972); the US Geological Survey's efforts, especially in California (e.g. Brown and Kockelman, 1983); the emergent practices in drylands (e.g. Cooke et al., 1982); and more specific codification of such problems as sediment control (e.g. US Department of Housing and Urban Development, 1969), and building regulations (e.g. Cooke, 1984).

A final problem concerns the cost-effectiveness of geomorphological advice. Unfortunately it is extraordinarily difficult to assess cost-effectiveness because many costs are confidential or difficult to determine, the responses to advice are not always clear, the success of advice accepted is rarely evaluated, and it is often difficult to disaggregate geomorphological advice from the other advice.

Some parallels exist between geomorphological investigations and their costs, and soil surveys, for which some data exist. For example, in Table 1.2 costs are presented (in £/ha at mid-1970s prices) for soil surveys carried out within integrated projects, for an exploratory (reconnaissance) survey, a general (low-intensity) survey, a medium- and a high-intensity survey. Value for money can only be judged against the contribution the end-product made in improving the management of the environment concerned. Absolute costs, however, do not appear to be large when compared with the overall project costs or the value of the project to the community it is intended to serve. Unfortunately this conclusion has to be a value-judgement for precise data on these latter aspects are not available.

The manner of survey also affects costs, and where comparative data exist they show, for example, that surveys in which aerial photography is available tend to be less costly than those where it is not (Fig. 1.13). Fig. 1.13 also shows graphically how survey costs tend to vary with the map scale at which data are recorded, and this is usually a function of survey intensity.

Some notion of costs for geomorphological studies can be derived from Table 1.2. If the number of days on any one of the projects is totalled and a day-rate fee applied then a measure of the professional costs can be obtained. To this needs to be added travel and accommodation expenses, but even so, the key question is: how else could the same results be achieved for an equivalent or lower cost? There are several examples of the cost-effectiveness of geomorphological investi-

TABLE 1.2 Comparative costs for soil surveys at different levels of intensity

Survey	Country	Cost (£sterling mid 1970s)	Area (ha)	Cost/ha (£/ha)
Reconnaissance	Thailand	10 000	5×10^6	0.002
Low	Indonesia	3 750	25 000	0.15
Low/medium	Thailand	50 000	133 000	0.37
Medium	Nigeria	34 400	37 000	0.93
High	Greece	6 640	1 000	6.64

Source: Western (1978).

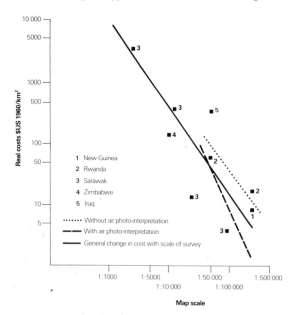

FIG. 1.13 Comparative costs of soil surveys with and without the use of air-photo interpretation at different survey scales (after Bie and Beckett, 1970 and Veenenbos, 1956)

gations within an engineering or planning project. For instance, Leighton's (1976) analysis of the ratio between the estimated losses due to landsliding and the *investigative* costs of *preventing* landsliding, based on a study in part of Los Angeles County, suggested an extremely beneficial ratio of 48.7:1, a conclusion that points strongly towards the advantages of geological and related surveys prior to slope development. In general there seems little doubt that geomorphological contributions are relatively cheap and rapid compared with other methods of terrain appraisal, and that they are in general cost-effective.

1.5 Conclusion

Geomorphology, the study of landforms, materials, and their related processes, is pertinent to many aspects of environmental management involving these phenomena. Within the main phases of planning, implementation, and evaluation in environmental management, geomorphology can contribute especially to hazard assessment, environmental auditing, resources evaluation, impact assessment, and post-development policy evaluation. Such contributions may be useful in most natural environments and especially at the loci of urban and regional development. Geomorphological contributions, often associated with work by other scientists such as engineering geologists, soil scientists, and hydrologists, require mainly mapping and monitoring of land-surface features in ways that are intelligible and useful to a wide range of clients who are not geomorphologists. Thus, the geomorphologist must be able to translate his scientific findings for the clients, and he must also be sensitive to responsibilities and views of those with whom he is working. Amongst those with whom he will come into contact are professional managers, including environmental administrators and technicians, engineers, planners, foresters, land lawyers, and, increasingly, the public, public pressure groups, and individuals who make significant use of land such as the farmer, the tourist, and the building contractor.

As a result, his reading material will need to extend beyond geomorphology, into such fields as economics, engineering, geography, planning, and even political sciences. In short, disciplinary boundaries have little meaning except in so far as they must be recognized in order to be crossed with caution.

2 Mapping Geomorphology

2.1 Introduction

In many environmental management situations the most useful contribution of the geomorphologist is the provision of a terrain map. This is particularly true where information is required concerning the distribution of landforms, soils and rock materials, or features created by surface processes. Such mapping has been used to considerable advantage on many engineering, planning, and land management projects (Tables 2.1, 2.2).

While the *concept* of making a terrain map is not a difficult one, the *task* of actually making one may in fact be so. Quite apart from the need for the person making the map to have the appropriate skills and experience, there are often several critical constraints and choices. Before mapping can begin, choices include the scale of mapping, what is to be incorporated, and how the final map is to be reproduced. Common constraints influencing what can be done include the availability of base maps at the required scale, the purpose of the mapping, the need for the map to provide a comprehensive or a selective statement of the geomorphology, and whether the map is to be printed and published in large quantities or as a small number of working documents.

In practice further constraints are imposed by access to the area to be mapped, and in some instances the map has to be created by air-photo interpretation, perhaps even without access to the field site itself. It is clear, therefore, that working practice greatly influences the nature of the map produced. For present purposes the discussion will concentrate on land-systems mapping (Sect. 2.2)

TABLE 2.1 Selected examples of projects in which terrain mapping has provided important information

Project	Country	Reference
Engineering		
Dharan–Dhankuta Road	Nepal	Brunsden et al., 1975; Hearn and Fulton, 1986
Taff Vale Trunk Road	Wales	Brunsden et al., 1975
Irrigation	Pakistan	Verstappen, 1970
Planning		
Land reclamation	Senegal	Tricart, 1959, 1961
Agricultural land	Venezuela	Tricart, 1966
Urban	Tenerife	Vallejo, 1977
National	Bahrain	Brunsden et al., 1979
Urban	Egypt	Ibrahim and Doornkamp, 1987
Urban	Poland	Klimaszewski, 1956, 1961
Urban	Italy	Fulton et al., 1987
Management		
Land use	Germany	Kugler, 1976
Erosion and conservation	Italy	Rao, 1975
Rural development	Lesotho	Schmitz, 1980

TABLE 2.2 Some applications of geomorphological mapping in environmental management

Regional land classification	Terrain classification used as basis for planning, engineering, or management scheme.
Situation mapping	Mapping used to define the general context of a specific project site. Mapping of active landform components and processes to detect those likely to affect a site from beyond the boundaries of a project area (e.g. landslides, floods, sand, or dust).
Site mapping	To define landform, materials, and processes as they affect planning, engineering, or management projects: especially for construction projects, damage assessment, engineering design. Typical engineering projects include road, harbour, and dam construction, and typically emphasis is given to hazards such as ground instability, flooding, aggressive saline soils, erosion, and deposition.

and geomorphological mapping (Sect. 2.3) as these are the two forms of terrain mapping most useful in environmental management.

Given the great variety of contexts in which geomorphological mapping may be used, it is undesirable to be too dogmatic about the precise methods to be adopted. For the purposes of some environmental management projects, particularly those which need to record the nature of large areas over a short period of time, it may be better to adopt more general mapping approaches such as that provided by land-systems surveys (Sect. 2.2) than the more detailed approach of geomorphological mapping as defined in Sect. 2.3. Surveys of the land-systems type seek to divide an area into morphological (or terrain) regions rather than to define individual landforms or slope characteristics. On the other hand, detailed mapping of specific processes, materials, and land-

forms, as in geomorphological mapping, may be essential for many projects. In the last decade or so, the context in which geomorphological maps are useful has been extended to include environmental geology mapping and engineering geology mapping (Sect. 2.5). They also have a role to play in terrain mapping for environmental impact assessments (EIA) and scenic evaluation (Sect. 2.6). Only a few examples of mapping are given in this chapter, other examples are provided to illustrate the particular themes of other chapters, and in the references provided.

2.2 Land-systems Mapping

Rural land planning in some of the larger countries, especially in the tropics, has been, and in some cases still is, handicapped by a lack of appropriate topographic, geological, and soil maps. The urgent need for a mapping programme was first tackled in Australia, and then adopted in several countries in Africa, Latin America, and Asia. The approach used, although having been conceived and developed by, for example, Bourne (1941), Wooldridge (1932), Veatch (1933), and Unstead (1933), came into prominence with the publication by the Commonwealth Scientific and Industrial Research Organisation (Australia) of *Summary of General Report on Survey of Katherine–Darwin Region, 1946* (Christian and Stewart, 1952). This was the first of more than 30 in a series of reports on land-systems mapping in the Australian territories. Mapping was based on the recognition of 'land systems' which have within them common terrain attributes different from those of adjacent areas.

Land systems may range in size from only tens of km² up to some hundreds of km². Within any one land system there is usually *a recurring pattern of topography, soils, and vegetation* Christian and Stewart, 1952). This threefold association is described as follows:

The topography and soils are dependent on the nature of the underlying rocks (i.e. geology), the erosional and depositional processes that have produced the present topography (i.e. geomorphology), and the climate under which these processes have operated. Thus the land system is a scientific classification of country based on topography, soils and vegetation correlated with geology, geomorphology and climate (Stewart and Perry, 1953, p. 55).

A land-systems map defines those areas within

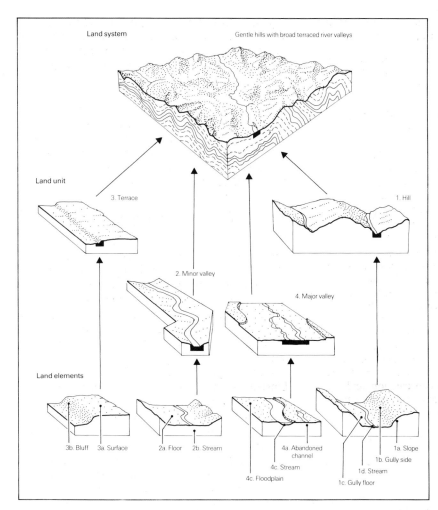

Land system — Gentle hills with broad terraced river valleys

Land unit
3. Terrace
1. Hill
2. Minor valley
4. Major valley

Land elements
3b. Bluff 3a. Surface 2a. Floor 2b. Stream 4a. Abandoned channel 1a. Slope 1b. Gully side 1d. Stream
4c. Stream 1c. Gully floor
4c. Floodplain

FIG. 2.1 Diagram to show the relationship between land system, land facet (or land unit), and land element (after Lawrance, 1972)

which certain predictable combinations of surface forms and their associated soils and vegetation are likely to be found.

In practice, whether the boundaries are being drawn in the field or from aerial photographs, the simplest criterion for distinguishing between land systems is topography. In addition, the interpretation of soils, geology, and both depositional and erosional processes comes most readily through landform analysis. That is why geomorphology is a key element in land-systems mapping, and indeed in most integrated resource surveys, especially at the rapid reconnaissance stage (Verstappen, 1966). Good descriptions of the land-systems approach to terrain mapping may be found in Christian (1957), Christian and Stewart (1968), Mabbutt (1968), Webster and Beckett (1970),

Lawrance (1972), Mitchell (1973), and Ollier (1977*b*). Recent practice is described by King (1987).

Every land system is divisible into smaller components (units and elements) (Fig. 2.1). However, in many of the Australian reports individual land units (sometimes called 'facets') are not usually shown on the maps, though their properties (including their soils and geology) are usually recorded by means of a three-dimensional diagram (Fig. 2.2), by a table of information (e.g. Table 2.3), and sometimes by reproducing a stereoscopic pair of aerial photographs which show a typical part of the system. Each land unit is normally (at least within semi-arid Australia) associated with a defined set of geological, pedological, vegetational, and sometimes hydrological conditions.

A. Distribution of the Napperby and Warburton land systems

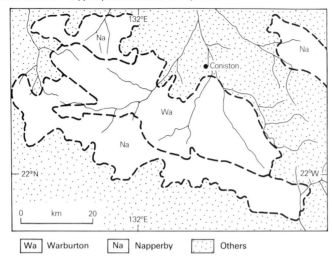

Fɪɢ. 2.2 Land systems near Coniston, Australia (after Perry, 1962)

| Wa | Warburton | Na | Napperby | :::: | Others |

B. Napperby Land System

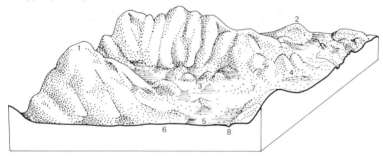

C. Warburton Land System

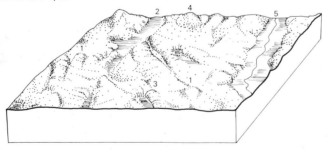

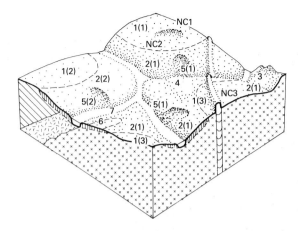

Engineering properties of some soils in the Kyalami Land System

Unit	Variant	Form	Soils and hydrology	Statistic	Liquid limit	Plasticity index	Linear shrinkage
1	2	Hillcrest, 0–2°, width 450–1800 m	Residual sandy clay with collapsing grain structure on granite (0–23 m). Ferralitic soil. Above groundwater influence except at depth.	X̄ S V	44.1 11.0 0.25	22.6 10.8 0.48	9.78 4.27 0.44
2	1	Convex side slope 2–12°, width 450–910 m	Hillwash of silty sand derived from granite (0.3–0.9 m) on granite schist or basic metamorphic rocks. Occasionally saturated.	X̄ S V	17.3 8.98 0.52	4.9 3.70 0.76	2.04 1.04 0.51
7		Alluvial flood-plain 1–3°, width 45–230 m	Expansive alluvial clays and sands (3–6 m) on granite schists and basic metamorphic rocks. Mineral hydromorphic soil. High watertable (data for black alluvium)	X̄ S V	58.5 14.7 0.25	37.1 13.7 0.37	13.7 10.3 0.75

x̄ mean
s standard deriation
v coefficient of variation

FIG. 2.3 Schematic block diagram of the Kyalami area, near Johannesburg, South Africa (after Brink *et al.*, 1970)

Supplementary data on climate, existing land uses, communications, settlements, or whatever was deemed to be appropriate for that area, were included in many of these reports. The reports provided the basis on which decisions could be made concerning land-use policy.

During the late 1950s and early 1960s, the (then) Ministry of Overseas Development (UK), through its Land Resources Division, was providing advice on land management and agricultural development to countries such as Lesotho, Nigeria, Tanzania, Botswana, and Gambia (Ministry of Overseas Development, 1970). Many of the studies in these countries involved large areas, and not surprisingly the Australian land-systems approach was normally adopted. Present policy within the ODA is to use the land-systems techniques to undertake rapid land resource surveys of large areas and to concentrate detailed effort on areas of higher agricultural potential (King, 1987).

Gathering the information both for the land-systems map and the definition of land units cannot always be achieved in the time available by field mapping alone. Great reliance therefore has to be placed on the interpretation of aerial photography (e.g. Webster and Beckett, 1970). Satellite imagery is sometimes used for an initial definition of land systems (e.g. Mitchell and Howard, 1978). Indeed, part of the fieldwork programme is often designed to check an air-photo interpretation map, though it also allows some additional site measurements to be made and both soil and rock samples to be collected for analysis.

TABLE 2.3 The Napperby and Warburton land systems

NAPPERBY LAND SYSTEM (2600 km²)

Granite hills and plains with lower rugged country in a strip from Aileron to west of Mt. Doreen homestead.
Geology—Massive granite and gneiss, some schist. Pre-Cambrian age, Arunta block, Mt. Doreen–Reynolds Range; Lower Proterozoic, Warramunga geosyncline.
Geomorphology—Erosional weathered land surface: hills up to 150 m high and plains with branching shallow valleys; less extensive rugged ridges with relief up to 15 m, and a dense rectangular pattern of narrow steep-sided valleys.
Water Resources—Isolated alluvial or fracture aquifers may yield supplies of groundwater. There are areas suitable for surface catchments.
Climate—Nearest comparable climatic station is Tea Tree Well.

Unit	Area	Landform	Soil	Plant community
1	Large	Granite hills: tors and domes up to 150 m high; bare rock summits and rectilinear boulder-covered hill slopes, 40–60%, with minor gullies; short colluvial aprons, 5–10%	Outcrop with pockets of shallow, gritty, or stony soils	Sparse shrubs and low trees over sparse forbs and grasses, *Triodia spicata*, or *Plectrachne pungens* (spinifex)
2	Medium	Close-set gneiss ridges and quartz reefs: up to 15 m high; short rocky slopes, 10–35%; narrow intervening valleys		
3	Medium	Interfluves: up to 7 m high and 0.8 km wide; flattish or convex crests, and concave marginal slopes attaining 2%	Mainly red earths, locally red clayey sands and texture-contrast soils, stony soils near hills	Sparse low trees over short grasses and forbs or *Eragrostis eriopoda* (woollybutt)
4	Medium	Erosional plains: up to 1.6 km in extent, slopes generally less than 1%		
5	Small	Drainage floors: 180–365 m wide, longitudinal gradients about 1 in 200	Mainly texture-contrast soils, locally alluvial soils and red earths	*Eremophila* spp.—*Hakea leucoptera* over short grasses and forbs; minor *Kochia aphylla* (cotton-bush)
6	Small	Alluvial fans: ill-defined distributary drainage; gradients above 1 in 200	Alluvial brown sands and red clayey sands	Sparse low trees over short grasses and forbs or *Aristida browniana* (kerosene grass)
7	Small	Rounded drainage heads: up to 180 m wide and 1.5 m deep on the flanks of unit 3	Red earths	Dense *A. aneura* (mulga) over short grasses and forbs
8	Very small	Channels: up to 45 m wide and 1.5 m deep and braiding locally	Bed-loads mainly coarse grit	*E. camaldulensis* (red gum)—*A. estrophiolata* (ironwood) over *Chloris acicularis* (curly windmill grass)

The most common scales for the production of land-systems maps are 1:1m to 1:500 000. Inevitably the desire to record greater detail, especially when the land-system approach was adopted for mapping smaller areas than was originally the case, led to the mapping of individual land units and elements (e.g. Scott *et al.*, 1985 in Papua New Guinea, who used a map scale of 1:100 000).

By the late 1960s land-unit mapping and the use of land units as the basis for data banks storing geotechnical information of value to engineers had been developed both in South Africa (Brink and Partridge, 1967; Brink *et al.*, 1968; Kantey, 1971; National Institute for Road Research, 1971) and in Australia (Aitchison and Grant, 1967; Grant and Lodwick, 1968; Grant 1968, 1972, 1974). A South African example is shown in Fig. 2.3. To arrive at the level of information shown on soil properties requires not only land-unit mapping but also the laboratory testing of soils. Since such tests can be costly it is clearly desirable not to allow the results to become inaccessibly buried within files. This data-bank approach allows such data to be conveniently stored and retrieved.

TABLE 2.3—*continued*

WARBURTON LAND SYSTEM (1554 km²)
Sparsely timbered granite plains in the north and north-west of the area, mainly near Coniston homestead.
Geology—Quaternary soils and and Tertiary 'deep weathering profile'. Overlying Pre-Cambrian granite, schist, and gneiss.
Geomorphology—Erosional weathered land surface: peneplain of selectively weathered rocks; open sub-rectangular pattern of shallow valleys.
Water Resources—Prospects generally poor but isolated fracture aquifers may yield supplies of groundwater. There are areas suitable for surface catchment.
Climate—Nearest comparable climatic station is Tea Tree Well.

Unit	Area	Landform	Soil	Plant community
1	Very large	Interfluves: up to 0.8 km wide and 4.5 m high; flat or slightly rounded stony crests with minor rock outcrops; short concave marginal slopes attaining 1%	Mainly red earths, locally red clayey sands and stony soils including texture-contrast soils	Sparse low trees. *A. aneura* (mulga) or *A. kempeana* (witchetty bush) over short grasses and forbs or *Eragrostis eriopoda* woollybutt)
2	Medium	Drainage floors: up to 270 m wide; flat, unchannelled central tracts and gently sloping margins	Presumably red earths	*A. aneura* (mulga) over short grasses and forbs or *Eragrostis eriopoda* (woollybutt)
3	Small	Short valley heads shallowly entrenched on the flanks of unit 1: unchannelled floors up to 460 m wide liable to shallow gullying; longitudinal gradients above 1 in 500	Texture-contrast soils	*Eremophila* spp.—*Hakea leucoptera* over short grasses and forbs or *Bassia* spp; or minor *Kochia aphylla* (cotton-bush)
4	Very small	Small hills and quartz reef ridges: up to 9 m high	Outcrop with pockets of shallow, gritty, or stony soil	Sparse shrubs and low trees over *Triodia pungens* (spinifex), *T. spicata* (spinifex), or sparse forbs and grasses. Far north: *E. brevifolia* (snappy gum) over *Triodia pungens* (spinifex) or *Plectrachne pungens* (spinifex)
5	Very small	Channels: wide, shallow and braiding		*E. camaldulensis* (red gum)—*A. estrophiolata* (ironwood) over *Chloris acicularis* (curly windmill grass)

Source: Perry (1962) (with metric conversion).

In the South African case a data bank was established, at CSIR in Pretoria, in which geotechnical data obtained by consulting engineers or contractors within the Republic could be stored under the reference number for the land unit from which they were derived. Under this scheme anyone who requires geotechnical data can interrogate the data bank as long as they know for which land unit the data are required. Obviously any data-bank scheme, such as this, benefits greatly from being computerized.

The Australian approach has been a little different in that a PUCE (Pattern-Unit-Component Evaluation) system was developed based mainly on local relief amplitude (relative relief) and dissection (an example is shown in Fig. 2.4). Other developments in land-systems studies include Speight's (1974, 1976) important demonstration that it is possible successfully to identify and delimit land units quantitatively using selected land-surface morphometric parameters.

Land-systems mapping provides an extremely simple, cost-effective, and versatile method for rapidly classifying large areas of relatively unknown territory, and providing a regional framework for environmental data collection and storage. For example, the first approximation of a terrain classification of the Wahiba Sand Sea in

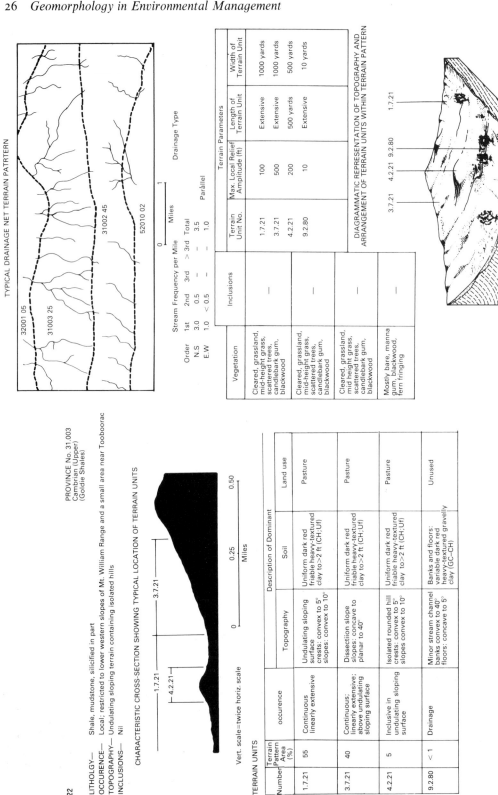

Fig. 2.4 Example of a terrain pattern in the Melbourne area, based on a survey for engineering purposes (after Grant, 1972)

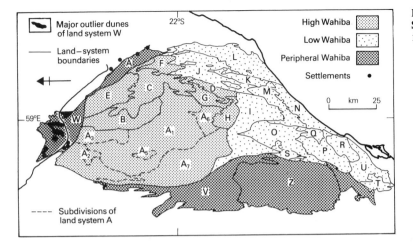

FIG. 2.5 Land systems of the Wahiba Sand Sea, Oman (after Goudie *et al.*, 1987). See also Plate 2.1

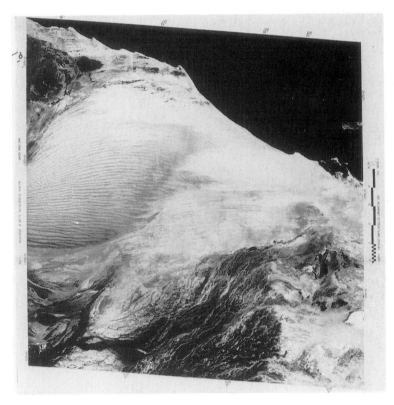

PLATE 2.1 The Wahiba Sand Sea, Oman, as revealed on a LANDSAT image, 1971

Oman (Fig. 2.5; Goudie *et al.*, 1987) was based on evidence of dune forms revealed on a satellite image (Plate 2.1): 26 major regions were distinguished in an area that had not previously been studied, and they provided the basis for systematic field sampling of geomorphological features, biological characteristics, and the patterns of nomadic activity.

But the land-systems approach has its weaknesses. It tends to be general, and has not always been successfully directed towards specific development purposes. It is, in its commonest form, both qualitative and subjective; and even when the definitions and descriptions are based on quantified criteria as in the work of Speight, the selection of criteria is still subjective, tending towards the quantifiable rather than the useful. Nevertheless, the apparent value of the approach is testified by its widespread international use.

2.3 Geomorphological Mapping

The aim of geomorphological mapping is to record information on surface form, materials (soil and rock), surface processes, and (in some cases) the age of landforms. As such it provides a basis for terrain assessment which is useful in the context of many environmental problems (Table 2.4). The most successful approach to such mapping is to combine field inspection with air-photo interpretation. There is no doubt that geomorphological mapping is both subjective and dependent upon the skill of the mapper. The subjectivity shows in the choices made over what to include, and skill (allied to experience) shows not only in the accuracy of plotting but also in the ability to recognize land features and their relationships. Nevertheless, the process of making a geomorphological map is an unrivalled way for the geomorphologist to become familiar with the landforms of an area. It also acts as a great stimulus to thought concerning both the relationships between forms, materials, and processes, and the manner of

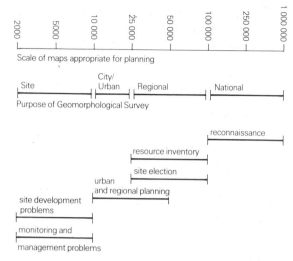

FIG. 2.6 Relations between scale and purposes of geomorphological mapping (from Cooke *et al.*, 1982)

landform development. Whether or not the geomorphological map is an appropriate document to place in the hands of engineers, planners, or other land managers is discussed later.

It can be argued that geomorphology provides the *best* basis for classifying terrain because, as mentioned in Chapter 1, landforms are often clearly displayed on aerial photographs and/or in the field; they can be classified hierarchically from smallest to largest; they cover the landscape continuously (vegetation and soils may be discontinuous); and many other parameters (such as soils) may have a defined and predictable relationship to landforms.

Some distinctions have to be made concerning mapping at different scales, often because different scales are appropriate for different purposes (Fig. 2.6). For convenience the subdivision used here is that adopted by the International Association of Engineering Geology (UNESCO, 1976) and A. Hansen (1984):

synoptic maps   1:100 000 and smaller
medium-scale maps   1:25 000–1:50 000
large-scale maps   1:10 000–1:5000
detailed maps   1:2000–1:5000
site plans   1:2000

(In the USA the published topographic maps at scales of 1:24 000 and 1:62 500 make them suitable as base maps for mapping at the medium scale.)

TABLE 2.4 Some applications of geomorphological mapping in planning and economic development

---

1. Land-use
   Territorial planning—regional area planning—conservation of the natural and cultural landscape

2. Agricultural and forestry
   Potential utilization—soil conservation—soil-erosion control—reclamation of destroyed or new areas—soil reclamation—drainage and irrigation

3. Underground and surface civil engineering
   Reconstruction and replanning of settlements—design of industrial installations—construction of communication lines—design of dams, reservoirs, canals, and harbours—shore protection—regulation of natural and artificial waterways

4. Prospecting and exploitation of mineral resources
   Geological survey—mining and exploitation—mining damage and evaluation—reclamation of mines—areas of landsliding and subsidence—maintenance and creation of new dumps for waste materials

---

Source: Demek (1972).

Clearly, as the scales become smaller (i.e. larger and larger areas are shown on the same size of map sheet), it is necessary either to eliminate some mapped elements or to show them in a more generalized way, or both. For example, at 1:10 000 it may be possible to map accurately and true to scale not only the location and dimensions of landslides but also some of their surface details. At 1:25 000 surface details probably cannot be shown; at 1:50 000, some of the smaller landforms, such as landslides, have to be omitted from the map; and at 1:250 000, generalizations have to be made, so for example 'areas prone to landsliding' may replace the mapping of individual landslides since they can no longer be shown true to scale. An alternative, at the 1:250 000 scale, would be to show the location of landslides by some form of symbol, although it would be drawn out of scale.

Thus, as scale changes so will both the cartographic style of the maps and their information content. Scale of mapping therefore takes on a singular importance, and the decision over scale determines to a very large extent all subsequent procedures and uses. For example, not only does scale influence what is shown and how it is shown, it also determines locational accuracy, the amount of effort required to include detail, the need for supplementary topographic surveying (more with detailed maps, usually none with synoptic maps), and final usefulness. The two most important determinants of map scale should be the ultimate purpose of the mapping and the complexities of the landscape to be mapped.

Despite several attempts to produce unified schemes, notably by the International Geographical Union (IGU), there is no commonly agreed approach to geomorphological mapping. The mapping schemes proposed by the IGU are set out in *Manual of Detailed Geomorphological Mapping* (Demek, 1972) and in *Guide to Medium-scale Geomorphological Mapping* (Demek and Embleton, 1978). The mapping symbols illustrated in Figs. 2.7 to 2.19 are based partly on Demek (1972) and partly on the scheme proposed by Verstappen and Zuidam (1968) and Verstappen (1970). Other schemes include those of the Geological Society's Engineering Group Working Party (Anon., 1972), and of Blachut and Müller (1966).

Different styles of mapping and different emphases on information content have emerged from one country to another. Gilewska (1967) provided an interesting and perceptive comparison of the approaches used in Hungary, France, Russia, and Poland, demonstrating how national trends in geomorphological thinking are apparent in the style of mapping. Other useful discussions are provided by Galon (1962) (with special reference to work in Poland), Tricart (1965) (the French approach), Doornkamp (1971, with a full bibliography), Barsch (1979), Cooke *et al.*, (1982) (drylands), Fleisher (1984) (US maps), and Verstappen (1983) (the practice followed at ITC, the International Institute for Aerial Survey, Enschede, The Netherlands, as well as a review of other approaches). Verstappen (1983) also provides a history of geomorphological mapping. Geomorphological mapping normally involves the mapping, together or separately, of one or more of the following: surface form, surface materials, surface processes, and landform age. Each of these will briefly be examined in turn.

(a) *Mapping surface form*

Surface form, as later chapters will show, is an important component of many geomorphological studies. Mapping surface form is usually based on a system of morphological mapping that depends on the recognition of the junction between slopes of differing steepness, cliff forms, and both the amount and direction of slope (Fig. 2.20). The map is constructed by recording the nature and position of slope junctions, and by placing a V-symbol on the steeper side of the line (and pointing downhill). Slope direction arrows are included and steepness of slope may be shown as a numerical value or as a shading (or colour) to identify a steepness class. The slope classes used may be those recommended in the IGU manual (Demek, 19172), i.e.:

$$0-2°, 2-5°, 5-15°, 15-35°, 35-55°, 55°,$$

or based on particular criteria relating to threshold values (Table 2.5) for specific land management purposes. Most engineers tend not to work in degrees, however, but use a gradient notation (e.g. 1 in 4), or percentage slope.

For an accurate representation of surface form the scale of mapping should be determined by the nature of the topography. In more complicated areas such as those which are intensely dissected, mapping might need to be at scales of 1:10 000. In areas of moderate dissection smaller scales may be sufficient to record the existing slopes. At scales

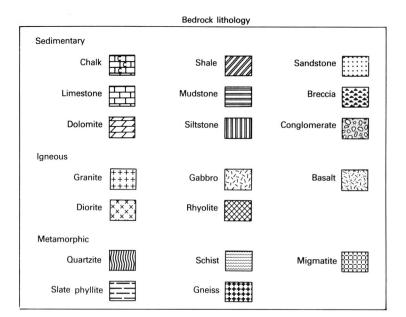

FIG. 2.7 Bedrock lithology: geomorphological mapping symbols

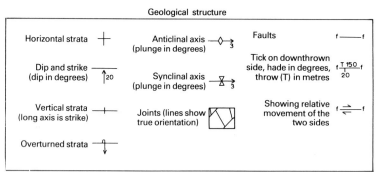

FIG. 2.8 Geological structure: geomorphological mapping symbols

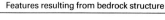

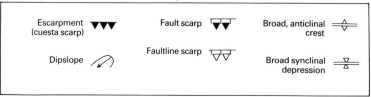

FIG. 2.9 Features resulting from bedrock structure: geomorphological mapping symbols

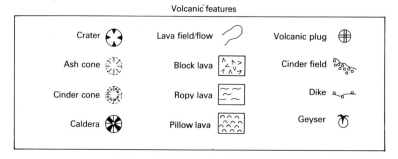

FIG. 2.10 Features of volcanic origin: geomorphological mapping symbols

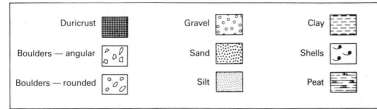

FIG. 2.11 Superficial unconsolidated materials: geomorphological mapping symbols

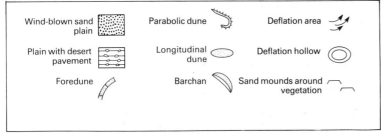

FIG. 2.12 Slope instability features: geomorphological mapping symbols

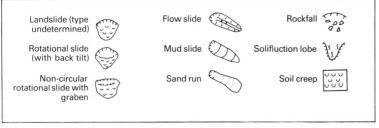

FIG. 2.13 Aeolian features: geomorphological mapping symbols

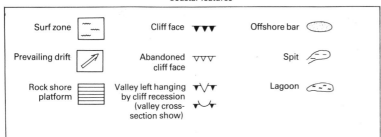

FIG. 2.14 Coastal features: geomorphological mapping symbols

Forms of permafrost areas, glacial and periglacial features

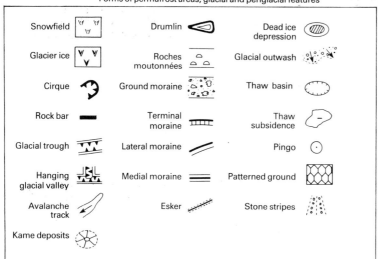

FIG. 2.15 Forms of permafrost areas, glacial and periglacial features: geomorphological mapping symbols

Forms of fluvial origin

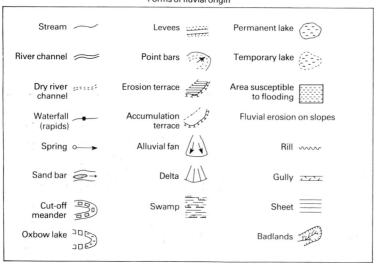

FIG. 2.16 Forms of fluvial origin: geomorphological mapping symbols

Karst features

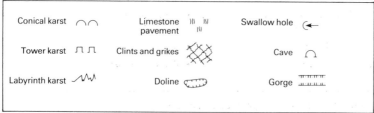

FIG. 2.17 Karst landscape features: geomorphological mapping symbols

Other major features

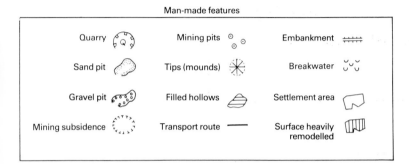

Planation surface	Rock wall	Piedmont
Residual hill	Pass (col)	Ridge

FIG. 2.18 Major features not included in previous figures: geomorphological mapping symbols

Man-made features

Quarry	Mining pits	Embankment
Sand pit	Tips (mounds)	Breakwater
Gravel pit	Filled hollows	Settlement area
Mining subsidence	Transport route	Surface heavily remodelled

FIG. 2.19 Man-made features: geomorphological mapping symbols

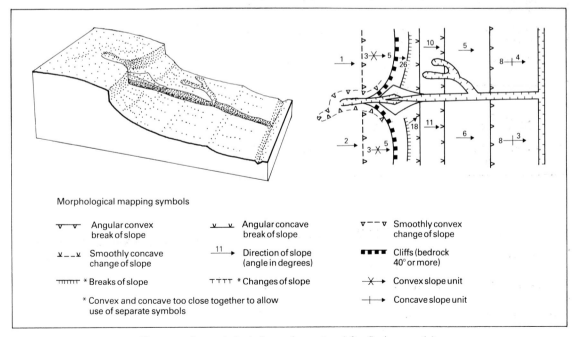

Morphological mapping symbols

▽—▽ Angular convex break of slope	⩒—⩒ Angular concave break of slope	▽ — — ▽ Smoothly convex change of slope
⩒ — — ⩒ Smoothly concave change of slope	——11——▶ Direction of slope (angle in degrees)	▪▪▪▪ Cliffs (bedrock 40° or more)
⊤⊤⊤⊤⊤ * Breaks of slope	⊤ ⊤ ⊤ ⊤ * Changes of slope	—✕—▶ Convex slope unit
		—┼—▶ Concave slope unit

* Convex and concave too close together to allow use of separate symbols

FIG. 2.20 A morphological mapping system (after Savigear, 1965)

TABLE 2.5 Critical thresholds of slope steepness for some practical purposes

Slope threshold steepness (degrees)	Examples of uses limited by slope steepness[a]
1	International airport runways
1	Main-line rail transport
	Local aerodrome runways
	Free ploughing
2	Major roads
	Agricultural machinery for weeding and seeding
	Constructional (land development) problems begin
	Soil erosion may be initiated
3	Housing, roads
5	Railways
5	Heavy agricultural machinery
	Large-scale industrial site development
10	Site development
	Standard wheeled tractor

[a] Unless specially adapted equipment is used and the site is treated with particular regard to its steepness.

Source: Modified from Crofts (1973).

less than 1:75 000 it is rarely possible to record meaningful slope discontinuities.

The mapping scale chosen depends on the final accuracy required which in its turn depends on the thickness of the lines drawn to represent slope boundaries. The breadth of the thinnest line that can be drawn on a map in the field is about 0.5 mm; if only pecks are added (see Fig. 2.20) the space required becomes 1.5 mm; and if an arrow is included with the appropriate curvature sign (cross or negative sign) then at least 3.0 mm, and frequently twice this space, may be required. These considerations alone may determine the mapping scale. For example, at a scale of 1:10 000, 1.5 mm on the map is equivalent to 15 m on the ground, so no feature less than 15 m across can be individually and accurately represented on the map by the smallest symbol allowed, namely a line with pecks on it. In practice it is not possible to consider the effects of pencil thickness every time a feature has to be recorded, and an appropriate mapping scale has to be chosen based on experience. This will be the scale that allows the representation of all features relevant to the problem in hand.

In any field-mapping technique there may be an element of subjectivity about the location of boundaries, as in the case of locating breaks and changes of slope on a morphological map. In general this subjectivity decreases as slope discontinuities become sharper and as locational reference points, such as field boundaries, increase in density. Where cultural reference features are either absent or not shown on the base map, then air-photos may have to be used in field plotting. In fact, aerial photographs are generally to be preferred to topographic maps, especially when landforms can be clearly seen upon them.

It has often been found useful to plot surface morphology on to a contour base map. The contours provide elevation data while the morphological mapping symbols show details of form that is usually not displayed by the contours alone. Such detail may provide essential clues to the nature of the landforms (e.g. the precise margins of a landslide, the location of sink-holes) that in an environmental management context are crucial pieces of information. The morphological map is also a good basis for constructing a map of slope steepness, and this may be of value to both planners and engineers when part of their task is to avoid ground too steep for certain purposes (Table 2.5). Thus a slope map thereby becomes a useful planning tool. It is also good practice to produce a morphological map, using the symbols shown in Fig. 2.20, before or at the same time as mapping either surface materials or indicators of surface processes. Indeed, the morphological map can be used as a base map for recording these other characteristics (Fig. 2.21). This approach was used in site mapping for sections of the Dharan–Dhankuta road in Nepal (Fig. 2.22).

Slope maps can be obtained in a variety of ways. These include: direct field measurement (e.g. with an inclinometer or by levelling), from contour maps (by analysing contour spacing), or photogrametrically (by measurement from stereoscopic vertical aerial photographs). Despite the inaccuracies involved, slope steepness maps are frequently derived from contours on published maps. Indeed, although the results may not be precise (in that contour data are only an imperfect reflection of ground form) they are often taken to be sufficiently precise for general planning purposes. Thus, a slope map (Fig. 2.23) produced as part of a mapping programme for the Torbay area, UK (Geomorphological Services Ltd., 1988) was derived from contours shown on the published

FIG. 2.21 Different uses of geomorphological mapping data and styles of presentation for the same land surface

Mudstone
Conglomerate
Sandstone
Mudstone
Siltstone
Sandstone
Sand
Gravel

A Morphological/Morphometric map

Morphological mapping symbols

ᴠᴠᴠ Convex break of slope
ᴠ_ᴠ Concave break of slope
—ᵥ—ᵥ— Convex change of slope
ᴠ_ᴠ Concave change of slope
⟶ Slope direction and angle
▾▾▾ Cliff > 45°
⌐┬┬ Convex and concave breaks of slope in close association
⊢⟶ Concave unit
×⟶ Convex unit
—— Contours in metres
• Spot height
⊙ Depth of incision

B Morphochronological map

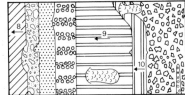

Bedrock succession

▨ Planation surface — mid-Tertiary
⬭ Conglomerate
▦ Sandstone
▥ Mudstone (highly weathered)
▯▯▯ Siltstone — late Pleistocene valley incision

Unconsolidated sediments

▨ River terrace and infill — Devensian
▨ River sand (Recent)
▨ Angular boulders — intermixed recent gravel & sand

Superficially disturbed

⟨⟩ Landslips — active
⟵ Dip

C Morphographic map

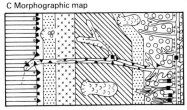

▤ Planation surface
▾▾▾ Cuesta scarp face
▪▪▪ Rock wall
▨ Scree-debris slope
▨ Pediment
▨ River terrace and valley infill
▨ Incised valley-side slope
⟨⟩ Landslides
⌐┬┬ Minor gully
○⟶ Spring   ■⟶ Waterfall   ⟿ Permanent stream

D. Morphogenetic/Dynamic map

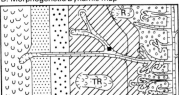

Dominant Slope Forming Processes

ᴜᴜ Soil creep and throughflow on planation surface
⋮ Frost weathering and rock fall from scarp
∴ Talus creep on scree debris
⟨⟩ Landslips on highly weathered mudstone — Active R Rotational TR Translational
▨ Potential instability on river terrace gravels
▨ Wash on terrace
⌐┬┬ Gully erosion
⌒ Actively eroding gully heads

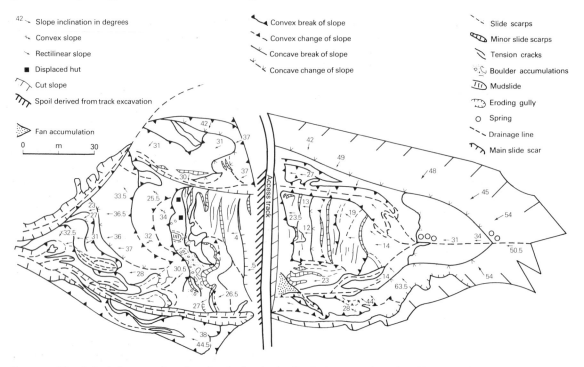

Slope inclination in degrees
Convex slope
Rectilinear slope
Displaced hut
Cut slope
Spoil derived from track excavation
Fan accumulation

Convex break of slope
Convex change of slope
Concave break of slope
Concave change of slope

Slide scarps
Minor slide scarps
Tension cracks
Boulder accumulations
Mudslide
Eroding gully
Spring
Drainage line
Main slide scar

FIG. 2.22 Morphological map used as a base for plotting the details of a deep-seated rotational landslide along the proposed route of the Dharan–Dhankuta road, Nepal (after Hearn and Fulton, 1986; reproduced by permission of the Geological Society of London).

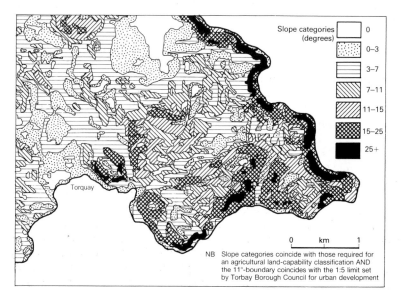

Slope categories (degrees)

	0
	0–3
	3–7
	7–11
	11–15
	15–25
	25+

FIG. 2.23 Segment of a slope map of the Torbay area (courtesy of Geomorphological Services Ltd)

Torquay

NB Slope categories coincide with those required for an agricultural land-capability classification AND the 11°-boundary coincides with the 1:5 limit set by Torbay Borough Council for urban development

Ordnance Survey 1:25 000 map. The slope classes were chosen to coincide with (i) an urban construction slope limit of 11°, and (ii) other slope values useful for defining within the rural areas grades of land capability for agriculture (see Sect. 2.6). Similarly, in the USA, in the Connecticut Valley (where 8° or 15 per cent is taken as the effective upper limit for building conventional houses and carrying out intensive urban land uses), a slope map was compiled from contour data (US Geological Survey, 1979).

Although the task of producing a slope map, using a template such as that published by Thrower and Cooke (1968) is possible, it is more cost-effective to use either photo-mechanical methods (see US Geological Survey, 1979) or computerized methods based on digital terrain modelling procedures. In the latter case the contours need to be digitized to magnetic tape, but once in that form it is possible to produce maps at a variety of scales and with a variety of slope classes by computerized means (as was done for the Torbay study mentioned above).

(b) *Mapping surface materials*

Surface materials occur either as solid rock or as superficial deposits (which include both weathered bedrock and transported sediments). These deposits are all *soils* to the engineer, though the pedologist and agriculturalist would restrict the use of the word 'soil' to the medium in which plants grow. The distinction between bedrock and regolith (i.e. the products of weathering and denudation) is often important (Chapter 12). The boundary between the two (the *weathering front*, Chapter 12) marks the junction of materials with normally quite different physical properties and different influences on the retention, passage, and chemical composition of water. Mapping lithology and structure in the field is described in standard textbooks in geology, such as Gilluly *et al.* (1968); Longwell, Flint, and Sanders (1969); and Barnes (1981). Mapping geology from aerial photographs is explained in Allum (1966), American Society of Photogrammetry (1960), and Drury (1987).

There are many distinctive properties of both bedrock and regolith that can be mapped in the field and recorded on a geomorphological map. Their relevance depends on the purpose of the map. A distinction can be drawn between those characteristics that are required to be known for most purposes (e.g. bedrock lithology) and those material properties that may only be required for special purposes (e.g. slope stability analysis requires a knowledge of shear strengths). The former would appear on a general geomorphological map while the problem-specific technical information would normally be shown on a separate special-purpose, or problem-oriented map.

Bedrock lithology is recorded in terms such as granite, limestone, sandstone, shale, and slate, but the list can be increased to include other frequently encountered rock types. Examples are given in Fig. 2.7 together with the symbols by which they are normally recorded. Also the general-purpose geomorphological map would show bedrock structures, such as fold axes and fault lines (Fig. 2.8), as well as the features which are the direct result of structure (Fig. 2.9), and those of volcanic origin (Fig. 2.10).

It is not possible to identify every special-purpose application of geomorphological mapping for which separate maps of selected properties of materials can be drawn. In the context of weathering, for example, it may be useful to portray not only the spatial distribution of the grades of rock decomposition as defined in Table 12.4 but also the types of superficial materials (Fig. 2.11). This has special application in many engineering problems (such as in a search for local construction materials and in the evaluation of foundation conditions). A more detailed examination of bedrock would include its subdivision on the basis of mineral composition.

The depth of the weathered mantle is of relevance in many specialized studies, but the data are not easily obtained except from isolated exposures of borehole records. Porosity and permeability are physical properties of material that have a strong influence on water penetration (infiltration) and thus on weathering. This type of measure, as with soil texture, shear strengths, liquid and plastic limits (see Chapter 5), and indeed any geotechnical data concerning the mechanical properties of materials belong to very specific problems and require mapping at a detailed scale. This type of mapping takes place at the site investigation stage of projects, such as along proposed routes or at a foundation site. The relevant characteristics in an engineering geology sense are incorporated in a special issue of the *Quarterly Journal of Engineering Geology* (Engineering Group Working Party, 1972) which describes a proposed legend for engineering geology maps, and in the UNESCO (1976) publication *Engineering Geology Maps: A Guide to their Preparation*. Later chapters demonstrate the role of both bedrock and soil assessments in relation to particular themes and environmental problems.

(c) *Mapping geomorphological processes*

It is rarely the case that an actual process is mapped. Much more usual is the mapping of either surface form or surface materials, or both, as a surrogate for the process itself (see Chapter 3). For example, *landsliding* (as a process) is not mapped but a *landslide* (as a feature betrayed by its form and the evident movement of soil or rock materials) may be. *Flooding* is not mapped, but the upper

limits of a past flood, or the margins of a flood plain, may be. In Fig. 2.22 both landslide features and features resulting from flooding are shown for an area in Nepal. The map *implies* process but actually shows a landform response.

Process maps therefore usually record an interpretation of forms and materials associated with a defined process, or they record other data, such as hydrological observations in the case of a flood, that relate to a process. This second approach is also exemplified by those maps which not only show the distribution of process-related features (e.g. wind-blown sands), but also show the nature of the process (e.g. wind roses) (Fig. 2.24).

In terms of environmental management an interpretation of the origin and processes affecting landforms is often critical to effective land plan-

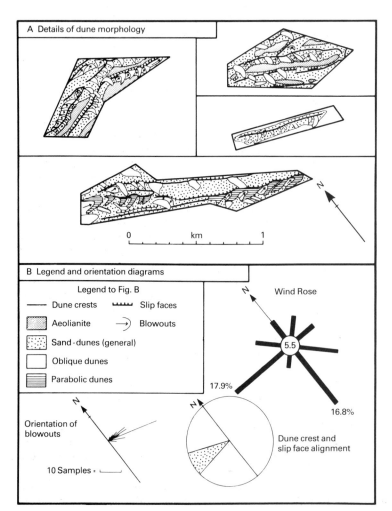

FIG. 2.24 Sand-dune characteristics and wind rose for a proposed development area in Dubai (after Cooke *et al.*, 1982)

ning. Thus, for example, wind erosion (Chapter 9) gives rise to specific landforms (Fig. 2.13). Under extreme conditions distinctive desert forms occur, such as sand-dunes, but in marginal areas smaller-scale forms (e.g. deflation hollows, sand-and-silt mounds around vegetation) may exist which indicate the general potential danger of wind erosion in the area. The mapping of these may indicate the extent of the area potentially suscep- tible to this danger. Soil erosion by water (Chapter 4) gives rise to distinct surface features. Active gully systems may be the first sign that a problem exists, and their location may betray whether they are caused by overland flow or saturated through- flow (Chapter 4). Mapping the extent of the problem is a desirable prerequisite for estimating the cost of remedial measures and for deciding upon the necessary and most effective control measures. In fact, mapping the fluvial features of any landscape is commonly an important part of its effective management. A set of symbols useful for this purpose is shown in Fig. 2.16. Such features include those relating to river-channel morphology (Chapter 6), areas susceptible to flooding (Chapter 6), and the various other fluvial components of a drainage basin.

UNESCO (1970) published a mapping legend of value in mapping hydrogeological aspects of terrain, but this legend has much in common with that normally to be found on a competent geomorphological map. The background informa- tion necessary for a hydrogeological map is that of bedrock and superficial materials as well as that of river characteristics, all of which would normally be identified on a geomorphological map. The nature of the geological and soil materials deter- mines groundwater storage characteristics. The definition of watersheds (catchment boundaries) is relevant to any projects planned within separate drainage basins. Variations in soil drainage pro- perties (e.g. as defined through an examination of soil texture) are often closely coincident with landforms identified on a geomorphological map. For the most part little additional fieldwork is necessary for the conversion of a geomorphologi- cal map into one showing hydrogeology. The legend for hydrogeological maps published by UNESCO (1970) also suggests that information on groundwater hydrology, hydrochemistry, boreholes, and wells should be included where these are known in sufficient detail.

As is shown in Chapter 5, no one condition can

be held to be responsible for slope instability, for instability usually results from a combination of factors occurring together. The results of instabil- ity can be mapped, but a prediction of those areas liable to future instability has to be based on an assessment of the spatial juxtaposition of a whole set of conditioning characteristics (Table 5.5). Prediction can be greatly assisted, however, by a geomorphological map which has been compiled to include all relevant factors likely to influence stability. In Fig. 2.25, for example, not only are landslides shown but also characteristics of ero- sion and bedrock lithology which relate to slope stability.

Coastal features (Fig. 2.14) tend to form a special group in that they are generally easy to identify, except where they have been abandoned by a relative lowering of sea-level and have become degraded under subaerial processes. Thus, raised beaches may become less easy to identify the older they are. Coastal processes are not always easy to map, but tidal ranges, the dominant wave direc- tions, and conditions of longshore drift are amongst the variables commonly of interest in coastal management.

Permafrost and periglacial features (Fig. 2.15) may be readily identified where there is little vegetation. Chapter 8 shows that such features are indicative of causative processes, and include ground hummocks formed by ground heaving, pingos which are associated with intrusive ice, polygonal patterns related to ice wedges, and flow forms due to solifluction. Such features can occur as fossil forms in temperate areas, and the importance of some of these to slope stability analysis is referred to in Chapter 8.

Glacial landscapes are subject to much landslid- ing, and also provide the sites for some sand and gravel deposits. They are also being increasingly used as tourist areas and for the development of recreational resources. In addition, they are fre- quently the setting for major civil engineering projects, such as dam construction. For all of these purposes advance knowledge of the terrain's attributes is important to its efficient and safe use (Chapter 8).

To complete the basic set of legends which can be used in geomorphological mapping, Fig. 2.17 illustrates the symbols that may be employed for karst areas (see also Chapter 14) while Fig. 2.18 includes some useful symbols for major denuda- tional forms. There may also be instances in which

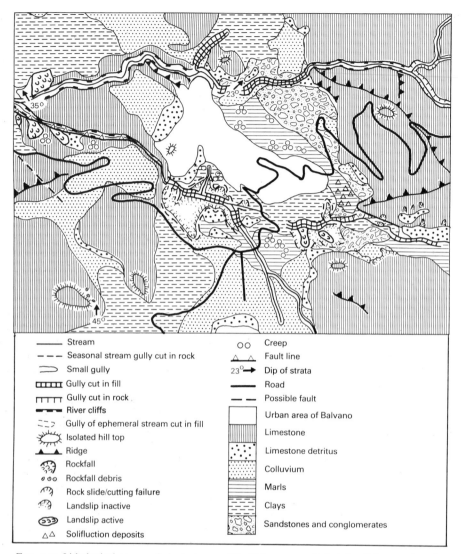

—— Stream	○○ Creep
- - - Seasonal stream gully cut in rock	△—△ Fault line
⊃ Small gully	23°➤ Dip of strata
⊞⊞ Gully cut in fill	—— Road
⊤⊤⊤ Gully cut in rock	- - Possible fault
▬▬ River cliffs	Urban area of Balvano
⁻⁻⊃ Gully of ephemeral stream cut in fill	Limestone
⌣ Isolated hill top	Limestone detritus
▲ Ridge	Colluvium
⌓ Rockfall	Marls
₀ ₀₀ Rockfall debris	Clays
⌒ Rock slide/cutting failure	Sandstones and conglomerates
⌒ Landslip inactive	
⊙ Landslip active	
△△ Solifluction deposits	

FIG. 2.25 Lithological-geomorphological map of the Balvano area, Italy (after Fulton *et al.*, 1987;
© 1987 and reproduced by permission of J. Wiley and Sons Ltd)

man-made features need to be recorded (Fig. 2.19).
A more comprehensive set of symbols is given in
Demek (1972).

(d) *Mapping the age of landforms*

Identifying the age of landforms is one of the most
difficult tasks in geomorphology unless specific
evidence is available (e.g. [14]C dates, fossil content,
or archaeological evidence). While in the case of
individual features it may sometimes be possible to
assign a specific age, it is rarely so in the case of

larger expanses of terrain. Specific dating becomes
possible, for example, in the case of an end-
moraine when historical (often map) records show
the date at which a glacier snout was at that
location. Elsewhere ^{14}C dating of organic material
may enable the age of a beach deposit to be
established. It is possible to date a landslide when
historical records, or the memory of local inhabitants, identify the moment of that event with some
precision. That is quite different, for example, from
being able to place an age on the whole slope

within which the landslide has taken place, or indeed on the plateau surface below which that slope may be cut.

Despite the subjectivity involved, many geomorphological maps, especially those produced in Europe, do attempt to assign dates to the origin of a land surface. Such an approach is always linked to a conceptual model of landscape evolution. In the European situation this is usually related to concepts of Quaternary landform evolution with special reference to periods of glaciation and climatic change. Problems arise, of course, when the evolutionary model is revised.

For environmental management purposes, knowledge of surface materials and active processes is usually far more important than knowledge of absolute age, except where recent events are likely to be repeated (e.g. flooding). The time-period of interest to the land-user is the time-period of the projected/intended life of the project. Of greatest relevance therefore are events that will happen at that site during the existence of the project. This requires knowledge about the magnitude and frequency of current geomorphological events. Thus if an event or feature is assigned to the present, attention is automatically focused again on processes.

2.4 Data Sources for Mapping

For most practical purposes geomorphological mapping requires a first-hand inspection of the site or problem in the field. However, there are also many occasions when other approaches can provide useful, sometimes unique, information. For example, remote sensing sources, ranging from satellite imagery to aerial photography, can and often do provide information not readily obtained by any other means. This is especially the case when they provide: (i) a view of areas not accessible in the field (in which case they are a substitute for field mapping); (ii) a historical record of former conditions (in which case they form a base-line against which present conditions can be compared); (iii) a synoptic view (or perspective) not available by any other means; (iv) data, by means of special sensors or films, in the non-visible part of the spectrum (in which case they supplement what can be mapped in the field or by means of ordinary aerial photography).

The essentials of remote sensing, as they relate to applied geomorphology, are presented below.

There are also other types of data that are valuable either as part of a mapping programme or as a complementary data set. Sometimes such data would be shown on the map, sometimes not. The former includes information on the thickness of superficial deposits (such as glacial drift), which may be shown as values at specific sites or as isolines. Alternatively such data might be shown as a table to accompany the map.

Another type of data that forms a useful addition to some maps relates to surface processes. For example, for an area of sand-dunes it is useful to include a wind rose on the map (Fig. 2.24). In a coastal area it is useful to include not only a wind rose but also information on the length of fetch influencing different beaches. On maps of river systems it is sometimes informative to provide inset diagrams of flow hydrographs.

(a) *Remote sensing imagery*

The technical aspects of remote sensing, including the variety of imaging platforms and sensors, are described in the American Society of Photogrammetry's (1975) *Manual of Remote Sensing*, by Sabins (1978), Lillesand and Keifer (1979), Townshend (1981), Curran (1985), Drury (1987), Harris (1987), and Mather (1987).

Geomorphological applications are examined in Verstappen (1977*b*) and Rosenfeld (1984); and many examples of the application of remote sensing imagery to mapping are described in *International Archives of Photogrammetry and Remote Sensing* (1986). Aerial photographic methods are described in the American Society of Photogrammetry's (1960) *Manual of Photographic Interpretation*, and Zuidam's (1985–6) *Aerial Photo Interpretation in Terrain Analysis and Geomorphological Mapping*. It is important to define the place of these techniques for geomorphology in environmental management. In order to do this it is useful to return to the four situations identified above.

(i) *Areas not accessible in the field* In remote areas some sites may be inaccessible in the field by virtue of distance or the physical difficulty of reaching them. In such cases, any available remote sensing imagery is useful, but aerial photography at an appropriate scale is especially valuable. In fact, since in any field-mapping programme it is rarely possible to see all sites on the ground, it is always useful to have aerial photography available. This concept is admirably illustrated by the examples given in *Applied Geomorphology: Geo-morphological Surveys for Environmental Development* (Verstappen, 1983).

(ii) *Historical records of former states and recording change over time* In particularly dynamic environments landform changes can occur between successive dates of acquisition of remote sensing imagery. Whether or not the imagery is useful in recording such changes depends on the magnitude and nature of the change; the sensors used to record the image; and the resolution and scale of the final image. The lower the spatial resolution of the sensors, the less likely they are to pick up small changes (Fig. 2.26). For this reason, the LANDSAT MSS (multi-spectral sensors with an effective pixel resolution of 80 × 80 m) is unable to record some types of changes, and hence is not an adequate historical record of a former state when the amount of change affects a land area less than 80 m in any one direction. In practice the change has to cover about three pixels (i.e. 240 m on the ground) before an interpreter can begin to treat the record with confidence. This means that successful applications of satellite imagery are restricted to large-scale events, such as the encroachment of more arid conditions (desertification) into parts of Africa, or tracing the passage of

volcanic events during the eruption of Mount St Helens (Rosenfeld, 1980).

To a lesser extent the same limitations apply to LANDSAT TM (thematic mapper) imagery, though effective pixel resolution is better (30 × 30 m). Thus on TM imagery it is possible to identify large landslides (which is not the case with MSS imagery), but not all landslides. In 1986 the French satellite system SPOT was launched with a capability of providing some imagery at 10 m resolution, and also with a stereoscopic image facility. Thus, even from space systems, technical improvements are allowing more and more details of surface features to be identified—which makes them increasingly useful both in remote areas and as historical data sources. In practice many geomorphologists still use aerial photography to record change over time, but it is frequently difficult to obtain aerial photographs of the area of interest, especially when it is in a country that is sensitive about military security. To that extent satellite imagery is useful because if it has been recorded it is normally also available on request. In these circumstances it may be the only remote sensing source available to supplement mapping.

(iii) *Synoptic view* One of the most valuable features of remote sensing imagery is that it

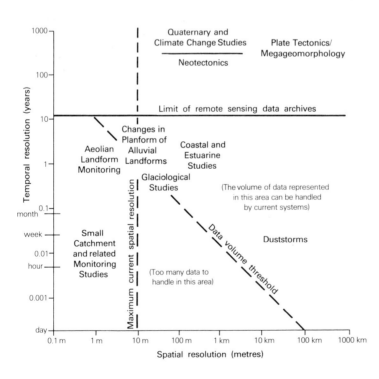

FIG. 2.26 Geomorphological applications of satellite remote sensing in relation to spatial and temporal resolution, and illustrating the factors limiting their suitability (after Millington and Townshend, 1984*b*)

provides a view from above showing the land surface in a manner very different from that obtained by standing on the ground. This alternative perspective can be very revealing. It often allows the patterns of features to be seen, and the spatial relationship of one set of features to be noted with respect to other features in a way that may not be possible from any other perspective. For example, a braided river pattern is usually best appreciated from above. The pattern of sand-dunes in a desert is often best seen from above, and their relationship to rocky hills, mountain fronts, natural depressions, and areas of high water-table, is often most apparent on vertical aerial photography or satellite imagery (Jones *et al.*, 1986). Fig. 2.27 shows directions of sand drift in a desert area derived from LANDSAT imagery; these directions are not obvious on the ground and are very difficult indeed to derive from field observations. Such association in space, and the patterns of the various landforms involved can often provide a valuable indication of the behaviour of the processes involved.

One important advantage of space imagery over aerial photography is that it can provide a single view of a larger area of ground. A typical LANDSAT MSS image, for example, is normally produced so that it shows an area of about 185 × 185 km. This allows a regional perspective on terrain and the pattern of landforms that is not available by any other means. It can also provide valuable ideas on landform development within the region, and if such images are mosaicked they can be used to study areas as large as the Sahara. Mainguet (1983), for example, used this approach in a study of sand-dune systems and wind-erosion features across the whole of the Sahara. McKee (1979) used a similar approach to study the sand seas of the world.

Associated with this aspect of satellite imagery is the fact that the pixel resolution suppresses detail and allows dominant forms and patterns to be well displayed. For regional studies this can be a great advantage. However, in environmental management there are very few projects that operate at this scale—though those that do are very notable (e.g. desertification and land-cover changes in Africa; major floods on the Ganges–Brahmaputra; extent of salinization of soils in the Indus Valley). In the United States published records of using space imagery for regional flood studies include those by Rango and Salomonson (1974) and by Williamson (1974).

(iv) *Data detected by special sensors (or film)* A

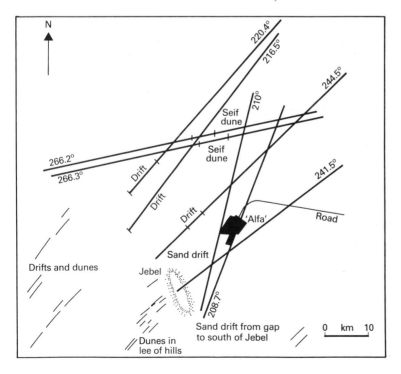

FIG. 2.27 Major dunes and sand drift directions as identified on LANDSAT imagery for an area in Saudi Arabia (after Jones *et al.*, 1986; reproduced by permission of the Geological Society of London)

major advantage of some remote sensing imagery is that data can be recorded outside the visible spectrum. On satellite and some aircraft this is done by sensors set up to record within a specified range of the electromagnetic spectrum (Table 2.6). The data are stored either by means of a digital record on magnetic (computer) tape, or photographically. In the case of camera images film may be used that records outside the visible spectrum (e.g. in the infra-red wavebands), as film negatives. The advantage of the digital record is that it is amenable to computer processing and thereby to a whole range of treatment that allow preprocessing (including geometrical rectification, registration, and resampling), data analysis (including statistical analysis, time difference studies, and image classification), and visual display, first on a computer graphics screen and then as a hard copy which resembles a photographic print.

While satellite imagery, and indeed aerial photography, is at its best in cloud-free areas with sparse vegetation (e.g. in desert areas where there may be only a limited need for applied geomorphology because of the sparse population), it may be of little value in cloudy areas or those covered in dense vegetation (or both). Special sensors, especially radar sensors, come into their own here in that they can penetrate both clouds and forests and provide an image which records terrain roughness. One form of radar imagery, side-looking airborne radar (SLAR), was used by MacDonald (1969) with success to map geological structures and the

drainage network of part of Panama State. Other examples of the geomorphological use of radar are provided by Rosenfeld (1984), Lewis (1976), McCoy (1969), Cannon (1973), McAnerney (1966), and Munday (1984). One difficulty with radar imagery is that it usually has to be specially flown. Only for Brazil is it available 'across the counter', although SEASAT imagery (obtained as L-band radar carried by satellites), and which can also be purchased on demand, may be useful for some purposes.

2.5 Mapping in Practice

Geomorphological mapping, in the context of environmental management, is carried out for three principal reasons: (i) to enable the geomorphologist to obtain a better understanding of the landscape before giving advice; (ii) to provide a map record of landscape characteristics relevant to the project in hand; and (iii) to provide an essential basis for derivative and special-purpose maps.

In these situations there is a tendency when not mapping individual components, such as surface form or materials, to map landforms according to a genetic classification.

One of the reasons for adopting this approach is that in practice the amount of time made available within the context of a project is often insufficient to make a painstaking detailed map of all aspects of the land surface. Instead the elements of significance to a project have to be identified and

TABLE 2.6 Recording characteristics of sensors

MSS LANDSAT 1, 2, 3	4, 5	Wavelength (microns)	Examples of features discernable on images
Band 4	Band 1	0.5–0.6	sediments in water, bathimetry
5	2	0.6–0.7	cultural features
6	3	0.7–0.8	vegetation, water boundaries, landforms
7	4	0.8–1.1	geological structures, landforms, water boundaries
TM Band			
1		0.45–0.52	water penetration, soils, vegetation
2		0.52–0.60	vigour of plant growth
3		0.63–0.69	vegetation discrimination
4		0.74–0.91	biomass, water bodies
5		1.55–1.75	moisture in soils
6		10.50–12.50	vegetation stress, soil moisture
7		2.08–2.35	geology, hydrothermal alteration

mapped quickly following as closely as possible the well-tried scheme shown in Table 2.7. Pushed too far, the dangers of working quickly can mean that insufficient attention is given both to details of a site and the number of sites is too few to provide a clear picture. Thus gross, and sometimes costly, errors can arise. On the other hand, a geomorphologist experienced in the environment concerned can gather a very large amount of relevant and useful data in a short amount of time, and as such his labours can be very cost-effective.

In practice, the geomorphological map produced, in order to understand what is going on in the project area, is seldom an appropriate document to pass on to the engineer, planner, or other end-user. *It is nearly always necessary to produce a new project-related document from the primary geomorphological map*, a process of translation referred to in Chapter 1. There are two main reasons for this: (i) the original map will often appear too complex to the non-geomorphologist and many of its features will be meaningless to him; and (ii) the map will carry data which are not relevant to his immediate concerns.

It is often necessary, therefore, to produce one or more *derivative* maps based on the geomorphological map but which are both clear and only carry relevant project-related information (see below). In many cases this reduces itself to either hazards faced by the project, or local resources available to the project. An example of the former

Table 2.7 Stages in a geomorphological mapping programme

1. Familiarization with problem (e.g. road construction)
2. Selection of mapping team
3. Desk study of background data (including maps)
4. Acquisition of air photos
5. Preliminary mapping procedure, mapping key, and programme
6. Planning mapping procedure and programme
7. Reconnaissance field work
8. Finalize mapping procedure and programme
9. Field mapping
10. Extrapolation of field mapping to unvisited/unsampled areas
11. Laboratory analyses
12. Cartography, including preparation of derivative maps
13. Report compilation
14. Presentation of results

would be a map of landslides, and of the latter would be a map of sand and gravel available for use as aggregate.

It is also useful to adopt the perspective of the discipline for which mapping is required. For example, the needs of a highway engineer will define the aims that should feature in a mapping programme designed for his use (Table 2.8). A similar case can be made for other disciplines.

It would be a mistake to see geomorphological mapping as an exercise unrelated to other methods of terrain investigation. For example, geomorphological maps provide a useful basis for extrapolating subsurface information gathered from trial pits or shallow boreholes. Investigations of ground conditions for urban planning in Suez, Egypt (Bush *et al.*, 1980) ran into interpretation difficulties while only trial-pit records existed. As soon as geomorphological mapping defined both the nature and distribution of the landforms of the area, the subsurface data could be sensibly interpreted with reference to a terrain framework. For

Table 2.8 Aims of geomorphological mapping for highway engineering

1. *Identification of the general terrain characteristics of the route corridor*, including suggestion of alternative routes and location of hazards
2. *Defining the 'situation' of the route corridor*, e.g. identifying influences from beyond the boundary of the corridor
3. *Provision of a synopsis of geomorphological development of the site*, including location of materials for use in construction and location of processes affecting safety during and after construction
4. *Definition of specific hazards*, e.g. landsliding, flooding, etc.
5. *Description of drainage characteristics*, location and pattern of surface and subsurface drainage, nature of drainage measures required
6. *Slope classification*—according to steepness, genesis and stability
7. *Characterization of nature and extent of weathering*—also susceptibility to mining subsidence and erosion
8. *Definition of geomorphological units*—to act as framework for a borehole sampling plan and to extend the derived data away from the sample points

Source: Brunsden *et al.* (1975).

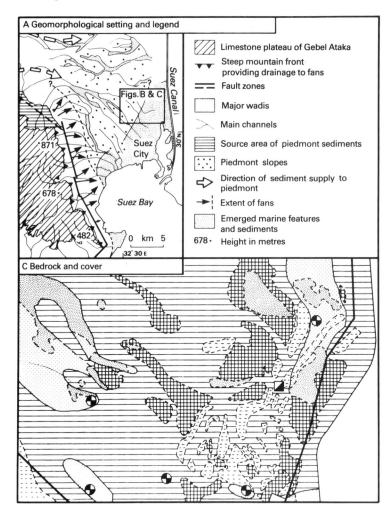

instance, through the mapping, a pattern of raised marine features (Fig. 2.28) (including former offshore bars and raised beaches) was identified and used as a basis for extrapolating trial-pit observations up to the appropriate boundary of each feature mapped. Thus, sense could be made out of the different soil observations made in each of the trial pits.

This approach can also be used the other way round. Once a geomorphological map has been prepared it can be used to decide on the location of trial pits or boreholes so that the smallest number of pits can be used to provide the maximum amount of subsurface information. This is achieved by making sure that there is a minimum appropriate number of observations within each of the main landform components. This approach

was adopted at a proposed airport site in Dubai, greatly reducing the number of boreholes and trial pits used compared with the number on an adjacent and geomorphologically very similar site where the sampling was based on a grid unrelated to geomorphology (Doornkamp *et al.*, 1979).

Geomorphological mapping has begun to feature prominently in both environmental geology mapping and engineering geology mapping, each of which has received considerable support and which has had growing success in influencing planning policy since the mid 1970s.

(a) *Environmental geology mapping*

The impetus for environmental geology mapping has come mainly from the United States. Its purpose (Robinson and Spieker, 1978) is to inform

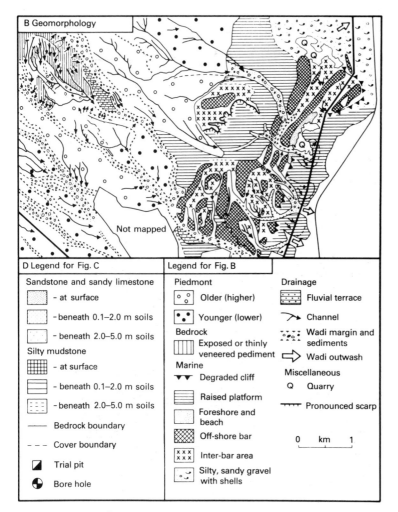

FIG. 2.28 Geomorphological survey of the Suez area, Egypt: application to bedrock and superficial sediments (after Bush *et al.*, 1980)

planners, policy-makers, or land developers so that they can: (i) forestall or relocate new developments in areas where lives or property might be in danger; (ii) propose appropriate design precautions in developments that cannot be placed elsewhere; and (iii) alert inhabitants of imperilled developments to seek protection through engineering or insurance. These aims are also often the aims of geomorphological mapping, and it is not surprising that many environmental geology mapping programmes contain a strong geomorphological mapping component. Although coincident in time with the development of environmental impact statements after the passing by Congress of the National Environmental Policy Act (1969) environmental geology mapping had an independent basis for development. Through the work

of the US Geological Survey in California (Brabb *et al.*, 1972; Taylor and Brabb, 1972; Nilsen *et al.*, 1979), for example, there was an increasing awareness amongst planners, engineers, and administrators, that the right geological information could form the basis for land zoning in advance of urban development. In particular, in California, this related to landslide hazards. Elsewhere in the United States environmental geology mapping was carried out at a variety of scales and for a range of development purposes, several of which are described in *Nature to be Commanded* (Robinson and Spieker, 1978).

In Britain, beginning with a study of the Glenrothes area (Nickless, 1982), about 20 environmental geology mapping studies have been carried out. Most have been funded by the

Department of the Environment and focus on areas extensively affected by mining (e.g. Glasgow, West Midlands), urban areas experiencing redevelopment (e.g. London docklands), or areas of urban expansion where there are competing pressures for land. In each of these studies, although to varying degrees, geomorphological maps are included (e.g. steepness of slope, sites of landsliding or subsidence).

The stated purpose in most environmental geology reports in the UK is to provide the layperson (i.e. one without a formal training in geology or geomorphology) with uncomplicated information about ground conditions. Hence, each map produced needs to have a simple legend and relate to a single purpose, such as risk of landsliding or extent of ground subsidence. To achieve this, there has to be a process of simplifying all the available data—beginning with inventory and passing through analysis and synthesis to presentation—a task which may lead to an over-simplification of reality, at least in the eyes of the scientist producing the map (e.g. Fig. 2.29). Such simplification tends to hide specific site details, and hence makes the maps unsuitable for site-specific use, but they are especially valuable for regional policy and planning purposes. The same information would need to be shown in much greater detail if required for engineering decisions, especially at the site level. For this reason some reports include two versions of the maps, one with more detail for a technical audience, the other, with little detail, for a non-technical audience.

(b) *Engineering geology mapping*

Mapping for engineering, and even town-plan-

ning, purposes has generally come under the heading 'engineering geology mapping'. Many examples of these maps are described in various issues of the *Bulletin of the International Association of Engineering Geology*, with a suggested mapping legend provided in *Bulletin 24*. In general these maps emphasize four themes (Doornkamp *et al.*, 1986): (i) geotechnical characteristics of surface materials; (ii) resources available for construction purposes; (iii) risk from earthquakes (seismic activity); and (iv) the sensitivity of the environment to change. All these themes commonly require geomorphological data to be included.

Amongst the most common topics to appear on these maps are: stratigraphy, lithology, bedrock structure, tectonic/seismic history, hydrology, geohydrology, geotechnical properties of materials, geomorphology, mining activity, land use, and agricultural soils. In many cases the studies include maps that show a zoning of the land in terms of suitability for development, reclamation potential, or appropriate engineering/planning practices.

The maps tend to have a considerable influence on policy for they are sometimes used as a basis for regulating permissible land uses, for prohibiting certain land uses, or even discouraging certain land uses through tax assessments or financial policies.

In France, engineering geology maps are used in the context of what is known as POS (les plans d'occupation des sols) and ZERMOS (zones exposés aux risques liés aux mouvement du sol et du sous-sol) (Porcher and Guillope, 1979). POS is linked to *Le Code de l'urbanisme et de l'habitation, Article R.128. 18 (1970)* under which planning

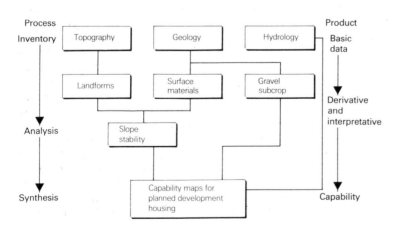

FIG. 2.29 An example of sieve mapping (after Robinson and Spieker, 1978)

documents must make clear the existence of geomorphological hazards such as floods, erosion, subsidence, and landslides. As such these documents establish codes of practice for construction or development (Champetier de Ribes, 1979). Fig. 2.30 shows the legend for a ZERMOS map with its qualitative classification of hazard.

In Britain there have been several attempts to introduce engineering geology mapping as a technique (Engineering Group Working Party, 1972, 1982; Dearman and Fookes, 1974), but the current practitioners appear to be few. Certainly there has, as yet, been no attempt to bring it into planning procedures.

In the United States engineering geology maps have been produced, frequently by abstracting information from published geological maps. Gardner and Johnson (1978) provided an example in the context of regional planning in which the geological maps for the area around Boulder, Colorado, were used to produce maps showing expandable clays, landsliding, flooding, and high groundwater, mining activity features which form the particular parameters of concern to engineers in *that* area. Further examples are given in Ferguson (1974) and in Robinson and Spieker (1978).

(c) *Derivative application maps*

Almost without exception the geomorphological map, as produced by a geomorphologist for his own purposes in coming to terms with a landscape, is not the right document to pass on to others concerned with environmental management. What has to be created from the original map is another which is designed to show those features specifically relevant to the management concern in that area. Thus, for example, a civil engineering construction project may require a map of existing landslides and of landslide-prone slopes which are currently near the limiting angle of stability. Other data, although intrinsically interesting to the geomorphologist, if shown on the map may only serve to confuse and to take attention away from the real issues facing the engineer. This is not to be confused, however, with the geomorphologist's need to record other data in order to understand why landslides (or potentially unstable slopes) occur where they do.

A distinction has to be made here between single-element maps and derivative maps. *Single-element maps* include all those maps that show one

ground property derived from the geomorphological analysis. Examples include maps of slope steepness (Fig. 2.23), location of landslides, and geotechnical properties of rocks and soils. Derivative maps on the other hand are based not only on data on geomorphological maps but also on analysis and interpretation of the data, and their presentation in a distinctive form of value to clients. *This process of translation is fundamental.* Examples of such maps include those showing ease of rock excavation, susceptibility to landsliding, appropriate locations for waste disposal, and almost all maps of geomorphological hazards (including, for example, flooding, salt weathering of foundations, and wind-erosion risk) and resources (such as aggregates). Such maps commonly adopt a qualitative scale or risk (e.g. 'high', 'medium', 'low'), as for example in Fig. 2.31 which shows a flood-hazard map derived from a geomorphological survey in Egypt (see also Chapter 3). Fig. 2.32 shows the location of aggregates for a region in Bahrain which is exclusively based on the antecedent geomorphological map field survey, and its interpretation.

2.6 Geomorphology in Other Mapping Contexts

(a) *Geomorphology and environmental impact assessment*

The environmental impact assessment process gained considerable momentum after the passing of the Natural Environmental Policy Act in 1969 in the United States, and is now accepted as part of both planning and environmental care by many countries as well as by the World Bank and agencies such as the US Agency for International Development (USAID). Within Europe the environmental impact assessment process was stimulated by the EEC Directive of June 1985 which made it mandatory for Member States to have appropriate legislation in place by June 1988. This legislation makes it obligatory for proposals for major development projects to be accompanied by an environmental impact assessment. Environmental impact assessment (EIA) has been described in many texts (e.g. Leopold *et al.*, 1971; Dickert and Domeny, 1974; Catlow and Thirlwall, 1976; Canter, 1977; Heer and Hagery, 1977; Munn, 1979; Bisset, 1980; Chapman, 1981).

The purpose of the EIA process is to determine

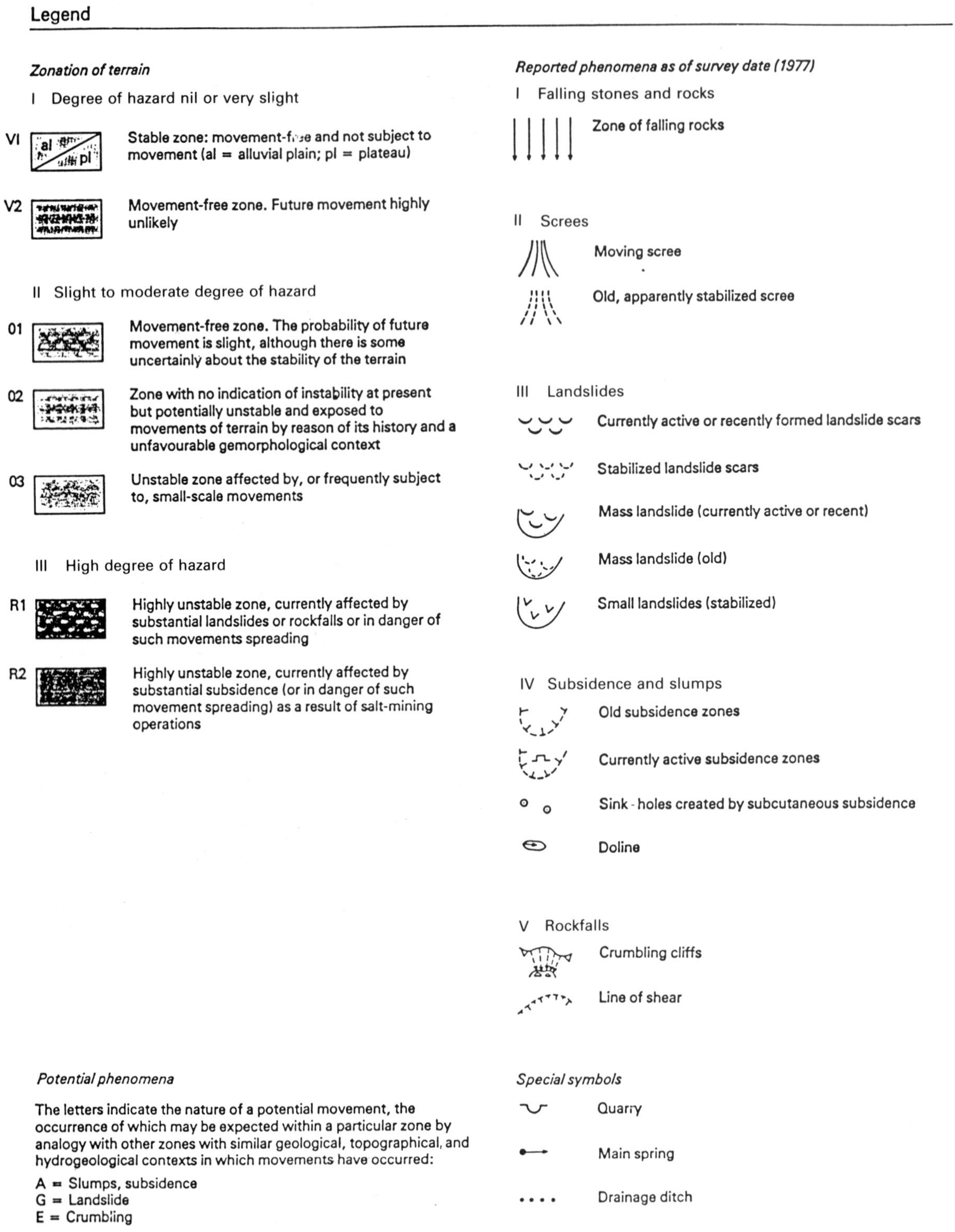

Legend

Zonation of terrain

I   Degree of hazard nil or very slight

V1 Stable zone: movement-free and not subject to movement (al = alluvial plain; pl = plateau)

V2 Movement-free zone. Future movement highly unlikely

II   Slight to moderate degree of hazard

01 Movement-free zone. The probability of future movement is slight, although there is some uncertainly about the stability of the terrain

02 Zone with no indication of instability at present but potentially unstable and exposed to movements of terrain by reason of its history and a unfavourable geomorphological context

03 Unstable zone affected by, or frequently subject to, small-scale movements

III   High degree of hazard

R1 Highly unstable zone, currently affected by substantial landslides or rockfalls or in danger of such movements spreading

R2 Highly unstable zone, currently affected by substantial subsidence (or in danger of such movement spreading) as a result of salt-mining operations

Reported phenomena as of survey date (1977)

I   Falling stones and rocks

Zone of falling rocks

II   Screes

Moving scree

Old, apparently stabilized scree

III   Landslides

Currently active or recently formed landslide scars

Stabilized landslide scars

Mass landslide (currently active or recent)

Mass landslide (old)

Small landslides (stabilized)

IV   Subsidence and slumps

Old subsidence zones

Currently active subsidence zones

Sink - holes created by subcutaneous subsidence

Doline

V   Rockfalls

Crumbling cliffs

Line of shear

Potential phenomena

The letters indicate the nature of a potential movement, the occurrence of which may be expected within a particular zone by analogy with other zones with similar geological, topographical, and hydrogeological contexts in which movements have occurred:

A = Slumps, subsidence
G = Landslide
E = Crumbling

Special symbols

Quarry

Main spring

Drainage ditch

FIG. 2.30 Key accompanying the ZERMOS map of the Lons-le-Saunier region at Poligny, France (Landry, 1979)

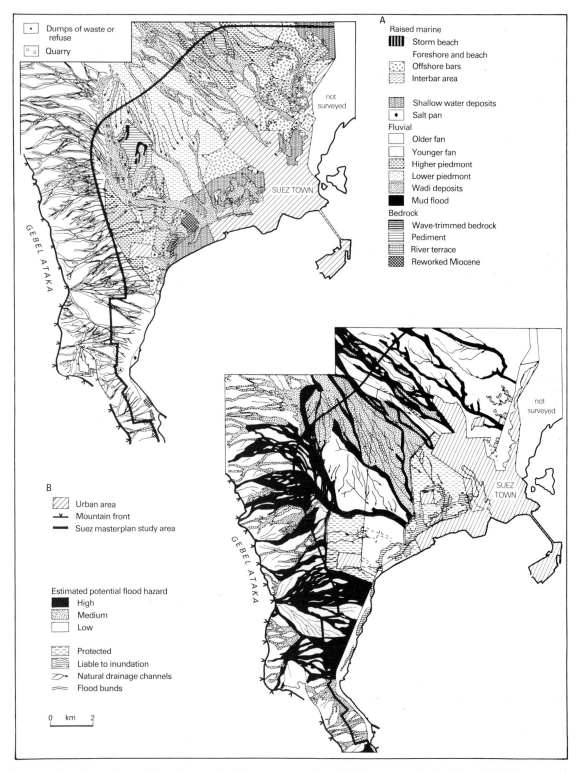

FIG. 2.31 Flood-hazard map of Suez, Egypt derived from geomorphological field mapping (after Bush *et al.*, 1980). Compare with Fig. 2.28.
(A) Geomorphological map (B) Flood-hazard map

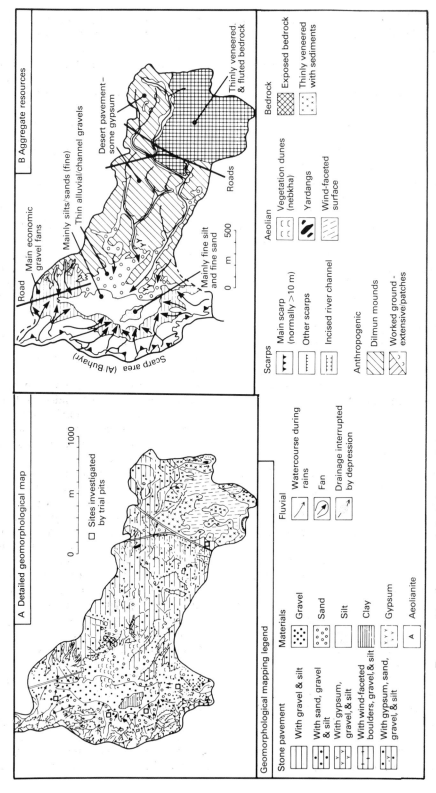

FIG. 2.32 Use of a geomorphological map to derive an aggregate resources map, Bahrain (after Doornkamp *et al.*, 1979)

the potential environmental, social, and health effects of a proposed development. Clearly the EIA includes a great deal more than just geomorphology, though 'environment' often includes geomorphological aspects. In particular, the geomorphology of a site (or area) may place constraints upon the proposed development (e.g. as revealed by flood-hazard zoning or by an assessment of slope instability); or the development, as proposed, may have an adverse effect upon geomorphological conditions (e.g. by increasing coastal erosion, or by adversely affecting stream flow). The geomorphological contribution to the EIA therefore consists of making an analysis of the landforms, surface materials, and processes within and adjacent to the project area, to see if they will either produce constraints or react unfavourably to the development. Such an analysis will almost certainly include geomorphological mapping, but it will also require the systematic analysis of both processes and materials along the lines described in other chapters. A geomorphological analysis can also help to define the character of the proposed project area (as is required by an EIA); to forecast any changes that will result from the proposed development; and to find alternatives that would reduce any adverse impacts upon the environment.

(b) *Geomorphology and land-capability classification for agriculture*

Land-classification schemes exist in order to grade land according to its quality for agriculture and pastoral activities. The criteria used are normally those of soils, relief, and climate as illustrated in Fig. 2.33 for Britain and Fig. 2.34 for the United States.

In many ways the classification systems are similar in kind to that of the land-systems approach described in Sect. 2.2. The agricultural land classification used in the United States, for example, recognizes a threefold hierarchy from smaller to larger groupings of (i) capability units, (ii) capability subclasses, and (iii) capability classes (Klingebiel and Montgomery, 1961). The capability unit resembles the land unit of a land system in that the soils in a capability unit are sufficiently uniform (i) to produce similar kinds of cultivated crops and pasture plants with similar management practices, (ii) to require similar conservation treatment and management under the same kind and condition of vegetative cover,

and (iii) to have comparable potential productivity. The capability unit is described as a grouping of soils that are nearly alike, and is based on an interpretation of soil data. The relevance of geomorphology can be assessed only when the basis for describing the soils is known. If site relief and drainage conditions are included in the soil classification then geomorphology becomes intimately involved. On the other hand if the soils are defined more on the basis of their internal physical and chemical properties then the relevance of geomorphology to the classification may be more remote.

At the subclass and class level of the United States capability classification, however, there is a great dependence on relief and drainage characteristics. Subclasses are defined according to their limitations for agricultural use and according to the hazards to which they are exposed. Four general kinds of limitations are recognized: (i) erosion hazard, (ii) wetness, (iii) rooting-zone limitations, and (iv) climate. Of these the first three are closely related to the geomorphology of an area. Subclasses are composed of groups of capability units.

At the most general level there are eight capability classes, which, under the United States system of notation, are numbered from I to VIII according to a scale of increasing severity of soil damage or limitations. Soils in class I to class IV under good management are capable of producing adapted plants, such as forest trees or range plants, and the common cultivated field crops and pasture plants, but some conservation practices become increasingly necessary through classes II–IV. The range of uses decreases through classes V–VII, and soils in class VIII do not return on-site benefits for inputs of management for crops, grasses, or trees without major reclamation. The particular nature of hazards and limitations is summarized in Fig. 2.34. Locally soil salinity may produce a limitation additional to those listed in this figure. Relief characteristics provide an important element in assigning land to a particular capability classification. Indeed, an initial classification of land on the basis of slope steepness and its liability to erosion or flooding can provide a first approximation of the eventual land-classification class based on soil characteristics as well. This is on the assumption, of course, that climate alone has not already determined the allocation to a particular class. In the United States an estimated 2 per cent of the

Class physical limitations	Texture (SOILS)			Structure			Rooting depth				Drainage					Stoniness		Slope (RELIEF)				Liability to erosion			Liability to flooding			General character (CLIMATE)					
	Good	Fairly good	Poor	Good	Fairly good	Poor	Good	Moderate	Restricted	Poor	Good	Moderate	Imperfect	Poor	Very poor	None	Many	Flat/gentle	Moderate	Steep	Very steep	None	Slight	Severe	None	Some	Frequent	Favourable	Slightly unfavourable	Unfavourable	Moderately severe	Severe	Very severe
1 None or minor	×			×			×				×							×										×					
2 Some		×			×			×				×	×						×				×					×	×				
3 Moderately severe			×			×			×				×	×						×			×						×	×			
4 Severe (i)										×				×	×		×		×				×			×					×	×	
5 Severe (ii)														×	×					×				×			×				×	×	
6 Severe (iii)										×					×		×			×				×			×					×	
7 Extremely severe										×					×		×				×			×									×

Note : In order to be assigned to the next poorest class an area must show one or more of the detrimental characteristics typical of that class
Source: Compiled from information by Bibby and Mackney (1969)

FIG. 2.33 A land-classification system for agriculture in the UK (after Bibby and Mackney, 1969)

Class physical limitations	Texture (SOILS)			Structure			Rooting depth				Drainage					Stoniness		Slope (RELIEF)				Liability to erosion			Liability to flooding			General character (CLIMATE)					
	Good	Fairly good	Poor	Good	Fairly good	Poor	Good	Moderate	Restricted	Poor	Good	Moderate	Imperfect	Poor	Very poor	None	Many	Flat/gentle	Moderate	Steep	Very steep	Moderate	Small	Severe	None	Some	Frequent	Favourable	Slightly unfavourable	Unfavourable	Moderately severe	Severe	Very severe
1 None or minor	Not specified				×		×				×							×				×			×			×					
2 Some					×			×					×					×					×			×		×					
3 Severe						×			×				×						×	×				×			×	×	×				
4 Very severe (i)										×										×				×			×				×		
5 Very severe (ii)															×		×	×									×				×		
6 Very severe (iii)															×		×			×				×			×					×	
7 Extremely severe										×					×		×				×											×	×
8 Absolute															×		×							×									×

Source : Compiled from information by Klingbiel and Montgomery (1961)

FIG. 2.34 A land-classification system for agriculture in the United States (after Klingbiel and Montgomery, 1961)

land falls in class I, classes II, III, VI, and VII occupy about 20 per cent each, class IV 12 per cent, class V only 3 per cent, and class VIII occupies 2 per cent of the United States mainland area (US Department of Agriculture, 1962; see also Young, 1973).

In Britain a similar land-capability classification system has been proposed (Bibby and Mackney, 1969). Here land is graded into seven classes also based upon the site relief, climate, and soil characteristics (Fig. 2.33). Apart from climate, all of the criteria used once more concern landform, materials, and process, and as such are partly geomorphological in character. The chemical properties of the soil may become significant in terms of limitations on its use for agriculture, but at the primary level of classification it is the physical properties of both soils and site that matter.

The mapping of land units, as in a land-systems survey, potentially appears to be of value at the reconnaissance level of a land-capability classification. These units can then be combined into capability subclasses, or classes, according to their susceptibility to erosion or their inherent limitations for agricultural use. These combinations would not equate with land systems, however, for land systems may contain land units of differing slope steepness and liability to erosion.

(c) Land-unit mapping and land classification for agriculture; an example from Tideswell, Derbyshire, UK

The land-classification system described by Bibby and Mackney (1969) has been applied to a limestone upland area near Tideswell, Derbyshire (Johnson, 1971). In general this is an undulating upland dissected by deep and steep-sided river valleys, occasionally with flat alluvial valley-floors. Altitudes range from 150 m OD in the valley-floors to 400 m on the higher parts of the limestone plateau. Near-vertical outcrops of massive Carboniferous Limestone occur on some of the steeper valley-sides and scree slopes (mainly grassed over) are common within the deeper valleys. There is no upland surface drainage, though isolated areas of ill-drained land occur where clays prevent rapid infiltration to the limestone bedrock. Plateau-top slopes are locally steep (up to 25°), but generally the upland slopes are less than six degrees.

A land classification of this area is of consider-

able interest because it is close to the upper limit for agriculture in the Peak District. This becomes apparent in the classification as minor changes in aspect introduce micro-climatic effects which change the classification category of otherwise continuous or similar slopes.

A portion of the land-use capability map of the Tideswell area is reproduced in Fig. 2.35A, together with the classification index (refer to Fig. 2.33). Climate provides an important limiting factor on the capability of the land for agriculture, but within this general context it is the factors of slope steepness, soil materials, and drainage which, either singly or in combination, provide the key to the class of land-use capability. Fieldwork was carried out, by means of soil-profile analysis and hand-augering, in order to compile a soils map of the area. The land-use capability classification arises from combining the soils information with agronomic, climatic, and topographic information (Johnson, 1971). It is admitted that some of the boundaries were arrived at subjectively, and that some arbitrary decisions were made.

If the purpose in mapping this area had been solely to define land-use capability then many of the boundaries and class characteristics could have been defined through land-unit mapping (as defined in Sect. 2.2). Fig. 2.35B shows the land units of the same area as Fig. 2.35A as derived from an air-photo interpretation with field checking. A comparison between these two maps shows a considerable degree of accordance in the boundary positions. However, the land-unit map carries most detail. A comparison of the keys to the figures indicates that many of the physical properties of the land-capability groups can be derived from a land-unit description produced as a basis for land-use capability classification. The significant advantage of the land-unit approach is that it involves less effort and a lower cost than standard field-mapping techniques. In addition, many of the land-unit boundaries coincide with soils boundaries, since relief is an important factor in soil formation (and profile characteristics) in this area. Thus it also provides a valuable guide to soil boundary locations at the reconnaissance stage in mapping soils.

(d) Geomorphology and mapping scenic quality

In recent years the numbers in Western society with a taste for scenery, a desire to protect it, and the ability to visit it have grown rapidly, and the

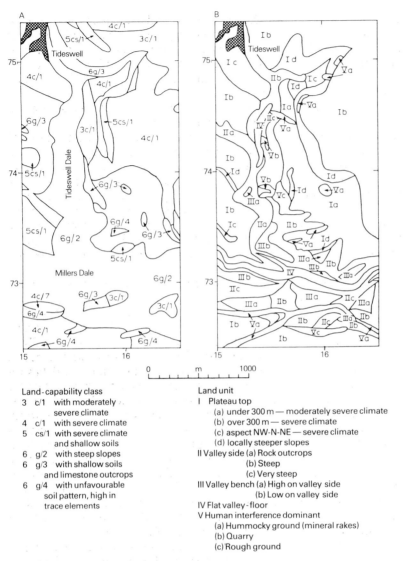

FIG. 2.35 (A) Land-use capability classification of a part of the Tideswell area, Derbyshire, UK, and (B) a land-unit subdivision of the same area (Johnson, 1971)

Land-capability class
3 c/1 with moderately
         severe climate
4 c/1 with severe climate
5 cs/1 with severe climate
         and shallow soils
6 g/2 with steep slopes
6 g/3 with shallow soils
         and limestone outcrops
6 g/4 with unfavourable
         soil pattern, high in
         trace elements

Land unit
I Plateau top
      (a) under 300 m — moderately severe climate
      (b) over 300 m — severe climate
      (c) aspect NW-N-NE — severe climate
      (d) locally steeper slopes
II Valley side (a) Rock outcrops
               (b) Steep
               (c) Very steep
III Valley bench (a) High on valley side
                 (b) Low on valley side
IV Flat valley-floor
V Human interference dominant
      (a) Hummocky ground (mineral rakes)
      (b) Quarry
      (c) Rough ground

demands for recreational landscapes has increased. But environmental managers, faced with the task of designating such areas, have had to come to terms with the problems of defining 'attractive scenery', and defining it in such a way that its aesthetic value can be compared realistically with the more easily quantified claims of the resource-users, such as mining interests, farmers, and developers. The problems are immense. Who can say, or who can deny, that a babbling brook is more valuable than a hydroelectric power-station? One thing is clear, however. In the appraisal of landscape attractiveness, the form of the ground

and the nature of geomorphological processes are normally regarded as being important ingredients (e.g. Crofts, 1975). It is therefore possible that geomorphological criteria may play a useful role in contemporary efforts to evaluate landscape for planning purposes.

The techniques that have been proposed for appraising landscape fall between two extremes. At one extreme, measurable components of landscape deemed to be representative or scenic quality are evaluated in quantitative or semi-quantitative terms. At the other extreme, emphasis is placed on the perception by 'consumers' of scenic quality;

'comprehensive' consumer responses are classified, perhaps on numerical scales, and individual components of scenery are not in themselves considered. Some techniques are only concerned with the evaluation of sites or particular views; others attempt to provide areal classification of terrain. Techniques may vary according to the scale of inquiry.

As Craik (1972) has observed, the results derived from these techniques yield several advantages to environmental decision-makers. They focus attention on criteria involving such important planning issues as landscape preservation and the routeing of power-lines. A regional description of landscape quality provides a useful context for making choices among specific sites and for evaluating land-use proposals. In addition, quantitative information may endow environmental factors with a status comparable to that of economic and social factors, although there is a danger that such status is quite spurious. And if the surveys are carried out from time to time, the rate of change of environmental quality can be monitored.

The literature on this field is extensive (e.g. Appleton, 1975; Crofts, 1975; Robinson *et al.*, 1976; Penning-Rowsell, 1981*a*). In the following brief discussion, approaches are reviewed which use geomorphological criteria and other landscape components to appraise either sites or regional contexts.

(i) *Leopold's method of site evaluation* A proposal to build a dam in Hell's Canyon on the Snake River, Idaho, was strongly opposed by conservation interests. Leopold (1969*a*, *b*) attempted to quantify aesthetic features of the proposed site, together with those of 11 other possible dam sites in Idaho and sites of renowned beauty elsewhere in the United States, in order to determine objectively the aesthetic quality of the Hell's Canyon site in the context of information from other sites. Leopold's approach was founded on the assumption that there is some benefit to society from the existence of 'unchanged' landscapes, that a unique landscape has more value to society than a common one, and that the unique qualities enhancing landscape value are those having some aesthetic, scenic, or human-interest connotations. All these assumptions, of course, are open to question.

Three types of factors were selected to represent the aesthetic qualities of a site: physical factors, biological factors, and human-use and interest factors. A total of 46 factors was identified (Table 2.9). The value of each factor at any one site was determined on a scale of 1 to 5: in some cases, the evaluation number was based on precise measurement (e.g. stream width), in others it was based on qualitative assessment (e.g. water condition), as shown in Table 2.9. The physical factors are, in general, the easiest to measure precisely. It should be emphasized that the list, though long, by no means comprehensively covers the factor that might be considered to compose the beauty of a site: for instance, important features such as smell, weather, and illumination are omitted. And there is rather little substantive evidence that the variables are in fact, either important, or equally important, in personal perception of scenic beauty by most Americans.

Site 'uniqueness' can be defined as the reciprocal of the number of sites sharing a particular evaluation number for a factor. For example, if all of 12 sites share the same number, the 'uniqueness ratio' for each site will be 1/12 or 0.08; equally, if only one site of 12 had a particular number, it would be 'relatively unique' and would have a uniqueness ratio of 1. The uniqueness ratios for each site can be summed, and the total can be compared with the totals for other sites. By simply adding factor-uniqueness ratios, each factor is given equal weight. At the same time, the total values do not indicate if a site is uniquely aesthetic or unaesthetic. In the case of 12 Idaho sites, for instance, Hell's Canyon (5) was unique in a positive (attractive) sense, and a site on the Little Salmon River (7) was unique in a negative (unattractive) sense (Table 2.10).

The data can be exploited further. For example, *valley character* and *river character* can be examined as two variables considered by some to be the most important in influencing a viewer's impression of a river-valley site.

Valley character may be, Leopold suggested, a function of *landscape scale*, the availability of distant views (*landscape interest*), and the degree of urbanization. The ranking of sites in terms of these variables on a scale of valley character can be achieved as follows (Fig. 2.36). The landscape scale is derived by plotting the height of mountains against valley width for each site, and projecting these points orthogonally on to a line sloping at 45° across the graph (thus giving equal weight to both variables). The values on this landscape scale

TABLE 2.9 Factors representing the aesthetic qualities of a site

Factor number	Descriptive categories	Evaluation numbers				
		1	2	3	4	5
	Physical Factors					
1	River width (m) at low flow	<1	1-3	3-9	9-30	>30
2	Depth (m)	<0.15	0.15-0.30	0.3-0.6	0.6-1.52	>1.52
3	Velocity (m per sec.)	<0.15	0.15-0.30	0.3-0.6	0.6-1.52	>1.52
4	Stream depth (m.)	<0.30	0.30-0.60	0.6-1.2	1.22-2.44	>2.44
5	Flow variability	Little variation		normal		Ephemeral or large variation
6	River pattern	Torrent	Pool and riffle	Without riffles	Meander	Braided
7	Valley height/width	≦1	2-5	5-10	11-14	≧15
8	Stream bed material	Clay or silt	Sand	Sand and gravel	Gravel	Cobbles or larger
9	Bed slope (m/m)	<0.0005	0.0005-0.001	0.001-0.005	0.005-0.01	>0.01
10	Drainage area (km²)	<2.59	2.59-25.9	25.9-259	259-2589	>2589
11	Stream order	≦2	3	4	5	≧6
12	Erosion of banks	Stable		Slumping		Eroding large-scale deposition
13	Sediment deposition in bed	Stable				
14	Width of valley flat (m)	<30.5	30.5-91	91-152	132-305	>305
	Biological and Water-Quality Factors					
15	Water colour	Clear colourless		Green tints		Brown
16	Turbidity (parts per million)	<25	25-150	150-1000	1000-5000	>5000
17	Floating material	None	Vegetation	Foamy	Oily	Variety
18	Water condition (general)	Poor		Good		Excellent
	Algae					
19	Amount	Absent				Infested
20	Type	Green	Blue-green	Diatom	Floating green	None
	Larger plants					
21	Amount	Absent				Infested
22	Kind	None	Unknown rooted	Elodea, duck weed	Water lily	Cattail
23	River fauna	None				Large variety

		1	2	3	4	5
24	Pollution evidence	None				Evident
	Land flora					
25	Valley	Open	Open w. grass, trees	Brushy	Wooded	Trees and brush
26	Hillside	Open	Open w. grass, trees	Brushy	Wooded	Trees and brush
27	Diversity	Small				Great
28	Condition	Good				Overused
	Human-use and Interest Factors					
	Rubbish and litter					
29	Metal ⎛ no. per	<2	2–5	5–10	10–50	>50
30	Paper ⎜ 30.5 m of	<2	2–5	5–10	10–50	>50
31	Other ⎝ river ⎠	<2	2–5	5–10	10–50	>50
32	Material removable	Easily removed				Difficult removal
33	Artificial controls (dams, etc.)	Free and natural				Controlled
	Accessibility					
34	Individual	Wilderness				Urban or paved access
35	Mass use	Wilderness				Urban or paved access
36	Local scene	Diverse views and scenes				Closed or without diversity
37	Vistas	Vistas of far places				Closed or no vistas
38	View confinement	Open or no obstructions				Closed by hills, cliffs or trees
39	Land use	Wilderness	Grazed	Lumbering	Forest, mixed recreation	Urbanized
40	Utilities	Scene unobstructed by power lines				Scene obstructed by utilities
41	Degree of change	Original				Materially altered
42	Recovery potential	Natural recovery				Natural recovery unlikely
43	Urbanization	No buildings				Many buildings
44	Special views	None				Unusual interest
45	Historic features	None				Many
46	Misfits	None				Many

Notes: <less than, > greater than, ≦ less than or equal to, /divided by
Source: Leopold, 1969*b* (with metric conversion).

TABLE 2.10 Summary totals of uniqueness ratios of aesthetic factor values, Hell's Canyon region

Site in Idaho	Aesthetic factors			Total
	Physical	Biological	Human interest	
1. Wood River, nr. Ketchum	3.06	2.92	5.09	11.07
2. Salmon River, nr. Stanley	3.73	2.66	4.61	11.00
3. Middle Fork, Salmon River, at Dagger Falls	3.53	2.81	5.53	11.87
4. South Fork, Salmon River, nr. Warm Lake	4.69	5.15	4.09	13.93
5. Hell's Canyon, Snake River	3.20	4.41	8.48	16.09
6. Weiser River, at Evergreen Forest Camp	3.75	3.67	3.75	11.17
7. Little Salmon River, nr. New Meadows	8.88	9.74	4.48	23.10
8. Little Salmon River, nr. Boulder Creek	3.26	4.24	6.28	13.78
9. Salmon River, nr. Riggins	3.28	2.65	4.32	10.25
10. Salmon River, at Carey Falls	3.30	3.96	7.05	14.31
11. French Creek, nr. Salmon River	3.43	3.06	5.46	11.95
12. North Fork, Payette River	2.84	3.70	3.67	10.21

Source: Leopold (1969*b*).

are then plotted against *scenic outlook* (degree of view-confinement, or availability of distant vistas), and the resulting points are projected, as before, at 45° on to a new scale of landscape interest. Landscape-interest values are then plotted against degree of urbanization, and the points are projected on to a *scale of valley character*. The data plotted on Fig. 2.36 are for the 12 Idaho sites—and the Hell's Canyon (5) stands out as the most wild and spectacular location.

River character can be determined in a similar way (Fig. 2.37). In general, grandeur of rivers can be attributed to size, apparent speed of flow, and the extent to which the water surface is broken. Thus the river width and depth data are plotted and projected on to a scale of river size, and this variable is related to the presence or absence of rapids, riffles, and falls, to produce a *scale of river character*.

The data on valley character and river character can be combined to produce a ranking of sites in terms of all the variables that go to make up the two scales. Fig. 2.38 shows the information of the 12 Idaho sites and for a selection of sites elsewhere in the United States. In terms of this sample of sites, Hell's Canyon is clearly outstanding, both locally and nationally.

Leopold's technique is attractive because it is largely based on many fairly easily measured factors, and it provides a clear, semi-quantitative result. But it has several drawbacks. Firstly, it

relates only to sites, and it is difficult to see how it could easily be extended to *areas*. Secondly, some important components of scene, especially transitory ones, are omitted. And finally, although many may agree with it, the scheme is based on the assumption that the criteria selected are important, and are equally important.

(ii) *Linton's method of regional appraisal* Linton (1968) developed a technique for the assessment of scenic resources in Scotland that provided an areal classification of landscape. He argued that the appraisal should be founded on the elements of scenery that influence our reactions to it, the spatial variations of the elements should be mapped, and the several categories mapped should be arranged in a hierarchy of value. In Linton's view, those elements of scenery that contribute fundamentally to its quality can be classified into two major categories: the form of the ground (*landform landscapes*), and the use made of the land by man (*land-use landscapes*).

Landform landscapes could be quantitatively described in terms of relative relief (differences in height between the highest and lowest points within unit areas) or a similar measure, using topographic maps as a data source. But, in Linton's view, this would be a fairly laborious process. Instead, using his own considerable experience of Scottish scenery, Linton substituted a subjective appraisal of relief features shown on 10-miles-to-the-inch maps (1:633 600). Six types

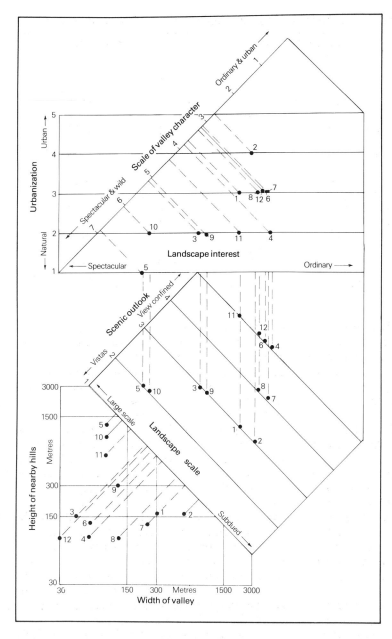

FIG. 2.36 Analysis of valley character at 12 sites in Idaho (after Leopold, 1969*a*)

of landform landscape were recognized, each based on the qualitative assessment of such criteria as relative relief, steepness of slope, abruptness of accidentation, frequency and depth of dissecting valleys in upland regions, and isolation of hill masses from their neighbours. Each type was given a point rating, the rating being based on the personal evaluation of the relative attractiveness of the six types. The scenic quality of landscape was generally reckoned to increase with relief, slope, etc., although the intervals between ratings were not always considered to be the same. Linton also appreciated that other natural features contributed to the attractiveness of landform landscapes; the most important of these is water, which was evaluated but not built into the final map because of scale problems. A fundamental problem in the areal classification of landscapes on the

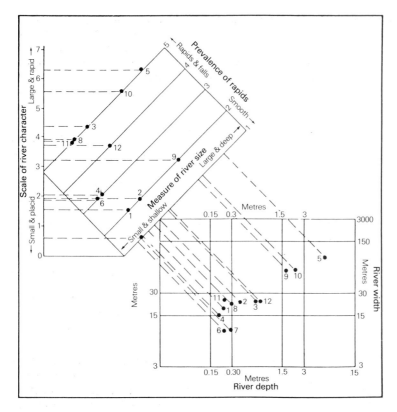

FIG. 2.37 Analysis of river character at 12 sites in Idaho (after Leopold, 1969a)

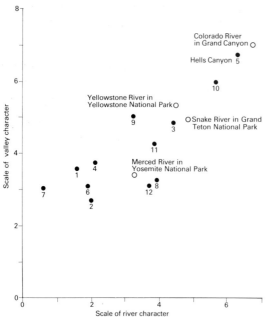

FIG. 2.38 Relations between river character and valley character at 12 sites in Idaho, and four other sites in the United States (after Leopold, 1969a)

basis of landform characteristics is that of defining the boundaries of units. In Linton's technique, the boundaries are determined subjectively.

Linton identified several land-use landscapes, delimited their extent using information on topographic maps, and gave each a point rating. In the classification, an urbanized and industrial landscapes lie at one extreme (−5) and are generally assumed to be, for example, ugly, dull, or depressing; at the other extreme, there are wild, lonely landscapes, too steep and remote for development (+6) and considered to be of major scenic value. Armed with the maps of landform the land-use landscapes, a composite map of scenic assessment based on the combination of ratings, can be produced.

Linton's technique clearly produces, relatively quickly and cheaply, a map of scenic value that is suitable for planning purposes. But as applied by Linton, the method has several limitations. In the first place, and despite claims to the contrary, it is subjective: significant landscape elements are subjectively selected, delineated, and rated, and perceptual responses to the elements are assumed.

Secondly, it takes no account of coastal scenery, of individually striking features or of water features. This limitation could perhaps be overcome by the judicious use of bonus points for areas with especially attractive sites. Thirdly, landscape and land-use ratings are equated without justification.

Crofts and Cooke (1974) modified Linton's method so that it could be applied to the analysis of any area in the UK using data derived solely from within 1-km grid-squares on Ordnance Survey maps at a scale of 1:63 360. Landform and land-use landscape maps for East Sussex were combined to produce a landscape evaluation map. This can then be compared with a map for the same region by Fines (1968) which is based on the subjective assessment of the totality of landscape quality by individuals and groups in the field. While the two distributions do differ, they both identify broadly the same areas of high and low quality. Given this similarity, it is arguable that the map-based method is quicker, cheaper, and more consistent and replicable than the qualitative field method. But both are open to serious criticisms of partiality, subjectivity, prejudice, and selectivity, criticisms that may undermine the value of any maps designed to assess scenic quality. At best, such maps serve pragmatically to focus attention on areas of potential aesthetic value that may need to be considered in the planning process. At worst, they ignore social attitudes, and disguise prejudice beneath the mask of semi-quantitative objectivity.

2.7 Conclusion

The purpose of terrain mapping, in whatever form, is to portray the form of the earth's surface, the properties of its soil and bedrock, and the nature of those processes which have been or are now operating in that area. Such maps potentially carry information of considerable value in land-use planning, hydraulic and marine engineering, civil engineering, agriculture, and land conservation. In the context of choosing geomorphological mapping schemes, Verstappen (1970) provided useful guidelines:

(i) the mapping system should be flexible, allowing the compiler discretion in the adoption of the symbols most appropriate for the area concerned;
(ii) the maps should be as simple as possible (to counteract cartographic problems and to avoid high printing costs);
(iii) the system should be applicable for mapping at all scales;
(iv) general geomorphological maps should be supplemented by special-purpose ('applied') maps;
(v) colours should be used to indicate major genetic landform units (rather than lithology or chronology as in some other mapping schemes).

The value of this last point is that in practice the colours will thereby define those areas which under a land-systems survey would normally be identified either as a land system or as a land unit. In either case compatible boundaries between these two approaches to land evaluation will become clear if this basis is adopted for the application of colours.

The most useful maps, in environmental management terms, are those compiled in the context of a specified problem. To that end the role of terrain mapping in environmental geology mapping, engineering geology mapping, and during an environmental impact assessment needs to be understood.

3 Monitoring Geomorphological Change

3.1 The Need to Understand Change

The previous chapter has shown that the recording of geomorphological data—especially in the form of maps showing what is where—is one fundamental contribution of the geomorphologist to environmental management. Geomorphological mapping is an essential prerequisite in many circumstances to problem appraisal and it is often a necessary precursor to evaluating the nature of change. But many contributions by geomorphologists to environmental management also arise initially from questions relating to the need to understand environmental trends in order to improve the success of management efforts. The following exemplify the commonest questions that are posed:

(a) What is the explanation of the present geomorphological situation, its problems, hazards, and resources? A specific version of this type of question might be 'Where do landslides occur and why?' This type of question involves explaining the present situation in the light of contemporary and/or historical evidence. It may also involve an explanation of past events and evolution from present evidence (*postdiction*).

(b) What are the nature and rates of geomorphological change at present? Or, similarly, how stable is the present landscape? A specific example might be 'How rapidly are building stones weathering in London?' Answers to such questions require an understanding of contemporary processes and the dynamic status of geomorphological systems, and they also depend on contemporary and historical evidence.

(c) What will happen geomorphologically if an area is developed, and how can the problems that are likely to arise best be overcome? Or, given the prevailing circumstances, where and

how can the landscape best be managed in order to avoid/ameliorate/prevent/control the geomorphological problems? A specific example might be: 'In a drainage network, in which valley-floors are potentially unstable and vulnerable to entrenchment, how can their entrenchment best be avoided?' *Prediction* is clearly a primary requirement here.

(d) What *were* the geomorphological consequences of environmental management of developments in an area? For instance, how effective were soil-erosion control strategies? Here again, the evidence of the present and the past is essential.

These questions have in common the need to understand change—past, present, or future—in geomorphology.

3.2 Difficulties of Understanding Geomorphological Change

The explanation of past events and the prediction of the future is extremely difficult for several fundamental reasons, as Schumm (1985) has shown.

(a) Geomorphological systems, even small ones, are usually *complex*, reflecting not only the interrelationships between causative variables (climate, geology, soils, vegetation, morphology, etc.), but also the impress of evolution that may span great periods of time (often including much of the Quaternary). In any landscape, this complexity is reflected in the fact that there may be 'eroding, stable, healing and potentially unstable landforms' (Schumm et al., 1984, p. 6), and that the specific forms likely to develop at any one location are difficult to predict.

This *innate complexity* provides the context for the impact of stimuli of change, giving rise to what Schumm (1973) has called *complex response* in which the same stimulus may generate different changes in different systems. These stimuli may be external to the system (e.g. climatic change) or internal to it (e.g. adjustment of erosion and deposition to change of slope).

Such complexity makes it difficult to answer a range of questions fundamental to understanding change and improving management. Fig. 3.1 shows four sequences of change to a geomorphological variable over time. Some of the important questions arising from such sequences that may require answers include the following. What kind of *equilibrium* or *disequilibrium* prevails when a change or group of changes were imposed on an area? (Was there, for instance, a steady state or a condition of dynamic equilibrium?) What are the *thresholds* of change that must be crossed before a system responds to the stimuli of change? What is the system's *reaction time* (the time between the beginning of disruption and the beginning of adjustment)? Is the *transition* across a threshold abrupt or slow? (For example, is adjustment at first rapid and then declining, as often seems to be the case (e.g. Graf, 1977)?) Is the adjustment *ramped* or *pulsed* (e.g. Brunsden and Thornes, 1979)? And what is the *magnitude* of change, and the *relaxation time* (the time of adjustment to change from one equilibrium condition to another)?

All these and related questions have applied relevance. For example, Gregory (1977, p. 5)

declared in the context of river channel changes that 'Applied research may ... be directed towards the identification of thresholds below which change can be constrained.' And Schumm (1973, p. 309), in the same fluvial context, concluded: 'The fact that, at least locally, geomorphic thresholds of instability can be defined quantitatively indicates that they can be identified elsewhere and then used as a basis for recognition of potentially unstable landforms in the field. This approach provides a basis for preventive erosion control.'

The problems of complexity and complex response are made worse by the intervention of human activity in the form of land-use change (such as afforestation, agriculture, grazing, fire, and urbanization), the timing and location of which is often difficult to determine in the past and impossible to predict in the future; and in the form and effect of management practices, especially engineering structures.

(b) Each geomorphological system is unique, at least in terms of its location and history, and certainly in the ways the various factors contributing to its state are integrated. This *singularity* is 'the key to the difficulty of short-term prediction' (Schumm, 1985, p. 12), for it means that each system reponds to change in distinctive ways and at different rates.

(c) The *sensitivity* of a system to change, which Brunsden and Thornes (1979) saw as a function of the temporal and spatial distributions of resisting and disturbing forces at the

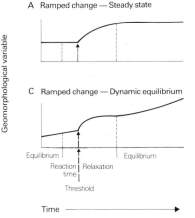
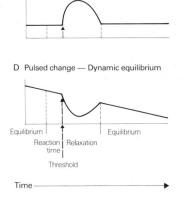

FIG. 3.1 Four theoretical types of response of geomorphological variables to stimuli for change (for explanation, see text)

earth's surface, is usually spatially variable and it may vary even for identical forms within a single system. For example, a series of apparently identical slope failures on a valley side may be in very different conditions of stability or instability at any one time. The sensitivity of landforms to stresses imposed internally or externally can be viewed as a ratio between the mean relaxation time (Fig. 3.1) and the mean *recurrence interval* of events that cause the change: Brunsden and Thornes (1979) refer to this as the *transient-form ratio*. If the ratio exceeds unity, the forms are transient and sensitive to change; if it is less than unity, the forms are relatively stable. The problem for understanding change is that the ratio is spatially and temporally variable, and the relaxation times and recurrence intervals may not be known sufficiently well for the ratio to be predictable.

(d) Closely allied to the notions of singularity and sensitivity, is the somewhat perplexing problem for the applied geomorphologist of the *principle of indeterminacy*. This refers to

those situations in which the applicable physical laws may be satisfied by a large number of combinations of values of inter-dependent variables. As a result, a number of individual cases will differ among themselves, although their average is reproducible in different samples. Any individual case, then, cannot be forecast or specified except in a statistical sense. The result of an individual case is indeterminate (Leopold and Langbein, 1963, p. 189).

For example, if discharge of water and sediment into a stream is to be increased, the engineer needs to know how the channel will respond but, as Leopold and Langbein (1963) pointed out, the channel may respond in many different ways through the interdependent variables of width, depth, velocity, slope, hydraulic roughness, and channel pattern. The precise form of adjustment will vary from place to place. In one sense, the principle is a confession of ignorance: with fuller understanding of the system, the precise response should become more predictable (e.g. Watson, 1969). But in another sense, indeterminacy confounds prediction for management. For example, a strategy for gully control often cannot be based on a knowledge of exactly where specific new gullies will be located.

(e) *Convergence* and *divergence* are two features of geomorphological systems that further inhibit the prediction and explanation of change. Convergence refers to the condition when 'different processes and causes produce similar effects' (Schumm, 1985, p. 11). Both mean that the observation of form is a dangerous guide to the nature of processes in the past, the present, or the future; but often form and deposits are the only available evidence for such interpretation (see Sect. 3.3). Schumm (1985) neatly illustrated the danger of divergence (Fig. 3.2A): because of the relationship between sediment yield, drainage density, and climate, a similar change of climate will have very different effects in arid, semi-arid, and humid regions. In this example, the divergence effects are predictable because the relationships are known; in many circumstances they are not.

(f) *Scales* of both space and time create further difficulties. In general, 'the shorter the time span, the smaller the space, and the more rapid the process the more specific can be the explanation and extrapolation' (Schumm, 1985, p. 5). These difficulties are particularly clear in the context of process monitoring. For example, because the magnitude, frequency, and duration of forces intermittently at work on the earth's surface are often unknown, intermittent sampling may fail to provide data that adequately reflect the trends (see Fig. 3.2B). In particular, the assessment of change in geomorphological systems is typically over a period of time much shorter than that of significant landscape changes (e.g. Church, 1980). Thus, as Fig. 3.2C shows, the sample periods may fail to identify the long-term mean, the linear trend, or the well-defined cycle; and in the case of the second sampling period in Fig. 3.2C, an exceptional sequence is recorded that could be misinterpreted as a short cycle. Church (1980) also emphasized the problem of *intermittency*, the tendency for the grouping of similar values over long periods of time that can lead to cumulative effects which ultimately cause a substantial deviation from the mean conditions (the so-called '*Hurst phenomenon*') and make statistical inferences from short periods of observations unrepresentative and even dangerously inaccurate.

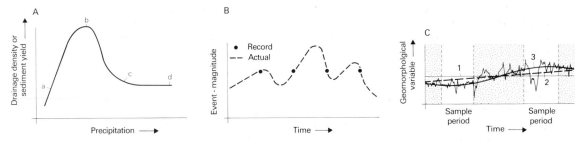

FIG. 3.2 (A) Relations between sediment yield and precipitation—with increases of precipitation, drainage density and/or sediment yield may increase to a maximum (a to b) or decrease (b to c), or remain constant (c to d) (after Schumm, 1985). (B) Inappropriate sampling of a geomorphological sequence may inadvertently fail to record its true pattern of change over time. (C) Inappropriate sampling periods in a sequence may fail to recognize long-term means, linear trends, or well-defined cycles; and the recording of exceptional events, as in the second period, may lead to mis-interpretations such as the recognition of a false short-cycle. (After Church, in R. A. Cullingford *et al.* (1980); © 1980, and reproduced by permission of J. Wiley and Sons Ltd.)

(*g*) All the preceding problems are further compounded by the fact that for many applied problems, answers to the geomorphological questions concerning change are required *quickly*, time for prolonged monitoring is not normally available, and existing *data are inadequate*. As Watson (1969, p. 493) said, 'If one knows all the conditions, laws, and factors for a given time, then one can determine for a system the values of all the factors prior to and subsequent to that time.' Unfortunately, in geomorphology as in most earth-science, many of the 'conditions, laws and factors' are either not known or are inadequately monitored.

Such fundamental problems undoubtedly explain why the geomorphological past is difficult to explain and the future difficult to predict. They mean that geomorphologists are usually unable to give as precise accounts of change to environmental managers as both would wish, and this is undoubtedly a major weakness in applied geomorphological research. But there is no need for despair. The questions *require* answers, and even approximate and appropriately qualified answers are likely to be more useful than none. Furthermore, as experience and understanding of geomorphological systems grow, so too does precision in explanation and prediction. And, not least, there are several important and extremely useful ways of trying to overcome the difficulties.

3.3 Approaches to Studying Change

Geomorphologists have adopted several strategies for explaining the past, understanding the present, and predicting the future, although the philosophical basis of some of the strategies is rather rarely analysed. In explaining the past, the traditional emphasis lies on interpreting the evidence of form and deposits and, where possible, dating the events responsible for them using a variety of relative and absolute dating techniques available for the study of the Pleistocene and more recent periods (e.g. Vita-Finzi, 1973). The strategy has much to do with hypothesis testing and, however it may be dressed up in contemporary jargon, it is rooted in fieldwork, and still owes much to the method of *multiple working hypotheses* (in which the research worker, acting independently of any one ruling theory, explores competing explanatory hypotheses until one generally acceptable explanation is derived (e.g. Chamberlain, 1897)).

In understanding the present and predicting the future, most of the strategies of value in applied work require some kind of *translation*, and the choice of strategy often depends heavily on the time available, especially for monitoring.

(a) *Evaluating change without time for monitoring: five major strategies*

(i) *Form and deposits* The commonest strategy involves *translation from the evidence of form and deposits to the prediction of process and change.* Rates of change may, for example, be determined from such evidence as the exposure of dated soil horizons at the ground surface (indicating erosion rates), the accumulation of sediments up-slope of fences (indicating rates of soil creep), and the destruction of dated inscriptions on tombstones (indicating weathering rates). Surface stability, often as important as change, may be revealed by such features as lichen growths, the thickness of weathering rinds, and soil-profile development. Dated or datable markers are very valuable in this context (Sect. 3.3(b)). Surrogates for process evidence include the *geometric characteristics* of geomorphological systems and landforms. For example, drainage catchment area, if used cautiously, can help to predict water and sediment discharge; dimensions of active dunes may be used to predict rates of dune movement; and drainage-basin shape, together with channel network topology, can help to predict the pattern of flood discharge. The relations between geometric properties may also be used. For example, when the relations between drainage area and valley-floor slope were matched to gullied and ungullied valley-floors in Piceance Creek Basin, Colorado (Fig. 3.3) it was possible to define the threshold slope that separates gullied from ungullied valley-floors, a threshold of predictive value (Patton and

Schumm, 1975). *Indicators of instability* are also of potential value for prediction. For example, distinctive patterns of ground cracking on slopes can be used to locate potential slope failures, the exposure or burial of groynes can indicate the nature of sediment movement along beaches, plant succession in dune fields can illuminate the progress of dune movement, and the curvature of tree trunks on slopes can betray soil creep. In this context, vegetation can be exceptionally valuable as an indicator of surface activity because it often responds quickly and visibly to geomorphological changes, especially on relatively unstable surfaces.

This general approach is well illustrated by the derivation of *hazard maps* of use to planners in areas without process monitoring data. In Fig. 2.31, for example, the flood-hazard intensity prediction for the alluvial plains of the Suez area, Egypt, is based *inter alia* on the precise interpretation and translation of a detailed geomorphological survey, using such features as catchment area (because discharge events on alluvial fans are normally proportional to catchment area); the relative ages of alluvial surfaces as revealed by iron and manganese oxide stains on surface pebbles ('desert varnish'), in the belief that the older surfaces are more stable and less liable to flood; and the distribution of recent mudflows (high-risk locations) (Bush *et al.*, 1980). In this study there were available no process or flood damage data of any kind. Such translations are often crude, qualitative, and imprecise; but they are often robust and useful to planners.

(ii) *From theory to site* A second approach is to *translate general theory into a realistic evaluation of conditions at a specific site*. An example of this is the application of Bagnold's (1941) principles of sand movement in air to the interpretation of sand movement at the site of a road development in Saudi Arabia and thus to the design of appropriate defensive measures (Redding and Lord, 1982; see Chapter 9). Such translations may appear simple, especially if the theory is as clear and fundamental as Bagnold's, but unfortunately theory is often inadequate and site conditions are unexpectedly complex. In this case, for instance, it is possible that the problem of local turbulence is inadequately covered by theory, and the knowledge of sand and dune activity may be incomplete.

(iii) *From laboratory to field* A third approach is the translation from *laboratory experiments and computer-based simulations to specific field condi-*

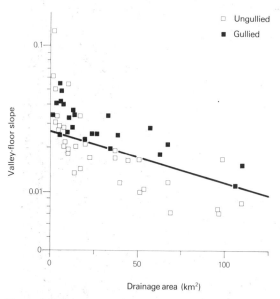

FIG. 3.3 Relations between valley-floor slope and drainage area in Piceance Creek Basin, Colorado, USA. The regression line defines the threshold between gullied and ungullied valley floors (after Patton and Schumm, 1975)

tions. In the case of laboratory experiments, it may be possible to isolate selected variables, but it is often very difficult to reproduce the complexity of natural conditions, to 'scale down' the system appropriately, and satisfactorily to reproduce the time-sequence on which natural processes operate. Thus, for instance, the difference is great between a small cube of rock subject to salt attack in the laboratory and a whole building of the same rock subjected to a number of different processes acting in various ways in a polluted city atmosphere. Nevertheless, 'hardware models' of geomorphological systems have been shown to be of predictive value. For example, the Hydraulics Research Station, Wallingford, UK, and the US Army Corps of Engineers Waterways Experiment Station, Vicksburg, Mississippi, USA, both use models of rivers and estuaries to predict the effect of proposed engineering works on erosion and sedimentation (Plate 3.1).

A computer-based simulation reproduces a natural system in digital form; once established it can be run cheaply at a faster rate than its counterparts in the real world, variables can be manipulated as desired, and change can be extrapolated into the future. For example, King and McCullagh (1971) simulated the evolution of a complex, recurved spit at Hurst Castle, Hampshire, UK. In this case, the form of the spit was thought to be related to different types of waves approaching it from different directions, the effect of deepening water to the east of the spit, and the proximity of the Isle of Wight. All of these variables were built into the model and the present form of the spit was accurately reproduced using an optimum combination of the different variables (Fig. 3.4). Once the present spit was established, further running of the computer model suggested that 'more recurves will be built on to the spit and that these will become longer and closer as the

PLATE 3.1 Model of the Mersey Estuary (photo courtesy of Hydraulics Research Ltd., Wallingford, UK)

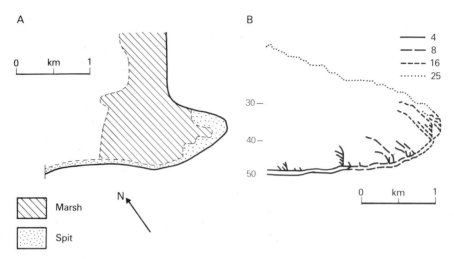

FIG. 3.4 (A) Map of Hurst Castle spit. (B) Computer-based simulation of the development of Hurst Castle spit through a number of stages and making assumptions about the variables involved and their values. Stage 16 represents the best fit, and stage 25 suggests a possible future plan of the spit (after King and McCullagh, 1971; reproduced by permission of the University of Chicago Press)

distal end of the main spit slows down its eastward movement as the depth becomes greater' (King and McCullagh, 1971, p. 37). In addition, it was possible to predict what would happen to the form of the spit if, for example, the wave regime or the depth of water were to change.

(iv) *The ergodic hypothesis* Space and time can be considered interchangeable under certain conditions, so that *differences observed between different locations in space at present can be translated into a temporal sequence of explanatory and predictive value* (e.g. Chorley and Kennedy, 1971). This translation is called the *ergodic hypothesis*. Brunsden and Kesel (1973) used the space-time transformation to describe the evolution of part of the Mississippi River bluff near Port Hudson, Louisiana, USA (Plate 3.2). The bluff was undercut by the river in 1722; thereafter it was progressively abandoned by the river so that its slopes have developed subaerially for different periods of time. Today, some slopes are actively eroded by the river (high-intensity zone), some are characterized by basal scour (intermediate-intensity zone), and some by basal aggradation (low-intensity zone). The profiles from these three zones can be arranged to represent a sequence of change over time (Fig. 3.5) that reflects changes in basal conditions. There is, for instance, an evolution from average slopes of 44° and cliffs in the high-intensity zone to average slopes of only 19.5° and concave profiles in the low-intensity zone. A

similar sequence was produced by Savigear (1952) for a cliff line in South Wales that had been progressively cut off from wave attack by the development of a spit. Profiles taken along the cliff at present reflect changes associated with the progressive withdrawal of the sea; it is then possible to arrange these profiles into a sequence that suggests the evolutionary progress of a single profile (Fig. 3.6A). Kirkby (1984) developed this model into a computer simulation model which incorporated a range of slope processes (including landslides) and limitations/thresholds. The model, once created, could then be run to predict the consequences of changing the values of any or all of the variables. The computer model is thus not only a means of refining explanation of Savigear's original model, it is also a way of predicting future change. Fig. 3.6B shows one example of Kirkby's analysis, in which a basal retreat of 20 mm/y for 0–8000 years is followed by basal aggradation for the remainder of a 10 000-year period: the profiles with the shortest period of wave attack had been degraded longest and had the lowest profiles. This particular example provides too rapid wave attack and the cliff does not survive long enough adequately to explain the observed situation.

(v) *From known to unknown locations* It may be possible *to predict conditions at an unmonitored location from known conditions at other monitored locations*. In this approach, the translation from one place to another is achieved through surro-

PLATE 3.2 Erosion of the Mississippi River bluff near Port Hudson, Louisiana

gates for process such as geometric variables. For example, soil loss at sample sites in a region may be correlated with explanatory variables that can be measured at *any* site in the region. Thus Bovis (1982) showed that soil loss at several sites in part of the Colorado Front Range, Colorado, USA, was closely related to the percentage of bare soil multiplied by the sine of slope angle (Fig. 3.7); from this graph it is possible to predict approximately the soil loss at any site within the same region by measuring slope and the proportion of bare soil. In this, as in many other studies (e.g. see Chapter 3), the statistical description becomes the basis for spatial prediction. Equally, *geomorphological and land-systems maps* may be used predictively. For example, if one type of land system (Chapter 2) is identified in several different locations, field conditions sampled in one example *should* form an adequate basis for predicting field conditions in another example. An illustration of the value of this approach is provided by Van Lopik and Kolb (1959). They mapped desert terrain conditions relevant to military purposes in an accessible area (Arizona, USA) where field tests of equipment were possible, and then were able to predict and assess militarily significant 'going conditions' for the same equipment in other inaccessible desert areas by identifying in those areas terrain analogues for the conditions in Arizona.

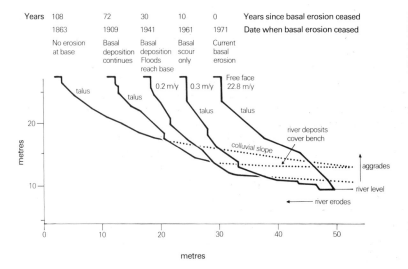

FIG. 3.5 Summary of slope evolution in a Mississippi River bluff near Port Hudson, Louisiana, USA (after Brunsden and Kesel, 1973; reproduced by permission of the University of Chicago Press)

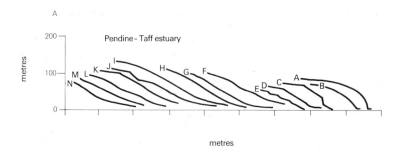

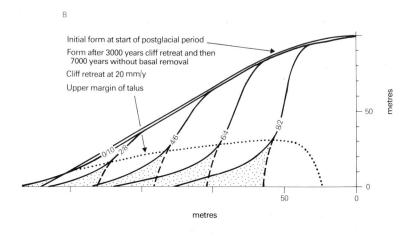

FIG. 3.6 Suggested sequences of cliff-profile development following retreat of the sea from the base of a cliff in South Wales based on (A) slope profiles (after Savigear, 1952) and (B) computer simulation, assuming 20 mm/y basal retreat from 0–8000 years, and basal accumulation for the remainder of a 10 000-year period (after Kirkby, 1984)

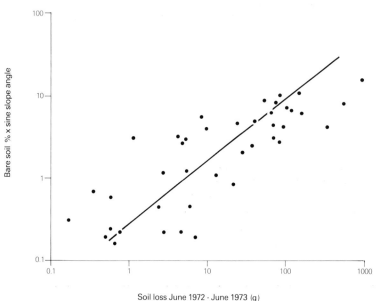

FIG. 3.7 Relations of annual soil loss to bare soil × sine of slope angle at erosion plot sites in the Front Range, Colorado, USA (after Bovis, 1982, reproduced by permission of Unwin Hyman Ltd.)

(vi) *The historical record* Finally, and very importantly, change can be evaluated for the recent past and for the near future using *historical sources*. Historical sources of great potential value are of two main types: those that directly record geomorphological change, and those that provide supporting environmental evidence of geomorphological change. Table 3.1 provides a list of such sources with specific examples for the USA. Amongst the most valuable are those surveys based on precise *instrumental work* (1, 2, 3, 8, and 9. Table 3.1), some of which now chronicle change back to at least the beginning of the nineteenth century. Of increasing value is the *photographic record* which can include evidence over a hundred years and can be used with caution to assess such features as biomass (e.g. Graf, 1979). More recently, the *remote sensing record* (Chapter 2) has become important, including as it now does *aerial photography* (sometimes dating from the 1920s) and *satellite imagery* (which began in the 1960s). The analysis of change using these sources depends heavily on *repeat imagery*. In the USA, the availability of repeat ground-based photography has been reviewed (e.g. Rogers *et al.*, 1984). Successive LANDSAT images have now been used to monitor such features as the movement of dust storms (e.g. Péwé, 1981). At present, however, successive aerial photographs of an area may be more useful than satellite imagery because they usually cover a longer time-span and have better spatial resolution. For example, Huntings Surveys Ltd. (1977) analysed aerial photographs taken in 1963, 1971, and 1976 of sand-dunes in Qatar to

TABLE 3.1 Documentary sources of geomorphological evidence in the USA

Geomorphological change	
1. Instrumental surveys	e.g. Metes and Bounds Survey e.g. US Rectilinear Land Survey
2. Topographic surveys	*Terrestrial* Topographic Maps (e.g. USGS since 1889) e.g. *Marine* Topographic Maps (e.g. US Coast and Geodetic Survey since 1830) *River and Harbour* Surveys (e.g. US Army Corps of Engineers) *Water Power* Surveys (e.g. 1880 US Census) *Road, Railroad, and Bridge* Surveys (e.g. Southern Pacific Railroad) *Reservoir* Surveys (e.g. US Army Corps of Engineers, US Dept. Agriculture) *Stream and Valley Sediment* Surveys (e.g. 1935–41, USDA) VIGIL Network
3. Soil degradation surveys	*Soil Surveys* (e.g. since 1899, USDA) *Erosion Surveys* (e.g. USDA, 1934) *Flood Control Reports* (e.g. US Army Corps of Engineers since 1935)
4. Air and ground photography	since 1840s
5. Travel and exploration accounts	
6. Newspapers and journals	
Supporting environmental evidence	
7. Land use	e.g. US *Census of Agriculture*, since 1810 e.g. USDA *Land-use reports*, since 1958 e.g. *Drainage and irrigation records* USDA since 1920
8. Climatic records	e.g. *Post Surveyor and Signal Service*, since 1840s e.g. *US Weather Bureau*, since 1880s
9. Stream/Sediment discharge records	e.g. *USGS Water Supply Papers*

Source: S. W. Trimble (1981) in part.

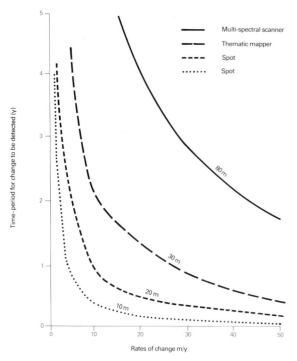

FIG. 3.8 Relations between the spatial resolution of remote sensing systems, rates of geomorphological change, and the time-periods for change to be detected (after Millington and Townshend, 1987*a*, reproduced by permission of Croom Helm Ltd.)

reveal an average annual movement of 'large ridge dunes' in a south-south-east direction of 7.4–7.5 m/y. There is in all such work a relationship between the time-period for change to be detected, the rate of change, and the spatial resolution of the imaging system (e.g. Fig. 3.8).

The use of historical sources in analysing geomorphological change involves many problems. Few sources were collected with geomorphology in mind, so that they are often inadequate and always require sensitive interpretation. Attention needs to be paid to the frequency and timing of observations (see Fig. 3.2B), perceptions and attitudes of observers, styles and conventions of presentation, the semantics of descriptions, the instruments (if any) used, and internal and external tests on accuracy and meaning are desirable (e.g. Hooke and Kain, 1982). In addition, there may be major methodological problems, such as those requiring the disentangling of the complex webs of cause and effect in the recent past.

Perhaps the most important recent developments in this approach are the advent of precise ways of dating the historical sedimentary record (e.g. using ^{210}Pb and ^{137}Cs dating techniques; see Table 3.2), and the establishment of some archival records that are long enough to capture the main trends of changes in many geomorphological systems. These two developments, taken together, can play a vital role in calibrating the nature of historical responses to change (e.g. Fig. 3.1), in evaluating the effectiveness of environmental management policies (e.g. Trimble, 1981), and in helping to predict future trends.

(b) *Evaluating change with time for monitoring*

(i) *Monitoring contemporary processes* To many, the best way to understand contemporary changes and the most secure way of predicting future changes is to monitor processes and landforms directly over periods of time sufficiently long to reveal significant trends. The problems of extending contemporary monitoring records beyond the period for which they are derived were discussed above (Sect. 3.2(*f*)). For most geomorphological work, monitoring takes two main forms. The first involves data collected systematically for other purposes but which are of value to geomorphology (*adventitious monitoring*). For example, many hydrological data, collected chiefly for water-supply purposes, have been shown to be invaluable in the study of fluvial geomorphology (e.g. Leopold and Langbein, 1963), not least because they have in places been collected for quite long periods and are usually based on standard, reliable, and quite expensive installations.

The second, and more important approach is through *field experiments* designed to illuminate specific geomorphological problems by 'a set of measurements, conducted under controlled field conditions' (e.g. Slaymaker, 1982, p. 11). Such experiments have flourished in recent years, especially in those situations that respond relatively quickly to change, and some have now been generating data for decades. For example, the VIGIL network (Leopold, 1962) and the use of instrumented catchments in association with the International Hydrological Decade, 1965–74 (Gregory and Walling, 1974) have provided abundant data. Slaymaker (1980) recognized three major experimental approaches: the monitoring of intentional and controlled interference with natural conditions (e.g. monitoring the effects of river diversions); selection of one landform, and the monitoring of its changes with respect to one or

TABLE 3.2 Techniques of dating the historical sedimentary record

Evidence	Useful time-scale (years)	Use	Relationship to documents
Caesium 137 and lead 210 radiometry	10–50	Absolute dates; sources of sediments	As corroboration
Lichenometry	10–500	Date of exposure of surface, e.g. frequency of flooding	Calibrated by documentary dating of objects on which they grow
Man-made structures	10–1000	Positional evidence	As a bench mark for measurements of movement
Tree-rings	10–7000	Date, climatic conditions, and events	Provides absolute date where documents unavailable, date and occurrence of events
Archaeological features	50–5000	Mark and date horizons; indicate climatic conditions and human activities	As direct evidence and as corroboration for documents
Morphology	5+	Location and nature of processes; relative dates of features	As corroboration
Sedimentary stratigraphy	5+	Evidence of processes; climatic and erosional conditions; marker horizons	As corroboration
Pollen analysis	100–100 000	Vegetation and climatic conditions, relative dates	To extend land-use documents

Source: Hooke and Kain (1982).

more processes (e.g. Washburn's (1967) monitoring of slope processes in Greenland); and the stratifying of landforms in terms of one selected variable (e.g. lithology or altitude) and monitoring change in two or more strata. The study of 'paired' catchments, in which one is urbanized and the other is not, or of 'before and after' studies in a single catchment (Gregory and Walling, 1974) illustrate this approach. The analysis of drainage catchments is particularly important. Such studies have been criticized because the catchments may be unrepresentative and the results are difficult to apply elsewhere, they are costly, they leak, and their integrated results conceal processes; but these problems can be overcome in some measure and they are outweighed by the fact that they are functional and often management units (e.g. Hewlett *et al.*, 1969).

All of these approaches normally involve either *repetitive* or *continuous* monitoring. The type of monitoring used, if there is any choice, depends on several considerations such as the length of

monitoring period (bearing in mind that change itself may be either continuous or intermittent) and the frequency of observations (see Fig. 3.2B), the cost and nature of the appropriate equipment, the difficulties of maintenance, protection, and site visits.

(ii) *Methods of monitoring* These can be classified into four main groups. *Topographic survey* is an invaluable source of landform data that is commonly repeated from time to time. Adventitious monitoring of topography can reveal subsidence due to water, oil, or mineral extraction (Chapter 5), peat extraction (e.g. Stephens and Speir, 1969; Fig. 3.9B), or neotectonic deformation (e.g. Burnett and Schumm, 1983; Fig. 3.9C). The repetitive survey of the surface of sediment in reservoirs in order to calculate rates of sedimentation and sediment yield (see Chapter 7) illustrates the use of topographic survey in studies designed specifically to solve geomorphological problems (Fig. 3.9E). *Reference markers* have been widely used to monitor rates of change. They are,

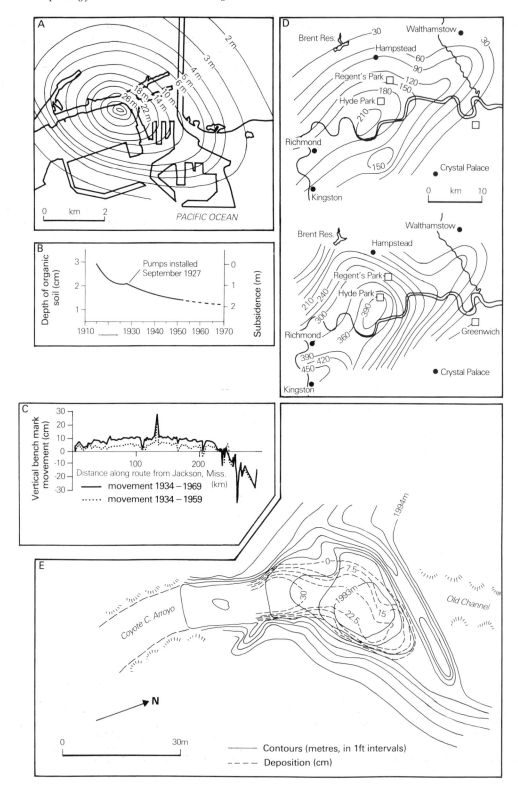

of course, essential for repetitive surveys, but other examples include the use of nails and washers to monitor the gain and loss of sediment on hill-slopes; the use of unconnected cross-sections on channels to chronicle channel change (e.g. Leopold, 1973); and the use of lichens which, because they grow at predictable rates, provide evidence of the age of the surfaces on which they live and can, for example, help to predict channel discharge and capacity (e.g. Gregory, 1976; Lock *et al.*, 1979). A

wide range of *monitoring equipment* has been devised specifically to chronicle geomorphological change (e.g. Goudie, 1981). Examples include land-surface compaction recorders (borehole extensometers, Fig. 3.10); the use of piezometers in the study of slope failure (Chapter 5); and the micro-erosion meter designed to monitor changes to the morphology of rock surfaces (Chapter 12; Trudgill *et al.*, 1981). A final category of monitoring method that often combines elements of the

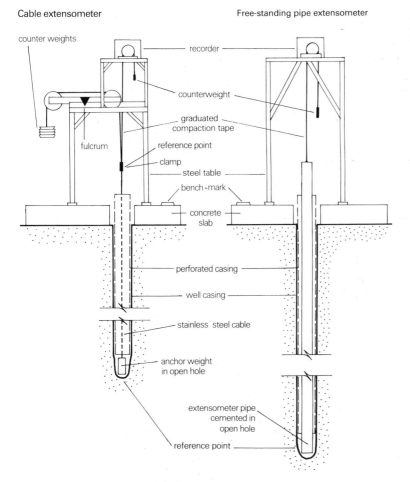

Cable extensometer

Free-standing pipe extensometer

FIG. 3.10 Surface-compaction recorders: left, cable extensometer; right, free-standing pipe extensometer (after Ireland *et al.*, 1984)

counter weights

recorder

counterweight

graduated compaction tape

fulcrum

reference point

clamp

steel table

bench-mark

concrete slab

perforated casing

well casing

stainless steel cable

anchor weight in open hole

extensometer pipe cemented in open hole

reference point

FIG. 3.9 Resurvey as a means of monitoring change: (A) Subsidence between 1928 and August 1971 in Long Beach, California, USA, arising from abstraction of oil in the Wilmington field (Port of Long Beach, 1971). (B) A periodically resurveyed profile near South Bay, Florida, indicates the normal pattern of subsidence on drained organic soils (after Stephens and Speir, 1969). (C) Vertical bench-mark movement along a National Geodetic Survey route between Jackson, Mississippi and New Orleans, Louisiana, USA (after Burnett and Schumm, 1983; © AAAS). (D) Subsidence in central London due to water withdrawal from the underlying acquifers in the Chalk; figures in mm (after Wilson and Henry, 1942). (E) Original topography of a dam and reservoir and contours of deposition, 1961–2, in a New Mexico arroyo (after Leopold *et al.*, 1966)

PLATE 3.3 Erosion plots to monitor runoff and soil erosion, near Ouagadougou, Burkina Faso

other three types involves monitoring based on markers and special equipment installed at field experiment sites. The monitoring of stone-tablet decay at field sites using such indices as weight loss and chemical and micro-morphological change to samples exposed on a standard carousel is one example (Chapter 12); the establishment of special erosion plots to monitor soil erosion within clearly delineated areas using devices for monitoring the transfer of water and sediment is another (Plate 3.3).

3.4 Conclusion

Monitoring change is fundamental to problem solving in applied geomorphology as is the mapping of landforms, surface materials, and processes. Geomorphologists commonly need to explain the past and the present and to predict the future and some understanding of the nature and rates of change is an essential prerequisite. Explanation and prediction is particularly difficult because of the complexity of geomorphological systems, and the related problems of sensitivity, singularity, indeterminacy, convergence, and divergence; the problems of scale and lack of data increase the difficulties. Geomorphologists adopt several strategies in overcoming these problems. Where prediction and evaluation are required without time for monitoring, they can translate from form and deposit to the prediction of process and change; from general theory to specific sites; from the laboratory and the computer to field conditions; from spatial to temporal variability; from known to unknown locations, and from the historical record into the future. Where time is available for monitoring contemporary processes, adventitious monitoring and specifically designed field experiments are commonly used. Amongst the most valuable methods are topographic survey, and the use of reference markers and a variety of specifically designed pieces of monitoring equipment. There is no single desirable approach to monitoring in applied geomorphology: each problem has to be tackled pragmatically in terms of its intrinsic characteristics, and the time and resources available.

4 Soil Erosion by Water

4.1 A Continuing Problem; a Burgeoning Literature

Loss of soil from land by rainfall and runoff has been a major environmental problem for much of the twentieth century. It creates a spectre of declining agricultural productivity and deteriorating rangeland that has haunted conservationists since it began to be exposed by George Perkins Marsh (1864) and others in the nineteenth century. And the natural loss of soil, often accelerated not only by agriculture and grazing but also by fire, urban development, construction activity, and off-road vehicles, can create additional downstream problems of sediment pollution in rivers (Chapter 7). Today, soil erosion by water is a serious and growing problem in many countries, including the 'old world' countries of Europe (such as the UK, Germany, Spain, Czechoslovakia, and the USSR), the 'new world' countries (such as the USA and Australia), and many developing countries, including China, Tanzania, Syria, and India (see e.g. Holý, 1970; Morgan, 1980; Zachar, 1982; Hodges and Arden-Clarke, 1986).

Because soil erosion by water is seldom visually dramatic, and usually only slowly affects productivity, the tasks of perceiving its seriousness, and then adequately measuring it and persuading decision-makers that action is required have been long and arduous. They have been marked by several milestones, including Jacks and Whyte's *Rape of the Earth* (1939), Bennett's influential *Soil Conservation* (1939), the FAO's *Soil Erosion by Water* (1965), and Eckholm's (1978) study, *Losing Ground*. Public awareness of the problem has certainly also been enhanced by unusually spectacular crises such as the dust storms in the 'Dust Bowl' area of the high plains of the USA (Chapter 9) and the ravages of erosion in the Tennessee Valley Authority during the 1930s. And farmers, faced with lost productivity that can

be directly attributed to soil erosion and cannot be offset by injections of pesticides and fertilizers, swell the ranks of those demanding action.

The responses in most societies threatened by soil depletion have four common elements: scientific efforts to understand and predict the nature of erosion dynamics; engineering and other investigations into methods of alleviating the problems; attempts to apply the results of research to land management; and reviews (as yet rather few) of the effectiveness of conservation programmes. The volume of literature arising from these responses fills a library, and continues to grow rapidly.

In many countries—and the United States provides the best example—literature on soil erosion usually comes from a variety of agencies. At a national level, the US Department of Agriculture, in addition to sponsoring research and experiment stations and soil- and water-conservation districts, also provides a huge range of publications such as agricultural information bulletins, agricultural handbooks, conservation research reports, and conference proceedings. At a state level, universities and other agencies are often directly involved in erosion research and providing farm advice. More locally, countries and cities may develop their own services and information, often in association with state or federal agencies, to control land use and to help improve land-use practices. Similarly in Australia, the Commonwealth Scientific and Industrial Research Organization (CSIRO) is a national agency with responsibilities for land-resources management, whereas some states (e.g. Tasmania and New South Wales) have soil-conservation authorities. This national-regional pattern is repeated in many other countries (e.g. Zachar, 1982), and is supplemented at an international level by UN/FAO publications.

The voluminous literature on soil erosion is published through three main channels: journals, conference proceedings, and books. Amongst the more important journals are the *Transactions of the American Society of Agricultural Engineers*, the *Journal of Soil and Water Conservation*, *Agricultural Engineering*, the *Proceedings of the Soil Science Society of America*, *Catena*, and *Earth Surface Processes and Landforms*. There are in addition many more parochial journals that include soil-erosion studies, such as *Agriculture* (UK Ministry of Agriculture, Fisheries and Food), the *Journal of the Australian Institute of Agricultural Science*, and in Czechoslovakia the *Agricultural Engineering* and *Water Management* magazines (in Czech). Amongst recent important conference volumes are *Soil Erosion: Prediction and Control* (Soil Conservation Society of America, 1977), *Assessment of Erosion* (De Boodt and Gabriels, 1980), *Recent Developments in the Explanations and Prediction of Erosion and Sediment Yield* (Walling, 1982), *Soil Erosion* (Prendergast, 1983), *Soil Erosion and Conservation* (El-Swaify et al., 1985), and *Hillslope Processes* (Abrahams, 1986). In addition, there are several recent books that review scientific aspects of soil erosion. These include Morgan's *Soil Erosion* (1979) and *Soil Erosion and Conservation* (1986), Zachar's (1982) *Soil Erosion*, and Jansson's (1982) *Land Erosion by Water in Different Climates*; and edited volumes *Erosion* (Toy, 1977), *Erosion and Sediment Yield* (Hadley and Walling, 1984), *Soil Erosion* (Kirkby and Morgan, 1980), and *Anatomy, Physiology and Psychology of Erosion* (Hallsworth, 1987).

The overwhelming majority of the studies published in journals, conference proceedings, and books are concerned with the science of soil erosion—with understanding the processes and predicting their location, magnitude, frequency, and duration, and with engineering and land-use methods of control. The fact that the scientific literature is so extensive reflects both the perceived seriousness of the soil-erosion problem and the pervasive belief that its solution lies in scientific research, for which extensive funds are made available. And yet, a fundamental question remains. Why is it that, after half a century of intense scientific enterprise, soil erosion remains such a serious problem in so many countries, including those where the research is being undertaken? The answer to this question will be considered below, but it has been addressed recently by Blaikie (1985) in *The Political Economy of Soil Erosion in Developing Countries* and in Blaikie and Brookfield's (1987) *Land Degradation*

and Society. The brief answer, and it is one of fundamental importance, is that soil erosion as a hazard is not simply a scientific problem; it is inextricably bound up with social and economic conditions. For example, it is one thing to invent conservation measures capable of controlling soil erosion; it is quite another to have them adopted and adopted successfully. It is thus also pertinent to ask how effective has the vast investment in soil-erosion research and soil conservation been?

4.2   The Nature of Soil Erosion

(a) *The soil-erosion system*

Soil erosion by water involves two important sequential events: the detachment of particles and their subsequent transportation. The two principal agents for this work are raindrops and flowing water. *Raindrop* (or *rainsplash*) *erosion* involves the detachment of particles from soil clods by impact and their movement by splashing. *Runoff erosion* largely concerns the transportation of loose material (much of it often prepared by raindrop erosion) by turbulent water flowing as unconcentrated flow in sheets, or as concentrated flow in rills or gullies, although some detachment of particles also occurs in runoff erosion.

Each of these two types of erosion comprises its own set of forces and resistances: some forces tend to promote particle detachment, for instance, and others tend to resist it. Put slightly differently, the nature of soil erosion depends on the relationship between the *erosivity* of raindrops and running water on the one hand, and the *erodibility* (i.e. detachability and transportability) of soil material on the other (Fig. 4.1). The two are not necessarily independent and may interact (e.g. Morgan, 1983). Much of soil-erosion research is concerned with measuring and comparing the variables that determine these forces in order to be able to predict the likelihood of erosion and reduce soil loss. In the following discussion, emphasis is placed on soil erosion of agricultural land. But it is important to remember that soil erosion may also be a problem, for example, on forest lands after a fire or tree-felling or where vegetation-covered land is laid bare, as by the practice of grassland burning in many savannah lands, or by site preparation for urban development.

(b) *Approaches to studying soil erosion*

The great majority of soil-erosion studies are

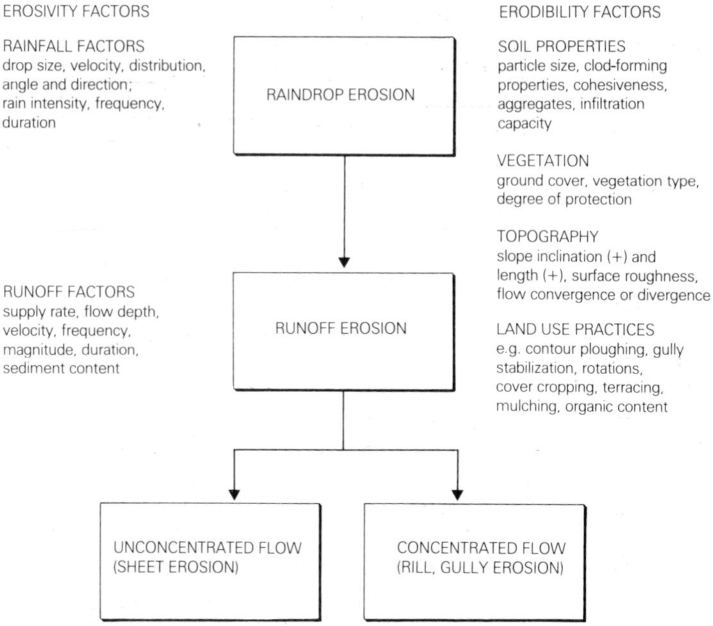

rain/runoff

so u

EROSIVITY FACTORS

RAINFALL FACTORS
drop size, velocity, distribution,
angle and direction;
rain intensity, frequency,
duration

RAINDROP EROSION

ERODIBILITY FACTORS

SOIL PROPERTIES
particle size, clod-forming
properties, cohesiveness,
aggregates, infiltration
capacity

VEGETATION
ground cover, vegetation type,
degree of protection

TOPOGRAPHY
slope inclination (+) and
length (+), surface roughness,
flow convergence or divergence

RUNOFF FACTORS
supply rate, flow depth,
velocity, frequency,
magnitude, duration,
sediment content

RUNOFF EROSION

LAND USE PRACTICES
e.g. contour ploughing, gully
stabilization, rotations,
cover cropping, terracing,
mulching, organic content

UNCONCENTRATED FLOW
(SHEET EROSION)

CONCENTRATED FLOW
(RILL, GULLY EROSION)

FIG. 4.1 The main factors in soil
erosion by water. In general, erosion
will be reduced if the value of an
erosivity factor is reduced and/or the
value of an erodibility factor is
increased, except for those factors
shown with a (+), where the reverse
is the case

measured

empirical and field based (e.g. FAO, 1977). One
approach of great importance involves the monitoring of soil loss from erosion plots of various
shapes and sizes for which various methods of
trapping sediment have been devised (e.g. De
Ploey and Gabriels, 1980; Zachar, 1982). Particularly widespread are the standard erosion plots
devised by the US Department of Agriculture,
which are c.22 m long and on a 5°-slope. Such
plots are used to monitor soil loss under different
conditions of erosivity, soil erodibility, land use,
and management practices (e.g. Wischmeier and
Smith, 1978). There are, in addition, numerous
other methods for monitoring soil loss from
slopes, including fixed-point levelling, calculation
of rill volume, photogrammetric techniques, and
erosion gauges (e.g. Zachar, 1982). Field experiments designed to explain some feature of the
soil-erosion system are widely practised by geomorphologists. For example, simulated rainfall is
sometimes used to determine erodibility (e.g.
Bryan, 1970; Lusby, 1977).

A second field-based, empirical approach
involves mapping of soil erosion. The techniques
used here include those described in Chapter 2: for
example, geomorphological mapping, land-systems mapping, and land-capability mapping,
based on remote sensing analysis (e.g. FAO, 1978;
Bergsma, 1980) and field assessment (e.g. Richter,

1980; Jansson, 1982). Mapping of relative hazard,
in which erosion is classified into a few qualitatively or semi-quantitatively defined categories, is a
common feature of erosion mapping (see Figs. 1.7
and 4.4). This approach is particularly useful in
providing a regional and/or national perspective
on soil loss to aid policy formulation, and in
facilitating extrapolation from erosion-plot observations to a broader management context.

Laboratory experiments and analyses also have
an important role to play in soil-erosion studies.
Experiments provide an alternative, and sometimes quicker approach for studying selected
aspects of the soil-erosion system under controlled
conditions such as the relation between raindrop
characteristics and erosivity, the assessment of
erodibility, and the effects of vegetation on erosion
(e.g. Morgan, 1985; Farres and Cousen, 1985;
Morgan et al., 1986). In addition, analysis of field
samples, for example to determine erodibility and
shear strength, adds a further dimension to
laboratory work.

One of the main objectives of the collection of
empirical data is to help the formulation of
explanatory theory and to improve prediction.
Many models have been developed (see the
following sections), but the parametric and the
deterministic are perhaps the most important (e.g.
Morgan, 1979). Parametric models are based on

identifying statistically significant relationships between variables that are assumed to be important through the generation of appropriate empirical data. The best-known model of this type is the Universal Soil Loss Equation (USLE, Sect. 4.5(e)). Deterministic models are based on mathematical equations that describe the processes in the system taking into account the laws of conservation of mass and energy. A well-known example of such models is that of Meyer and Wischmeier (1969) in which the soil-erosion system is structured in terms of its four major dynamic components and the movement of sediments by each component is calculated across a slope profile (Fig. 4.2). Models of this sort, which work within the mass-balance framework, seem likely to dominate predictive modelling of soil erosion in the future (e.g. Kirkby, 1980).

(c) *Natural and accelerated erosion rates*

Earth scientists normally draw an important distinction between 'natural' or 'geological' soil erosion, and soil erosion 'accelerated' by human activity. The distinction is fundamental. Geological erosion is the rate at which the land would normally be eroded without human disturbance. Accelerated erosion is the increased rate of erosion that often arises when 'natural' conditions are modified by various land-use practices, such as compaction of soil by machinery, up-and-down slope cultivation, removal of field boundaries, and long periods of fallow without vegetative protection. Accelerated erosion is really the problem of soil erosion that is central to environmental management. Attempts to reduce accelerated erosion to a natural rate or even less, or perhaps preferably to a rate comparable to the rate of soil formation, constitute the practice of *soil conservation*.

This distinction immediately raises several fundamental questions. What is the natural rate of erosion at a locality? Is that rate an acceptable target for conservation practices? Or, what should the target rate of erosion be for a given land use? Should the 'natural' or target rate be comparable to the local rate of soil formulation?

'Natural' rates of erosion vary enormously with controlling conditions such as climate, vegetation, soils, bedrock, and landforms. This variability is clearly revealed by studies of sediment yield based on the analysis of suspended sediment load (see Chapter 7). These studies, however, may include a variability altered to an unknown extent by human activity. Another approach to estimating natural rates is based on measuring soil loss from small erosion plots planted with close-growing vegetation. Such empirical approximations in the United States suggest that 'natural' rates are commonly less than 753 kg/ha/y. An alternative method would be to determine the actual rates of soil formation and to use these as the targets for soil conservation, but despite some efforts here (e.g. Vanoni, 1975), little progress has been made. Thus, target rates for conservation practices are largely a matter of opinion. For many practical purposes, acceptable target rates are commonly fixed between 2510 and 12 550 kg/ha/y.

The patterns of accelerated erosion, reflecting a combination of natural and land-use factors, differ from the patterns of erosion under natural conditions. Many surveys of erosion have been made, and most of them reflect contemporary, and thus accelerated erosion patterns. Surveys based on field assessment of erosion, as reflected for ex-

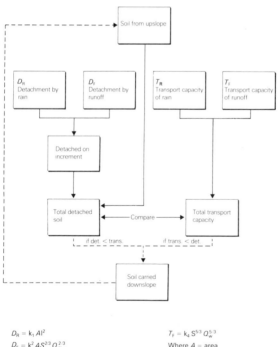

$$D_R = k_1 A I^2$$
$$D_F = k_2 A S^{2/3} Q_w^{2/3}$$
$$T_R = k_3 S I$$

$$T_F = k_4 S^{5/3} Q_w^{5/3}$$

Where A = area
I = rainfall intensity
S = ground slope (sin $\emptyset$)
Q_w = runoff

FIG. 4.2 Flow chart for the model of the processes of soil erosion by water (after Meyer and Wischmeier, 1969)

ample in the truncation of soil profiles, probably provide the best empirical maps. Thus erosion surveys based on soil-profile analysis in the United States in the 1930s and since have led to national and local maps of erosion (Fig. 4.3).

To *predict* patterns of potential accelerated erosion requires a suitable explanatory model. For example, Kugler (1976) produced a map of erosion potential in East Germany (Fig. 4.4) by combining values of slope gradient, land use, and soil texture into a simple qualitative classification. This model is based on empirical observations that suggest erosion is likely to be most severe on steeper,

cultivated slopes with loamy soils (especially those on loess) and clay loams. To develop such models requires a clear understanding both of the soil-erosion system and the relative importance of the variables that influence it (see Sect. 4.5).

Often field surveys are expensive and time-consuming, and alternative cheaper approaches to estimating the spatial variability of erosion are desirable, especially in developing countries. Mill-ington *et al.* (1982) provided one economical method. A grid-square system is used to record on a computer those characteristics related to soil erosion for which data are normally *already*

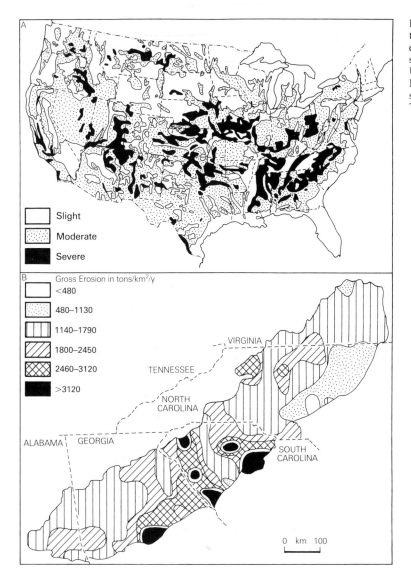

FIG. 4.3 (A) Degree of topsoil loss in the USA. Moderate erosion involves estimates of 25–75% of topsoil lost; severe erosion, more than 75% (after US Dept. Agriculture, 1957). (B) Depth of gross erosion in the southern Piedmont Plateau (after Trimble, 1974)

A

Slight

Moderate

Severe

B   Gross Erosion in tons/km²/y

<480

480–1130

1140–1790

1800–2450

2460–3120

>3120

VIRGINIA

TENNESSEE

NORTH CAROLINA

ALABAMA   GEORGIA

SOUTH CAROLINA

0   km   100

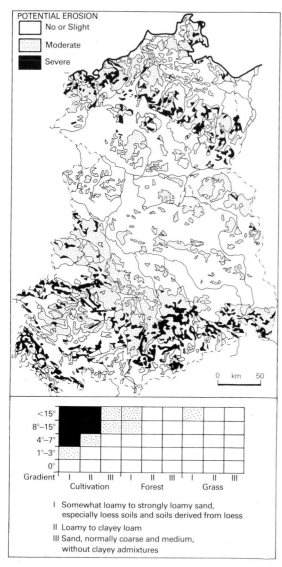

POTENTIAL EROSION
☐ No or Slight
▦ Moderate
■ Severe

0   km   50

Gradient	Cultivation			Forest			Grass		
	I	II	III	I	II	III	I	II	III
<15°									
8°–15°									
4°–7°									
1°–3°									
0°									

I Somewhat loamy to strongly loamy sand,
   especially loess soils and soils derived from loess
II Loamy to clayey loam
III Sand, normally coarse and medium,
   without clayey admixtures

FIG. 4.4 Potential sheet and rill erosion in East Germany (after Kugler, 1976, from Jansson, 1982)

available from archives and remote sensing imagery: rainfall, topography, soil, land use, and vegetation. The map of relative erosion potential in part of Sierra Leone (Fig. 4.5) is based on multiplying for each unit area, values for slope, soil erodibility, rainfall erosivity, and drainage density. The map of actual erosion loss was based on the same index divided by the length of bush fallow (i.e. the shorter the fallow, the greater the erosion). This method is simple and cheap; it also must be considered to be very approximate.

4.3 Raindrop Erosion

As research progressed in the 1930s it was realized that the amount of soil in runoff increased rapidly with raindrop energy, and it was noted that erosion could be very greatly reduced by preventing raindrop impact. Since that time, the detachment and movement of soil particles as a result of raindrop impact has come to be recognized as a fundamentally important and often initial phase of soil erosion (Fig. 4.6). It is no exaggeration to say that in certain circumstances as much as 90 per cent of erosion on agricultural land may result from this process. In addition to particle dispersion by detachment and movement, raindrops also tend to compact surface particles, especially by reorienting them to create a surface crust, thus promoting surface runoff. Research on this subject has been reviewed by Ellison (1947) and others (e.g. Mutchler and Young, 1975; Kirkby and Morgan, 1980).

The nature of particle detachment and movement is, of course, a reflection of the relations between the characteristics of the rainfall and the characteristics of the soil and ground surface. The major properties of rainfall of importance to its erosivity are drop mass, size, size distribution, direction, rainfall intensity, and raindrop terminal velocity. From these variables, the kinetic energy [0.5 mass (velocity)^2] and momentum (mass × velocity) can be determined.

Raindrop size can be calculated in several different ways (e.g. Hudson, 1971). One common method is based on catching a sample of rain in a dish containing flour. Each raindrop creates a pellet which, when dried, has a predictable relationship to the size of the drop that created it. In future it may be practicable to determine raindrop sizes from dual-polarization weather radars (Hall *et al.*, 1980).

The median size of raindrops increases with rainfall intensity for low- and medium-intensity falls in the form:

$$D_{50} = aIb \qquad (4.1)$$

where

D_{50} = median size of raindrops (mm),
I = intensity (e.g. mm per hour),
a, b = constants.

For high-intensity rainfalls, drop size declines slightly (Hudson, 1971), and maximum drop sizes appear to be in the order of 5–6 mm diameter.

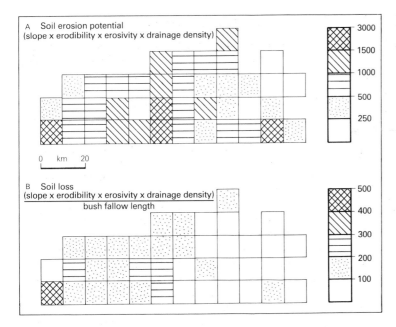

A Soil erosion potential
(slope x erodibility x erosivity x drainage density)

3000
1500
1000
500
250

0   km   20

B Soil loss
(slope x erodibility x erosivity x drainage density)
bush fallow length

500
400
300
200
100

FIG. 4.5 Computer-generated maps of soil erosion potential and soil loss for an area in Sierra Leone, based on the analysis of parameters derived from available data for 10 × 10 km grid-squares (after Millington *et al.*, 1982)

The impact of velocity of raindrops is also related to drop size. A free-falling raindrop accelerates under the influence of gravity until the gravitational force is equal to the frictional resistance of the air, when the drop will be falling at its terminal velocity. The relations between drop size, terminal velocity, and the distance of fall required for 95 per cent of terminal velocity to be attained under laboratory conditions are shown in Fig. 4.7. It is clear from this graph that, because fall distances to maximum fall velocity are so short, most drops will strike the surface at their terminal velocity. Air turbulence and winds will affect the terminal velocity of raindrops in natural rain.

Once size, terminal velocity, and intensity of drops are known, momentum and the kinetic energy of rainfall can be calculated by summing the values for individual drops. Rainfall intensity can easily be derived by recording rain-gauge data, and this intensity may be compared with calculated kinetic energy (Fig. 4.7). The kinetic energy of rain varies, of course, from storm to storm, and therefore in order to obtain a reasonable average annual figure (or similar general statement) attention must be given to the frequency and duration of storms as well as their intensity.

Having established the nature of raindrop impact, the next step is to determine the relations between rainfall and soil detachment and trans-portation. Several empirical studies describe these relations.

Multivariate analysis of erosion-plot data in the United States led to the adoption of a compound parameter that explains most satisfactorily soil loss in terms of rainfall (e.g. Smith and Wischmeier, 1972; Bergsma, 1981). This parameter, known as the EI_{30} index, is the product of kinetic energy, E and intensity, I, where I_{30} is the 30-minute intensity (i.e. the greatest average intensity in any 30-minute period during a storm). The spatial pattern of this index for the USA is shown in Fig. 4.8. Two comments are required on the index. First, it requires continuous and detailed rainfall records. Where these are not available, alternative approximations are possible. For example, in calculating R (the annual erosivity), Wischmeier (in Bergsma, 1981) found that:

$$R \simeq P_t \times 24hP_{max} \times 1hP_{max} \qquad (4.2)$$

where

$$P_t = \text{average annual precipitation,}$$
$$24hP_{max} = \text{the maximum 24-hour precipi-tation with a recurrence interval of 2 years,}$$
$$1hP_{max} = \text{the one-hour maximum precipi-tation with a recurrence interval of 2 years.}$$

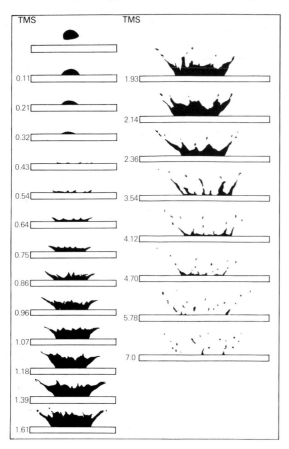

FIG. 4.6 The impact of a raindrop on a saturated soil paste, based on high-speed cine photography. Time-scale in milliseconds (after Ghadiri and Payne, 1980; © 1980 and reproduced by permission of J. Wiley and Sons Ltd.)

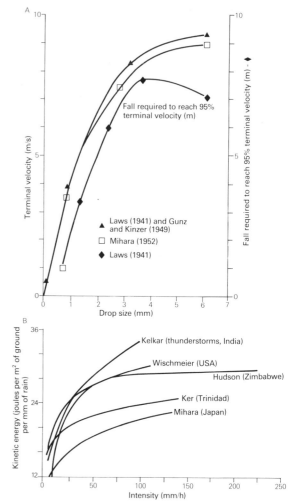

FIG. 4.7 (A) Relations between drop size, terminal velocity, and fall required to attain 95% of terminal velocity (after Smith and Wischmeier, 1962, from which the data sources may be obtained). (B) The relation between kinetic energy and intensity of rainfall, from studies by various workers in different localities (after Hudson, 1971)

Secondly, R does not always necessarily correlate well with soil erosion in areas outside the USA. Other indices have therefore been developed (Bergsma, 1981). For example, in preparing an iso-erodent map of West Africa (Fig. 4.8B), Roose (1980) found that R correlated well with 5–10 year average annual rainfall, and that in much of the region this provides an alternative, simpler index. Similarly, Morgan (1980) found that the $KE > 10$ index (the kinetic energy of all rains falling at intensities equal to or greater than 10 mm/h for 10 minutes) was the most suitable for the assessment of erosion in the UK (Fig. 4.8D). The FAO have modified Fournier's p^2/P index (Chapter 7) in developing an erosivity index of world-wide applicability (Arnoldus, 1980).

Soil splash also varies with the detachability and transportability of the soil and the vulnerability to splash of the surface involved. For example, soil loss is greatly reduced if the soil is protected by vegetation, and the degree of protection varies according to the type of cover. Similarly, the ease with which soil clods can be destroyed by raindrops will depend on factors affecting soil consolidation, such as clay, stone, base-, and organic-matter content. Again, the slope of the surface is important. If the surface is horizontal then splash erosion may break up the soil and move particles

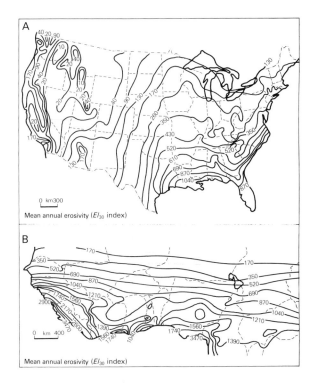

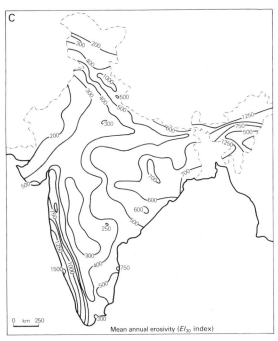

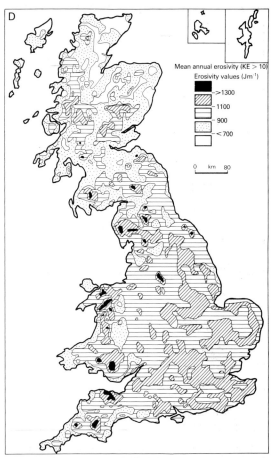

FIG. 4.8 (A) Mean annual erosivity in the United States estimated by the EI_{30} index. Conversion from US units to nearest tens of metric units through multiplication by 1.735 (from Wischmeier and Smith, 1978). USDA(SCS), 1975. USDA recommended that 610 should be the maximum EI_{30} value used in the USLE in the United States (after Wischmeier and Smith, 1978 and Jansson, 1982). (B) Mean annual erosivity, EI_{30} in West Africa. Conversion from US units to nearest tens of metric units through multiplication by 1.735 (from Roose, 1976 and Jansson, 1982). (C) Mean annual erosivity in India. EI_{30} in metric units (from Babu *et al.*, 1978). (D) Mean annual erosivity (KE > 10) in Great Britain (after Morgan, 1980)

through the air, but there will be no overall loss of soil from the field, despite the fact that individual particles may be splashed up to 60 cm into the air, and may be moved up to 1.52 m horizontally. Ellison (1947) showed, however, that there can be net erosion simply by rainsplash erosion on a sloping surface. For example, on a 10 per cent (6°) slope he found that 75 per cent of soil splash was *down-slope*. Ellison also showed that maximum splash occurs shortly after the surface is wetted and thereafter decreases, probably because the surface film of water increases in thickness (thus reducing impact on soil particles) and because the most easily eroded particles are soon removed.

The process of soil splash, which attacks the whole exposed surface uniformly, produces several types of damage. Firstly it detaches particles from clods, destroys soil structure, and prepares debris for transport by runoff. In certain circumstances, small pedestals of soil may be formed. Secondly, it transports debris in splash and results in soil loss from sloping surfaces. Thirdly this erosion elutriates the soil—it removes clays, humus, and other soil nutrients and leaves an impoverished soil behind. A fourth problem is that raindrop impact may disperse surface clay particles and lead to the formation on drying of a crust which reduces infiltration capacity, thus promoting surface runoff and the second type of water erosion, runoff erosion.

4.4 Runoff Erosion

Since the theory of runoff erosion began to be developed in the 1930s, two major explanatory models have been described and elaborated. These are briefly reviewed below.

(a) *Horton's model—overland flow*

In an argument of impeccable logic and impressive elegance, Horton (1945) formulated a theory of surface runoff and erosion which provided a sound basis for an understanding of the causes and control of runoff erosion and for much subsequent research (e.g. Dunne and Aubry, 1986).

According to Horton, runoff does not occur immediately rain falls on a bare soil surface. If the soil is unsaturated, water will infiltrate into the ground at a rate determined by soil structure, soil texture, vegetational cover, biological structures in the soil, soil-moisture content, and condition of the surface. *Infiltration capacity* (f_p) (the maximum sustained rate at which a particular soil can transmit water) varies during a rainstorm—initially it may be quite rapid (f_o), but with time it declines to a constant value (f_c). This decline is due to 'rain packing'—in-washing of fine material, swelling of colloids, and a breakdown of the surface soil structure. It is this minimum infiltration capacity which is predominant during most long rainstorms. If the rate of precipitation exceeds this value of infiltration capacity, water will begin to accumulate in and on the soil and runoff may result. If the rain continues for a period t_R, then:

$$f_p = f_c + (f_o - f_c)e^{-kt_R} \qquad (4.3)$$

where k is a constant and e = 2.71828 (the base of Naperian logarithms). As f_p decreases with time during rain, so runoff increases. In practice it is difficult to measure either the infiltration capacity or the runoff properties of a drainage basin at any one moment in time. This is because in most basins there are considerable variations in the infiltration characteristics of the soils that it contains. At the same time there are continual changes taking place in the amounts of precipitation being stored by the ground.

Infiltration is also controlled by the conditions that exist prior to the onset of rain. An earlier rainfall may have left the soil partially saturated. In addition, infiltration varies with the seasons as these govern the state of the vegetation, the condition of the land if agriculture is being practised, and the temperatures which in turn control evaporation rates. Areas with vegetation have a higher infiltration capacity than barren areas, for the vegetation retards surface flow, the roots make the soil more pervious, and the ground is shielded from some of the effects of raindrop impact, so that the amount of soil packing is reduced.

There are several methods for arriving at infiltration values. For example, field measurements can be made on experimental plots of the amount of runoff resulting from a known quantity of artificially induced rainfall; then from the relationship Precipitation = Infiltration + Runoff, an approximate value for infiltration can be obtained. (The amounts intercepted by vegetation and held in storage on the ground surface during the experiment are assumed to be small.)

The difference between rainfall intensity and infiltration capacity is called the supply rate (σ). Once there is a positive supply rate, water will begin to collect in the storage area provided by ubiquitous surface depressions. The next stage is for water to overflow between depressions and for a thin layer of water to develop. Ultimately, after a critical thickness of this layer is exceeded the water will begin to flow.

Initially the flowing water will have insufficient energy to pick up and transport soil, but as it proceeds down-slope its force will increase to the point where it exceeds the resistance of the surface material and erosion can begin. Up-slope of this critical distance (x_c) there is a zone of no runoff erosion where only raindrop erosion can occur. It is clearly desirable to calculate where x_c will occur on a slope, as it is here that gully erosion may begin.

The force per unit area exerted by flow parallel with the soil surface, F, can be defined by the DuBoys formula (in Imperial units) for shear stress:

$$F = w \frac{d}{12} \sin \theta \qquad (4.4)$$

where

F = eroding force in pounds per square foot,
w = weight of a cubic foot of water,
d = depth of overland flow (in),
θ = angle of slope.

Of the variables in this equation, the most difficult to determine is d, depth of flow. Horton showed that this depth of flow at a point any distance from the top of the slope can be described (in Imperial units) as follows:

$$d_x = \left(\frac{\sigma n x}{1020} \right)^{3/5} \left(\frac{1}{s^{0.3}} \right) \qquad (4.5)$$

where

d_x = depth of flow at a given distance from the slope divide (in),
σ = supply rate (in/h),
n = factor describing surface roughness,
x = distance (ft) down-slope,
s = tangent of slope.

A value of d_x can therefore be substituted into Equation 4.4 to give an expression of the total eroding force at x:

$$F = \frac{w}{12} \left(\frac{q_s n x}{1020} \right)^{3/5} \left(\frac{\sin \theta}{\tan^{0.3} \theta} \right) \qquad (4.6)$$

where F = eroding force (lb per sq ft), and q_s = runoff intensity (in/h), being equal to σ for steady overland flow.

It remains to determine x_c, the critical down-slope distance where runoff erosion begins. This can be done using Equation 4.6, substituting R_i (the threshold value of resistance of the surface) for F, making $x = x_c$, assuming $w = 62.41$ lb per cu ft, and solving x_c, thus:

$$x_c = \frac{65}{q_s n} \left(R_i \frac{\tan^{0.3} \theta}{\sin \theta} \right)^{5/3} \qquad (4.7)$$

That is to say, the width of the zone of no runoff erosion is inversely proportional to the intensity of runoff and surface roughness, and directly proportional to the 5/3 power of resistance. For slopes of less than 20°, the width of the belt of no runoff decreases with increase in slope. Naturally, raindrop erosion can be, and usually is, effective in this watershed zone.

One of the most important features of this demonstration is that it clearly identifies those variables that determine runoff erosion and which therefore must be modified in order to reduce erosion. The list of variables is worth repeating: rain intensity and infiltration capacity (i) as expressed in supply rate (σ) or runoff intensity (q_s); length of overland flow; slope (θ); and surface roughness (n).

Once the critical threshold has been crossed, flowing water will begin to incorporate sediment and, all other things being equal and assuming uniform material, the erodibility of sediment can be expressed in terms of the relations between particle size and flow velocity (Fig. 4.9). A point to notice on this figure is that the most easily eroded particles are between 0.1 and 0.5 mm (medium and fine sand), and that higher velocities are required to transport smaller as well as larger particles. The curve for settling velocity, below which deposition will occur, is also shown on Fig. 4.9.

Normally runoff will be in rills (defined as channels that can easily be removed by ploughing) or in gullies. As water is concentrated into rills its depth and velocity increase, and detachment and transport of particles both increase as a result (Plate 4.1). Erosion proceeds largely by the water-and-sediment mixture scouring bottoms and sides of the rills, by waterfall erosion at the head of channels, and by mass movement into the channels (through, for example, slumping and the

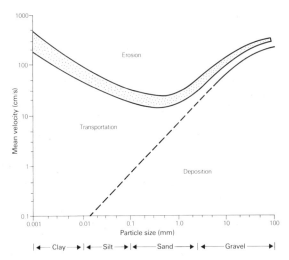

FIG. 4.9 Relations between particle size, erosion velocity, and settling velocity for uniform sediments (after Hjülstrom, 1939)

weathering of soil). The initial rill system usually develops rapidly, with some rills enlarging faster than others, until a simpler, more stable pattern is produced. Such rill and gully erosion is common only on vegetation-free surfaces, such as those in semi-arid and arable lands or where the vegetation cover has been seriously disrupted. The initial pattern may be guided by already-established plough lines and cattle paths, especially where they have a down-slope orientation.

Horton's model took no account of rainsplash. In an attempt to integrate this Dunne and Aubry (1986) undertook field experiments in Kenya which showed that some rills created by runoff became filled and eradicated by rainsplash. Their results indicate that the initiation and maintenance of rills depends on the balance between erosion in sheetwash causing incision, and rainsplash which tends to smooth the surface.

(b) *The saturated throughflow model*

As an alternative to the overland flow model the saturated throughflow model has been developed in which emphasis is placed on the movement of water down-slope through the upper soil horizons (Kirkby, 1969).

The velocity of groundwater moving through soils (throughflow) is slow (about 20 cm/h) compared with that of overland flow (about 27 000 cm/h) (Kirkby and Chorley, 1967). Velo-

PLATE 4.1 A case for conservation: developing rill erosion in a vineyard near Bratislava, Czechoslovakia (photo courtesy of D. Zachar)

city of throughflow, and its discharge, vary with the permeability of the material through which it passes. Darcy's Law (Fig. 4.10) can be applied to this situation whereby:

$$Q = \frac{k(h_t - h_b)A}{L} \qquad (4.8)$$

where

> Q = discharge,
> h_t = hydraulic head at the top,
> h_b = hydraulic head at bottom (both h_t and h_b are measured with piezometers),
> A = cross-sectional area of soil column,
> L = length of column of soil,

and

$$k = \frac{k'\rho g}{\eta}$$

where

> k' = permeability,
> ρ = density of water,
> g = acceleration of gravity,
> η = viscosity of water;

and also:

$$\text{velocity of flow} = \frac{Q}{A} = ki$$

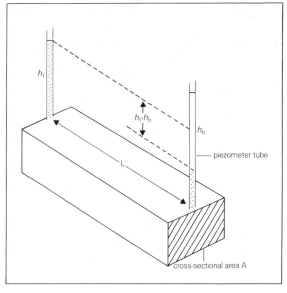

FIG. 4.10 The parameters used in Darcy's Law

where

$$i = \frac{(h_t - h_b)}{L}, \text{ the hydraulic gradient.}$$

Since k can be calculated for any one soil, and since h_t, h_o, and L *can* be measured, it is thus possible to calculate Q. An important assumption has to be made, however, which is that the soil is uniform in its permeability characteristics. No drainage basin will ever contain soils which are all of the same permeability. In order to calculate Q, for any basin, it is necessary to discover the variations in permeability that exist. In practice a first estimate of variations in permeability would come from an analysis of soil textures. Permeability is very low in unweathered clays, low in fine sands, silts, and stratified clays, higher for mixed sands and gravels, and high for sorted gravels.

Kirkby (1969) evaluated the mathematics of the throughflow model in the following way. Throughflow, travelling through soil-pore spaces, never attains a steady state, and its discharge per unit contour length (q_τ) is given by:

$$q_\tau = (P - f_*)vt \qquad (4.9)$$

where

> P = rate of surface percolation,
> f_* = rate of infiltration at base of the permeable soil,
> v = velocity of throughflow,
> t = time elapsed.

In order to calculate v, Kirkby combined Darcy's Law, restated as:

$$Q = z \cos \theta \left\{ K \cdot m \cdot \sin \theta - D \cdot \frac{\partial m}{\partial x} \right\} \qquad (4.10)$$

and the continuity equation, which states the difference between inflow and outflow must be accompanied by changes in moisture content:

$$\frac{\partial m}{\partial t} + \frac{\partial Q}{\partial x} = i \qquad (4.11)$$

where

> Q = down-slope discharge measured in a horizontal direction,
> x = distance down-slope,
> K = soil permeability,
> D = soil diffusivity,
> θ = surface angle of slope,

m = soil moisture content,
z = thickness of soil layer,
i = rainfall intensity,
t = the time elapsed.

Under conditions of high moisture content of the soil (i.e. with throughflow) the soil moisture is constant (i.e. does not vary with time and $\partial m = 0$), and thus equation 4.11 becomes:

$$\frac{\partial Q}{\partial x} = i. \qquad (4.12)$$

In other words near-saturated soils respond very rapidly to changes in rainfall intensity, even before overland flow begins.

From the above it can be seen that the important controls on throughflow are soil characteristics, distance down-slope, angle of slope, and the intensity of the rainfall. The attainment of a fully saturated state occurs most readily in soils adjacent to streams, on concave slopes, in hollows, or where the soils themselves are either thin or impermeable. This process generally involves an up-slope extension of the existing channel system (Kirkby, 1969) and is independent of distance from the hill crest, which is an important control in the case of overland flow.

As shown in the discussion of the Horton model for overland flow, the erosive force of runoff increases both with distance down-slope and the angle of slope. This means that on a hillside with a convex–concave profile the erosive force approaches its maximum on the steepest, and often central, part of the profile. Most erosion then takes place just below this region of steepest gradient, and it is here that gullies are initiated and channel flow begins.

Overland flow and throughflow are likely to be two extremes of a continuous sequence of possible conditions for gully development, with overland flow more common in semi-arid areas and throughflow more common in humid areas. However, if any group of gullies adopts either of the preferred sets of locations, this will indicate which of the two causes of gully formation is likely to be dominant. The control appears to be vegetation— where it is low, or removed, and the proportion of bare surface is increased, Hortonian-type flow is likely to dominate.

(c) *Sheet erosion, rills, and gullies*

When surface runoff is initiated it may be in sheet form or, where the irregularities of the surface demand it, in the form of filaments of flow in a complex, braided pattern. Rills are often seen as a development from such flows; they are ephemeral and discontinuous features that can develop rapidly during a runoff event. Many rills form at a critical distance down-slope, but they may also occur near to the bottom of a slope as suggested by the saturated throughflow model (Morgan, 1979). Various studies suggest that on many slopes, much of the sediment is removed in rills, where the erosive effect of flow is relatively concentrated.

Gullies may arise from the progressive development of rills into more permanent, larger ephemeral channels on slopes. But they may also form independently. Those arising from surface flow commonly follow a development such as that shown in Fig. 4.11, in which initial breaks in vegetation cover act as loci of erosion that grow and coalesce, and subsequent growth is largely by headwall retreat, bank collapse, and scouring by flows in them (Morgan, 1979). In some cases, subsurface erosion can lead to the creation of soil pipes which may eventually collapse to initiate surface gullies. Such features are common in loessal soils (e.g. Parker, 1963). In addition, many gullies can be initiated where surface flow is concentrated as a result of surface disruption by human activity—for example, off-road vehicle tracks and irrigation ditches can act as loci of erosion, and barriers to flow, such as walls and fences, may also concentrate flow (e.g. Webb and Wilshire, 1983).

Various attempts have been made to predict accurately rates of gully development but with only limited success because of the complexity of the problem. For example, the US Department of Agriculture's Soil Conservation Service used the following equation to estimate the average annual rate of gully advance, R:

$$R = (5.25 \times 10^{-3}) \cdot A^{0.46} \cdot P^{0.20} \qquad (4.13)$$

where

R = is in metres,
A = drainage area above gully (m^2),
P = the total rainfall from 24-h rains equal to or greater than 12.7 mm for a time-period, converted to an average annual basis (in mm) (Mitchell and Bubenzer, 1980).

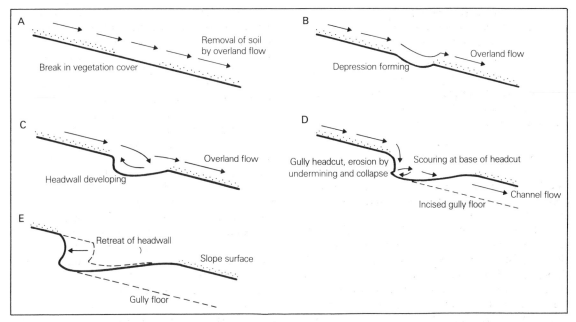

FIG. 4.11 Stages in the development of a hillside gully (after Morgan, 1979; © 1979, and reproduced by permission of the Longman Group Ltd.)

4.5 Variables in the Soil-Erosion System

From the preceding discussion it will be apparent that soil erosion is largely controlled by variables which relate to climate, topography, soil characteristics, vegetation, and land-use practices. The climatic variables, it has been shown, can be summarized by the rainfall index (EI_{30}). The role of soil characteristics, the effect of vegetation, topographic variables, and land-use practices must be examined in a little more detail before comprehensive relations between soil loss and explanatory variables can be presented in the form of a universal soil-loss equation.

(a) *Soil erodibility*

The erodibility of soils depends on several soil properties, notably those that affect detachment and transportability by rainsplash and runoff, and those influencing infiltration capacity. Of crucial important are aggregate stability, the texture of soil particles and related chemical and organic characteristics, and the strength of the soil. Aspects of each of these properties have been used to develop indices of erodibility. Each index requires field tests, or laboratory tests on field samples, or both.

One approach has been to use experimentally validated indices based on soil properties as predictors. These have the advantages that they can often be measured relatively quickly, routinely, and require relatively little special equipment in the laboratory. Such indices generally fall into one of two categories (e.g. Bryan, 1976): those that measure soil-dispersion properties, and those combining soil-dispersion properties with a measure of soil-transmission properties. An early example of the former is Middleton's (1930) dispersion ratio: the amount of silt and clay in an undispersed sample compared with that in a dispersed sample, expressed as a percentage. This ratio is based on the assumption that only dispersed material can be eroded, and Middleton found it to be usually above 15 per cent for erodible soils. However, it makes no allowance for the dispersion of previously undispersed material by raindrop impact, and it does not accurately reflect the erodibility of soils with a high sand content. An example of the second group is that developed by Chorley (1959) in which:

Index of erodibility =
(mean shearing resistance × permeability)$^{-1}$.

He found shearing resistance to be significantly

related to soil density, range of grain size, and soil-moisture content. This index has been criticized on the grounds that shearing resistance is more important in gravity-controlled mass movements of material than in erosion dominated by raindrop impact and surface runoff. But recent studies suggest that shear strength of soils may be a potentially valuable predictor (e.g. Singer *et al.*, 1978; Pall *et al.*, 1982). Other indices also involve the proportion of silt and clay in soils as they fundamentally influence aggregation properties. Bryan (1976) evaluated several of these and concluded that simple measures of aggregate stability (such as the percentage weight of water-stable aggregates >0.5 mm) were the most efficient, because they combine notions of entrainment resistance and runoff generation.

A second approach to measuring soil erodibility, which has been widely adopted, is to determine empirically in the field the actual soil lost under controlled conditions for given values of rainfall erosivity. The best-known of these indices

is the erosion index, K (e.g. Wischmeier and Smith, 1978). For this index, standard bare-soil erosion plots are required, having a length of c.22 m and a slope of 5°. Measured soil losses are then related to each unit of erosivity, determined by the EI_{30} index. Such studies require time and are expensive, but the data they provide have been used to develop a nomograph for predicting K on the basis of measurements of only particle-size distribution (percentage silt + fine sand, percentage sand and organic matter, soil structure and permeability; Fig. 4.12).

(b) *The effect of topography*

From the previous discussion of raindrop erosion and models of runoff erosion it is clear that topographic variables of importance in water erosion include surface slope, surface length, and surface roughness. Various estimates of the relation between soil loss and slope have been made.

In an early study, Zingg (1949) showed that erosion per unit-area, Q_s, was proportional to

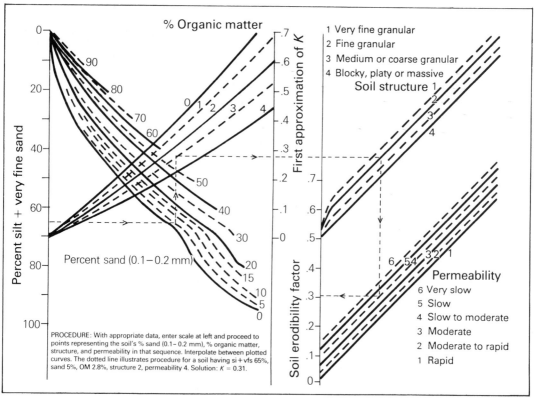

FIG. 4.12 Nomograph for determining the soil-erodibility factor, K, for the USA mainland soils (after Wischmeier and Smith, 1978)

slope angle and slope length. At the sites he studied in the United States, the relationship was

$$Q_s \propto \tan^{1.4} \theta L^{0.6} \qquad (4.14)$$

where θ = slope angle and L = slope length. The values of the exponents will vary from location to location depending on the relationship between slope characteristics and other variables in the soil system (e.g. Morgan, 1979).

In the general predictive model developed by Wischmeier and others (see below), slope length (L) and inclination (S) are regarded as two separate factors. Length is the distance from the point of origin of overland flow to the point where slope decreases sufficiently for deposition to begin or to where runoff reaches a defined channel; gradient is expressed in per cent (e.g. Wischmeier and Smith, 1978), LS is the predicted ratio of soil loss per unit-area of a given field slope, to that from the 22 m/9 per cent standard erosion plot under otherwise identical conditions, assuming uniform gradients (Fig. 4.13).

(c) *The role of vegetation*

Vegetation cover is perhaps the greatest deterrent to soil erosion for it tends to protect the surface from raindrop impact, reduce the amount of water available for runoff by consuming it and by improving infiltration capacity, and (by increasing surface roughness) decrease the velocity of runoff. In addition, plant roots help to keep the soil in place. A grass cover is often the most efficient defence against soil erosion by running water. Thus, in general, vegetation tends to reduce both runoff and erosion and vice versa (e.g. Thornes, 1985). Aspects of vegetation that determine this relationship include canopy density, height, and degree of cover, root density, mulch, and water consumption. Fig. 4.14 shows the general relations between climate and soil erosion by both water and wind (see also Chapter 9). The main focus of interest in the context of environmental management is in *changing* vegetation cover and the effect of different crops on soil loss.

'Natural' vegetation may be modified in many

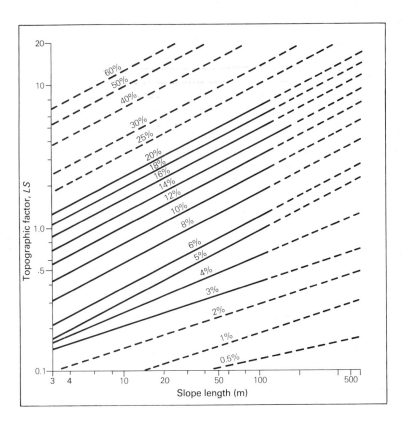

FIG. 4.13 Slope length and gradient factor *LS*, for use with the Universal Soil Loss Equation (after Wischmeier and Smith, 1978)

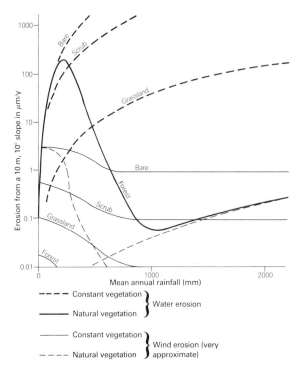

FIG. 4.14 Estimated rates for soil erosion by wind and water as a function of rainfall and vegetation cover (after Kirkby, 1980. © 1980 and reproduced by permission of J. Wiley and Sons Ltd.)

ways—for example, by cutting of forest or ploughing of grassland to create agricultural land, by forest and grassland management practices, and by fires. The case of fire is instructive, because forest fires are a common, usually man-made hazard that can affect many areas, especially in semi-arid lands. Thus, in the chaparral and forested mountains of southern California, it is possible to recognize a fire-induced cycle of erosion (Fig. 4.15). One distinctive feature of this sequence of erosional changes is the creation of water-repellent soils through the coating of soil particles with hydrophobic organic substances during a fire. This change reduces infiltration capacity and serves to accelerate erosion. An impression of the initial acceleration, and subsequent subsidence of erosion rates in the San Gabriel Mountains of Southern California following a fire can be gained from Fig. 4.15B.

In a given locality, where other variables are held constant, it can readily be demonstrated that different vegetation cover strongly influences soil loss. For example, Table 4.1 illustrates the effect of different crops on erosion within standard erosion plots in the United States. The high soil loss from row crops arises partly because they require the preparation of a seed-bed which tends to break up the soil, and partly because planting in rows leaves a relatively high proportion of bare soil.

The beneficial effects of vegetation cover are not confined to living plants; mulches of crop residue, such as chopped maize stalks, provide protection from raindrops and, as they decompose, their residues temporarily improve soil aggregation.

The relations between plants and soil erosion are considerably more subtle than these general examples suggest. For example, the protective effect of vegetation varies with the time of year—the protection afforded by seedlings is naturally less than that provided by mature crops, and the greatest vegetative protection may not coincide with the most destructive rains. Similarly, certain plants may encourage rainfall to drip from leaves or flow down stems, thus perhaps locally

TABLE 4.1 Soil loss from silt-loam soil 1933–42 (USA) (Plots: 9% slope, 22 m long)

Cropping system	Soil loss $kg \times 10^2$ per ha
Continuous maize (unfertilized rows up and down slope)	861.82
Continuous maize (as above, but on subsoil)	1179.33
Rotation	
a. Maize	408.23
b. Oats	226.79
c. Cover	113.40
Continuous lucerne	2.27
Continuous blue grass	0.68

Source: FAO (1965).

A

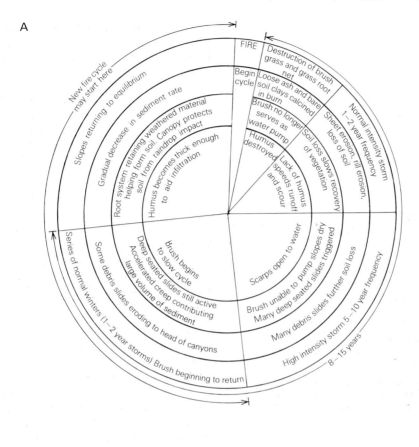

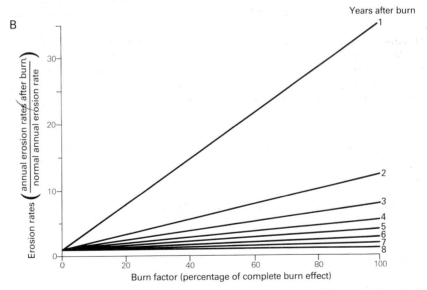

FIG. 4.15 (A) The 'fire-induced' sediment cycle: a summary of common consequences of a fire in southern California mountains (after Smalley and Cappa, 1971). (B) Relations between proportion of watershed burned and the ratio of annual erosion rate after a fire to normal annual erosion rate in the Los Angeles region of the San Gabriel Mountains (after Rowe *et al.*, 1954)

increasing raindrop impact on the one hand and concentrating runoff on the other.

The vegetation effect can be assessed in terms of crop-management practices. In developing a general equation for predicting soil loss, for example, Wischmeier and Smith (1978)'s *cover and management factor, C,* represents the ratio of soil loss from a specific cropping or vegetal cover condition to soil loss from a ploughed, continuous fallow condition with the same soil and topography. In general the ratio will be reduced as the percentage of vegetative canopy cover increases and as mulch cover increases. Other factors influencing *C* are ploughing type, frequency, and timing; and residual effects in a soil of plant roots, residue, and changes in soil structure and composition. An example of *C* values from the United States is shown in Table 4.2.

(d) Conservation practices

If the rate of soil erosion can be enhanced by increased erosivity or erodibility, or by slope length/steepness and vegetation/land-use changes, the rate of erosion can be reduced by reversing some or all of these factors. The aim of soil-erosion control is to reduce soil loss so that soil productivity is economically maintained. In systems terms, the aim is to restore equilibrium in the soil system so that the rate of loss is similar to the rate of regeneration. At present, tolerable rates of loss have to be estimated, and clearly they will vary from place to place because both rates of erosion and the economics of control will vary. In the United States, losses of about 1.1 kg/m²/y seem to be acceptable for deep, medium-textured, moderately permeable soils (Mitchell and Bubenzer,

1989), and rates of 2510–12 550 kg/ha/y to be generally acceptable (e.g. Kirkby, 1980).

Soil-conservation practices fall into three main groups—crop-management practices, supporting erosion-prevention practices, and practices designed to restore eroded land. Each practice relates to the manipulation of one or more of the variables in the water-erosion system. In the following brief summary, the relations between practice and principle are emphasized. For greater detail on this subject the reader is referred to the works of FAO (1965), Hudson (1971), and Morgan (1986).

(i) *Crop-management practices* As much erosion control must be carried out on land in agricultural use, it is clearly desirable to exploit the beneficial effects of vegetation and healthy soil in the form of intelligent cropping practices appropriate to the particular environment. The use of a legume or grass crop in rotation at least one year in five, for instance, often gives a high degree of protection from raindrop erosion. At the same time it bestows additional advantages by providing for a period of soil recuperation, improving soil structure and nitrogen content, and increasing soil organic content (thus tending to reduce soil detachability). Similarly, the judicious application of fertilizers and manures may not only help to improve crop yields but may also encourage soil conditions that decrease detachability and increase infiltration capacity. A third technique involves the planting of 'cover crops'. Cover crops may be grown when the main crop has been harvested in areas where the growing season is long enough to sustain them. Such crops serve to protect the soil at times when they would be exposed to rainfall erosion; in

TABLE 4.2 Soil loss ratios for selected crops

Cover, sequence, and management	Productivity kg/m²	Soil loss ratio, percentage for crop-stage period					
		F	SB	1	2	3	4[a]
First-year corn after grain and legume hay, spring turn plough, conventional tillage, residue left	0.6+	8	22	19	17	10	14
Small grain in disked row-crop residue, after one year corn after meadow, residue left		—	12	12	11	7	2
Grass and legume meadow	0.7+						0.4

[a] Crop-stage periods: F—Rough fallow; SB—Seedbed; 1—Establishment; 2—Development; 3—Maturing crop; 4—Residue or stubble.

Source: Wischmeier and Smith (1978); Mitchell and Bubenzer (1980).

addition they may be ploughed in to provide a beneficial 'green manure'. An alternative approach is to 'interplant' protective crops of grass or a legume between the main crop if the latter tends to be associated with soil loss. 'Mulch tillage' involves the covering of the surface with crop residues (such as grain straw or stubble). This practice provides protection from raindrop impact when the field would normally be bare, and at the same time the residues hold water, retard surface flow, and promote the formation of erosion-resistant aggregates. These are only a few examples of crop-management practices which are designed to protect the soil by maintaining or improving soil structure and using the protective effect of vegetation to advantage.

(ii) *Supporting erosion-control practices* The most important supporting measures of erosion control are 'contour' farming', 'strip cropping', and 'terracing'. Contour farming involves planting rows of crops and using farm machinery along the contours of the land (normal to the direction of slope, e.g. Plate 4.2). This simple device—which is not as widely used as might be expected, especially in view of evidence that it is often more economical of effort—disrupts surface flow and reduces its velocity by increasing surface roughness, restricts the *loci* for rill development, and, in some circumstances, conserves water. Contour farming is most effective on medium slopes and on deep, permeable soils. It also has its dangers, for if soil ridges are breached, water may be concentrated in the breaches, giving rise to large gullies.

'Strip cropping' consists of creating alternating strips of crops and grass or legume parallel to the contours. The dense and complete cover in the sod strips serves to trap sediment carried from crop strips, filter runoff from up-slope and reduce its velocity, increase infiltration rate, and protect the soil from raindrop impact. Normally the strips form part of a crop-rotation system. Strip cropping is especially appropriate on slopes too steep

PLATE 4.2 Sensible management: soil conservation by terracing, contour planting in orchards, S. Moravia, Czechoslovakia (photo courtesy of D. Zachar)

for terracing (e.g. 6–15 per cent, or 3.5°–8.5°), and on farms where forage is an important part of the economy. The width of strips varies with ground slope. For example, recommended strip widths on soil with fairly high water intake are as follows (FAO, 1965, p. 92):

Percent slope	Slope in degrees	Strip width (m)
2–5	1–3	30–33
6–9	3.5–5	24
10–14	5.5–8	21
15–20	8.5–11.5	15

Terracing normally requires the creation by earth-moving equipment of an embankment (with or without a channel up-slope of it) parallel to the contours. Most terraces reduce slope gradient, break the original slope up into shorter units, conserve soil moisture, and remove runoff in a controlled fashion. They also allow a field to be given over to a single crop. The spacing of terraces is related to ground slope, in the same way as strip-crop widths, and will be constant for given soil and climatic conditions. Of critical importance in the design of efficient terraces is the geometry of the drainage channel which removes runoff from the terrace base: this must not be susceptible to erosion and should be able to accommodate the largest runoff likely to occur in a reasonable time-period (such as a decade). The problem of drainage-channel design is complex, but its solution rests on well-established and fundamental hydraulic principles (e.g. Chow, 1964; Henderson, 1966).

(iii) *Restoration of gullied land* The removal of runoff from terraces usually requires the construction of waterways, and if these are inappropriately designed they can easily become gullies. Natural waterways can also become gullied by increased runoff from areas of poor land-use practice. In both of these cases, measures are required to remove the gullies and to restore the drainage channels. The most commonly used method is to cover the waterway with grass or to encourage natural vegetation. A second approach is to convert the gully into a stable artificial channel with dimensions appropriate for the discharge of water. A third method is to reduce water supply by conservation practices in the tributary lands. An alternative is to eliminate flow from the gully by diverting it into an artificial channel. A fifth method is to reduce erosive flow velocities in gullies by building structures such as spillways and weirs which dissipate the flow energy, and by creating stable channel sections between the structures.

(iv) *A support practice factor (P)* In evaluating the factors for a general model of soil erosion, Wischmeier and Smith (1978) developed a support practice factor, P, which is the ratio of soil loss with a specific support practice to the corresponding loss with up-and-down slope cultivation. The main practices that might reduce the value of P include ploughing and planting along contours, contour strip cropping, and terracing.

(e) *The Universal Soil Loss Equation (USLE)*

Once the major variables affecting soil loss have been isolated and evaluated, it should be possible to derive a general equation that can efficiently predict total soil loss in terms of them. By far the most important equation for general soil loss prediction by rainsplash and runoff is the Universal Soil Loss Equation (USLE) which states (Wischmeier and Smith, 1978; Mitchell and Bubenzer, 1980):

$$A = R \times K \times LS \times C \times P \qquad (4.15)$$

where

A = average annual soil loss,
R = rainfall erosivity factors ($\Sigma EI_{30}/100$),
K = soil-erodibility factor,
LS = slope length–steepness factor,
C = cropping and management factor,
P = conservation practice factor.

The nature and derivation of each of these factors has already been described in the previous paragraphs. The equation is not as universally applicable as its title suggests (Hudson, 1980). Wischmeier and Smith (1978) made clear that its main purpose is to predict cropland losses by sheet and rill erosion and to provide guides for selecting adequate erosion-control practices for farm fields and small construction areas. In particular it allows the determination of those land-use and management practice combinations that will provide the selected level of erosion control. For example, in the equation, its tolerance level can be substituted for A (Equation 4.15) and the equation can then be solved in terms of C and P. The USLE is also useful in estimating watershed sediment yield, provided the watershed is subdivided into areas for which representative values for the factors can be calculated.

Because it is based on erosion-plot data and cropping practices in the United States, the equation is likely to be most accurate in that country. Indeed, it should only be applied elsewhere with caution, especially if factor values are not established (e.g. Hudson, 1980). Application of the equation to other areas has led to its refinement (e.g. SLEMSA—Soil Loss Estimator for Southern Africa, Elwell, 1984); Osborn *et al.* (1976) have suggested that it could be improved in some semi-arid areas at least, by including a channel factor to account for sediment loss in gullies.

Kirkby (1980) has noted several potential and real problems with the USLE. For example, the derivation of a value of A does not allow for any sort of non-linear reactions between the factors. Similarly 'the overland flow generated in a small quadrat is equal to rainfall intensity *minus* infiltration rate. Since maximum infiltration rate depends on soil rather than rainfall properties, it is plain that neither overland flow runoff, nor the soil transported by it can be expressed as a rainfall factor *multiplied* by a soil factor, as is demanded by the USLE' (Kirkby, 1980, p. 9).

Such comments serve as a warning against the belief that the USLE, robust, useful, and as widely applied as it is, actually embodies a full understanding of the soil-erosion system in a wholly satisfactory way. In fact, there are still many problems in the physical system that remain unsolved. For instance, the relative importance of and interaction between rainsplash and flow erosion is not fully understood or effectively predicted; and the relations between erosion and sediment supply require further examination (Kirkby, 1980).

4.6 Economic and Productivity Implications

(a) *Introduction*

The previous discussion of the soil-erosion system has been concerned mainly with the loss of soil from agricultural land and with measures for preventing it. The consequences of such soil loss vary greatly from area to area: here, there may be a problem of silting and loss of feed, fertilizers, and seed; there, sprinkler irrigation may be prevented because it exacerbates raindrop impact and raindrop erosion. These consequences and those of the appropriate control measures are reflected in soil production and the economics of farm management, but the precise economic implications of soil erosion are extremely difficult to disentangle from the complex variety of conditions which affect a farmer's budget sheet.

(b) *The EPIC model*

A fundamental question in soil-erosion studies is the effect that soil erosion has on soil productivity. To predict this relationship requires more than knowledge of the physical system; it also requires an understanding of the economic and technical environment. In the United States, and arising from the requirements of the Soil and Water Resources Conservation Act 1977, a computer-based simulation model called EPIC (Erosion–Productivity Impact Calculator) has been devised to allow prediction of the relationship between erosion and soil productivity, both locally and nationally (e.g. Williams *et al.*, 1984; Williams, 1985). The model includes numerous physically based components for simulating erosion and plant growth: hydrology (surface runoff, percolation, subsurface flow, drainage, evapotranspiration, irrigation, and snow-melt); weather (precipitation, temperature and radiation, wind); erosion by water and wind (using modified versions of the USLE for both types); nutrients (e.g. nitrogen, phosphorus); soil temperature; and tillage. A single crop-growth model for simulating productivity of a range of crops was included. The crop budgets and economics are calculated using components from the Enterprise Budget Generator, in which inputs are *fixed* (e.g. depreciation, interest or return on investment, insurance, various taxes, and capital improvements) or *variable* (e.g. machinery repairs, fuel and other energy, lubricants, seed, fertilizer, pesticides, labour, and irrigation water). The model is only in the early stages of its development, but it is already able to produce reasonable productivity predictions.

(c) *Economic consequences: examples from Australia*

In a series of studies, Molnar (e.g. 1964) has attempted to measure the loss of production and the decline of land values due to soil erosion by means of economic models, experiments, and farm surveys in Australia.

It can be argued that in order to maintain output from a soil that has been eroded an increase in input will be required. Such change is illustrated graphically in Fig. 4.16A, where X_2 would need to be increased to X_3 to maintain output on an eroded soil. This model was tested experimentally in New South Wales and Victoria by comparing production of wheat in a wheat–fallow rotation on uneroded plots, and on plots from which 7.6 cm or 15.42 cm of topsoil had been removed (to simulate sheet erosion). All plots received 20.41 kg of superphosphate with the wheat. The results are

shown in Table 4.3. Clearly these experiments indicate that since inputs are the same regardless of yield obtained, there is less return on capital and labour on land where the yield is lower.

Differences in the value of annual production due to differences in erosion and soil types in two areas of Victoria can also be determined approximately by farm surveys, as Table 4.4 shows.

It is important to realize that economic implications of soil conservation must be viewed in the light of the *economic aims* of the conservation measures. Molnar pointed out that both in

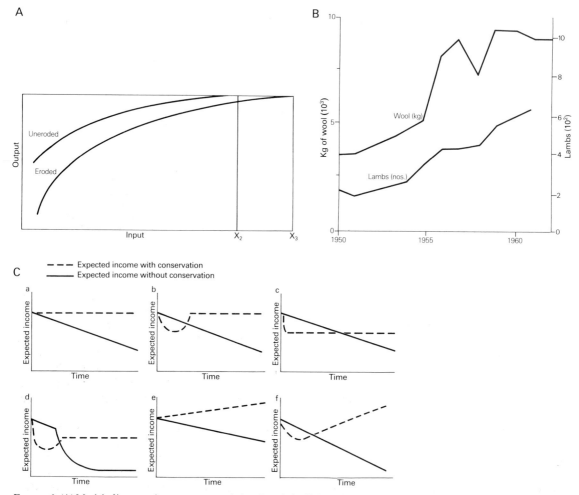

FIG. 4.16 (A) Model of input and output on uneroded and eroded soils for soils with similar surface and subsurface conditions (after Shroder, from Molnar, 1964). (B) The effect of soil conservation on production for a farm at Armstrong, Victoria, Australia (after Garside, from Molnar, 1964). Conservation works began in 1951. Between 1951 and 1963 the improved pasture increased from 88 ha to 264 ha, the number of sheep increased from 700 to 2000, and wool cut per ha increased from 1.78 kg to 4.99 kg. (C) Relations between expected incomes and time on land with and without conservation practices, using various assumptions (after Barlowe, from Molnar, 1964)

TABLE 4.3 The effect of loss of topsoil on wheat yield and protein content at Wellington and Gunnedah, New South Wales, and on soil nitrogen at Gunnedah

Depth of soil removed (cm)	Wellington (1955–63 av.)		Gunnedah (1955)		
	wheat yield (kg/ha)	wheat protein (%)	wheat yield (kg/ha)	wheat protein (%)	soil nitrogen (%)
0	1325	11.3	1450	10.8	0.16
7.6	965	10.6	990	8.4	0.12
15.42	800	10.2	620	9.1	0.10

Source: Molnar (1964), with metric conversion.

TABLE 4.4 Differences in the value of annual production due to differences in erosion and soils at Dookie and Coleraine, Victoria

Locality	Years of data	Difference in value (in Australian £) of production per acre (0.4 ha.) between:	
		(a) Uneroded and worst eroded land	(b) Best and worst soils
Dookie	1951–2	£1 9s. 6d.	—
Coleraine	1954–5	£2 16s. 0d.	£5 0s. 0d.

Note: These data must be used with caution because of several problems of putting a cash value on soil quality and erosion. Values for different years need to be adjusted to a common base, and are in any case only of relative significance; the figures are understated because the quality of lambs could not be valued (e.g. lambs may not have been kept as long on eroded land); the analysis is not based on complete information (e.g. cereal production at Dookie was not considered); other effects of erosion, such as increased working hours involved in cultivating eroded land, were also not considered.

Source: Molnar (1964).

Australia and in the USA conservation measures are in general supposed to halt the decline in production and it is not normally assumed that an *increase* in production will follow. Economic models designed to describe production trends need to recognize the possibility that income may be maintained, increased, or decreased. Some of the relations between expected income and time, using different assumptions, are shown in Fig. 4.16C. In *a*, conservation practices restore income to the level before erosion began; *b* shows a temporary initial drop in income; in *c* income is eventually stabilized at a lower level than before the onset of erosion; and *d* corresponds to *b*, except that the 'stable' income is lower than initial income. In *e* and *f*, the alternative assumption is made that income will eventually rise. It is difficult to validate these models because the interesting variables cannot easily be isolated from other aspects of the farm economy. One approach is through the comparison of 'before' and 'after' incomes from adequate farm records. For example, Fig. 4.16B implicitly suggests the effect of soil-conservation works, begun in 1951, on a farm at Armstrong, Victoria. The analysis indicates a clear trend, but unfortunately it is based on incomplete records and shows only gross production.

Accelerated soil erosion arises from mismanagement. It is therefore important to know why mismanagement occurs. In Australia, sheet erosion occurs most frequently on land that has been overstocked in successive years. Molnar pointed out that overstocking can arise from ignorance, or because farmers have been unable to earn an adequate income with fewer stock. Reasons for the latter situation may have complex origins in lack of capital, high interest rates, drought, adverse price conditions, too small farms, or inappropriate farm location and boundaries. None of these problems is necessarily the responsibility of the farmer. Responsibility for the solution of soil-erosion problems must surely rest in part with government agencies.

4.7 Why Is There Still a Problem?

Decades of research into the problems of soil erosion have led to a substantial improvement in understanding of the physical erosion system and to the recognition of many techniques for avoiding or reducing soil loss. And yet soil erosion remains a fundamental environmental problem in many countries. Why? The reason, as Blaikie (1985) has discussed in detail in his important book, is that the 'soil erosion problem' is less one of physical understanding and conservation techniques, and more one of the failure of *programmes* and *policies*, all of which must face economic, social, and political realities. It is true that soil-conservation policies have had some successes, locally and nationally, in several countries, including the United States, China, Jamaica, and the Republic of South Africa. But the evidence of failure is accumulating, even in the United States, and in many countries of Europe and the Third World.

The reasons for failure of soil-conservation policies are numerous. Blaikie (1985, pp. 64–71) listed five of them:

(a) Conservation techniques do not conserve soil in practice because of technical failures through inadequate or misapplied research.
(b) Conservation techniques do not fit into agricultural and pastoral practices and are therefore not applied by farmers or pastoralists.
(c) Conservation is hampered by existing land-tenure conditions.
(d) There is a lack of participation by land-users in government-sponsored conservation.
(e) There are fundamental institutional weaknesses for enforcement, participation, farming, etc.

Blaikie pointed out that failure is often related to the imposition of policies 'from above' by governments, or international agencies, remote from the problems of the land-user and ignorant of the fact that policies may require unacceptable social changes and inappropriate technical adjustments.

He identified, for example, in developing countries a 'colonial model' and its neocolonial successors that blame the problem on 'lazy, ignorant and backward land-users', link erosion with 'over-population', seek 'local land-users' involvement in a market economy, and impose solutions without local support. He also recognized contradictions between most conservation policies and the limitations and objectives of foreign aid (Table 4.5). Even where adoption is achieved its rate of introduction may be inadequate to overcome the rate of degradation and population growth, and the growing need for capital, especially amongst small landholders. Closely allied to this failure is the failure to recognize that soil erosion is strongly enmeshed with the whole social and economic fabric, and the aim of reducing soil loss, even if it is locally accepted, not only requires knowledge of the methods, but also requires access to such things as fertilizers, seeds, credit, and appropriate assistance. Blaikie (1985) showed convincingly that soil erosion in developing countries is thus a *sympton*, a *cause*, and a *result* of underdevelopment.

There are signs that these causes of failure are beginning to be recognized. Increasingly efforts are being made to adjust policies to local practices, to involve land-users themselves, and to focus the study of soil erosion locally and on its possible economic and political dimensions as well as on its physical manifestations; and to integrate the problem of soil erosion more fully into the broader problem of rural development.

4.8 Conclusion

The advent of process models of soil erosion by water, and of predictive models such as the USLE and EPIC, mark substantial progress in the understanding of the soil-erosion system. There remain large areas of ignorance among scientists: for example, too little is known about 'tolerable soil loss' in different circumstances, basic empirical data are scarce in many countries, the USLE and other predictors are improperly applied in places, and several fundamental, important theoretical problems, such as the relative roles of rainsplash and runoff in soil loss, remain. It is nevertheless possible substantially to reduce soil loss. The largest remaining problem is undoubtedly to achieve this objective by formulating and success-

TABLE 4.5 Contradictions between foreign aid and soil conservation programmes

Essential elements in conservation programmes	Objectives and limitations of foreign aid projects programmes	Result
Long-maturing	Measurable benefits within three to five years	Emphasis on short-term and often peripheral objectives to soil conservation
Diverse, timely, and highly co-ordinated inputs	Inability to deal with a large number of line ministries	Either disorganization, or attempts to set up independent foreign-staffed implementation agencies
Outputs diffuse, diverse and difficult to quantify	Quantifiable benefits need to be predicted at proposal stage for purposes of justification	Concentration on physical and often less important objectives of conservation
Implementation deeply involved in sensitive political issues	No overt interference in internal political affairs of recipient country	Acute problems of implementation
In-depth analysis of political economic circumstances	Short-term consultants with the necessity to be tactful on political feasibility	Project documents full of rhetoric and technical details
Sustained political will at central government level	Short-term consultancies (1–3 years usual)	Uneven and uncertain back-up of project/programme after aid finishes

Source: Blaikie (1975).

fully implementing policies that are acceptable to land-users. To achieve such success requires much more than a knowledge of process and engineering solutions and where the problem occurs. It also requires a detailed understanding of why so many societies are faced with soil-erosion problems and how they can best be persuaded to conserve their soil resources. That problem is not new, but it has been ignored and it is bound to be high on the soil-erosion agenda in the future.

5 Slope Failure and Subsidence

5.1 Introduction

Almost every year a major disaster is caused by landsliding. In many landslides deaths exceed a hundred, and at times are in the thousands (Table 5.1). In 1985, volcanic activity on the Nevado del Ruiz in Colombia caused rapid ice-melt near the mountain summit, which in turn sent large volumes of water down the mountain side. This water mixed with the volcanic ash to create debris flows (or *lahars*). These debris flows overwhelmed the town of Armero and killed an estimated 22 000 people (Fig. 5.1). Such dramatic events attract world-wide attention. In addition, there are also hundreds of smaller, less newsworthy cases of slope failure, some of which are no more than a nuisance, but they can cost money either in terms of the damage they do or through the remedial works they may require.

The largest known landslide in the world, the Saidmarreh landslide in the Zagros Mountains of Iran, occurred more than 10 000 years ago. It was 15 km long and slid 20 km away from its source (Watson and Wright, 1969). This slope failure was not a disaster, despite its size, because it affected nobody. As with all natural hazards, a disaster only occurs if slope failure actually affects people and property. Landslides in uninhabited areas or ancient landslides are of academic interest only, unless new developments, such as a civil engineering project, reactivates such a landslide; sometimes an engineering project can create a new and damaging landslide (Plate 5.1). Notable amongst these is the landslide into the Vaiont Dam, Italy, in 1963 which created a wave within the reservoir waters so large that it caused the dam to burst. The resulting flood drowned over 2000 people in the village below. The landslide itself was caused by the water-level of the reservoir being lowered too rapidly. This had left a perched water-table within unsupported reservoir side slopes, thus creating instability in the slope which resulted in a landslide (Kiersch, 1964).

The costs, in monetary terms, resulting from landslides are unknown. Within the United States alone they have been estimated to exceed $1000 million per annum (Schuster, 1978; US Geological Survey, 1982), while annual losses in Italy are said to be about $1140 million (Arnould and Frey, 1977).

Policy reactions have taken place where landslides commonly have a human impact. In Brazil, it was found that urban expansion led to landsliding as houses were constructed on the higher, steeper slopes around Rio de Janeiro, especially where forests had to be cleared to make way for construction. Legislation was therefore introduced to prevent this type of expansion. The Forest Law of 1959 made it illegal to carry out construction work above a specified level. This helped to maintain the stabilizing influence of the forests and to prevent construction across springs. A second law, the Law for Licence of Construction in Uneven Terrain, passed in 1967, regulated construction on steep slopes or wherever instability was possible. It demanded that the contractor should obtain proof of slope stability before building construction was allowed to commence (Barata, 1969).

Elsewhere, such as in parts of the United States and Western Europe, development is preceded by hazard mapping that includes the identification of slopes that have failed or are near to failure. Such mapping and assessments are built into the legislation or planning procedure with varying degrees of rigour. However, the general trend is to take more and more account of slope failure and the potential for slope failure in advance of a construction programme. This is also true in site investigation practice in connection with civil engineering work.

The greatest contribution made by geomor-

TABLE 5.1 Some major disasters caused by landslides

Place	Date	Type of landslide	Est. max speed	Impact
Goldau, Switzerland	2 Sept. 1806			457 people killed
Elm, Switzerland	1881		44 m/s	115 people killed
Java	1919	Debris flow		5100 killed, 140 villages destroyed
Kansu, China	16 Dec. 1920	Loess flows		10 000+ killed
California, USA	31 Dec. 1934	Debris flow		40 killed, 400 houses destroyed
Kure, Japan	1945			1154 killed
SW of Tokyo, Japan	1958			1100 killed
Ranrachirca, Peru	10 June 1962	Ice and rock avalanche		3500+ people killed
Vaiont, Italy	1963	Rockslide into reservoir	50 m/s	about 2600 killed
Aberfan, UK	21 Oct. 1966	Flowslide?	8.8 m/s	144 people killed
Rio de Janeiro, Brazil	1966			1000 killed
Rio de Janeiro, Brazil	1967			1700 killed
Virginia, USA	1969	Debris flow		150 killed
Japan	1969–72	Various		519 died, 7328 houses destroyed
Yungay, Peru	31 May 1970	Earthquake-triggered debris avalanche–debris flow	133 m/s	up to 25 000 killed
Chungar	1971			259 people killed
Hong Kong	June 1972	Various		138 killed
Kamijima, Japan	1972			112 killed
S. Italy	1972–3	Various		about 100 villages abandoned affecting about 200 000 people
Mayunmarca, Peru	25 Apr. 1974	Debris flow	39 m/s	town destroyed, 451 killed
Mantaro Valley, Peru	1974			450 killed
Mt. Semeru	1981			500 killed
Yacitan, Peru	1983			233+ killed
W. Nepal	1983			186 killed
Dongxiang, China	1983			227 killed
Armero, Colombia	Nov. 1985	Lahar		about 22 000 killed

phologists to the study of slope failure is in the recognition of how landslides have failed, what caused the failure, and where else failure could occur. Such prediction is difficult, and comes only from a clear understanding of the nature and mechanisms of any landsliding that has already taken place in an area. Inevitably, therefore, the geomorphologist must be concerned with a study of past and recent failures, of the geological and hillslope conditions that allowed failure, and of the fundamental causes of that failure. He must be able to map and record not only existing landslides but also those hillslope parameters that determine whether or not further failure is likely to occur. Increasingly, this also requires a good understanding not only of natural processes (e.g. river undercutting) but also of human influences on slopes (such as those that occur with different types of engineering works including dam construction, road building, urban expansion, and forest clearance).

The literature that covers the range of knowledge required is vast, and embraces not only geomorphological texts but also engineering ones. Amongst the most informative of recent texts are *Landslide: Causes, Consequences, and Environment*

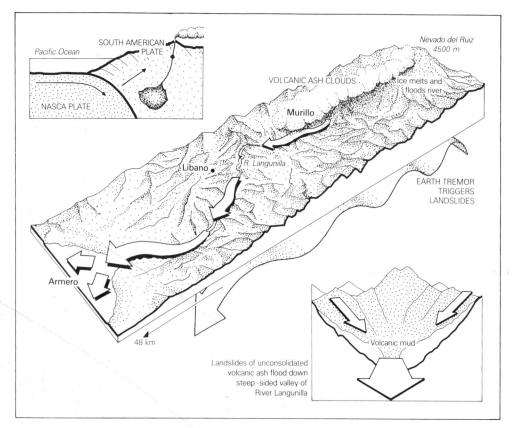

FIG. 5.1 The lahar that overwhelmed Armero, Colombia (after a drawing by Peter Sullivan, 1985; reproduced by permission of the *Sunday Times*)

by Crozier (1986); *Hillslope Stability and Land Use* by Sidle *et al.* (1985); *Landslide Hazard Zonation* by Varnes (1984); *Slope Instability* edited by Brunsden and Prior (1984); *Landslides—Analysis and Control* edited by Schuster and Krizek (1978); the chapter on mass movement by Brunsden in Embleton and Thornes (1979) *Process in Geomorphology*; *Rockslides and Avalanches* (Voight, 1978); *Landslides and their Control* (Zaruba and Mencl, 1969); while much useful information and ideas are still to be found in what is rapidly becoming a classic study, *Landslides and Related Phenomena* written in 1938 by C. F. S. Sharpe. Important journals in this field include *Geotechnique*, *Engineering Geology*, the *Quarterly Journal of Engineering Geology*, and that of the Soil Mechanics and Foundations Division of the American Society of Civil Engineers. Many case-studies are also published in the *Bulletin of International Association of Engineering Geology*.

This chapter identifies some of the more important factors that are likely to contribute towards landsliding. This basic understanding is necessary before an appreciation can be acquired of techniques of landslide prediction. Many geomorphological studies are based upon, or use, morphological classifications of landslides and, as will be shown, landslide form is often a very good guide to the cause and mechanism of hillslope failure.

5.2 Classification of Landslides

Features demonstrating slope movement range in size from small terracettes, indicating soil creep, up to large landslides. For present purposes soil creep features, because of their small size, are ignored.

Several landslide classifications exist. These have been reviewed by Hansen (1984) and Crozier (1986). The one adopted here is based on Varnes (1978) and is illustrated in Fig. 5.2. In this

PLATE 5.1 Main trans-Pennine road destroyed by landsliding at Mam Tor, Derbyshire, UK (photo courtesy of J. F. Winn)

classification the types of slope movement are identified as falls, topples, slides (rotational and translational), lateral spreads and flows, as well as complex slides in which two or more types of failure are present. The distinction between failures in bedrock and failures in soils is an important one for the nature and behaviour of the materials involved in each case are very different. A study of the illustrations (Fig. 5.2) shows that the form of the landslide tends to betray the nature of the movement. Form is often a critical diagnostic feature of a landslide. Landform features that aid in the recognition of landslides are listed in detail by Rib and Liang (1978), and are reproduced in Varnes (1984).

(a) *Falls*

Rockfalls may occur where a steeply sloping rock-face consists of well-jointed rock. Rockfalls tend to be generated in three stages. The first sees the creation of the cracks from which the block eventually breaks. These cracks become enlarged and the fall takes place because support has been removed from the base of the rock. The crack enlargement process arises mainly from pressures related to the freezing of water or the growth of plant roots, or directly as a result of gravity. Loss of support at the base may be the result of erosion by glaciers, the sea, rivers, differential weathering of a softer layer that lies beneath the capping rock, or human interference. Rockfall debris may be removed by processes working at the base of the slope, in which case further falls may occur, or it may accumulate to form a talus footslope helping to protect the slope from further undercutting.

Soil falls can occur wherever a soil bank or

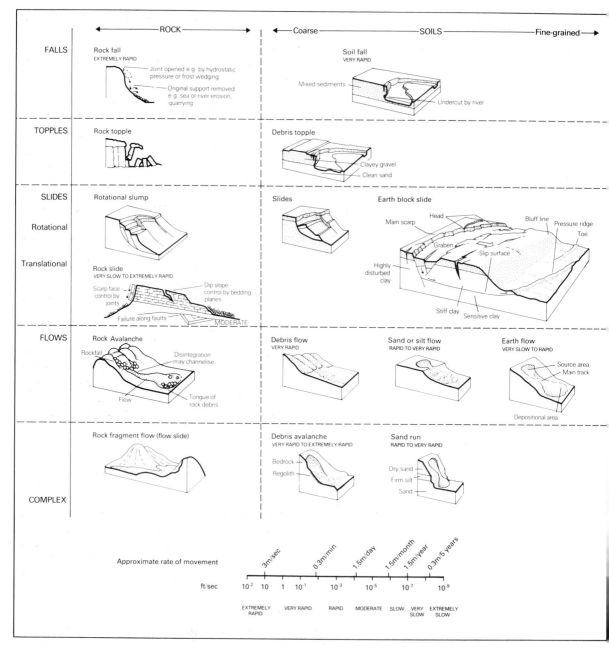

FIG. 5.2 Classification of landslides (after Varnes, 1978)

terrace face is undercut. The unstable unit is initially separated from the parent cliff by tension cracks, before its abrupt collapse as shear failure occurs at the base of this unit.

Many smaller falls tend to occur near the surface where the effects of pressure release and of pressure build-up through water freezing can be most effective.

Catastrophic rockfalls were involved in the large landslides at Elm, Switzerland, which, in 1881, killed 115 people; the Huascaran failure which, on 31 May 1970, turned into a debris flow and destroyed Yungay in Peru, killing about 25 000 people; the prehistoric failure on the slopes of Blackhawk Mountain, where rock descended 610 m in free fall before crashing into the valley below; and the Frank slide in Alberta, Canada, where in 1903 the mining settlement of Frank suffered a loss of 76 inhabitants.

Once a rockfall reaches the ground it may roll on, for a short distance, if small volumes of rock are involved. With volumes in excess of 0.01 km³, or so, the mass may turn into a flowslide (as defined by Rouse, 1984) and travel on for several kilometres at speeds of up to 480 km/h (133 m/s). In the case of the Frank slide, 90 million tons of rock ran out more than 1.3 km in 100 seconds, a speed of 13 m/s.

(b) *Topples*

A forward toppling motion can occur in either rocks or soils (Fig. 5.2) usually as a result of wedging behind the falling mass. The toppled material may then move on as a flow or as a slide.

(c) *Slides*

Rotational slides (or slips) involve a turning movement that often leaves an upper surface, on the failed mass, inclined back into the hillside. *Translational* slides involve a down-slope movement along a more or less inclined (planar) surface. Rotational slides occur with a concave-upwards curved failure plane (rupture surface) passing through a thick and relatively homogeneous bed of clay or shale (Hutchinson, 1968). They are more deep-seated than translational slides. Movement tends to be rotational on this slip surface.

Although the rupture surface tends to be semicircular, this form is seldom perfectly achieved since its shape is influenced by original structures such as faults, bedding planes, and joints. The back-tilting at the head of the mobile mass forms hollows in which water can collect. This tends to drain down the rupture surface and maintain permanently high pore-water pressures. By this means there is a tendency for movement to be renewed.

Translational slides include rock slides, blockslides, and debris slides, and can be used as a universal term to cover lateral spreads as well (Hutchinson, 1968). Translational slides move on shear planes which are roughly parallel to the ground surface. The depth to the slip surface is generally less than one-tenth of the distance from the toe to the rear scarp of the slide (Hutchinson, 1968). Many of these slides in bedrock or in soils suffer shearing along a marked discontinuity, such as the junction of a stronger and a weaker bed. In consolidated materials the failure plane may be kept near the surface because of the general tendency within these beds for shear strengths to increase rapidly with depth.

When a large mass is moved laterally it generally does so on a slip surface commonly formed by a bedding, cleavage, or joint plane, frequently occupied by a filling of material having a low shear strength. The landslide types which occur on steep slopes (15–40°) in unconsolidated materials tend to behave as a more or less cohesionless mass, becoming distorted with movement. The actual details of the distortion and the depth of the slide depend on the nature of the slide debris. Sand runs involve dry, cohesionless granular material experiencing only shallow movement, whereas clays, being more cohesive, tend to move in less distorted units. The trigger mechanism for the movement of slides in the unconsolidated materials frequently appears to be a heavy rainfall which creates high pore-water pressures. When distortion takes place the movement may include flow as well as sliding. The distinction between these two types of movement is not always clear, and many slope failures involve both sliding and flowing.

Slides are also the most common form of slope failure on the sea-floor (Prior and Coleman, 1984), and these have been particularly well documented for the Mississippi delta (Roberts *et al.*, 1983). The largest area of submarine slides occurs as a complex of slides off the coast of South Africa (the Agulhas slide) which lies between longitude 22° E and 30° E, and between latitude 36° S and the coast. It occupies nearly 80 000 km². Many submarine slides occur on much gentler slopes than their onshore equivalents.

(d) *Flows*

Catastrophic flows in bedrock have been described above under 'falls' as 'flowslides'. Flows in soils are also important. These range from debris flows in coarse granular material to mudflows in predominantly clays. A critical component of any flow in soils (as distinct from that involving rock) is its water content, and this may be reflected in the final form of the failure (Fig. 5.2). However, all flows resemble the movement of viscous fluids (even those in dry materials), and usually show a sharp boundary (slip surface) between the moving mass and the ground across which it is moving.

(e) *Complex*

Complex failures involve several types of movement within one event. Typical combinations include rockfalls and debris avalanches, rockfalls and rock flowslides, rotational slides, and earth (often mud) flows.

(f) *Active and inactive landslides*

It is crucial for land-management purposes to know whether a landslide is active or inactive. Certain criteria have been identified which allow the distinction between these two states to be made based on surface evidence (Table 5.2). However, such a distinction is rarely simple, and arbitrary decisions often have to be made. For example, is a landslide that happened 5 (or 10, 15, or 50) years ago an active landslide? The term is probably best reserved for slope failures that have shown movement in the recent past *and* are expected to show movement again in the near future. Other landslides may be recent rather than active and can be considered to be historical in that they can be assigned to some past moment or period in time. The precise form of such a classification does not matter too much for environmental management (especially engineering) purposes. What does matter, however, is that landslides are recognized for what they are. Inadvertent mismanagement, such

as an excavation into the toe of an inactive (or even unnoticed) landslide, may cause it to become active again, sometimes with devastating consequences.

(g) *Avalanches*

Avalanches in snow and ice form a distinct category of slope failure because of the medium involved. To understand them is important because of the increasing pressure of development upon alpine lands. Interest is growing in the mapping of avalanches and avalanche-prone sites, in the determination of the return period of avalanching, in a fuller understanding of their behaviour, and in predicting where and when they will occur (La Chapelle, 1977; see also Chapter 8).

5.3 Forces That Lead to Slope Failure

Landsliding takes place when slope materials are no longer able to resist the force of gravity. This decrease in (shear) resistance may result either from internal or external causes. Internal causes usually involve some change in either the physical or the chemical properties of the material (rock or soil) or its water content (Table 5.3). External factors which lead to an increase in shear stress on the slope usually involve a form of disturbance that may be either natural or induced by man (Table 5.4).

It is useful in this context to think in terms of two

TABLE 5.2 Features which distinguish active from inactive landslides

Active	Inactive
Scarps, terraces and crevices with sharp edges	Scarps, terraces, and crevices with rounded edges
Crevices and depressions without secondary infilling	Crevices and depressions infilled with secondary deposits
Secondary mass movement on scarp faces	No secondary mass movement on scarp faces
Surface-of-rupture and marginal shear planes show fresh slickensides and striations	Surface-of-rupture and marginal shear planes show old or no slickensides and striations
Fresh fractured surfaces on blocks	Weathering on fractured surfaces of blocks
Deranged drainage system; many ponds and undrained depressions	Integrated drainage system
Pressure ridges in contact with slide margin	Marginal fissures and abandoned levees
No soil development on exposed surface-of-rupture	Soil development on exposed surface-of-rupture
Presence of fast-growing vegetation	Presence of slow-growing vegetation
Distinct vegetation differences 'on' and 'off' slide	No distinction between vegetation 'on' and 'off' slide
Tilted trees with no new vertical growth	Tilted trees with new vertical growth above inclined trunk
No new supportive, secondary tissue on trunks	New supportive, secondary tissue on trunks

Source: Crozier, 1984.

TABLE 5.3 Factors leading to a decrease in shear resistance

---

1. Materials
   beds which decrease in shear strength if water content increases (clays, shale) (e.g. when local water-table is artificially increased in height by reservoir construction, or as a result of stress release (vertical and/or horizontal) following slope formation)
   low internal cohesion (e.g. consolidated clays, sands, porous organic matter)
   in bedrock: faults, bedding planes, joints, foliation in schists, cleavage, brecciated zones, and pre-existing shears

2. Weathering changes
   weathering reduces effective cohesion (c'), and to a lesser extent the angle of shearing resistance (ϕ')
   absorption of water leading to changes in the fabric of clays (e.g. loss of bonds between particles or the formation of fissures)

3. Pore-water pressure increase
   higher groundwater table as a result of increased precipitation or as a result of human interference (e.g. dam construction)

TABLE 5.4 Factors leading to an increase in shear stress

---

1. Removal of lateral or underlying support
   undercutting by water (e.g. river, waves), or glacier ice
   washing out of granular material by seepage erosion
   man-made cuts and excavations
   drainage of lakes or reservoirs

2. Increased loading (external pressures)
   natural accumulations of water, snow, talus
   man-made pressures (e.g. stock-pile of ore, tip-heaps, rubbish dumps, or buildings)

3. Transitory earth stresses
   earthquakes
   continual passing of heavy traffic

---

elements: (a) those conditions that prepare the slope materials for failure, and (b) those forces that actually cause it to fail. These may not be the same as each other. For example, tropical weathering may produce a deep mantle of soil that remains in place on a slope until the slope is undercut by a river (or human activity), and the soil is no longer able to remain at rest on the undercut slope. Unless the soil had been prepared by weathering,

the effect of river undercutting would have been much less.

In general terms, the stability of a slope may be defined by a factor of safety F_s, where:

$$F_s = \frac{\text{the sum of forces resisting slope failure}}{\text{the sum of disturbing forces}}$$

or

$$F_s = \frac{\text{shear strength}}{\text{shear stress}}$$

If $F_s > 1.0$ stability is likely (but nervousness exists while $F_s < 1.2$), but if $F_s < 1.0$ instability exists (or is imminent).

The calculation of F_s depends upon the on-site measurement of the geotechnical properties of the slope materials or the analysis of rock or soil samples in the laboratory (see Petley, 1984). These geotechnical properties tend to identify the state of preparation, as defined above, and may fail to consider factors (such as river undercutting) which may be present but have not yet transmitted their influence to the site being investigated. It is in such situations that a geomorphological study can make a significant contribution to the assessment of slope stability, by identifying those character-istics of the situation of a slope that are likely to have an external influence upon its stability. Geotechnical site investigations are costly and are usually carried out only in the context of an engineering project where slope failure will affect the structures being built. The main geomorpho-logical contribution at this stage is to define, on the basis of visible evidence, those sites that appear to be most likely to fail (or show a history of past failures) as a basis for guiding the precise location of sampling and testing in any geotechnical site investigation.

5.4 Rocks, Soils, and Landslides

Strata composed of cohesive rock are not prone to landsliding in the same way as less cohesive beds and regolith. In the former the presence of weak elements, such as lineaments (including joints, faults, or bedding planes) or bands of weaker material are critical in determining the overall ability of the bedrock to resist landsliding. Less cohesive materials (e.g. dry sands, clays) tend to respond to instability conditions by shearing or deformation within themselves.

(a) *Bedrock materials*

Landsliding in coherent, predominantly unweathered bedrock is not caused by the lack of any inherent strength in the rock itself but is due to the presence of internal weakness, such as joint planes and faults. Whatever the causes of a failure in bedrock the actual plane of movement will very largely be determined by the position of the weakest zones (e.g. joints).

One of the important controls of bedrock instability is the angle at which the major joints intersect the ground surface. Some examples are shown in Fig. 5.3, and illustrate the importance of the relation between the angle of friction (at which under gravity sliding would occur) along the joint plane (ϕ_f) and the inclination of the joints (α). If $\alpha < \phi_f$ the slope could stand at any angle up to 90° (Fig. 5.3), but if $\alpha > \phi_f$ then gravity would induce movement along the joint plane. It is seldom that bedrock displays only one set of joints. Examples C–E in Fig. 5.3 illustrate the relations between ϕ_f, α, and ϕ_c when a secondary set of joints is also present.

The angle of friction is determined by a number of different characteristics along the discontinuity (here referred to under the term 'joints', but this could apply also to faults, fissures, bedding planes, and shear planes). Other contributing characteristics include the continuity of the joints within the rock mass, the roughness of the joint planes, and their undulations (waviness). In some instances the most important control on the stability of the bedrock is not so much the joints themselves but the weathering products (or in the case of a fault, the breccia) which may fill them. This gouge material may separate two bedrock surfaces from each other, and slope stability will depend on the shear strength of the gouge rather than on the plane-to-plane contact of two bedrock surfaces. In some cases the gouge–bedrock interface may have even lower shearing strengths than the gouge itself. With respect to the presence of gouge material on failure, four situations are possible (Piteau, 1970): (i) sliding plane passes entirely through gouge, and shear strength is dependent upon gouge material only; (ii) sliding plane passes partly through gouge, partly through joint wall, and shear strength is made up of contributions from both; (iii) gouge is very thin, and only marginally modifies the shear resistance of the joint plane; and (iv) no gouge, and shear strength is entirely that of the joint plane.

The orientation of joints is also important in determining the passage of water through the bedrock. The movement of water is related to the degree to which joints are continuous or form an interlocking network. Free-water flow prevents the build-up of locally excessive water pressures, and thus decreases the likelihood of instability. However, the underground continuity of joints is usually difficult to determine.

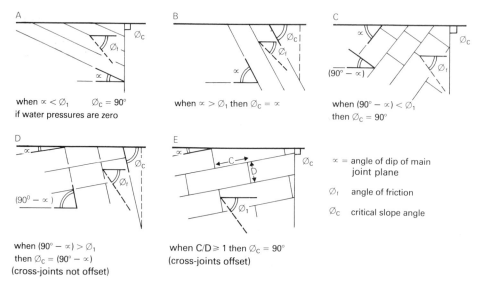

FIG. 5.3 Some possible relationships between joint planes in bedrock and slope stability (after Terzaghi, 1962)

Within joints it is not the quantity of water but the water pressures that are important, for as the latter increase so the shear resistance decreases. Since bedrock frequently contains joints running in several directions, providing predetermined routes for water movement, they exhibit anisotropic and non-homogeneous flow properties. In this respect they differ from soils, unless the joint spacing is close and there is a random hydraulic connection between the joints.

The shear strength of a joint plane is related also to bedrock mineral composition. For example, quartz and calcite increase their resistance to sliding if they become moist, but talc-like minerals are less resistant to sliding when they are wet. Weathering along the joint plane can have important effects. Montmorillonite (a clay mineral) has a low shear strength, and swells exceptionally on wetting, making for potentially unstable conditions. Soluble minerals, such as rock salt, gypsum, limestone, or dolomite can be dissolved and cause a decrease in shear resistance with time as they are removed along the joint plane (Piteau, 1970); conversely they may increase shear strengths if they are deposited within joints.

Three categories of rock slope failure are illustrated in Fig. 5.4. These show how joints (or any other physical discontinuity) can control the shape of both local (case A) and large-scale (case B) failures. If these discontinuities are very close together they allow deep and intense bedrock weathering to occur, degrading the original bedrock strength, changing its deformability properties and permeability characteristics, and producing a complex of (i) residual soil, (ii) weathered rock, and (iii) unweathered rock. Some zones have a major influence on groundwater flow, and may cause excess pore-water pressure to build up, thus leading to instability (case C). Once movement has taken place, in any type of bedrock failure, the residual strengths will be much lower than the initial resistance to shearing, and subsequent movements occur with less inducement. The observation of an initial small displacement is thus critical in the assessment of bedrock slope stability.

Some materials that are classified by geologists as bedrock, such as clay beds (e.g. London Clay), may behave in a similar manner to soils, and their stability may depend on criteria similar to those examined below for soils. Clays tend to be distinctive both in their physical properties and their behaviour on sliding. For example, stiff fissured clays may have a shear resistance of between 10 and 20 tonnes/m² which is their peak strength as measured by laboratory tests on small samples. However, if a cut is made into clay (e.g. by excavation, or by an incising stream) this value may fall to as little as 3 tonnes/m². This is because, as the cut develops, stresses in the clays are relaxed

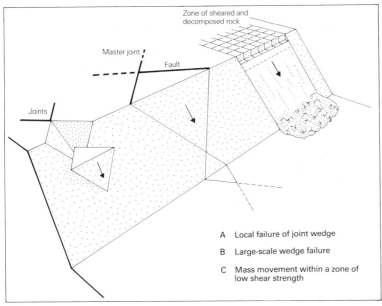

Zone of sheared and decomposed rock

Master joint

Fault

Joints

A   Local failure of joint wedge

B   Large-scale wedge failure

C   Mass movement within a zone of
     low shear strength

FIG. 5.4 Three types of bedrock failure. (A) Local failure of joint wedge. (B) Large-scale wedge failure. (C) Mass movement within a zone of low shear strength (after Patton, 1970)

and joints open up into which rain can pass. This in turn increases the pore-water pressures in the clay, which can build up to high internal values over the years, independent of the season of the year or the dampness of the clays at the surface. In addition, the water can cause swelling, clay blocks to break up, and shear resistance to decrease. As soon as it becomes equal to the average shear stress on a potential surface of sliding, the slope fails (Terzaghi, 1950). Such failures can occur long after the cut was made. For example, the banks of a railway cutting through stiff, fissured, Weald Clay at Sevenoaks, Kent, UK, suffered sliding 70 years after the cut had been excavated.

In an engineering situation, such as in an excavation for a cutting, it is important to distinguish between 'short-term', 'intermediate', and 'long-term' failures. 'Short-term' failures take place soon after excavation and before the water content has changed. 'Intermediate' and 'long-term' failures occur largely as a result of the swelling that follows excavation and the reduction in stress. Because of the low permeability of many clays, although not those containing silt and sand layers, the swelling of the clay towards a new equilibrium pore-pressure is a comparatively long process that could well occupy the 70 years of the Sevenoaks Weald Clay example. As swelling and water content increase so shear strength decreases (see below). If sliding occurs during the swelling process it is classified as an 'intermediate' failure; if it occurs at or near the end of the swelling period it is a 'long-term' failure. Laboratory measurement of the shear strength of materials in this situation overestimates actual shear-strength values. This is largely because the small laboratory sample does not adequately represent the physical controls of failure (e.g. fissures running through the material) in the slope itself.

The time-lags between slope exposure and subsequent failure in clays appear also to be related to slope steepness. Skempton (1948) found that in highly colloidal, stiff, fissured London Clay, a vertical slope with a height of 6 m may stand for several weeks before sliding takes place. A slope of 1 in 2 (25°), but of the same height, tends to fail after 10 to 20 years. Progressively gentler slopes take longer to fail, so that a slope of 1 in 3 (18°) is likely to remain stable for up to 50 years. The steepest natural slopes in London Clay are of the order of 1 in 6 (8°), at which relative stability is achieved.

(b) *Soils*

The shear strength of a soil is the maximum available resistance that it has to movement. This shear strength (S) is a function of the friction at grain-to-grain contact, and is related to the amount of grain interlocking, which increases with the angularity of the material, and the density of packing (consolidation) of the grains. These control the angle of shearing resistance (ϕ'). In addition, shear strength is related to the effective pressure transmitted between particles (σ') and their effective cohesion (c').

For normally consolidated soils, with no internal cohesion (such as sands and gravels):

$$S = \sigma' \tan \phi'. \qquad (5.1)$$

For overconsolidated soils (i.e. subject to overburden pressure) and displaying some effective cohesion (such as in clays):

$$S = c' + \sigma' \tan \phi' \qquad (5.2)$$

where S is the shear strength at the potential slip surface (which will tend to be curved concave-upwards).

These values are illustrated in Fig. 5.5. If water, under pressure u, is present in the pores of the soil then σ will be reduced (i.e. $\sigma' = \sigma - u$) for a saturated soil; the relationships are more complex if the soil is only partly saturated and either air or gas bubbles are present.

When shear stress rises to the value of shear strength then displacement occurs between two parts of the soil body, usually along a well-defined rupture plane (slip surface). Once movement has taken place along a slip surface there is a significant decrease in the shear strength of the material at the slip surface, for a given effective pressure, and subsequent movements will take place with lesser induced stress. Fig. 5.5 illustrates how, after movement, the residual values of ϕ' (ϕ'_r) are lower in both cohesionless and cohesive soils, while in the latter the effective cohesion is reduced to a residual value (c'_r). Any further stability calculations using equations 5.1 and 5.2 thus need to use ϕ'_r and c'_r values after movement has taken place. The difference between peak (before displacement) and residual values for these two parameters varies with the texture of the material, generally increasing with the clay content (Skempton, 1964). Once movement has taken place the clay particles become reoriented at the slip surface,

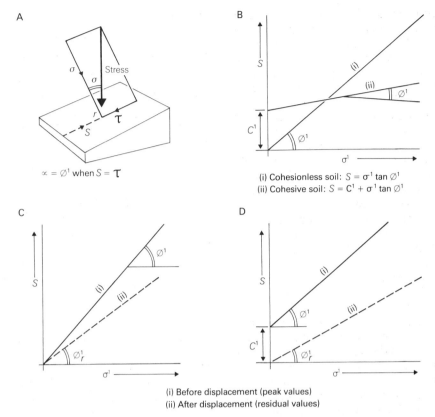

A

$\alpha = \emptyset^1$ when $S = \tau$

B

(i) Cohesionless soil: $S = \sigma^1 \tan \emptyset^1$
(ii) Cohesive soil: $S = C^1 + \sigma^1 \tan \emptyset^1$

C

D

(i) Before displacement (peak values)
(ii) After displacement (residual values)

FIG. 5.5 Some fundamental principles of soil mechanics which affect slope stability analysis. (A) Stress at a point can be divided into two components, the pressure normal to the slope (σ) and the shear stress (τ) which operates in the same plane but in the opposite direction to the shear strength (S). When $S = \tau$, the angle α = angle of shearing resistance (ϕ'). (B) shear strength (S) against effective pressure (σ') for cohesionless and cohesive soils (ϕ' may or may not be greater for the one than the other according to local conditions). (C) In cohesionless soil, the relation between peak and residual values of ϕ' and c'. In (C_i), $\phi'r$ is due only to dilation (which involves a decrease in density and an increase in water content) relative to density in the ϕ' state. In cohesive soil (D), ϕ'_r is due to dilation plus a loss of strength resulting from the reorientation of platey clay particles

tending to parallel each other and the direction of movement, thus making further movement much easier. In addition stiff clays may show an increase in water content associated with a dilation of the soil within the zone of shearing.

It was suggested above that stratified clays behave like soils rather than like more coherent bedrock. Skempton (1964) found that on a stable slope the resistance to sliding offered by the clay along the slip surface (i.e. its shear strength, S) is given by:

$$S = \bar{c}' + (\sigma - u) \tan \bar{\phi}' \qquad (5.3)$$

where $\bar{c}'$ = cohesion intercept (see Fig. 5.5B) and $\bar{\phi}'$ = angle of shearing resistance. $\bar{c}'$ and $\bar{\phi}'$ are average values of c' and ϕ' around the slip surface

and both of these are expressed in terms of the effective stress. Based on this relationship the average shear stress ($\bar{\tau}$) may be defined as:

$$\bar{\tau} = \frac{\bar{c}'}{F_s} + (\sigma - u) \frac{\tan \bar{\phi}'}{F_s} \qquad (5.4)$$

where F_s is the factor of safety.

In practice, on natural slopes there is often a large variation in soil properties such that the values of c' and ϕ' obtained on any one slope may cover a broad range sufficient to produce variations in F_s values from below 1.0 to well above 1.0. This makes a precise judgement of the actual level of stability very difficult. Nevertheless, this type of approach to the study of slope failure, through an analysis of the geotechnical properties of the materials, is routine within many engineering projects. Further expansion of this approach can be found in texts such as *Rock Mechanics and Engineering* (Jaeger, 1972), *Soil Mechanics* (Craig, 1978), *Soil Mechanics* (*SI version*) (Lambe and Whitman, 1979), *Fundamentals of Soil Behaviour* (Mitchell, 1976), the British Standards Institution's BS1377 *Methods of Testing Soils for Civil Engineering Purposes* (1975), and *Geotechnical Engineering* (Lee *et al.*, 1983). A summary of methods of stability analysis is provided by Morgenstern and Sangrey (1978) and by Graham (1984), which takes the reader well beyond the point reached above.

5.5 Geomorphological Assessments of Slope Failure

The main contributions of geomorphology to the assessment of slope failure are through: (i) an understanding of the evolutionary history of the landforms of an area; (ii) an eye for recognizing landslides (especially the more subdued, perhaps fossil, forms); and (iii) methods for defining the susceptibility of an area (or a site) to potential instability.

The remainder of this review is concerned with these and related aspects of a geomorphological contribution to landslide assessment.

(a) *The relevance of geomorphological history*

In general, slopes prone to landsliding have been left in that state by previous influences upon them. These influences may be more or less recent, but to understand the potential behaviour of a slope requires an understanding of its history. For example, it can be very important to recognize areas of previous slope steepening by glacial erosion, perhaps followed by glacial deposition for, as in the case of the landslides in Snoqualinie Pass, Washington State, USA, in August 1953, this can often provide an explanation of why slope failure has taken place. During the Pleistocene a glacier cut a steep slope into highly fractured graywacké. The glacier snout then retreated and a landslide began, the toe of which was subsequently removed by a readvance of the glacier later in the Pleistocene. With the next and final retreat of the ice the slide was partially buried and left supported by a 3-m thickness of glacial till. The resulting slope remained stable until excavation at its base by road works removed much of the supporting

material. This reactivated the instability and severely damaged the construction scheme (Ritchie, 1958).

Another legacy from past processes which has caused frequent engineering difficulties is that of solifluction features (see also Chapter 8). Under periglacial activity, solifluction is a common form of slope instability with the seasonal thawing and saturation of the surface debris layers causing mass movement. Movement can take place in one of three ways: (i) as viscous flow, when the soil is extremely wet; (ii) by sliding on a shear plane; and (iii) as freeze–thaw movements bringing about the down-slope creep of unsorted materials, and which can operate on slopes of as little as 2°. Stability returns both as the periglacial climate is replaced by warmer climates with no seasonal freeze–thaw and less extreme groundwater conditions, and as the toes of the solifluction lobes (or sheets) come to rest at the foot of slopes. Reactivation of the sliding type of movement (where a shear surface of low shear strength already exists) may begin if excavations decrease support from below by re-steepening any part of the lobe (especially the toe), or by loading the slope to increase the pressure exerted from above on the solifluction material. Considerable trouble was caused during the construction of the Sevenoaks bypass in south-east England because fossil solifluction forms were reactivated by loading with a bank and then by excavation into the lobe (Weeks, 1969; Higginbottom and Fookes, 1970).

Around Bath, England, Lias clays are overlain by limestones. In Pleistocene times cambering, valley bulging, and mass movement by sliding took place under periglacial conditions (see Chapter 8); the cambering taking place on the interfluves where the more coherent beds now curve down-slope more steeply than the general dip, and the bulging of less coherent materials (e.g. clays) occurring in the valley floors where periglacial activity caused these materials to be thrust upwards and contorted. As later urban development occurred some movement was reactivated, and now old slips are stabilized (mainly by drainage) before being used as building sites. However, slopes exceeding 15° generally remain unstable, and have been designated as public open spaces within the city boundary (Kellaway and Taylor, 1968). Shallow movements still take place on the slopes around Bath, especially during heavy rains.

It is imperative, therefore, that in evaluating any area for proposed engineering works due consideration should be given to the recognition not only of present-day landslides but also to instability which took place in the past. In either case an understanding of slope failure must come from an appreciation of the geomorphological history of the area.

(b) *Geomorphological characteristics of landslide-prone slopes*

Landslide-prone slopes tend to possess certain distinctive features, including various combinations of the following: steep slopes, high slopes, concentrated soil water (high pore-water pressures), thick, deeply weathered soil cover, undercutting of the base of the slope, and weak (incoherent) material outcropping below stronger (more coherent) material; and if the area is also subject to either intense rainstorms or earthquake activity (or both) then the potential for landsliding is even higher.

Of increasing significance, however, is the role of human activity. This may be through excavating at the base of a slope and thereby not only creating a steep slope but also removing some of the basal support; or it may be through diverting too much water into a slope (e.g. by bad drainage design), or through clearing protective vegetation which would normally have resisted shallow slope failures not only through the binding effects of roots but also through the absorption and evapotranspiration of water drawn out of the soil. All of these characteristics, whether natural or human-induced, can be linked to the factors controlling shear resistance (Table 5.3) and shear stress (Table 5.4), through which their role in slope failure can be worked out.

In terms of environmental management in general, and engineering projects in particular, there is an important geomorphological task in identifying and recording landslide-prone areas. Whichever of the following techniques are used, they all require the recognition and recording of the main characteristics of landslide-prone slopes. In some cases local factors may be particularly important, and the skilled geomorphologist will spot these and recognize their significance.

(c) *Identifying landslide-prone areas*

The usual starting-point for identifying landslide-prone areas is to record (probably by marking on a map) the position and character of any existing landslides. This does two things: it shows where landslides could be reactivated (if inappropriate engineering took place), and it allows those sites where failure has already taken place to be studied in order to discover what causes landslides *in that area*. The search is then on for sites having similar properties but which, for whatever reason (and this has to be found out), have not yet failed. The task is to identify those sites close to the threshold of movement (in engineering terms these are the slopes where F_s (the factor of safety) is close to unity).

To assist in this task various techniques have been developed. The most useful of these, which are reviewed below, include: site analysis using pro forma check lists; grid-based recording of slope properties (which are then subjected to statistical analysis or to sieve mapping techniques); landslide (inventory) mapping; geomorphological mapping; and landslide susceptibility (hazard) mapping. These, and other techniques, are the subject of reviews by Schuster and Krizek (1978) and by Hansen (1984*a, b*).

(i) *Pro forma check lists* In cases where sufficient data sources are available (and these include topographic maps, aerial photographs, geological maps, and the opportunity for some field investigations) a comprehensive pro forma of the type shown in Table 5.5 may be useful. This allows a systematic consideration to be given, at each site under investigation, to the main factors which are known to be associated with landsliding. By carefully assessing the state of the relief, drainage, bedrock, and regolith, the incidence of earthquakes, legacies from past processes, and man-made features, the likelihood of landsliding can be assessed. If checking from this list reveals gaps in the available information then any further work can be directed at filling in these gaps.

The check list (Table 5.5) has been designed so that by placing an X in the appropriate places not only are the main causes of instability considered in turn, but the more the X signs align themselves to the right the more the slope concerned is approaching an unstable state. Such a list has the advantage that if only one or two Xs occur to the left, this identifies those parameters which should not be aggravated if landsliding is to be avoided. It is not realistic to supply on such a check list absolute values for each category since the significance of a particular value depends on its local

TABLE 5.5 A check list for sites liable to large-scale instability

Relief	More stable		→	Less stable
Valley depth	Small □	Moderate □	Large □	Very large □
Slope steepness	Low □	Moderate □	Steep □	Very steep □
Cliffs	Absent □			Present □
Height difference between adjacent valleys	Small □	Moderate □	Large □	Very large □
Valley-side shape	Spur □	Straight □	Shallow cove □	Deep cove □
DRAINAGE				
Drainage density	Low □	Moderate □	High □	Very high □
River gradient	Gentle □	Moderate □	Steep □	Very steep □
Slope undercutting	None □	Moderate □	Severe □	Very severe □
Concentrated seepage flow	Absent □			Present □
Standing water	Absent □	Present at local base level □	Present slowly draining □	Present rapidly draining □
Recent incision	Absent □	Small □	Moderate □	Large □
Pore-water pressure	Low □	Moderate □	High □	Very high □
BEDROCK				
Jointing density	Low □	Moderate □	High □	Very high □
Joint openings	Small □	moderate □	Wide □	Very wide □
Direction of major joints (faults or bedding planes) with respect to steepest slopes	Away □		Normal □	Towards □
Amount of dip out of slope (or steepness of joint and or fault planes)	None □	Small □	Moderate □	Large □
Joint gouge	Hard □			Soft □
Strong beds over weak beds	Absent □			Present □
Degree of weathering	None □	Small □	Moderate □	High □
Compressive strength	High □	Moderate □	Low □	Very low □
Coherence (particularly of lower beds)	High □	Moderate □	Low □	Very low □
SOILS (incl. drift materials)				
Site	Flat □	Gentle slopes □	Moderate slopes □	Steep slope □
Coherent over incoherent beds	Absent □			Present □
Depth	Small □	Moderate □	Large □	Very large □
Shear strength	High □	Moderate □	Low □	Very low □
Plastic limit	Low □	Moderate □	High □	Very high □
Liquid limit	Low □	Moderate □	High □	Very high □
EARTHQUAKE ZONE				
Tremors felt	Never □	Seldom □	Some □	Many □
LEGACIES FROM THE PAST				
Fossil solifluction lobes and sheets	Absent □	Rare □	Some □	Many □
Previous landslides	Absent □	Rare □	Some □	Many □
Deep weathering	None □	Slight □	Moderate □	Much □
CLIMATE				
Rainstorms	Low intensity □			High intensity □
Snow cover melt	Slow □			Rapid □

TABLE 5.5—*continued*

HUMAN FEATURES

Excavation-depth	None ☐	Small ☐	Moderate ☐	Large ☐
Excavation-position	Top of slope ☐			Bottom of slope ☐
Reservoir	Absent ☐	Small ☐	Moderately deep ☐	Very deep ☐
Drainage diversion across hillside	Absent ☐			Present ☐
Lowering of reservoir level	None ☐	Small ☐	Moderate ☐	Rapid ☐
Loading of upper valley side	None ☐	Some ☐	Moderate ☐	Large ☐
Removal of vegetation	Slight ☐			Extensive ☐

context. A 9°-slope in clays may be potentially unstable while a slope three times as steep in a more coherent material (e.g. sandstone) may be extremely stable.

Each category on this list should be considered in turn in the context of the preceding discussions on soil and rock mechanics, and in terms of the techniques which can be used for measuring their appropriate properties. In many instances, such as in the examination of relief, drainage, structure, lithology, and the search for signs of past conditions of instability, much can be gained from a study of aerial photographs. This can save a considerable amount of expensive field time not only by directly providing information but also by identifying those sites worthy of the most detailed examination in the field.

Relief. The steeper a slope the more liable it is to be unstable. Steep slopes tend to predominate in areas of deep valleys, as do bedrock cliffs, which are potential rockfall sites. Slope steepness may be obtained from contour maps, together with the other measures of relief, although as a data source such maps can be notoriously unreliable. However, if a general guide as to steepness is required the scales for determining slope from contoured maps such as those provided by Thrower and Cooke (1968) are useful. A quick assessment of the mean steepness of a slope can be obtained from a measure of valley width (W) and valley depth (D), since $\tan^{-1} D/0.5W = \bar{\theta}$, the mean angle of the valley-side slope. If the slope is not straight, however, then its steepest portions will be in excess of the value $\bar{\theta}$.

The significance of recording the height difference between adjacent valleys is the fact that one of the most likely locations for landslides is the lower end of a spur between a higher and lower stream. Groundwater seepage may take place laterally through the spur, especially if the bedrock is fractured or jointed. This leads to instability on the valley side of the lower stream. Similarly the recognition of coves, or embayments, in the valley side (as seen in plan) is significant in that higher pore-water pressures are likely at such sites since they are water-receiving sites from the slopes around.

Drainage. High drainage densities are a sign of such things as impervious strata, high rainfall, little vegetation, and active stream incision, all of which may tend to increase the likelihood of mass movement. Similarly, steep river gradients generally indicate a phase of active and rapid incision, unless the river is in a 'steady state'. Rockfalls, earth-falls, and sliding can arise directly from such incision, but they can also occur because of undercutting by a stream whose predominant forces are directed laterally rather than vertically downwards. Lateral undercutting can be achieved by rivers of low gradient in floodplain channels, as well as by mountain torrents.

Seepage from a hillside (e.g. along a spring line) can produce seepage erosion in fine sands and silts by a drag effect which takes with it individual soil particles, or a seepage pressure may be generated within the ground materials (especially in cohesionless water-bearing beds) so that particles are carried outwards from the slope; eventually undermining results, and slope stability is lost (Terzaghi, 1950; Hutchinson, 1968). Seepage sites may be identified through the recognition of the associated features such as belts of damp ground, incipient gullies, spring-sapped hollows, or the collapse of natural underground pipes. In clays, however, it is unlikely that water will be seen at the surface as the rate of evaporation tends to be greater than the rate of flow out of the slope.

Groundwater pressures (including pore-water pressures and water pressures in joints) are critical in many slope stability problems, as much of the above discussion has shown. These are usually measured with piezometers (Lambe and Whitman, 1979), by which a piezometric water surface is defined for use in stability calculations.

Bedrock. Joint density, or the density of other discontinuities in the rock (such as bedding planes or faults), will determine the size of falling blocks. Joint directions and inclinations with respect to the orientation and steepness of the ground surface have a direct bearing on the overall stability of the slope. These all need to be measured in the field, though sometimes aerial photographs can provide valuable indications of actual conditions. One approach to effective field recording is through the identification of a structural region within which joint spacing is more or less homogeneous, but which differs from that of the neighbouring parts. Methods for field description of joints are given by Piteau (1970), and Terzaghi (1965). Other important data to be collected on a field survey include the nature of bedrock lithology, and the coherence, compressive strength, and degree of weathering of the beds. Compressive strength is the load per unit-area under which a rock fails by shearing or splitting. If clays (weak beds) underlie more coherent beds (strong beds), then the greater the thickness of clay that is exposed the more likely it is that instability will arise.

Regolith (soils and drift materials). The site occupied by regolith materials, or the outcrop position of non-coherent beds such as clays, have an important bearing on stability conditions, the steepness of the slope being an important factor in predicting the likelihood of landslides. However, much also depends on the particular relationship to other beds, and to rivers, slope drainage, and human interference.

The susceptibility of materials to failure is in part also related to their physical condition, for this very largely determines their shear strength. Relevant properties include soil texture, structure, coherence, grain shape, relative density, permeability, the porosity of granular soils, and the void ratio in cohesive soils. In engineering terms, the Atterberg limits (liquid and plastic limits) may need to be known. These limits are based on the observation that a fine-grained soil can exist in any one of four states depending upon its water content. A soil is *solid* when it is dry, but on the addition of water the soil passes through the *semi-solid*, *plastic*, and *liquid* states as the water content increases. The water content at the stage when the soil is passing from the semi-solid to the plastic state is known as the *plastic limit*, while the water content between the plastic and liquid states defines the *liquid limit*.

$$L_I = \frac{W_n - PL}{LL - PL} \qquad (5.5)$$

where

W_n = natural water content,
PL = plastic limit,
LL = liquid limit.

LI expresses the water content of a soil in a dimensionless way. Each of these physical parameters and its measurement or derivation is discussed in volumes on soil mechanics such as Terzaghi and Peck (1948) and Lambe and Whitman (1979).

One other important factor is the stress history of the materials, and this may be intimately related to their geological and geomorphological histories. For example, London Clay has been overconsolidated by the pressure of overlying deposits which, together with large amounts of the London Clay itself, have been removed by erosion. Overconsolidation leaves a clay with a much lower water content, at a specified depth, than in a normally consolidated clay at the same depth. In a normally consolidated clay c' of Equation 5.3 equals zero, but in an overconsolidated clay, c' has a finite value (see Fig. 5.5C) which is a result of the overconsolidation. In addition overconsolidation leaves the clay with much higher lateral stresses than occur in normally consolidated clays. These lateral stresses are not relieved to any great extent by the removal of the overburden. Relief may only be found by slope failure (progressive failure) when these lateral stresses exceed the shear strength of the clay. Of additional importance is soil chemistry, for certain minerals (e.g. the clay mineral montmorillonite) are more likely to be associated with slope failure than others.

Earthquakes. Landslides are more likely if the area occurs within an earth-tremor belt (see Chapter 13).

Legacies from the past. Evidence of previous mass movement, such as solifluction lobes and sheets, and old landslides, indicates that slope failure could occur again. In particular nothing should be done to load the top of the slope or to take away support from the bottom of the slope. If an engineering structure is to cross the old landslide, then very careful slope stability analyses will need to be made, and appropriate precautionary measures taken during construction. Where deep weathering has taken place the soils may be particularly prone to movement, especially if they lie on steep slopes, or have been subject to deforestation. Borehole records, or advantageous sections (e.g. in cuttings) may be the only sources of information about the location and depth of weathering products.

Climate. When water arrives rapidly within a soil body (whether by rainstorm or snow-melt) there is a tendency for high pore-water pressures to build up, followed by slope failure.

Human features. Each of the items listed has been discussed above in its appropriate context. If a large number of X symbols has been entered in or near the right-hand column in the check list (Table 5.5) then every endeavour should be made to ensure that human modification of the land surface keeps all other Xs well to the left, or instability could result.

The proforma suggested here (Table 5.5) could be used at either the regional or the local scale. It does not attempt to seek absolute data, only data classed on an arbitrary scale. For some studies this may not be enough, and may not fulfil adequately the needs of a survey which requires precise slope and/or landslide data. A proforma approach is also particularly appropriate if information is to be stored in a computerized data base. Examples of proformas that have been used for collecting field information are given in Varnes (1984).

(ii) *Data collection on a grid basis* In theory it is possible to take a topographic map of an area, place a grid across it (thus creating cells), and to record for each cell each of the landscape properties listed on the proforma (Table 5.5). This would be a daunting task, and is usually impracticable, especially in the context of an actual engineering project, because neither the time nor the data exist. Thus it is necessary, when using this approach, to restrict data collection to certain key parameters. In practice *the* key parameter varies from place to place, but research has shown that a high level of

prediction can be based on relatively few parameters.

In California (Brabb *et al.*, 1972) as well as in Calabria, Italy (Lucini, 1973), a combined analysis of slope and bedrock provided a good basis for identifying landslide-prone sites. An easy way of looking for critical combinations is to produce two grids over the study area. On one the cells are coded (by colours or numbers) according to slope steepness; on the other the outcrops of particularly weak strata are marked. By laying one grid over the other (e.g. on a light table) the cells having a high slope steepness coincident with the identified rock outcrop can be singled out as being most likely to be prone to instability. This can be confirmed by comparing the analysis with a map of actual landslides within the area. Although a simple approach, the idea can be extended to as many parameters as are thought to be sensible. It can also be refined to include more than one rock type and more than one classification of slope steepness. The technique can also be extended to include more parameters. De Graff and Romesburg (1980, see Fig. 5.6) included slope aspect as a third component in their study, but in some areas it might be more meaningful to include information on either soils or hillslope drainage. The latter is especially important where water in the slope is the fundamental component that determines slope failure, and it often is. In the approach shown in Fig. 5.6 the technique is used to obtain, for each landscape cell, an index of landslide susceptibility. However derived, such indices tend to be arbitrary, have no absolute meaning, and cannot be transfered from one area to another. Each scale of index-values tends to be unique to the area for which it is derived. Matters are further complicated because even within one area distinctions have to be made between a susceptibility to soil failure and a susceptibility to deep-seated failures in rock.

It would be encouraging to be able to say that the matrix assessment approach is a rapid, easily operated, and meaningful way of defining landslide-prone areas. In reality examples to date tend to suggest that it is time-consuming, requires appropriate computer facilities for operation when several parameters are involved, and the results are at best ambiguous and it is usually difficult to assign a precise meaning to them. More information is still likely to come from geomorphological mapping which seeks to define those slopes most

Bedrock map

Slope map

Aspect map

Landscape
susceptibility
matrix

Susceptibility class map

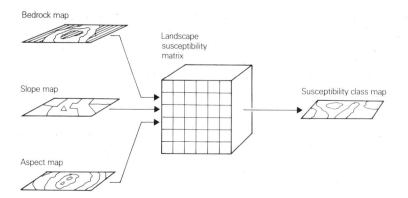

FIG. 5.6 The process followed to create a landslide susceptibility class map (after De Graff and Romesburg, 1980; reproduced by permission of Unwin Hyman Ltd.)

likely to be vulnerable to failure not only because of their site characteristics but also because of the situation in which they occur (see below).

(iii) *Landslide inventory mapping* Landslide mapping may be based on the grid cells, as defined for regional geomorphological data collection, whereby every cell that contains a landslide is identified in a particular way. This is useful in the context of cell-by-cell comparisons. However, this approach only provides an approximation to reality which can be overcome by mapping the precise boundaries (extent) of the landslide. Such

landslide mapping can also include a classification into landslide type (e.g. by reference to Fig. 5.2).

Mapping is normally carried out from aerial photographs, in the first instance, followed by field checking on the ground. Landslide inventory maps have been compiled at various scales. Some examples are given in Table 5.6.

(iv) *Geomorphological mapping* Although geomorphological maps would be expected to record the presence and character of landslides, as Chapter 2 has shown, they also carry other information. In the case of a geomorphological

TABLE 5.6 Examples of landslide inventory mapping

Country	Scale of mapping	Details	Source
Czechoslovakia	1:1 million	Based on 12 000 cases recorded in 1962–3	Nemcok and Rybar (1968)
USA (California, e.g. various counties)	1:24 000	One of several different types	Campbell (1980)
Wales, UK (South Wales Coalfield)	1:50 000	579 landslides	Conway *et al.* (1980)
Italy (Calabria)	1:400 000	Deep-seated landslides and lithology	Sorriso-Valvo (1984)
Italy (Lattarico)	1:25 000	Landslides plotted on geological map	Carrara and Merenda (1974)
USA	1:24 000	Landslides and slopes susceptible to landsliding	USGS Map MF-685B illustrated in Schuster and Krizek (1978)
Sicily	1:10 000	Distinguishes between active and dormant. Identifies landslides of 'great economic or social importance'	Agnesi *et al.* (1984)
Nepal	1:50 000	Part of soil degradation study	Fort *et al.* (1984)

map drawn as part of a landslide study much of the additional data would tend to have a bearing on the occurrence of the landslides, or would portray the characteristics of the slopes on which the landslides occur (Table 5.7). An example is shown in Fig. 5.7.

Geomorphological maps also form the basis of a hazard mapping system used in France known as ZERMOS (Zones exposées à des risques liés aux mouvements du sol et du sous-sol). Amongst the hazards included are landslides (see Fig. 2.30).

(v) *Landslide susceptibility (hazard) maps*

TABLE 5.7 Examples of geomorphological maps of landslide-prone terrain

Location	Scale	Observations	Source reference
Nepal		Specifically to identify landslides likely to affect a road alignment	Brunsden *et al.* (1975)
Italy	1:25 000	Includes information on erosion as well as landslides. Type and degree of activity of landslide shown.	
	1:10 000	Type and activity of sliding	Canuti *et al.* (1987)
Czechoslovakia	Range between 1:10 000 and 1:2000	Landslides continue to damage property and services. Map needed for land management	Malgot and Mahr (1979)
Switzerland	1:10 000	Landslides a hazard to farming and tourist industry	Kienholz (1978)
Austria	1:10 000	Landslide recognition important for land-use decisions	Rupke, De Jong *et al.* (1983)
Japan	1:50 000	Morphometric map of large-scale features, including landslides	Hatano *et al.* (1974)

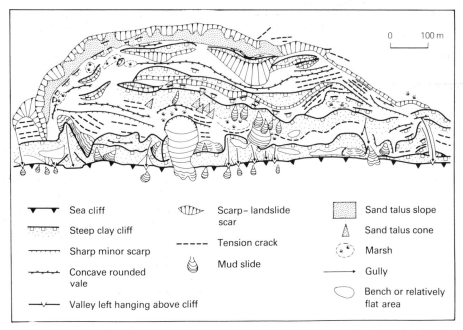

FIG. 5.7 Geomorphological map of a landslide area, Dorset, England (based on Geological Society Engineering Group Working Party, 1972; original mapping by D. Brunsden)

Landslide susceptibility maps go one stage beyond an inventory map, or even many geomorphological maps, in that they define a tendency towards instability in addition to slopes that have already failed. A typical slope classification in such cases is that shown in Table 5.8. A landslide susceptibility map expresses a judgement about how close a slope is to an instability threshold. As such it predicts the likelihood of a landslide. This is both difficult and contractually a heavy responsibility. When such a map is produced in the context of a major civil engineering project it can lead to very costly decisions being made. For example, an incorrect judgement which claims that a slope is near to failure may lead to unnecessary (and often expensive) geotechnical subsurface investigations. An error in judgement the other way may give the engineer a false sense of security which can become as shattered as the damaged or destroyed engineering structure. Clearly, landslide susceptibility maps should be drawn only with the greatest caution, and then only by those with long experience in the geomorphological analysis of landslide-prone slopes.

Such a geomorphologically based map should identify those sites where landslide-generating influences are strongest (e.g. slope undercut by a river having a weak outcrop at its base). However, such 'identification' is inevitably built around the judgement of the person carrying out the mapping. There is much to be said, therefore, for the systematic cell-by-cell approach of data-gathering–so as to make sure all slopes are included— dovetailed to the geomorphological mapping approach which provides a much stronger sense of the dynamic properties of the landscape (e.g. erosion, existing landslides, pattern of steep slopes, outcrop of weathered strata).

When a landslide hazard map is produced for use by planners it is often helpful to simplify the map and to colour or shade it into zones of different landslide susceptibility. Many different zoning schemes have been produced (Table 5.9). For example, Table 5.10 shows (in abbreviated form) the scheme used in the Grindlewald area of the Swiss Alps (Kienholz, 1978). The scheme is phrased unambiguously and does not demand any specific geomorphological knowledge of the user. This is the only successful way of imparting some kinds of geomorphological knowledge into environmental management.

As Carrara (1984) recognizing in his review of the aims and methods of landslide hazard mapping, there is no one standard approach. Different methods have been developed to suit the nature of a particular area or of a particular project. A fuller review by Varnes (1984) provided short examples of some of the different approaches used, and other examples are given in the US Geological Survey (1982). Not only are hazard assessments carried out based on experience and judgement, but some attempts are also based on numerical analysis. In some cases attempts are made to devise a hazard index (see Sect. 5.6) from quantitative values assigned to individual slope properties. In others a statistical approach is adopted in order to evaluate the influence of each landslide-determining factor in producing an actual or potential hazard (Carrara, 1984; Table 5.11).

More successful perhaps is the approach

TABLE 5.8 Slopes classified according to their stability characteristics

Class I	Slopes with active landslides. Movement may be continuous or seasonal
Class II	Slopes frequently subject to new or renewed landslide activity. Triggering of landslides results from events with recurrence intervals of up to five years
Class III	Slopes infrequently subject to new or renewed landslide activity. Recurrence intervals greater than five years
Class IV	Slopes with evidence of previous landslide activity but which have not undergone movement in the preceding 100 years Subclass IVa: Erosional forms still evident. Subclass IVb: Erosional forms no longer present—previous activity indicated by landslide deposits
Class V	Slopes which show no evidence of previous landslide activity but which are considered likely to develop landslides in the future. Landslide potential indicated by stress analalysis or analogy with other slopes
Class VI	Slopes which show no evidence of previous landslide activity and which by stress analysis or analogy with other slopes are considered stable

Source: Crozier (1984).

TABLE 5.9 Examples of landslide-hazard maps

Location	Scale	Observations	Source references
Italy (Calabria)	1:10 000	Five risk classes superimposed on geomorphology and geology	Dumas *et al.* (1984)
France	1:25 000	Four risk classes and map of existing hazard sites	Meneroud and Calvino (1976)[a]
Switzerland	1:10 000	Four risk classes combined with geomorphology	Kienholz (1978)[a]
Czechoslovakia	1:10 000	Three risk classes plus engineering geological zones	Malgot *et al.* (1973)[a] Mahr and Malgot (1978)[a]
Italy	data in cells of 200 × 200 m	Four hazard classes	Carrara *et al.* (1978)[a]
Canada	1:25 000 and 1:50 000	Slope classes based on calculated F_s values	Klugman and Chung (1976)[a]

[a] Reported in Varnes (1984).

TABLE 5.10 Hazard-zonation categories in the Grindelwald area of the Swiss Alps

Hazard category	Description
3	Houses are destroyed and people are in danger from landsliding (or avalanches)
2	Houses in little danger, but areas between houses may experience some landslides (or avalanches), hence people may be in danger
1	Houses in very little danger, slight but infrequent danger to people outside the houses
0	No known danger

Source: Kienholz (1978), simplified.

described by Brabb *et al.* (1972) for their study of slope stability in San Mateo County, California. They used a sequence of mapping and quantitative techniques as follows:

1. Area of each rock outcrop measured.
2. Area of landsliding within each outcrop measured—by superimposing a landslide inventory map over the geological maps.
3. Rock types ranked in order from those with greatest percentage of landslides to those with lowest percentage from which the degree of susceptibility to landsliding was assessed.
4. A map of slope steepness was superimposed on the geology and landslide maps to determine the associations between them.
5. Hazard classes were defined on the basis of susceptibility, in geological terms, and slope

TABLE 5.11 Definitions of hazard, vulnerability, and risk

Natural hazard	(H): the probability of occurrence, within a specified period of time and within a given area of a potentially damaging phenomenon (e.g. landslide)
Vulnerability	(V): the degree of loss (or damage) resulting from H, of a given magnitude, expressed on a scale from 0 (no damage) to 1 (total loss)
Specific risk	(R_s): the expected degree of loss due to a particular H. Expressed as $H \times V$
Elements at risk	(E): the population, properties, economic activities, including public services etc. at risk in a given area
Total risk	(Rt): the expected number of lives lost, persons injured, damage to property, or description of economic activity due to a particular H. Expressed as $R_s \times E$
Thus	$Rt = E \times R_s = E \times H \times V$

Source: Carrara (1984).

steepness in terms of the association of particular slope classes with landslides.

One feature of the resulting map is that a single boundary line can enclose, and hence include in one class, resistant rocks on steep slopes and weaker rocks on gentler slopes.

All landslide hazard predictions need to be revised, however, if there is a significant shift in land management. In many parts of the world it is man's effect on hillslope that either increase or decrease the potential for landsliding. Such induced man-made changes are normally far more rapid than most natural changes.

5.6 Landslide Hazard Assessment and Legislative Responses

If an area is prone to landsliding, several alternative reactions are possible. These range from ignoring the fact to avoiding the site completely. In between these two extremes are the situations where landslide-prone areas are already occupied (e.g. parts of California, mountain slopes of southern Italy), or where there are plans to occupy the area after introducing landslide control measures. In areas already occupied it is sometimes possible to introduce remedial measures that will reduce the risk of slope failure. However, situations will continue to arise where (i) the losses sustained from landslides have to be carried by someone, and (ii) legislation is required to prevent (or at least restrict) the unwise use of landslide-prone slopes.

Losses can be carried by (i) the individual; (ii) some form of rehabilitation aid, and (iii) funds designed to bear such losses (e.g. private insurance policies, and in New Zealand, the Earthquake and War Damage Fund, which can cover damage by landslides). In the United States the 1968 National Flood Insurance Program was amended in 1969 to include payment for damage caused by mudslides associated with a flood.

Land planning policies, however, are concerned with future developments, and if landsliding is to be taken into account they must rely on some form of landslide susceptibility study. Such planning policies, usually in the form of zoning and building ordinances, exist, for example, in the Los Angeles area, where increasingly rigorous laws have been adopted since the first controls were introduced in October 1952 (Fleming *et al.*, 1979). These con-

trols involved slope grading, and required geological (including geomorphological) and engineering advice to be incorporated in the design, construction, final inspection, and certification of the use of a new site. These practices appear to have produced a significant reduction in landslide losses (Cooke, 1984) and the US Geological Survey (1982) has embarked upon a continuing programme of landslide damage reduction.

In San Mateo County, California, landslide hazard maps are used to define the density of house building which is allowed in the different land-use zones (Atwater, 1978). Normal densities of one house per 2 ha is reduced to one per 16 ha where landslides occur. Other US examples are given in US Geological Survey (1982).

In Tasmania a landslide hazard scheme has been established (Stevenson and Sloane, 1980). Landslide susceptibility analyses through geomorphological and geological investigations are used to establish stability classes which are mapped (in Tasmania at scale of 1:5000 or 1:10 000). Supplementary geotechnical investigations may also be carried out. A decision is then made, in the case of areas of marginal stability, either to warn the public (including intending builders and purchasers) or to restrict development under the Local Government Act of 1973. This may even lead to building being prohibited. The effects in either case are to control development in those areas thought to be hazardous, and thereby to protect the landscape and to prevent losses. Additional effects include the modification of planning schemes and an influence on land values.

All of the above concerns can be incorporated in a hazard zonation programme (Table 5.12) which leads to appropriate policy action by planners and/or government. The main problem lies in transferring a landslide hazard awareness from the geomorphologist to the planner and legislator.

5.7 Land-Surface Subsidence

(a) *Introduction*

Landslides and related slope failure involve both lateral and vertical movement, and are therefore only found on inclined surfaces. In a sense, ground-surface subsidence may be considered as the limiting case of slope failure, in which movement is vertically downwards under the influence of gravity, and usually limited lateral movement is a consequence of it.

TABLE 5.12 Phases in a landslide hazard zonation programme

1. Identify the local, regional, or national concern
2. Define the extent of the area of concern
3. Identify areas of active or past landsliding
4. Identify slopes which are potentially unstable
5. Assess what public utilities, buildings, or property are at risk
6. Determine the threat to lives
7. Provide an evaluation of the potential economic social and environmental costs of landsliding
8. Relate the assessment of landslides and sites of potential failure to their impact and produce a hazard zoning map
9. Define the need for further investigations
10. Indicate, in the short term, the need for avoidance, prevention, or correction of landslides
11. Incorporate the results in planning and legislation

TABLE 5.13 Causes of ground-surface subsidence

Natural
1. Tectonic deformation and volcanic activity
2. Imposition of heavy loads (e.g. lakes): isostasy
3. Compaction, arising from vibration during earthquakes of from the superimposition of sediments
4. Desiccation of fine-grained sediments
5. Oxidation of organic soils
6. Subterranean solution of limestone, salts
7. Thawing of permafrost

Human-induced
8. Compaction due to wetting of dry, alluvial sediments (hydrocompaction)
9. Compaction due to the imposition of heavy loads (e.g. dams, buildings)
10. Compaction due to land drainage, and biochemical oxidation of organic soils
11. Compaction due to vibration of sediments
12. Compaction due to melting of permafrost
13. Deformation due to the creation of lakes
14. Removal of subterranean solids (e.g. coal, salt)
15. Removal of subterranean fluids (e.g. oil, water, and gases)

Source (in part): Carbognin (1985).

The subsidence system may be described very simply. The ground surface is maintained at an equilibrium which represents a balance between various stresses that are applied to the materials of the earth's crust and the resistance of those materials. The equilibrium can be disturbed so that the surface subsides, as a result of a relative increase in downward-directed stresses, a decrease in the strengths or support of crustal materials, or a combination of both. For subsidence to occur, the balance between local stresses and material characteristics must be disturbed.

The nature of subsidence will reflect the nature of the changes responsible for disruption, and the ways in which materials respond. The causes are many and varied; the responses likewise. As a result, subsidence may be local or regional; rapid or gradual; it may involve fracture or deformation; it may or may not be predictable, avoidable, or reversible.

The major causes of subsidence are shown in Table 5.13. Here, a basic distinction is drawn between natural and human-induced causes. In the context of the physical principles behind the causes, a further distinction is possible between externally imposed changes that increase downward-directed stresses and material density (Table 5.13: 1, 2, 3, 4, 5, 7, 8, 9, 10, 11, 12, 13), and those internal changes that decrease material strength and support (Table 5.13: 6, 14, 15). The geomorphological manifestations of subsidence include such large regional phenomena as subsidence basins, and a host of small features such as pits,

ponds, hummocky terrain, fissures, tension cracks, terracettes and bluffs, and ponded drainage. More significantly, ground-surface subsidence can cause serious damage to human structures, not least because some of its major manifestations are in urban areas. Table 5.14 summarizes the principal areas that are affected. Damage can include cracked roads, broken utility lines, buildings rendered uninhabitable, fractured well-casings, disrupted irrigation ditches, and flooding. The areas affected may range from a few square metres, to whole regions, such as the Central Valley of California, where over half the area (*c.*13 500 km²) has suffered from groundwater withdrawal and hydrocompaction (e.g. Ireland *et al.*, 1984). Several major cities are facing serious problems today. For example, the historic city of Venice is threatened by flooding as a result of both tectonic deformation and groundwater withdrawal, and the increasing amplitude of tidal storm surges (e.g. Ghetti and Batisse, 1983); in the city of Long Beach, California, the sustained extraction of oil has created major subsidence problems, especially in the port (e.g. Port of Long Beach, 1971); in Mexico City, widespread surface disruption and subsidence of up to 9 m is caused by locally

TABLE 5.14 Major human-induced land subsidence areas

Name of locality	Maximum subsidence (m)	Area affected (km²)	Cause	Damage ($ millions)
San Joaquin Valley, California	9.0	13 500	a	100
Houston–Galveston, Texas	2.75	12 170	a	>1000
Eloy–Picacho, Arizona and adjacent area	3.6	8700	a	several million
Tokyo area, Japan	4.6	2400	a	225 from 1957 to 1970
Nobi Plain, Japan	1.5	800	a	
Po Valley, Italy	3.0	780	d	
Santa Clara Valley, California	3.9	650	a	>25
Baton Range, Louisiana	0.5	650 exceed 5 cm subsidence	a	>1
Sacramento Valley, California	0.7	500	a	
Osaka, Japan	3.0	500	a	tens of millions
San Joaquin Valley (south-west), California	5.0	>500	b	
Lake Maracaibo, Venezuela	3.9	450	c	35 up to 1976
London, England	0.35	450	a	
Niigata, Japan	2.6	430	d	tens of millions
Saga Plain, Japan	1.2	400	a	523 from 1960 to 1979
Venice area, Italy	0.14	400	a	
Debrecen, Hungary	0.42	390	a	
Savannah, Georgia	0.15	330 exceed 2 cm	a	
Las Vegas, Nevada	1.7	300	a	several million
Raft Valley, Idaho	2.8	260	a	
Taipei, Taiwan	1.9	230	a	
Mexico City, Mexico	8.7	225	a	>500
New Orleans, Louisiana	0.8	150	a	
Victoria–Gippsland, Australia	1.6	102	f	
Saxet oilfield, Texas	0.93	92	c	
Wilmington, Long Beach, California	8.8	78	c	200
Chocolate Bayou oilfield, Texas	0.53	40	c	
Visonta, Hungary	0.5	40	a	
Haranomachi City, Japan	2.0	25	a	
Goose Creek oilfield, Texas	1.0	10	c	
Baldwin Hills, California	3.0	5	c	25
Wairakei, New Zealand	4.8	1.3	e	

Note: a = groundwater wells; b = hydrocompaction from surface water; c = oil and gas wells; d = methane in water wells; c = geothermal; f = dewatering coal mines.

Source: Coates (1983).

variable withdrawal of groundwater (e.g. Poland and Davis, 1969), and in Houston-Galveston, Texas, USA, where there has been subsidence due to groundwater withdrawal for many years, Holzer (1984a and b) estimated that surface faulting due to groundwater withdrawal is alone responsible for millions of dollars' worth of damage to property.

The geomorphologist's contribution to the problems of subsidence are similar to those described for landsliding: mapping the extent and monitoring the progress of subsidence, and relating the results of these surveys to causes, to predicting the location and magnitude of future changes, and to management plans. The methods of mapping and monitoring subsidence generally

Slope Failure and Subsidence

TABLE 5.15 Measured rates of subsidence of organic soils for specific sites in different areas

Location of site	Annual subsidence rate (cm/y)	Cumulative subsidence (cm)	Time-period (y)	Average depth to water-table (cm)
California Delta (2 sites)	2.5–8.2	152–244	26	
Louisiana (estimated)	1.0–5.0			
Michigan	1.2–2.5	7.6–15	5	
New York	2.5	150	60	90
Indiana	1.2–2.5	7.6–15	6	
Florida Everglades (2 sites)	2.7	147	54	90
Netherlands (2 sites)	0.7	70	100	10–20
	1.0–1.7	6–10	6	50
Ireland	1.8			90
Norway	2.5	152	65	
England	0.5–5.0	325 (by 1932)	84	
		348 (by 1951)	103	
Israel	10			
USSR (Minsk bog)	2.1	100	47	

Source: Stephens *et al.*, 1984.

depend on techniques for recording precisely relative changes to surface elevation. Detailed repetitive field surveying by levelling using precisely located bench-marks is the commonest approach (e.g. Fig. 3.8), but the use of compaction recorders (Fig. 3.9) and marker posts (Plate 5.2) is common (see Chapter 3). In the following brief discussion, only a few examples illustrating the variety of the subsidence problem will be discussed.

The literature on subsidence tends to be disseminated in specialist publications concerned with specific causes (e.g. oil extraction and coal-mining journals), but the most useful general references include the *International Association of Hydrological Sciences Publication 121*, *Reviews in Engineering Geology, 2* (1969), *Man-induced Land Subsidence* (Holzer, 1984a), and UNESCO's *Guidebook to Studies of Land Subsidence due to Groundwater Withdrawal* (Poland, 1984), and its earlier two volumes on *Land Subsidence* (UNESCO, 1969), as well as the short review of land subsidence world-wide by Carbognin (1985).

(b) *Removal of fluids*

The removal of fluids, especially water and oil, are the most important causes of land-surface subsidence. The ultimate cause is the same for both fluids: extraction results in the reduction of fluid pressure in the underground reservoir, which leads directly to an increase of *effective stress* (or grain-to-grain stress) so that compaction results. The principal of effective stress, attributed to Karl Terzaghi (in Terzaghi and Peck, 1948), states that the *total overburden load* (geostatic pressure or total stress, P) of a vertical column of unit cross-section is sustained partly by the *neutral stress* (hydrostatic pressure), U_w, of the fluid in the porous medium, and partly by the *effective stress* (the intergranual pressure, or grain-to-grain load), p^1, (Poland, 1984; Carbognin, 1985; Poland and Davis, 1969). Thus

$$P = U_w + p^1 \qquad (5.6)$$

Effective stress in this context is composed of two separate stresses. The first is *gravitational stress*, caused by the weight of the overlying deposits. The second is *dynamic seepage stress* exerted on the grains of sediment by the viscous drag of vertically moving interstitial fluid. The effect of these two stresses is additive, and together they increase as fluid pressure is reduced, and they lead to a reduction of *void ratios* (ratios between volume of voids and volume of solids), and to changes in the mechanical properties of the deposits (e.g. Lofgren, 1968).

As several authors have indicated, the nature of subsidence depends on certain variables in the system, of which the most important are the effective stress and its increase, the nature of the deposits (especially their compressibility, lithology, geochemistry, and thickness), the time the deposits have been subjected to increased stress, and whether or not the increased stress is being applied for the first time. A further general consideration is whether the fluids are confined (e.g. artesian conditions) or unconfined: most serious subsidence is associated with confined conditions.

A basic distinction, of considerable practical importance, is between elastic and non-elastic compaction, a distinction which depends largely on the nature of the deposits. Many reservoir sediments respond as *elastic* bodies, in which stress and strain are proportional, independent of time and reversible. Deposits of coarse sand and gravel respond as elastic bodies. That is to say, when fluid pressures fall, compaction is immediate, and if fluid pressures are restored, expansion follows. *Non-elastic* compaction results from permanent rearrangement of the granular structure of the reservoir sediments. Such is the case in fine-grained clay beds, which may occur adjacent to or within the principal fluid-storage sediments. These beds are highly compressible, but the adjustment of pore pressure is slow, time-dependent, and permanent.

The compaction of sediments due to fluid withdrawal is predictable, given an understanding of the variables involved (e.g. Helm, 1984). One approach is to plot past subsidence over time, and to extrapolate using an appropriate curve. A second approach is theoretical and based on laboratory tests and the application of fundamental physical laws. The third is to link measured subsidence with related field-derived variables (such as material characteristics). Fig. 5.8, for example, shows relations between the rate of subsidence and the depth to water in Osaka, Japan, in which

$$du_z/dt = c(h_o - h) \qquad (5.7)$$

where c is an empirical constant, h_o is the critical value of hydraulic head, h where subsidence ceases (Poland, 1984 after Wadachi). Other approaches have been reviewed by Helm (1984). Martin and Serdengecti (1984) have indicated that the prediction of large-scale subsidence due to oil extraction is very difficult as it results from special circumstances such as large pressure decline in shallow, highly compressible reservoirs.

Water and oil extraction subsidence not only share common physical characteristics: both are also relatively recent. Subsidence due to oil extraction was first noted in the Goose Creek oilfield of Texas in 1925, and subsidence arising from groundwater exploitation was recorded in the Santa Clara Valley, California, in 1933. It is no accident that the phenomenon is recent, for it arises largely from the novelty of oil extraction and the rapid rate at which water use has grown, especially in industrialized cities (several Japanese cities are seriously affected, for instance) and in areas of extensive agricultural irrigation. In addi-

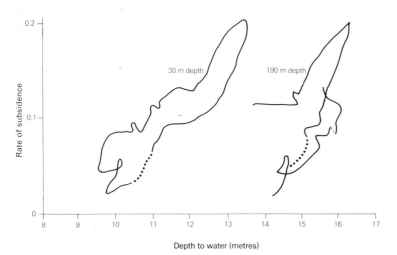

FIG. 5.8 Wadachi's correlation of water levels and observed rates of subsidence (after Poland, 1984)

tion, the recent extraction of geothermal fluids has created subsidence, for example in Wairakei, New Zealand, since 1950, and in the Geysers, California, since 1960 (Narasimhan and Goyal, 1984). All of these changes, of course, have been made possible by the contemporaneous development of appropriate equipment for exploitation. But there are differences between the consequence of oil extraction and water extraction. In the first place, groundwater is normally withdrawn from reservoirs that are shallower, have greater porosity and permeability, and are more extensive than those from which oil is extracted. The reduction of fluid pressures is also usually much less in groundwater reservoirs than in oilfields: the former may be no more than 13 atmospheres of pressure, whereas the latter may be as high as 275 atmospheres (Poland and Davis, 1969). As a result of these differences, subsidence in oilfields tends to be greater and more localized than in areas of groundwater exploitation..

Fig. 5.9 exemplifies the relations between oil extraction and subsidence; Fig. 5.10 shows the effects of water extraction on subsidence. In both examples, management responses have led to reductions in the rate of subsidence and even to rebound: in the first, subsidence has been reduced by injecting water into the oil reservoirs and by restricting extraction; in the second, controls on water extraction have been imposed.

(c) *Removal of solids*

Coal is by far the most important material the extraction of which is responsible for surface subsidence. In the USA alone, subsidence from this cause affects 800 000 ha in 30 states, whereas other underground mining affects only some 70 000 ha (Gray and Bruhn, 1984). The nature of coal-mining subsidence is strongly influenced by the type of underground mining. A useful distinction is between *room-and-pillar* methods (in which coal pillars are left in place at least temporarily to restrict or delay collapse) and *longwall mining* (in which coal is extracted along an advancing front and the explored area is either permitted to collapse or is sustained by being packed with rubble to reduce subsidence) (e.g. Institution of Civil Engineers, 1972; Gray and Bruhn, 1984). The principal components of mining subsidence are shown in Fig. 5.11, based on conditions in Germany's Ruhr coalfield. The consequence of coal removal are predictable, as the National Coal

Board's (1975) *Subsidence Engineer's Handbook* reveals. Prediction is based commonly on establishing a relationship between the amount of subsidence, and such controlling variables as the thickness of the coal seam and its depth, the location and size of cavities, and the nature of the overburden. For example, the NCB (1975) has established empirically derived curves relating the amount of subsidence to, for example, the width and depth of extraction so that subsidence can be predicted for areas where conditions are similar to those from which the original data were collected.

As Fig. 5.12 shows, surface displacement may involve vertical displacement, tilting, and horizontal displacement. The precise form of those changes depends crucially on the width of extraction: the *critical* width is that at which the centre of the subsidence trough will reach its maximum possible value. Subsidence does not occur instantly, but in general it occurs mostly while the active mine-face is within the critical area. Subsidence may be considerably delayed or even avoided altogether where pillars are left (e.g. Fig. 5.13). Horizontal displacements by coal removal may be either compressional or, often more importantly, tensional. They are approximately proportional to the amount of vertical subsidence and inversely proportional to mining depth. They can cause serious damage to buildings, utilities, bridges, roads, and other structures. Much depends on the relations between the strain (the lengthening or shortening per unit length) and the length of the structure (e.g. Fig. 5.14). In buildings, for instance, cracks and jamming doors and windows are an early sign; as the stresses increase, damage becomes more severe, service pipes may fracture, and floors begin to slope, until the buildings ultimately become unsafe. Horizontal and vertical displacements also create distinctive landforms, including compression bulges, tension cracks, pits, and depressions; in addition, surface drainage may be disrupted, often creating areas of flooding. The initial appearance of such features provides early warning of future hazards, but in most mining areas such hazards are also predictable on the basis of mining information.

(d) *Other forms of subsidence*

The major natural causes of subsidence (Table 5.13) include the widespread problems of subterranean solution of limestone (Chapter 14); the more localized thawing of permafrost (Chapter 8);

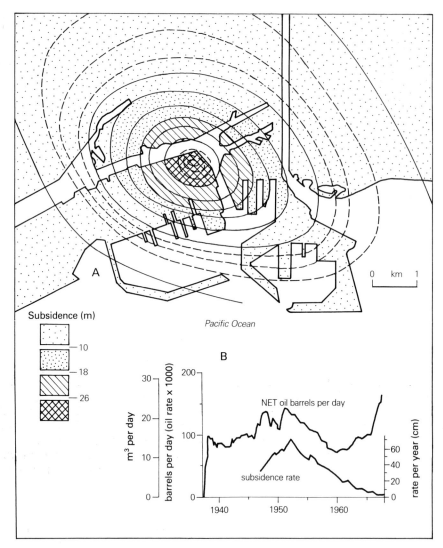

Subsidence (m)

Pacific Ocean

FIG. 5.9 (A) Subsidence of Long Beach area, California, between 1928 and 1971 (Mayuga and Allen 1969). (B) Rate of oil production and subsidence rate in the Wilmington oilfield (after Mayuga and Allen, 1969). Subsidence in the Wilmington oilfield, in Long Beach, California, now exceeds a maximum of 10 m between 1928 and the present. Rates of subsidence generally change in harmony with rates of extraction (except for slight increases with stress relief following earthquakes). The total volume of oil recovered between 1928 and 1962, 1397 million barrels, compares with subsidence of *c*.500 million barrels within the 60-m subsidence contour (i.e. *c*.39%). Associated damage includes flooding damage to oil field structures. Successful responses include increasing fluid pressure by recharge with water and flood protection works (e.g. Mayuga and Allen, 1969; Poland 1969)

and crustal deformation that may be rapid and violent (as in the Alaska earthquake of 1964) or slower and less dangerous (as in the North Sea Basin), as discussed in Chapter 13. Of the other hazards, four deserve review because they can also be caused by human-induced changes: wetting-and-drying of alluvial sediments (*hydro-compaction*); the biochemical oxidation of organic soils; subsidence caused by surface loading; and desiccation of clays.

(i) *Hydrocompaction*

In some areas, moisture-deficient, unconsolidated, low-density sediments have sufficient dry strength to support considerable effective stresses without

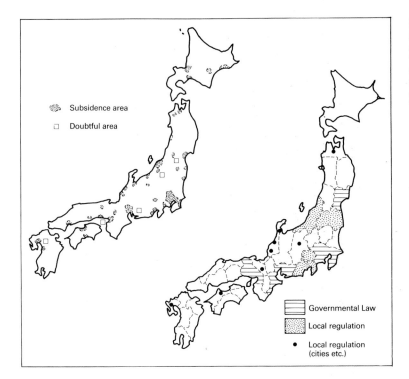

FIG. 5.10 Subsidence due to groundwater withdrawal now affects 40 areas in Japan, covering over 7380 km². Various regulations have attempted to control withdrawal, so that in some areas rebound occurs, creating new problems (after Yamamoto, 1977)

compacting. When such deposits are wetted for the first time, by percolating irrigation water for example, or when the overburden load is significantly increased, the intergranular strength of the deposits is weakened, rapid compaction occurs, and ground-surface subsidence follows (Lofgren, 1969).

The main requirement for this kind of subsidence is therefore low-density, unconsolidated, moisture-deficient, surface sediments. It is satisfied principally in those arid and semi-arid lands where there are either wind-blown loess (wind-blown silt deposits) or certain alluvial sediments which are above the water-table, are not normally wetted

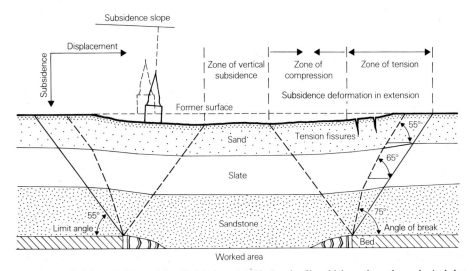

FIG. 5.11 Components of mining subsidence (after Wohlrab, 1969). The 'angle of break' depends on the geological characteristics of the rock formations. The 'limit angle' is defined by the line joining the limit of collapse to the limit of surface subsidence

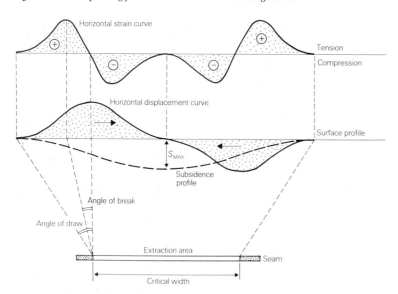

FIG. 5.12 Strain (*S*), horizontal and vertical displacement, and critical width of extraction associated with subsidence (after Gray and Bruhn, 1984)

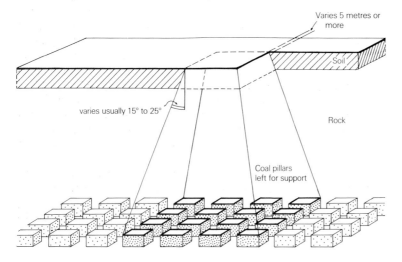

FIG. 5.13 Partial extraction for surface support (after Gray and Bruhn, 1984)

below the root zone, and have high void ratios. Areas of such sediments where hydrocompaction has been described are mainly in the drylands of the United States and the USSR.

In the creation of alluvial fans in drylands, a layer of sediment may be deposited in a flood and then rapidly dried. The dried layer may have many voids, such as intergranular openings held in place by clay bonds, bubble cavities, and desiccation cracks. Nevertheless, it may have sufficient strength to withstand the imposition of another increment of sediment in the next flood, except

perhaps for slight compaction of the upper part as the result of wetting by percolating water from the second deposit (Fig. 5.15). When such a deposit is first wetted by percolating irrigation water, the strength of the dry material is reduced (for example by the weakening of clay bonds), the void ratio is reduced, and the surface subsides. In these circumstances, the amount of subsidence depends mainly on the overburden load, the moisture conditions (such as depth of water penetration), and the amount and type of clay in the deposits. For example, Bull (1964*a*) showed in the south-western

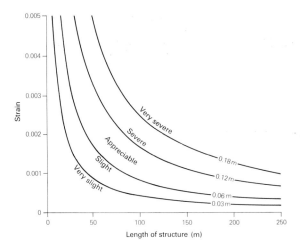

FIG. 5.14 Relationship between damage to length of structure and horizontal ground strain (after National Coal Board, 1975)

Central Valley of California that hydrocompaction increases with overburden load and that maximum compaction occurs at a clay content of about 12 per cent. Total amount of compaction here may be over 4.5 m. In areas of loess there is considerable lithological and structural variety, but the low-density sediment normally has a high void ratio, with voids often consisting of intergranular openings sustained by clay bonds, clustering of silt aggregates and tubular channels, and it responds to wetting by compacting (Lofgren, 1969).

Bull (1964a) described in detail the nature and consequence of hydrocompaction in 21 000 ha on certain alluvial fans of western Fresno County, California. Subsidence is quite variable, being greater for example where water penetrates more deeply or more continuously, such as along canals and ditches. Often field surfaces become quite irregular, and subsidence cracks may develop, e.g. along unlined canals and ditches or in irrigated gardens. The consequences of subsidence in Fresno County are numerous. To the farmer, subsidence in fields is a major problem, because canal-irrigation water accumulates in hollows and some plants are flooded, and water supply is reduced to plants on higher ground. It cost as much as $10 per acre (1964) to relevel the land. The farmer also faces the problem of damage to well-cases, pipelines, ditches, and canals, with the attendant difficulties of leakage and repair. Re-elevation of these structures, or even pumping may become necessary. One partial solution for the irrigator is to adopt a sprinkler system, which avoids the necessity for careful relevelling and distributes water uniformly (but which is nevertheless itself susceptible to subsidence). To the engineer, subsidence brings local problems of damage to buildings, roads, pipelines, power-transmission lines, and canals. For example, subsidence along a 2.5-km length of cement-lined canal caused damage that to repair required as much as $5000 a year for sand and gravel alone.

The most effective method of preventing hydrocompaction and its associated problems is to compact the vulnerable sediments by wetting before development.

(ii) *Organic deposits* Many organic soils subside when they are drained either because of relatively rapid *densification* (arising from loss of buoyancy, shrinkage from drying, or compaction from tillage) or from the usually slower *loss of mass* by

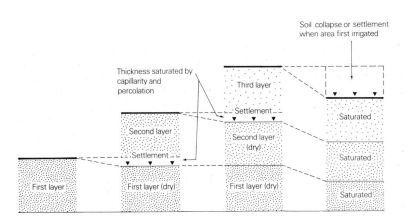

FIG. 5.15 Development of hydrocompaction in relatively strong, dry, alluvial fan deposits (after Roberts and Melickian, 1970)

biochemical oxidation in the presence of aerobic bacteria (and by burning, hydrolysis and leaching, wind erosion, and peat mining) (Stephens *et al.*, 1984). In general, rates of subsidence are positively correlated with groundwater depth; in addition, the rate of oxidation is a function of temperature and the type of peat. One prediction equation is as follows:

$$S_T = (a + bD)e^k(T - T_0) \qquad (5.8)$$

where

S_T = biochemical subsidence rate at temperature, T,
D = depth to water-table,
e = base of natural logarithm,
k = reaction rate constant,
T_0 = threshold soil temperature where biochemical action becomes perceptible,

a and b are constants (Stephens *et al.*, 1984).

The problems of 'organic subsidence' arise in such areas as the Sacramento–San Joaquin delta in California, the Florida Everglades, certain Dutch polders, near Minsk in the USSR, and the Fens in England. Table 5.15 exemplifies the rate of subsidence in these and other areas. The peatlands of the fens in Huntingdonshire, UK, were drained in 1850. A fixed datum point (Holme Post, Plate 5.2) was established in 1848, against which peat wastage could be recorded. Careful, detailed archival research by Hutchinson (1980) resulted in a precise chronology of surface lowering over 130 years, in which the phased nature of change is clearly seen to be associated with pump installations near by (Fig. 5.16). Hutchinson concluded that in the early years, shrinkage was probably the main process, but that biochemical oxidation became dominant later.

(iii) *Surface loading subsidence* This arises from the imposition of man-made features on materials capable of compaction or plastic flowage. Such subsidence is responsible for the 'Leaning Tower' of Pisa. New York's La Guardia airport has subsided as a result of artificial loading by more than 2 m in 25 years. On a larger scale, the creation of lakes can lead to crustal deformation, possibly associated with earthquakes (Nikonov, 1977).

(iv) *Desiccation of clays* In periods of drought, desiccation of clays, a process that is often exacerbated by tree roots (e.g. Cutler and Richard-son, 1980), can cause shrinkage and subsidence which seriously damages buildings, roads, etc. In south-east England, for example, the number of insurance claims for subsidence-related problems is said to have risen by about 50 per cent above the average in the exceptionally dry summer of 1983 (*The Times*, 15 September 1984), and the claims arise mainly from the London Clay, Weald Clay, and other clay areas.

5.8 Conclusion

The damage caused by landslides can range from the catastrophic to minor. Globally the total cost of landslides is unknown, but they are so widespread that seldom can they be ignored in environmental management.

There are well-tried methods available for the geotechnical analysis of slopes in general and landslides in particular. These are assisted by the fact that landslide form is often a clear indication of the physical manner of failure. However, the science of predicting where landslides will occur or defining which slopes are close to a failure threshold, is still in its infancy. It is also the part of landslide studies where geomorphology has the most to contribute. It is through a better under-standing of hillslope processes, as set in their geological context, their space and time contexts, that such advances will be made. In some cases no sense will enter the analysis until the role of man in modifying slope forms and processes has been correctly identified.

In many occupied areas, landslides form a hazard to both people and their property, as well as to a nation's infrastructure (e.g. major high-ways, dams, and electricity supply). Landslide hazard assessment thus becomes an important contribution to the evaluation of environmental risk that should feature in both planning and policy decisions (including legislation). The link between the physical and human aspects of this scenario is shown in Fig. 5.17. Interestingly, human activities feature here as though they were a geomorphological process, and in many cases they are.

The management of land subsidence is likely to involve at least three strategies. The first is to predict the location and nature of subsidence before it occurs, and to use this knowledge to plan the safe location of future surface developments

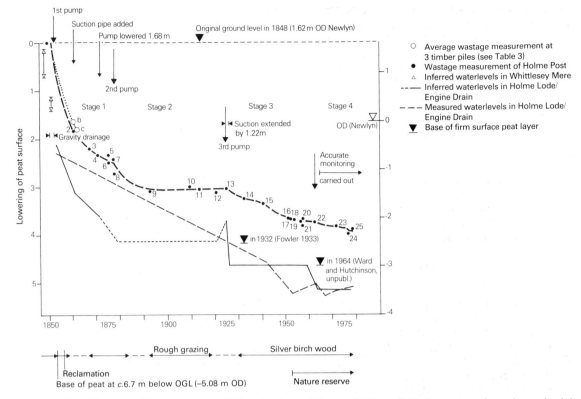

FIG. 5.16 Plot of lowering of the peat surface at Holme Post from 1848 to 1978. The available data on pumping and water levels in Holme Lode/Engine Drain are also shown (after Hutchinson, 1980)

PLATE 5.2 Holme Post, in the peatland fens of Huntingdonshire, UK, *c.*1910–13, *c.*1928, and 1978 (photos courtesy of J. N. Hutchinson, with acknowledgements to Cooper Square Publishers, New York, and the Huntingdon County Record Office)

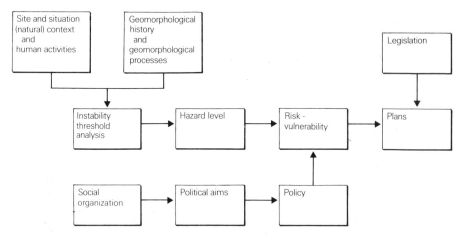

FIG. 5.17 Landslide-hazard assessment and planning policy (after Panizza, 1987; © 1987 and reproduced by permission of J. Wiley and Sons Ltd.)

and to avoid subsidence beneath established developments. The second is to ameliorate the effects of subsidence through efforts to retain surface levels (for example, by room-and-pillar mining), or to reduce the amount of subsidence or restore surface levels (for example, by underground storage in the case of solid removal, and by recharging of reservoir rocks in the case of oil and water extraction). Third, the effects of subsidence on surface developments can be reduced by careful planning and design (e.g. Institution of Civil Engineers, 1972) and by protection works (such as those required to reduce flooding, of which the Lower Thames Barrier is a good example).

6 Rivers and Floodplains

6.1 Introduction

The management of rivers involves many disciplines. Particular responsibilities, however, attach to such titles as hydraulic engineer, irrigation engineer, design engineer, public health inspector, and so on. Geomorphologists share an interest in rivers but tend to view rivers as only one of the process elements involved in landform change, whatever the time-scale. Since this includes both a medium- and a long-term perspective this allows geomorphologists to develop an understanding of rivers which is somewhat different from that of professional river managers who have to be much more concerned with immediate problems.

Of considerable relevance to all that is said below is the fact that river management is sometimes complicated by piecemeal administration. In the UK the situation was improved in 1974 when, as a result of the Water Act 1973, ten Regional Water Authorities, based largely on boundaries drawn along watersheds, replaced a more fragmented system of administration and control. The aim was to allow Water Authorities to control water 'from rainfall to the tap'. The existing pattern of water supplies in the UK is described by Gregory (1980). Further, as yet unknown changes may occur if 'privatization' of these authorities takes place.

More complicated administrative conditions can exist, as in some parts of the USA, where fragmentation of responsibilities prevails. For example, responsibilities for river management in the Santa Clara Valley, California, are held not only by the South Santa Clara Valley Water Conservation District, but are shared with the Santa Clara County Parks and Recreation Commission, and the Santa Clara County Flood Control and Water Conservation District.

Generally within the United States, however, the Constitution divides federal responsibility in the water field between the legislative, executive, and judicial branches. There is a similar subdivision within the individual states. The Constitution also distinguishes between federal and state-authority responsibilities. Most of these responsibilities lie with the states, but federal powers are reserved for water policies involving international or interstate relations. This includes the power to regulate interstate commerce. The actual use of rivers may in practice involve many organizations. To give one example, the use of rivers in the United States for navigation primarily involves planning, construction, and operation of the project by the US Army Corps of Engineers. However, since water use is usually planned for multiple purposes, any one project might also involve the US Bureau of Reclamation, the US Coast Guard, or the US Geological Survey.

Since, geomorphologically speaking, a river system functions as a process–response system within which a change in one part can have an effect on many others, its management is best conceived in a 'whole basin planning' sense. This has already been realized in many parts of the world. For example, the Maas and the Rhine, which flow through several countries of Western Europe, are administered as a unified whole, to the general benefit of all along its course.

As in all environmental management situations, the geomorphology of rivers and floodplains must be seen within the administrative framework that exists. This varies considerably from country to country.

The aim of this chapter is to concentrate on the contribution that a geomorphological assessment of rivers can make to their management. The general geomorphological concern with form–materials–process and time is modified here to a discussion of form–process–response. The forms of river channels, especially those flowing within an alluvial bed, are defined in cross-section plan and profile. These all have a relationship to river processes, and the processes themselves generate certain responses within the river channel. Flooding is an end-member of the continuous set of possible responses within the river-fluvial system. Since flooding has such a large human impact it is dealt with separately in Sect. 6.3.

Given the very large amount of work that has been done on rivers and floodplains, this account has to be selective. It must concentrate on those aspects which are related to environmental management.

The very wide interest in rivers and river management has led to a large literature within each of the relevant disciplines. As far as geomorphological publications are concerned, recent contributions include Richards's concentrated synthesis *Rivers: Form and Process in Alluvial Channels* (1982), Morisawa's general account *Rivers* (1985), *Rivers and Landscape* (Petts and Foster, 1985), *Fluvial Forms and Processes* (Knighton, 1984), *Rivers* (Petts, 1983), *Gravel-bed Rivers* (Hey *et al.*, 1982, and *River Channels* (Thornes, 1979a), with a discussion on rivers contained in more general texts such as *Water in Environmental Planning* (Dunne and Leopold, 1978), and *The Fluvial System* (Schumm, 1977).

Two examples of books on river management are: *Impounded Rivers: Perspectives for Ecological Management* (Petts, 1984); and *The Colorado River: Instability and Basin Management* (Graf, 1985).

A sign of an increasing interest in rivers is the appearance of two new journals: *Regulated Rivers: Research and Management* (Vol. 1, No. 1, January 1987) and *Hydrological Processes* (Vol. 1, No. 1, December 1986) (both published by J. Wiley). These join the existing list of relevant English serials including: *Journal of Hydrology*; *Water Resources Research*; *Journal of the Hydraulics Division, American Society of Civil Engineers*; *Proceedings of the Institution of Civil Engineers*; *Sedimentology*; *Sedimentary Geology*; *Transactions of the American Geophysical Union*; *Earth Surface Processes and Landforms*; *Water Supply Papers, United States Geological Survey*; *Bulletin of the International Association of Scientific Hydrology*; *Water Resources Research*; and the *Hydrological Sciences Bulletin*.

Searching through these sources shows that there have been changes in approach to river studies by geomorphologists. Gregory (1979a) has represented this in diagrammatic form (Fig. 6.1) as a change from the study of processes (for their own sake) to a study of river channel change and, more recently, applications of this knowledge to river engineering and management. The latter is exemplified, according to Gregory (Fig. 6.1) by integrated basin planning, the design of drainage systems, gully control, channelization, the design

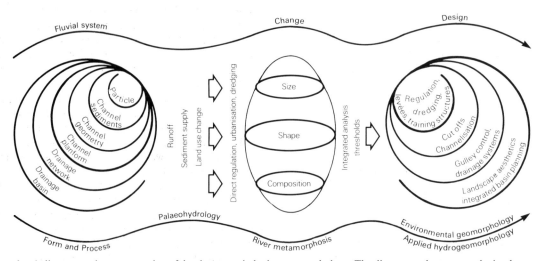

FIG. 6.1 A diagrammatic representation of developments in hydrogeomorphology. The diagram endeavours to depict the way in which studies of process in the fluvial system in the 1960s (left) have been complemented by studies of change (centre) and by applications (right) (after Gregory, 1979a)

of training structures, dredging, or river regulation. In fact, applied hydrogeomorphology exists whenever and wherever man has a significant interface with river systems.

6.2 River Channels

Rivers normally flow in channels cut either into bedrock or in recent (usually alluvial) sediments. From a management point of view the distinction is important. Rivers in bedrock tend to follow a stable course, those in alluvium have a strong tendency to change both their position and their behaviour. Failure to understand the natural behaviour of alluvial river systems can lead to unfortunate, even damaging, consequences especially if artificial changes are introduced into the system.

The basic component of alluvial river channel geomorphology is an understanding of channel form and how this is used to derive other channel parameters, including predictions of flow characteristics. Channel form can be defined in terms of plan shape, long-profile, and cross-section.

(a) Alluvial channels in plan

Natural river channels flowing across alluvium are seldom straight but tend to meander. Schumm (1963) classified meandering channels (Fig. 6.2) as either tortuous, irregular, regular, transitional, or straight. For the most *tortuous channels* meander bends are deformed, and the smoothness typical of the ideal meander curve is absent. *Irregular meanders* are irregular only with respect to the smoothly curved regular meander pattern, and may appear to consist of a meander pattern of low amplitude and wavelength superimposed on a larger pattern. Hjülstrom (1949) suggested that in such cases the smaller meanders may be related to periods of low perennial flow, while the larger occur in response to higher flows related perhaps to the mean annual flood.

The *regular pattern* is the one most amenable to quantitative analysis because it approximates to a regular wave-form in plan. The *transitional pattern* is characterized by very flat curves which may also approach a regular wave-like oscillation in plan. The *straight pattern* is almost unknown, but for this classification it is defined as the channel which has only minor bends showing no regularity.

It is well known that the meandering process leads to changes in channel position. Such changes

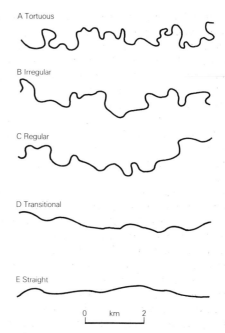

FIG. 6.2 Classification of rivers according to their meandering characteristics: (A) White River, nr. Whitney, Nebraska; (B) Solomon River, nr. Niles, Kansas; (C) South Loup River, nr. St Michael, Nebraska; (D) North Fork Republican River, nr. Benkleman, Nebraska; (E) Niobrara River, nr. Hay Springs, Nebraska (after Schumm, 1963)

may be localized (e.g. through the formation of meander cut-offs, or even through gradual meander migration as illustrated in Fig. 6.3). These changes should be anticipated in the use of any alluvial river channel as they may occur on a short as well as medium time-scale. Less easy to predict, a least in detail, are the major shifts in river position that can occur on some of the world's largest rivers. The mouth of the Hwang Ho (Yellow River), China, for example has been known to change its position by several tens of kilometres between the beginning and the end of one large flood. Indeed, at least nine major shifts in the mouth of the Hwang Ho have taken place since 802 BC (Gregory, 1979a).

The fact that such changes have taken place implies that a threshold of some sort has been exceeded. In this case a threshold related to sediment deposition during flooding. Critical thresholds in geomorphology are difficult to define, but they are crucial in terms of sound environmental management. Despite these difficulties Schumm and Beathard (1976) showed that,

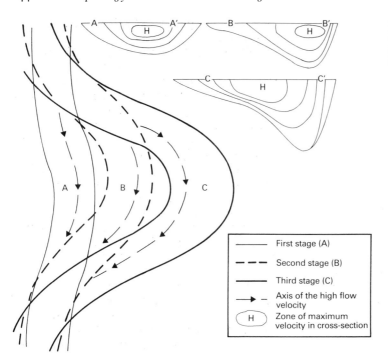

FIG. 6.3 Changes in channel plan form, cross-section, and flow characteristics in an alluvial river (after Hickin, 1974)

Legend:
—— First stage (A)
– – – Second stage (B)
—— Third stage (C)
→ – Axis of the high flow velocity
(H) Zone of maximum velocity in cross-section

in semi-arid lands, if channel sinuosity is compared with valley-floor gradient an indication is obtained of the variations in channel stability that may exist.

Since meanders have a wave-form in plan they can be defined in terms of measured properties such as wavelength and amplitude (Fig. 6.4B), which themselves have defined relationships to other river characteristics (see Sect. 6.2(e)).

(b) *Alluvial channels in cross-section*

The cross-sectional form of alluvial channels can change very rapidly both in space and in time. Adjacent sections may be very different from each other, and the cross-sectional form at any one location may change over very short periods of time. The measurements normally used to define channel cross-sections are shown in Fig. 6.4A. This illustration is somewhat misleading in that channel-bed form is usually irregular and not smooth, as the diagram suggests.

From these dimensions certain parameters are derived which are used in hydrological estimates. For example, channel capacity is usually defined as the product of channel width and the mean channel depth (based on several depth measurements across the river):

$$c = w\bar{d} \qquad (6.1)$$

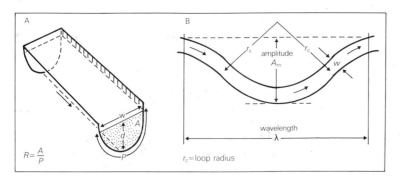

FIG. 6.4 The geometric properties of river channels: (A) cross-section, (B) plan, where w = width, p = wetted perimeter, A = cross-sectional area, A_m = meander amplitude, λ = meander wavelength, r_c = loop radius, and s = slope

where

  c = channel capacity,
  w = channel width,
  $\bar{d}$ = mean channel depth.

Hydraulic radius (R) is an often used parameter defined as:

$$R = c/p \qquad (6.2)$$

where p = wetted perimeter (see Fig. 6.4A), and is often deemed to be more or less equal to channel width.

River discharge (Q) is defined as the product of river velocity (strictly mean velocity, $\bar{v}$) and channel capacity:

$$Q = c\bar{v}, \qquad (6.3)$$

which shows how river discharge can be related to channel dimensions. These simple examples are indicative of how channel *form* is linked to channel *processes*, a subject which will be touched upon again in Sect. 6.2(d).

(c) *Alluvial channels in long profile*

In general a river channel displays a concave profile from source to mouth. Any departures from this will tend to be temporary, within the lifetime of the river, and will need to be explained if their significance is to be understood. The geomorphological significance of a convex section, or sharp step in the long profile may vary from an evolutionary explanation (e.g. rejuvenation knick-point) to a physical one (e.g. resistant element in the channel bed) to a process one (e.g. increase in discharge as a result of the inflow from a tributary which causes an increase in bed erosion and hence channel deepening).

An important geomorphological principle is contained in the concept that a concave-upward profile, for the river channel, implies proximity to a graded state. Such a state implies a balance between the forces of both water and sediment movement, within the system, and the geometry of the channel. At any one moment in time there may be a steady state along any one portion of a river channel when input of water and sediment equals output over a prescribed period of time. Such a steady state is time-independent in that at any time in the river's history and at any locality along its profile, steady-state conditions may occur. However, unless a change in climate or geology intervenes, as grade is approached so more and more of the long profile of a river will approach a steady state.

Relationships between drainage-basin morphometry and fluvial processes, and between channel forms and fluvial processes are based on the concept that dependent relationships exist between form, process, and materials, so that under steady-state conditions empirical predictive relationships can be established. A section of a channel in steady state will lose its fine balance between input and output if major controlling components (e.g. climate, sediment supply, or human structures) change.

Concepts of channel form relationships to processes are central to geomorphology, and their geomorphological evaluation is crucial in many management situations (see Sect. 6.2(i)).

(d) *Processes in alluvial channels*

Processes in alluvial river channels, in the final analysis, reduce to those of (i) erosion, (ii) transport, and (iii) deposition. Each of these is associated with specific geomorphological conditions, and an understanding of the physical principles involved in each case is important when it comes to making environmental management decisions, whether in terms of planning, management, or engineering.

(i) *Processes of erosion* As river flow occurs potential energy (Equation 6.4) is converted into kinetic energy (Equation 6.5):

$$E_p = mgh \qquad (6.4)$$

where

  E_p = potential energy,
  m = mass of water,
  g = gravitational acceleration,
  h = height above base-level.

$$E_k = 1/2\ mv^2 \qquad (6.5)$$

where

  E_k = kinetic energy,
  m = mass,
  v = velocity of flow.

It is this kinetic energy which enables the river to erode and do its work. This erosion may take place on the bed or on the channel walls, and requires turbulent flow conditions. The power of the water to dislodge material on the stream bed or bank is related to the forces generated and their general efficiency in overcoming the forces resisting movement. The force tending to produce movement is known as the shear stress (τ). The shear stress that can be generated on the bed of a river (τ_0) is a function of the specific weight of the fluid (γ), which for clear water is 1000 kg/m³ (or 62.4 lb/ft³), and both the hydraulic mean radius (R) and the slope (S) of the channel. These latter are defined in Fig. 6.4 (where R = cross-sectional area ÷ wetted perimeter):

$$\tau_0 = \gamma RS. \tag{6.6}$$

τ_0 is also known as the *unit tractive force*, in that it is a measure of the drag (shear) generated per unit wetted area by the flowing water.

Thus minimum bed erosion occurs when channel slope (S) is low, and the wetted perimeter is large compared with the cross-sectional area of the channel. These constraints are important, for example, in the design of an irrigation channel in which erosion has to be kept to a minimum.

Turbulence is generated by a number of factors which include the roughness of the river bed. This roughness is usually represented by Manning's roughness coefficient, n, which is used in the Manning formula for the determination of mean river velocity ($\bar{V}$) and is given as:

$$\bar{V} = \frac{1.486}{n} \cdot R^{2/3} \cdot S^{1/2} \tag{6.7}$$

where

$\bar{V}$ is in ft/s,
R is in ft,
S is the rate of loss of head per foot of channel (channel slope).

In practice this means that for a channel of given dimensions and gradient, velocity of flow increases as roughness decreases. Values for n have been assessed for different bed conditions. A useful publication is that by Barnes (1967) in which detailed variations in bed materials and form are illustrated by means of photographs alongside their n values. Some *appropriate* values for n related to different channel-floor materials are: sand $n = 0.02$; gravel $n = 0.03$; cobbles $n = 0.04$; boulders $n = 0.05$. Mountain streams with rocky beds have values of n between 0.04 and 0.05.

The movement of material along the bed of a river channel can be studied through a series of empirical relations which revolve around an assessment of *Shields entrainment function*, F_s, and the *particle Reynolds number*, R_e^*. The latter indicates the relation between the forces of inertia and those of viscosity.

Shields entrainment function (F_s) is given by:

$$F_s = \frac{V^{*2}}{gD(S_s - 1)} \tag{6.8}$$

where

g = acceleration due to gravity,
D = material (particle) diameter,
S_s = sediment specific gravity,
V^* = shear velocity (i.e. shear stress on channel bed ÷ fluid density)$^{1/2}$,

F_s can also be expressed as:

$$F_s = \frac{RS}{D(S_s - 1)}. \tag{6.9}$$

For any one portion of a river channel all of the items on the right-hand side of the equation can be measured and F_s calculated.

The particle Reynolds number (R_e^*) can also be related to measurable properties in that:

$$R_e^* = \frac{(\gamma RS)^{0.5} \, D\rho_f^{0.5}}{\mu} \tag{6.10}$$

where ρ_f = fluid density and μ = fluid viscosity. These two values (F_s and R_e^*) are of greatest interest when they are plotted against each other on double log. axes (Fig. 6.5). This shows that the threshold of sediment transport occurs at the minimum values of F_s and when R_e^* increases beyond a value of 6. By the time $R_e^* > 400$ and $F_s = 0.056$ turbulent flow is fully developed and material is removed from the channel bed. Since R_e^* must exceed 400 for full turbulence to exist a reordering of Equation 6.10 allows (γRS) to be stated as:

$$(\gamma RS)^{0.5} = \frac{400\mu}{D\rho_f^{0.5}}. \tag{6.11}$$

Since μ and ρ_f are constants for water at defined temperatures and both D and S are known for the channel concerned, R can be calculated, and related to channel cross-sectional area in order to establish the likely wetted perimeter (Fig. 6.4) for the channel when the movement takes place of sediment of a specific size.

Of the various contributing parts of a river basin to the sediment load carried by the river, a high

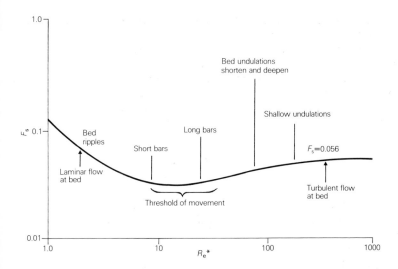

FIG. 6.5 Bed characteristics and flow conditions in terms of Shield entrainment function (F_s) and the particle Reynolds number R_e^*

proportion is derived from the channel walls. As with any inclined slope, the material of which the channel walls are composed (assuming that it is not bedrock but sediment or regolith) in the short term will remain stable as long as the angle of slope (θ) does not exceed the angle of repose (ϕ) of the material. On a river channel bank, unlike a hill slope, there is not only the gravitational force ($W \sin \theta$), exerted on each particle (where W is the weight of the particle) but also the lateral force generated by the flowing water. This force is proportional to τ_o, the shearing force on the channel bed. The resistance to movement is provided by:

$$\text{Resistance} = W \cos \theta \tan \phi. \qquad (6.12)$$

In this situation τ_o is related to τ_c (the critical shear stress required to move a particle) according to the relationship:

$$\frac{\tau_o}{\tau_c} = \sqrt{\left(1 - \frac{\sin^2 \theta}{\sin \phi}\right)}. \qquad (6.13)$$

On a side slope:

$$\frac{\tau_o}{\tau_c} = 0.75 \, \gamma ds \qquad (6.14)$$

where

γ = specific weight of the fluid,
d = vertical depth of flow,
s = channel slope.

The sediment from the channel banks combines with that washed in from the valley slopes and eroded from the channel bed to give the total sediment carried by the river. However, its actual release from the channel banks is a function *inter alia* of the size and shape of the material, its packing characteristics, the steepness of the bank, the depth of river flow, and the channel gradient.

In terms of environmental management one of the most important areas for concern is that where river channels merge upstream with gully systems that form the surface manifestations of soil erosion. The boundary between 'river channels' and 'erosional gully systems' is impossible to define. The principles of bank and bed erosion, defined above, apply in both cases. However, as is shown in Chapter 7, soil erosion has a human and land-management dimension which may be far more important than some of these physical laws.

(ii) *Processes of sediment transport* To be effective erosion must be linked to sediment transport, and transport is linked to both flow velocities and turbulence. Most rapid flow tends to occur farthest away from the bed and banks. However, turbulent flow generates velocity variations within this general set of conditions so that high velocities generated by turbulence can increase erosion of channel bed and walls. Such turbulent effects make it difficult to apply the empirical calculations involving critical shear stress defined above, and estimates are often much at variance with reality (Richards, 1982).

Even the simple equation: $Q = c\bar{v}$ (Equations 6.3 above) is difficult to apply without error for the measurement of depth required to calculate channel capacity (c) is both difficult and subject to inaccuracy, not least because depth can vary as bed sediments shift. In addition there are problems with the measurement of velocity, not only because velocity varies within any one cross-section but also because velocity changes as water-level rises or falls. This latter variation is in part defined by the Chezy equation:

$$V = C_h \sqrt{(RS)} \qquad (6.15)$$

where

R = hydraulic radius,
S = the energy gradient (which is approximately the slope of the water surface),
C_h = a resistance factor (normally indicated as C—the Chezy coefficient but here designated C_h.

The comparable formula (widely used in the US) is the Manning equation (Equation 6.7). The amount and calibre of sediment load transported by rivers is a function of many controls. These can be divided into two types:

(i) physical laws of association between sediment load and channel flow characteristics, and
(ii) human modifications or actions within the drainage basin containing the river channel.

Amongst the physical constraints shear stress is probably the most important. For example, a relationship can be defined between the shear stress required to move a particle (τ_0) and the size of that particle. If it is assumed that the specific gravity of the sediment (S_s) is 2.65 and that the kinematic viscosity (i.e. fluid viscosity ÷ fluid density) is 1.2×10^{-5} ft^2/s, then Fig. 6.6 illustrates the theoretical relationship between τ_0/γ as represented by RS, and material size (D).

From this diagram it is therefore possible to arrive at a theoretical measure of the size of material that should be mobile for a given hydraulic mean radius (R) and channel slope (S). Conversely, and as important, is the possibility of measuring the diameter of channel load in order to discover what the minimum value of R must have been to bring the material to that point. This is on the assumption that there will have been no appreciable change in S since the material was deposited. If it is desirable for practical purposes to ensure that the bed load in question does not move again, then enough water has to be diverted from the channel to keep R below a specified maximum

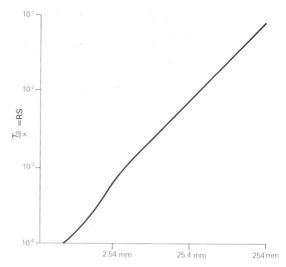

FIG. 6.6 The size of bed-load material mobile in a river channel for specific combinations of hydraulic radius and channel slope. τ_0 unit-tractive force; γ = specific weight of water (modified from Henderson, 1966)

value, namely that value necessary to move the bedload for the given value of S. The appropriate maximum value of R can be estimated from Fig. 6.6.

Other equations have also been defined for the relationship between sediments and channel characteristics. For example, Rubey (1952) suggested that when the form (width–depth) ratio of a channel is constant:

$$SF = \frac{kL^a \bar{D}^b}{Q^c} \qquad (6.16)$$

where

S = graded slope,
F = optimum form ratio (the depth–width ratio which gives to a stream its greatest capacity for traction),
L = amount of load (through any cross-section per unit time),
$\bar{D}$ = average diameter of bedload,
Q = discharge (through any cross-section per unit time).

Morisawa (1968) modified this relationship by including a product (n) on the left-hand side of the equation, where n is channel-bed roughness. In effect the equation implies that changes in either the load (L and/or D) or discharge (Q) will lead to changes in either slope and/or channel form, as well as influencing bed roughness. This illustrates the dangers of considering any one, or even any pair of channel parameters in isolation, as each has to be assessed in the context of the whole system in which it operates.

Expertise in hydraulic engineering has made it possible to isolate those parameters upon which design decisions can be based. For example Lacey (1929–30) based the design of irrigation channels on a regime theory expressed by:

$$\frac{\bar{V}^2}{R} = 1.324 \, f_{vr} \qquad (6.17)$$

$$R^{1/2} \cdot S^{2/3} = 0.0052 \, f_{rs} \qquad (6.18)$$

$$P = 2.67 \, Q^{1/2} \qquad (6.19)$$

where

$\bar{V}$ = mean velocity of flow,
R = hydraulic radius of channel,
S = channel slope,
P = wetted perimeter,
Q = discharge,
f_{vr} and f_{rs} are silt (sediment) factors.

From these predictive equations the known parameters can be used to estimate the others, or channel characteristics can be designed to cope with prescribed silt factors. However, as in many similar design situations, there are serious constraints upon the use of these equations. For example, they only apply to channels that have a bed of loose material of the same type as that being moved along the bed. In addition the channels should also be in equilibrium; and these equations may only be applicable to those areas in India and Pakistan for which they were originally developed.

(iii) *Processes of deposition* The empirical equations used to describe various states of channel-bed erosion can also be used to detect the likelihood of deposition taking place. If it is necessary for engineering purposes to induce the deposition of material above a certain size (e.g. to avoid damage to machinery) then this can be governed by channel shape and gradient characteristics, so that, for example R_e^* is kept well below 400 (Fig. 6.5).

Alluvial channel deposits have been found to adopt specific forms according to described criteria (Simons and Richardson, 1961). A sequence of six channel-bed forms has been recognized, each

related to different stream velocities (Fig. 6.7). Linked to each there is also a value for the Manning roughness coefficient (n) as used in Equation 6.7. Observation of the alluvial bed forms can therefore lead to an estimate of the likely extreme values of n, and these, for a channel having prescribed R and S values (Equation 6.7), can be used to estimate the likely range of stream velocities before the bed forms will change. Reference is also made in Fig. 6.7 to a Froude number (Fr), where:

$$\mathrm{Fr} = \frac{\bar{V}}{(gd)^{0.5}} \qquad (6.20)$$

in which

$\bar{V}$ = mean river velocity,
g = acceleration due to gravity,
d = vertical depth of water flow.

If Figs. 6.5 and 6.7 are compared it will be seen that ripples occur at R_e^* values well below those required for transport under turbulent conditions. At high velocities ripples give way to dunes and bars as turbulence begins. The upstream migration of anti-dunes implies sediment movement and is probably little short of the R_e^* values required for erosive conditions. Field observation of bed forms in alluvial channels can sometimes be used, therefore, to estimate values employed in hydraulic engineering.

(e) *River (water) discharge*

The most important common link between erosion, transportation, and deposition is the amount of river water being carried by the channel. In terms of environmental management river discharge is crucial to many planning and engineering decisions. Special attention is given below (Sect. 6.3) to flooding, which is only an extreme form of discharge. In many semi-arid, savannah, and monsoon lands periods of low flow are just as important in terms of human interactions with the river channel. These extremes are, of course, in part at least a function of climate, and variations in climate can produce extreme conditions as the so-called 'desertification' process in Africa has shown.

While climatic variability has served to make difficult any valid estimate of future water supplies within river channels, there have nevertheless been several attempts to use some of the channel parameters defined above to estimate river dis-

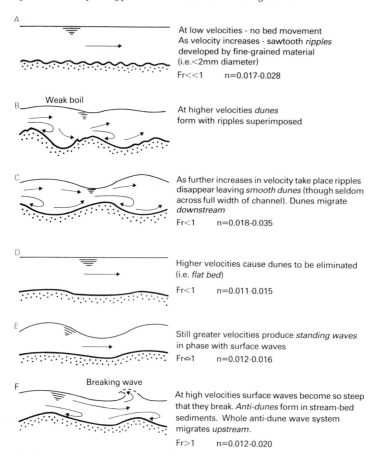

A

At low velocities - no bed movement
As velocity increases - sawtooth *ripples*
developed by fine-grained material
(i.e.<2mm diameter)

Fr<<1   n=0.017-0.028

B   Weak boil

At higher velocities *dunes*
form with ripples superimposed

C

As further increases in velocity take place ripples
disappear leaving *smooth dunes* (though seldom
across full width of channel). Dunes migrate
downstream

Fr<1   n=0.018-0.035

D

Higher velocities cause dunes to be eliminated
(i.e. *flat bed*)

Fr<1   n=0.011-0.015

E

Still greater velocities produce *standing waves*
in phase with surface waves

Fr≏1   n=0.012-0.016

F   Breaking wave

At high velocities surface waves become so steep
that they break. *Anti-dunes* form in stream-bed
sediments. Whole anti-dune wave system
migrates *upstream*.

Fr>1   n=0.012-0.020

FIG. 6.7 Stream-bed forms in alluvial
channels, their roughness (*n*) and
relation to the Froude number (Fr)
(modified from Simons and
Richardson, 1961)

charge values. This may be particularly important in areas for which there are no discharge records and yet an agricultural, industrial, or other development scheme needs such information. The fact that variable flows, especially in the short term, may not match the predictions made is often seen as inevitable.

In this section consideration is given to methods of estimating the likely amounts of discharge when no gauging stations are present because these methods all depend on geomorphological parameters. The equally important factors relating to the frequency and duration of specific flows are discussed in relation to flooding in Sect. 6.3. River discharge has figured in several of the equations given in the preceding sections. Each of these allows discharge to be estimated from other more readily measured characteristics.

Leopold and Maddock (1953) demonstrated simple relations between channel dimensions, velocity of flow, and discharge:

$$w \propto Q^{0.5} \qquad (6.21)$$

$$d \propto Q^{0.4} \qquad (6.22)$$

$$V \propto Q^{0.1} \qquad (6.23)$$

where

w = channel width,
d = channel depth,
V = velocity of flow,
Q = discharge.

(Note that the powers sum to 1.0, as the product of w, d, and V is Q). Fig. 6.8 illustrates the relations between Q and each of w, d, and V for the Missouri–Mississippi rivers, and for the Kansas River. In these cases, d increases faster than w with Q along the Missouri–Mississippi, but vice versa on the Kansas River.

Working on the premise that mean annual discharge (Q_m) is related to channel slope (S) by a power law:

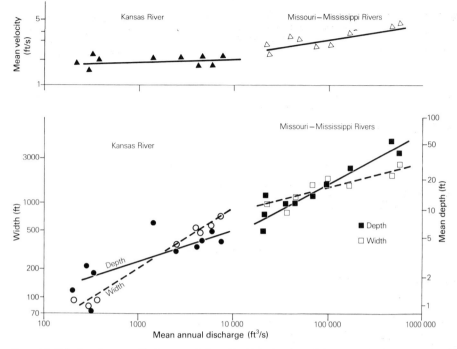

FIG. 6.8 Relations between discharge and channel width, depth, and river velocity for the Kansas River and the Missouri and Mississippi Rivers (after Langbein, 1962)

$$S \propto Q_m^a \qquad (6.43)$$

Carlston (1968) found that the value of a varied significantly with the character of the channel. On sections which were in a steady state, and where the channel was cut into an alluvial bed, a could be defined and the correlation between S and Q_m was good. In ungraded sections there was no significant correlation between S and Q_m. Thus on the graded portions of Red River, Tennessee River, and Delaware River a varied between -0.55 and -0.93. For the mountain reaches, or sections of a valley along which bedrock is highly variable, channel slope and mean annual discharge are uncorrelated. Recognizing this significant distinction between graded and ungraded portions of a river, it is possible from the relationship given above to assess the influence of S on Q_m in particular cases.

Empirical relationships have also been suggested between discharge (Q) and the measurable geometry of alluvial channels (including meander wavelength (λ), meander amplitude (A_m), and channel width (w)) (Fig. 6.5). For example Albertson and Simons (1964) reported that:

$$\lambda = k_1 Q^{1/2}, \qquad (6.24)$$

$$A_m = k_2 Q^{1/2}, \qquad (6.25)$$

$$w = k_3 Q^{1/2} \qquad (6.26)$$

where $k_1 = 29.6$, $k_2 = 84.7$, and $k_3 = 4.88$, but each of these may vary from locality to locality. Such statements only define general tendencies and may not be applicable to specific situations. More complex relationships can be derived by including more varied information about the meandering channels. For example Schumm (1967) showed that:

$$\lambda = \frac{1890 \, Q_m^{0.34}}{M^{0.74}} \qquad (6.27)$$

where

λ = meander wavelength,
Q_m = mean annual discharge,
M = percentage of silt and clay in channel perimeter (which is representative of the river load).

Clearly, channel meanders should not be considered independently of their material properties. Similarly, Schumm (1969) showed that channel width also is related not so much to Q_m but to the ratio of Q_m and M:

$$w = 2.3 \frac{Q_m^{0.38}}{M^{0.39}}, \qquad (6.28)$$

and likewise:

$$d = 0.6\ M^{0.34} \cdot Q_m^{0.29}, \qquad (6.29)$$

where d is channel depth. Variations in channel dimensions with constant discharge are probably attributable to changes in the calibre of the sediment load. For example, the width–depth ratio of alluvial channels appears to be determined by the nature of the sediment transported through the channel. A high width–depth ratio is associated with large bed-material load, and vice versa.

An alternative approach to the estimating of discharge is through examining the morphometry of the drainage basin concerned, rather than concentrating on channel characteristics. For example, discharge may be thought of as being proportional to some function of basin areas. Leopold and Miller (1956) tested the general equation:

$$Q = jA^m \qquad (6.30)$$

for parts of central New Mexico and found that the flood discharge equalled or exceeded in 2.3 years ($Q_{2.3}$) was closely estimated by:

$$Q_{2.3} = 12A^{0.79}. \qquad (6.31)$$

Values for j and m vary from one area to another, although m tends to remain in the range 0.5–1.0. Topographically similar areas nevertheless prob-

ably give rise to very similar j and m coefficient values.

Discharge can be more closely predicted by including additional measures of basin morphometry, and also by including a climatic parameter. Patterson (1970) analysed the available data for Arkansas and found that:

$$Q = 0.082\ A^{1.02} \cdot P^{0.75} \cdot E^{0.06} \qquad (6.32)$$

where

Q = mean annual discharge (ft³/s),
A = basin area (sq miles),
P = mean annual precipitation (in inches) minus 30,
E = mean elevation of the basin above sea-level.

The omission of E from the equation increases the standard error of estimated values of Q by only 0.5 per cent, thus indicating the dual importance of both the amount of precipitation and the size of the catchment.

In practice river discharge cannot be recalculated at a frequent time interval (say hourly) because the necessary measurements would, in most cases, be too time-consuming. Instead, reference is made to 'rating curves' which allow discharge at a particular cross-section to be estimated from river height (Fig. 6.9). In practice this also provides the single most important parameter in a flood-warning scheme, namely rise in river level.

(f) Sediment discharge

The erosion and sediment transport processes described above produce a sediment load which is referred to as sediment discharge passed or at a

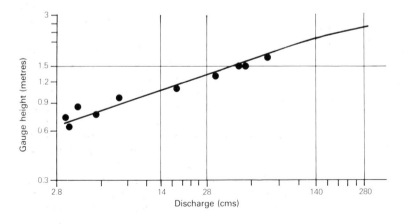

FIG. 6.9 Rating curve for the New Fork River at Boulder, Wyoming (after Dunne and Leopold, 1978; © 1978 W. H. Freeman and Co.)

particular point. Excess sediment can cause much damage. It may bring about reservoir siltation to such an extent that the reservoir never fulfils its expected economic life. Sediment can choke irrigation canals, and it may build up within harbours or estuaries. Frequently, any one of these may imply an immense economic loss.

In many development projects it is necessary to predict likely amounts of sediment discharge. It is possible simply to follow Lane (1955) and assume that bed load (Q_s) and sediment size (D) are related to water discharge (Q_w) and channel gradient (S):

$$Q_s D \simeq Q_w S. \qquad (6.33)$$

Schumm (1969) reconsidered Lane's suggestion and concluded that:

$$Q_s \simeq \frac{w\lambda s}{dP} \qquad (6.34)$$

where

Q_s = bed-material load,
w = bankful width,
λ = meander wavelength,
s = gradient,
d = maximum channel depth,
P = sinuosity (ratio of channel length to valley length).

In many cases, however, water and sediment discharges can be independent of each other. For example, climatic fluctuations, changes in land use, river regulation, and diversions can significantly modify the balance between water discharge and sediment load. Human interference with the system is often the primary cause of high sediment loads.

Schumm showed how the relationships in Equation 6.34 could be used to predict human-induced changes. For example, a change in bed-material load at constant mean annual discharge causes a change in channel dimensions (w and d), wavelength (λ), slope (s), and sinuosity (P). Deforestation or clearing land for housing development can significantly increase Q_s, and thus induce downstream changes in these other parameters.

Schumm (1969) also showed that:

$$Q_w Q_t \simeq \frac{w\lambda F}{P} sd \qquad (6.35)$$

where

w, λ, P, s, and d are as in Equation 6.34,
Q_w = water discharge,
Q_t = percentage of total sediment load that is bed load,
F = width–depth ratio.

Thus he linked water and sediment discharge in one empirical statement which can be applied to situations such as that of dam construction whereby water and sediment are cut off so that Q_w and Q_t are decreased for all channels below the dam. This results in downstream decreases in channel width, wavelength, and width–depth ratio, and an increase in sinuosity. Channel gradient would probably decrease with the increase in sinuosity and depths might also increase.

(g) River channel change—river metamorphosis

The study by Schumm (1969) was concerned with the ways in which river channel changes may be brought about. The identification of change can be achieved by (i) measurement, (ii) historical methods, (iii) dating techniques, and (iv) space-time substitution techniques (Gregory, 1977). Measurements are usually only available for short periods of time; historical methods rely on the availability of suitable sources (such as old maps or aerial photographs) (Gregory, 1979a); dating techniques depend on the preservation of datable materials (such as ^{14}C or pollen); and space-time substitutions (the ergodic approach) depend on valid comparisons between an unmodified river and the river which is being investigated.

Natural changes in channel position, or of dominant flow paths within a braided system, are not unusual in some rivers; abandoned meanders are a natural part of a dynamic alluvial valley system; braided channels on alluvial fans change their position and relative dominance from one flood to the next.

These natural processes may be important when a channel is relied upon to provide a continued water supply. Such was the case, for example, with the Kosi River which supplies irrigation water to farmland on the northern slopes of the Ganges plains (Fig. 6.10). In the early 1970s the river began to shift its course away from the entrance to the irrigation feeder canal, leaving the whole irrigation system acutely short of water (Doornkamp, 1982b). Geomorphological analysis of aerial photography demonstrated that the sediment banks between the braided channels were fre-

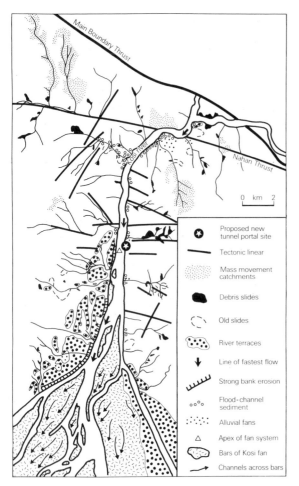

FIG. 6.10 Geomorphology of the Kosi study area

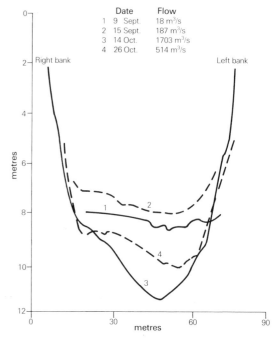

Date		Flow
1	9 Sept.	18 m³/s
2	15 Sept.	187 m³/s
3	14 Oct.	1703 m³/s
4	26 Oct.	514 m³/s

FIG. 6.11 Changes in channel cross-profile during the passing of a flood. Sept.–Dec. 1941, San Juan River, nr. Bluff, Utah (after Leopold *et al.*, 1964; © 1964 W. H. Freeman & Co.)

quently changing and causing the channel flow to vary greatly from one post-monsoon period to the next. Had this been recognized sooner, the irrigation intake canal could have been placed within the Chatra Gorge upstream of the braided river section.

At a more detailed level changes in the cross-section of any portion of an alluvial channel are normal. For example, large changes were observed with the passing of a flood on the San Juan River (Fig. 6.11). The arrival of the flood, with an increase in discharge from 179 m³/s to 1857 m³/s, caused a rise in channel-bed level, but scouring soon began. This deepened the channel and changed its cross-profile. As the flow decreased from 1687 m³/s to 509 m³/s sedimentation was

renewed and the channel cross-profile was once again modified (Leopold *et al.*, 1964).

Measurements of both channel characteristics and stream-flow characteristics are thus time-dependent. Stream flow also varies with the position in the channel at any one cross-section. This variability must be recognized in the field measurement of stream and channel properties and in making predictions based upon these concerning flow or sediment discharge. Data acquired in the past about channel conditions at a point may not remain constant for future site development plans.

There is a close link between channel plan changes and changes in both cross-sectional form and velocity patterns. Hickin (1978) shows how changes in meander form (Fig. 6.3) lead to changes in these other two components.

River meanders frequently change their form and their position. In this sense, they are unstable not only on the geological time-scale but also on the human time-scale.

While, in general, it may be that (following Henderson, 1966) meander wavelength (λ) and loop radius (r_c) are related to channel width (w) (see Fig. 6.4) by:

$$\frac{\lambda}{w} = 7 \text{ to } 11 \qquad (6.36)$$

$$\frac{r_c}{w} = 2 \text{ to } 3, \qquad (6.37)$$

it does not necessarily follow that these limits define an equilibrium condition.

(h) *Classification of alluvial channels*

Rivers cut into alluvium display great irregularities which make their classification difficult. Nevertheless, the classification proposed by Schumm (1977) (Table 6.1) provides an indication of how this may be achieved. Schumm relates the relative amounts of river load to the nature of channel stability, making this a functional classification rather than a morphologically based classification.

(i) *Problems of channel management*

Channel management is usually a component of a broader issue, such as the use of the river for navigation, irrigation, or within a flood control scheme. Although general relationships appear to exist between certain parameters of channel form, water and sediment discharge, and channel behaviour, there is always an element of uncertainty about the response of the channel to a particular action.

Examples of major attempts at channel management are numerous and ongoing. Control of the Mississippi, for example, began in 1699 with the building of levees as individual landowners tried to protect their land from flooding. In 1879, this work was taken over by a newly formed Mississippi River Commission who tried to bring systematic order into random endeavours at river control. Continual management of the Mississippi reduced an average flood frequency of once every 2.8 years, up to 1931, to only one flood between 1956 and 1969 (Stevens *et al.*, 1975).

Other forms of river control include the building of dams for water supply and electric power generation, channelization (see Brookes *et al.*, 1983) and the straightening of river courses, irrigation diversions, and the abstraction of large amounts of water for use in industry. Each of these 'controls' lead to modifications in the natural river system. Perhaps the best-documented of these are those modifications arising out of the building of dams.

The effects of dam construction (Petts, 1979) differ according to whether a site is above, at, or below the reservoir created (Fig. 6.12). Above the reservoir there is a rise in local base level for groundwater, and rivers draining into the reservoir deposit their load both within their beds near the reservoir margins and as a delta within the

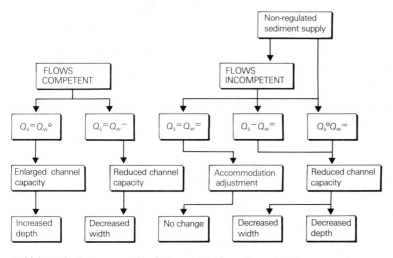

FIG. 6.12 Potential morphological adjustments of river channels, downstream from reservoirs (after Petts, pers. comm.)

Non-regulated sediment supply

FLOWS COMPETENT

FLOWS INCOMPETENT

$Q_s = Q_w{}^{\circ}$ $Q_s = Q_w{}^{-}$ $Q_s = Q_w{}^{=}$ $Q_s{}^{-}Q_w{}^{=}$ $Q_s{}^{\circ}Q_w{}^{=}$

Enlarged channel capacity   Reduced channel capacity   Accommodation adjustment   Reduced channel capacity

Increased depth   Decreased width   No change   Decreased width   Decreased depth

$=$ Major reduction
$-$ Minor reduction
$\circ$ No change

Q_s Sediment load (quantity and size)
Q_w Discharge (fluid magnitude and frequency)

TABLE 6.1 Classification of alluvial channels

Mode of sediment transport and type of channel	Channel sediment (M) (%)	Bed load (% of total load)	Channel stability		
			Stable (graded stream)	Depositing (excess load)	Eroding (deficiency of load)
Suspended load	> 20	< 3	Stable suspended-load channel. Width–depth ratio less than 10; sinuosity usually greater than 2.0; gradient relatively gentle	Depositing suspended-load channel. Major deposition on banks causes narrowing of channel; initial stream-bed deposition minor	Eroding suspended-load channel. Stream-bed erosion predominant; initial channel widening minor
Mixed load	5–20	3–11	Stable mixed-load channel. Width–depth ratio greater than 10, less than 40; sinuosity usually less than 2.0 and greater than 1.3; gradient moderate	Depositing mixed-load channel. Initial major deposition on banks followed by stream-bed deposition	Eroding mixed-load channel. Initial stream-bed erosion followed by channel widening
Bed load	< 5	> 11	Stable bed-load channel. Width–depth ratio greater than 40; sinuosity usually less than 1.3; gradient relatively steep	Depositing bed-load channel. Stream-bed deposition and island formation	Eroding bed-load channel. Little stream-bed erosion; widening channel predominant

Note: Sinuosity is the ratio of channel length to valley length.

Source: Schumm (1977).

reservoir itself. Higher groundwater may have an adverse effect in that in certain circumstances it can produce slope instability (see Chapter 5). Deposition of sediment decreases the gradient of the river bed and increases its height, thus increasing the risk of flooding. Deposition of sediment within the reservoir decreases its water-holding capacity, and in an area of high rates of soil erosion may even render the construction of the dam uneconomic.

Downstream of the dam rivers carry a reduced sediment load, and in an attempt to acquire a new load rivers immediately below a dam can produce large amounts of channel erosion. The net effect of dam construction tends to be to decrease maximum discharges, and the effect of this is to reduce channel sizes. Much depends on whether the flows are competent or not to transport sediment (Fig. 6.12). Former flood lands become vegetated, and this may extend into channels whose full capacity is no longer used. As tributaries come in and the contribution of water from side slopes and groundwater sources increases so the effects of dam construction tend to decrease the further downstream observations are made.

In particular cases channel adjustments may take on different forms. Channels can respond by changing their slope, roughness, width, depth, or form in plan in order to acquire a quasi-equilibrium state (Langbein, 1964). The response may come as a change in any one or in any combination of these.

In terms of river discharges a dam will reduce flow downstream and even cut flood peaks. River straightening has the opposite effect, especially if it is accompanied by an enlargement of the channel. Here water velocities are increased and the frequency of high discharges may increase.

Changes in channel behaviour may also result from changes in land-use practices on the hillslopes of the drainage basin. For example, deforestation may lead to higher runoff rates, increased soil erosion, and hence an increase in both peak channel flows and sediment load. Wolman (1967) has produced a model of river responses to land-use changes in the north-east USA since 1700 which identifies deforestation as a cause of increased sediment transport over 100 years or so, and urbanization and especially construction as a major source of increased sediment, but over a very restricted time-period (Fig. 6.13).

Channel management extends well beyond these topics into problems associated with such things as land drainage, irrigation schemes, and channel dredging for navigation. Gregory (1979b) defines some of these (Table 6.2) in terms of this relationship to channel behaviour at the local and regional scale, and in terms of the geomorphological analyses that need to be carried out. Some of these problem areas have received detailed consideration elsewhere. For example, Fookes *et al.* (1985) deal with details of road and hillslope drainage in a Himalayan environment, and Schick (1974) dealt with road drainage on alluvial fans. A context for the study of all river channel changes was provided by Gregory (1977), in an introduction to a volume devoted to several papers on this theme. There is no doubt that much of this work

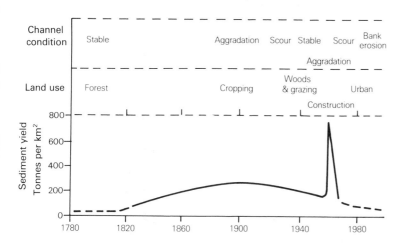

FIG. 6.13 Models of river-channel adjustment. Changes of sediment yield since 1780 in the north-east USA associated with man-induced alterations of land use and induced alterations in river-channel stability (from Wolman, 1967)

TABLE 6.2 Channel management and geomorphology

Subject of attention	Application	Possible implications requiring attention by hydrogeomorphologist		
	Engineering	Design aspects	Local impact	Regional impact
Drainage basin Irrigation schemes	Planning of irrigation scheme including pattern, size, shape, and slope of canals and watercourses and relation to pre-existing drainage	Knowledge of regime equations used in design and evaluation of alternatives. Extent to which associated channelization is necessary	Water-table, groundwater, erosion, and sedimentation effects	Effects of changed flow frequency on main channels in river system. Effects of channelization throughout drainage basin
Gully control	Control measures for specific gullies and for treatment of area	Evaluation of alternatives in relation to rate of gully development established by empirical studies	Adjustments including scour immediately downstream	Adjustments on trunk streams, scour and erosion of banks
Land drainage	Installation of types of drain, pattern of drainage	Knowledge of efficiency of alternative designs and patterns	Increase of runoff immediately downstream. Change in local water balance	Possible adjustment of channels in basin downstream of drained area
Rural road drainage	Types of road drainage, culverts	Basis for estimation of culvert dimensions	Roadside gully development	Metamorphosis of channels downstream from entry point of road drainage to stream system
Urban stormwater drainage	Installation of system of stormwater drains of specific size, pattern and with particular storage capacity	Method of estimating design discharge for urban drainage system. Knowledge of internal transmission including that from building down pipes to stormwater sewers	Erosion and deposition at discharge points. Nature of building operations. Aesthetic value of system	Downstream effects of increased or decreased peak discharge of water and of sediment

TABLE 6.2—*continued*

Subject of attention	Application	Possible implications requiring attention by hydrogeomorphologist		
	Engineering	Design aspects	Local impact	Regional impact
Channel reach				
River gravel exploitation	Selection of extraction methods	Location of suitable sites. Extraction in relation to access	Effects on flood plain water table, sediment availability to main river channel	Downstream effects. Amenity and aesthetic value
River diversion	Institution of diversion channel	Basis for design of size or diversion channel	Adjustment of channels experiencing increased and decreased flows	Effects downstream from diversion
Cutoffs	Building of spur channel	Size and character of channel	Effects within reach. Time for adjustments to occur	Downstream effects
Channelization	Widening, deepening, straightening, clearing, bank strengthening, dyking	Knowledge of water and sediment discharge in relation to alternative strategies	Effects of changes of flow frequency immediately downstream	Influence of hydrological regime changes in main stream channels
Construction on flood plain	Building in zoned belts	Basis for estimation of flood plain discharges. Significance of flood plain levels in relation to flows	Change of drainage sequence and rate	Downstream effects. Aesthetic degradation
Specific locations				
Dam construction and power generation	Building of dam in relation to selected design discharge	Flood frequency, flood routing analysis. Scour below dam	Aggradation upstream. Scour below dam aesthetic effects	Downstream adjustment of main channels
Bridge building	Construction with piers to allow passage of design flood	Flood frequency analysis: scour around bridge piers	Confined channel pattern, adjustment of local erosion	Cumulative effects downstream. Translation of channel pattern inhibited
Dredging for navigation	Depth, frequency, method of sediment removal	Rate of accretion indicating need for dredging	Loss of armoured layer, change of local sediment transport	Sediment supply and velocity effects transmitted downstream

Source: Gregory, 1979b.

has been stimulated by Schumm's ideas on river metamorphosis (Schumm, 1969), many of which are brought together in Schumm (1977).

6.3 Floodplains and Flooding

(a) *Introduction*

Discharge of water and sediment in rivers varies greatly in space and time. Discharge is normally confined below the banks of channels, but occasionally the channels are unable to contain the discharge, and water and sediment spill on to and move across the adjacent surfaces. Adjacent to perennial rivers these surfaces are usually (but not always) *alluvial floodplains*, which are created by the fluvial system specifically to accommodate the larger, less frequent flows. In certain areas where flow is ephemeral, floods often spread across the surfaces of *alluvial fans* (Plate 6.1). As floods commonly pose a hazard to people and their activities it is important to emphasize at the outset that they are a natural phenomenon and are only a problem where people have elected to use areas susceptible to flooding and where they have induced flooding that would otherwise not have occurred.

Yet flooding is both a serious hazard and one that is becoming more serious in some countries. In the USA, for example, average annual damage caused by floods up to 1970 was about $1.2 billion, about 10 per cent of the population is exposed to flood threat, and deaths occur at an average

annual rate of two per million at risk (Burton *et al.*, 1979; White, 1975a). And, despite the fact that over $6 billion has been spent on flood control since the federal Flood Control Act of 1936, annual property damage due to flooding is actually increasing (Fig. 6.14). In the Indian subcontinent, crop losses from flooding may run at more than $300 million per annum (Hewitt, 1983). Without doubt , on a world perspective, flooding is not only a major geomorphologically-related environmental hazard, it is also a major resource, especially in the developing world, where it may contribute significantly, for example, to soil fertility and the maintenance of productive wetlands (Drijver and Marchand, 1985). Indeed, there is strong evidence of a contemporary resurgence of interest in wetlands—many of which are sustained by flooding—because of their importance as regulators of flooding and water regimes in general, as distinctive habitats (for migrant birds, for example), as resources of characteristic, useful, and scientifically interesting fauna and flora, and as sites of recreational value (e.g. Horwitz, 1978). Conversion, for agricultural and urban development, threatens the wetlands and is a major cause of environmental concern, for example in North Africa and Canada (Fig. 6.15).

Large areas which are prone to flooding include deltas such as those of the Nile, Ganges, Mississippi, Huang He (Yellow River), and Rhine rivers. The fact that they are so to varying extents is now largely the result of engineering works. The High

PLATE 6.1 Flash floods in deserts: often unexpected and extremely hazardous, as here in the Wahiba Sands area, Oman

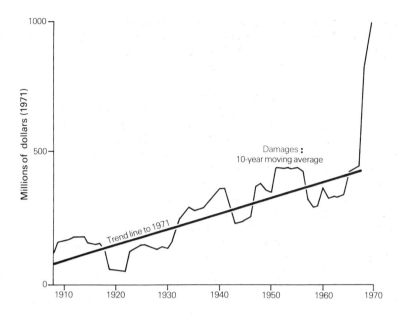

Fig. 6.14 Flood losses in the USA—
10-year moving averages (damage
data adjusted to 1971 values using the
wholesale price index) (after Smith
and Tobin, 1979; © 1979 and
reproduced by permission of the
Longman Group Ltd.)

Aswan Dam, completed in 1970, traps the flow of the Nile and prevents it from flooding the delta area each year, as was previously the case. Both the Mississippi and the Rhine deltas are protected, not by a storage dam, as in the case of the Nile, but by engineering works along the river channels. Artificial levées, or dykes, have been built and sluicegates inserted to control river levels. In the case of the Rhine engineering work has now, through a series of projects along the North Sea coast, almost completely prevented tidal effects from reaching into the delta. Neither the Ganges nor the Huang

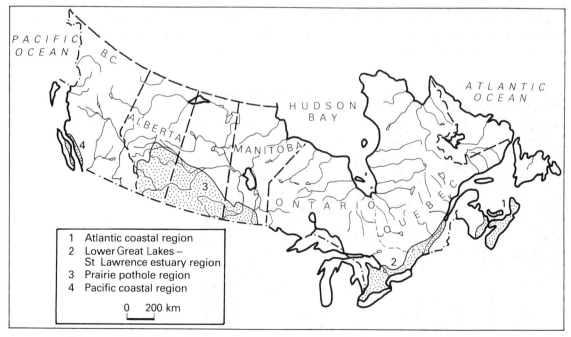

Fig. 6.15 Regions of significant wetland conversion in Canada (after Environment Canada, 1986)

He are as well controlled and flooding on these deltas can be catastrophic, especially if high flood waters are coming down-river as coastal storm surges occur.

Historically, flood protection work has not always been successful. It may introduce unwanted side-effects, as is happening in the Nile (Din, 1977) or it may prove to be unable to withstand certain conditions. In 1953 Holland experienced its worst floods of the recent past as a result of coastal storms, while in 1861 it suffered bad floods as a result of high flood waters coming down the Rhine. Recent geomorphological and stratigraphic mapping by Berendsen (1986) has provided an interesting record of dyke bursts which led to flooding in the Bommelerwaard. Dykes have been constructed in this area since the thirteenth century in an attempt to restrict flooding by the rivers Waal and Maas. To a large extent they succeeded but dyke bursts (especially that of 1861) have taken place, and even now new work is going on to both broaden and raise the height of some of the dykes which appear to be vulnerable to future bursts. Detailed mapping has shown that dyke bursts are associated with the position of sand lenses that passed under the original dyke. High pore-water pressures created during high river flows cause the water to be transmitted under the dyke at these points, and the sand is forced out creating holes under the dyke and allowing the flood waters to burst through.

Geomorphologically it is interesting to note that many of these sand deposits coincide with the position of former river channels, and some of the bursts coincide with these old channel positions and away from immediate contact with the main channels of the Waal and Maas. The palaeo-drainage provides one of the clues to predicting where bursts are likely to occur. The response to this threat of flooding has been not only to strengthen the dykes but also to raise the ground on which some of the settlements within the polder areas occur.

The lessons from this case-study are clear. Controlling flooding is possible, but the geomorphological system has to be adequately understood before effective control can be achieved. The Dutch have a reputation for understanding and controlling the Rhine–Maas–Waal system. Some of them would argue that it has come about more by historical experience involving trial and error rather than by knowing from the outset how the system operates. The work of Berendsen (1986) has shown that through mapping based on sedimentological and stratigraphic observations (by means of 10 000 boreholes in this case) a geomorphological statement can be made that explains why dyke bursts occur, and hence helps to define the protective measures that need to be taken in those places where palaeo-river channels cross the position of a dyke.

Flood-management analysis, prediction, control, crisis management, and rehabilitation, are normally in the hands of national, regional, or local river- or water-administration and other authorities. One aspect of the problem, indeed possibly part of its explanation, is the fact that different aspects of management are often disseminated amongst a complex hierarchy of agencies. Fig. 6.14 illustrates this complexity within the County of Los Angeles. Fig. 6.16 shows the structure for Britain. A second aspect, often ignored by scientists concerned with flooding, are the often numerous and varied reasons why people choose to live in flood-hazard zones, and the ways in which they perceive the problems (e.g. Burton *et al.*, 1979). To understand and manage successfully the flood problem requires an understanding of the physical problem; but it also requires an understanding of the threatened communities.

The literature on flooding is extremely extensive, and includes several recent, major contributions, such as those by Mayer and Nash (1987), Dunne and Leopold (1978), Newson (1975), Penning-Rowsell and Chatterton (1977), Penning-Rowsell *et al.* (1986), Smith and Tobin (1979), and Ward (1978). Chow's *Handbook of Applied Hydrology* (1964) is still a basic reference work. The Natural Environment Research Council's *Flood Studies Report* (1975) and Water Authority land-drainage surveys are national sources for the UK. In North America, many specific flood events are systematically recorded, sometimes as a statutory responsibility. In this context, the publications of the US Army's Corps of Engineers, and the US Geological Survey are particularly valuable sources of data (e.g. Cooke, 1984). Flood studies are published in many journals of which the most important include the *Journal of Hydrology*, *Water Resources Research*, the publications of the International Association of Hydrological Sciences, the *Transactions of the American Geophysical Union*, the *Journal of the Hydraulics Division* (American Society of Civil Engineers),

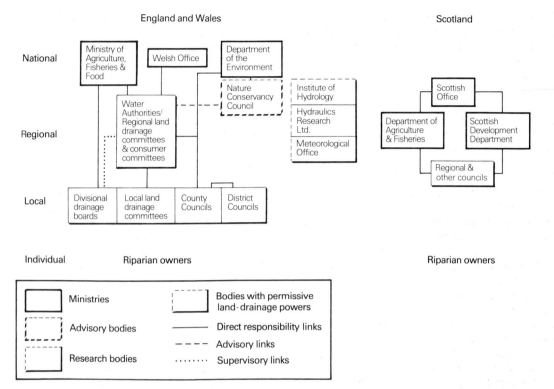

FIG. 6.16 A simplified organizational structure for flood alleviation and land drainage in Britain (after Penning-Rowsell, pers. comm.)

and publications of the engineering societies in most countries.

The purpose of the following review is to focus briefly on the geomorphological context of flooding and on some of the management responses to it.

(b) *Floodplains and fans*

A river floodplain results from the storage of sediment within and adjacent to the river channel. Two principal processes are involved. The first is the accumulation of sediment, often coarser sediment, within the shifting river channel. Sediment is commonly deposited, for example, on the slip-off slopes on the inside of meander bends to produce point-bars. As the river migrates in the direction of the outside of the bend, the point-bar grows and the floodplain deposit is augmented. Much of the sediment is only temporarily stored in a point-bar and it may be moved further downstream from time to time. This type of within-channel accumulation which can occur at any point along the valley, is mainly associated with below-bankfull

discharges. Secondly, suspended sediment carried by overbank discharges across the valley floor may settle and provide a further increment of flood-plain sediment, either generally over the flooded surface or, occasionally, locally along the channel margins to form levées. Where floodplain sediments comprise both coarse and fine material, most of the coarse fraction is the result of deposition by lateral accretion within the channels, and some of the fine material may result from overbank accretion; where the floodplain sediment is composed largely of fine material it is likely that most of it will be deposited within a channel.

The frequency with which bankfull discharge and flooding occurs is fairly similar for different rivers, even though they may be located in contrasted environments. The recurrence interval for bankfull flow is in the general range of 0.5–2 years, with 1.5 years a common value (e.g. Leopold *et al.*, 1964; Richards, 1982). The similarity of the recurrence interval in contrasted environments and the fact that channels normally do not become progressively deeper as floodplain deposition continues strongly imply that channels are adjusted to accommodate discharge generated within the watersheds for much of the time, that the floodplain is adjusted to transmit larger flows for the remainder of the time, and that the two features are functionally related to each other. Floodplains may be defined and mapped in various ways (e.g. Wolman, 1971). To the geomorphologist, the floodplain is an area characterized by a distinctive suite of forms and deposits; to the hydrologist it is the area inundated by flood events of particular magnitudes and frequencies; and to planners and lawyers it may be an area defined by statute. A particularly useful distinction on many floodplains, from the point of view of hazard problems, is that between the *floodway* and the *floodway fringe*. The floodway is that area of the floodplain, usually marginal to the main channel which is necessary to transmit a selected flood. Here, flood damage to structures is likely and inappropriate land-filling and flow concentration could increase flood levels. Thus attempts are commonly made severely to restrict developments and to permit only such activities as agriculture or recreation. The floodway fringe on the other hand may be suitable for certain developments, such as raising ground level or constructing flood-proofed buildings. The distinction is formally recognized by the US Army Corps Engineers and other agencies in the USA for flood-zoning purposes.

Table 6.3 lists some of the main techniques for mapping flood-prone areas in ascending order of precision, from relatively low-cost, qualitative appraisal of topographic features, soils, and vegetation, through the analysis of specific records of occasional floods, to systematic, quantitative, and generally more expensive studies of regional floods and specific flood profiles (Wolman, 1971). Well-monitored rivers allow discharges to be related to the statistical frequency of floods. Rang *et al.* (1987) show how these can then be related by mapping to areas likely to be inundated at each frequency.

Alluvial fans are fairly common landforms in arid and semi-arid areas and they are usually found where ephemeral flows from mountains spread out on to adjacent plains (e.g. Cooke and Warren, 1973). Individual fans vary enormously in their dimensions, but they are normally cone-shaped in plan, focusing on an apex near the mountain front. Fans frequently coalesce to form composite alluvial slopes. Deposition on fans arises from changes in the hydraulic geometry of flows as they leave the major feeder channels from the mountains—the flows increase in width, decrease in depth and velocity, and often lose water by infiltration into permeable alluvial deposits, thus causing deposition. Flooding on alluvial fans can occur in two principal locations: along the margins of the main supply channels, and in the depositional zones beyond the ends of the supply channels. Fig. 2.31, shows the predicted flood zones on an alluvial-fan sequence in the Suez area, based on the interpretation of geomorphological and sedimentary evidence. There are several problems associated with such flood zones that make the task of management difficult. Firstly, because flow on fans is ephemeral, and at most times there is no flow, the likelihood of flooding is often minimized or ignored by communities. Secondly, flow in alluvial-fan systems may vary from 'river' flows with low sediment concentration to highly viscous debris flows capable of moving large boulders. Thirdly, the channels followed by floods can change from time to time—for example when a channel becomes blocked with debris—and thus the loci of entrenchment and deposition and hence the hazard zones may vary in space and time. The nature of alluvial-fan flooding is therefore of great interest to farmers working agricultural land on fans, to those concerned with building and maintaining lines of communication across them, and to residents in communities built on them.

Geomorphological contributions to the flood-hazard problem commonly focus on the mapping of flood limits, and on defining the degree of hazard with flood zones, often in the context of preparing maps for insurance and planning purposes (e.g. Wolman, 1971). For example, many such maps are now being produced (e.g. Parker, 1981) in the USA (under the National Flood Insurance Program; US Dept. Housing and Urban Development, 1974), Canada, and many states in Australia; in England since 1973 flooding has been included within land-drainage surveys on maps at a scale of 1:25 000 (e.g. Wessex Water Authority, 1979). A second geomorphological contribution concerns predicting the nature of flooding based on an understanding of the causes of flows, and the conditions that intensify them.

A number of physical flood characteristics are critical to evaluating the flood problem (Table 6.4). Most of these can be graphically summarized

TABLE 6.3 Techniques of mapping areas subject to flood

Method	Principles	Principal methodological drawbacks	Approximate costs[a]
Physiographic	Topographic features: correlation of flood levels on the floodplain; return period, 1–2 years	Inadequate correlation Topographic form and flooding Omission of backwater effect	$1–4/mile of channel (estimated)
	Terraces: stepped topography		
Pedologic	Soil development Stratification	Distinguishing colluvial and alluvial soils;	$1–4/mile (estimated)
	Drainage	Terrace soil similarities Indistinct association soil and flooding	
Vegetation	Distinctive vegetation assemblages	Inadequate correlation	Unknown
	Vegetation form related to high water	Assemblages of species with flooding	
	Microvegetation related to high water	Soil moisture and flood effects undifferentiated Plant deformation not correlated with specific flood height	
Occasional flood	Aerial photos, remote sensing of floods	Records unavailable	$200/quadrangle ($4/mile converted)
Highest of record	Historic records, recorded flood profiles	Errors in spatial transposition	
Major	Regional stage frequency relations		
Recent major	Topography from stereoscopic air photos	Subtle topographic variation	—
Regional flood of selected frequency	Regional stage frequency relations	Errors in spatial transposition	$4/mile (estimated):$1.50–4/mile
	Regional physiographic relations and generalized hydraulic computations	Variability of hydraulic conditions Omission of backwater effect	
Flood profile and backwater curve	Definition of flood profile from high-water marks or detailed hydraulic computations	Detailed topographic information required	$400–1000/mile (includes topographic mapping)

[a] Does not include preparing the map, prices for 1971.

Source: Wolman, 1971.

TABLE 6.4 Critical physical characteristics of floods

---

  1. Depth of water
  2. Duration of inundation
  3. Area inundated
  4. Velocity of flow
  5. Frequency-recurrence relations
  6. Lag time (or flood-to-peak interval)
  7. Seasonality
  8. Peak flow
  9. Rate of discharge increase and decline
10. Sediment load
11. Total flood runoff volume

---

in two fundamental flood documents—the *flood frequency curve* and *flood hydrograph* (Fig. 6.17). The recurrence interval, *I*, the period of time in which a given event is *likely* to be equalled or exceeded once, is calculated using the simple formula:

$$I = \frac{n+1}{m} \qquad (6.38)$$

where n = number of years of discharge record, and

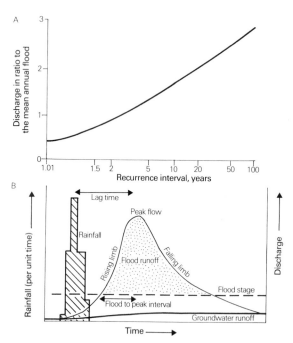

FIG. 6.17 (A) Typical regional flood-frequency curve, and (B) a typical flood hydrograph, showing several important flood characteristics

m = rank number of an individual item in an array. The rank values are flood magnitudes, such as annual peaks (measured in terms of discharge or depth of flow), or number of flood peaks above a chosen base. A major problem in calculating recurrence intervals in some areas is that appropriate historical records may not be available. Each recurrence interval is plotted against the appropriate measure of magnitude on semi-logarithmic (or probability) graph paper. From such graphs, statements relating to the *statistical probability* of flood events can be made. For example, the mean of a series based on annual peaks—the mean annual flood ($Q_{2.33}$)—usually has a recurrence interval of 2.33 years. Or it can be said that a flood of given magnitude is likely to occur, on average, once in every n years. This does not mean, of course, that such a flood can be forecast for a particular year. Put another way, a flood with a 10-year recurrence interval has a 10 per cent chance of recurring in any year. It should be noted that the technique described here is only one of several ways of arriving at flood-frequency curves (e.g. Chow, 1964).

In several studies (e.g. Dunne and Leopold, 1978) it has been noted that recent rainfall patterns do not conform to past behaviour. This makes the use of historical data an uncertain basis for predicting future recurrence intervals. It is conceivable, indeed it is likely, that short records will give different results from longer records. This approach must therefore be used with care.

Because discharge is a product of mean depth, width, and velocity of flow, and because its relations to characteristics such as area of inundation and sediment load can be precisely described, curves can be drawn relating these variables to their frequency of occurrence. Magnitude-frequency graphs are useful for many reasons, but perhaps the most important is that they provide a yardstick by which the hazard and the cost of preventing it can be assessed. The flood hydrograph describes the change of discharge during a flood for a single station. An example (Fig. 6.17b) shows several important flood characteristics, notably peak flow, total runoff, and the rate of discharge rise and fall (rising and falling limbs); when the storm rainfall is plotted on the same graph, lag time can be determined, for example, in terms of the time between 50 per cent of precipitation and 50 per cent of runoff volume; and when flood stage is defined, the flood-to-peak interval or 'time of rise' is also clear. As with the frequency-–magnitude plots, it is possible to construct hydrographs relating many specific flood characteristics to time-depth and area of inundation, for example.

Each of the important flood characteristics can be explained in terms of interrelated factors, of which three groups are fundamental: (i) transient phenomena, such as the nature of the rainstorm, and other changing features such as the evaporation rate and soil moisture conditions preceding the storm, (ii) permanent features, including especially basin characteristics (area, shape, etc.), the drainage network properties (such as density and length of streams), and the nature of the drainage channels (for example, slope, roughness, width, and depth), and (iii) the land use within the basin that may be either permanent or transient. Each of these groups requires brief examination.

(i) *Transient phenomena* There are several circumstances that may cause a flood (Fig. 6.18). The first is a sudden release of water, resulting perhaps from the thawing of a snow cover, or from blockage of flow by ice jams. (The breakage of

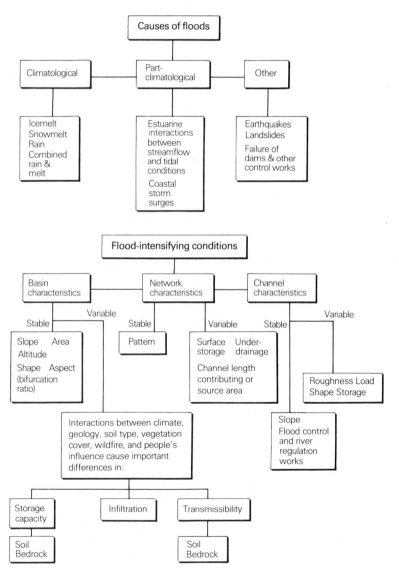

FIG. 6.18 Causes of floods and flood-intensifying conditions (after Ward, 1978)

dams, of course, has a similar effect.) The thawing process is commonly responsible for spring floods in middle and high latitudes: initially meltwater in a snowfield freezes as it percolates downwards; as melting continues, water begins to accumulate in the snow; eventually melt-water reaches the ground surface, and when most of the snow has become slush there follows a relatively rapid flushing of water into the drainage system.

The second, and the most common, cause of flooding is heavy precipitation, a phenomenon normally associated with depressions, hurricanes, thunderstorms, and other low-pressure systems, all of which are adequately described in standard textbooks (e.g. Barry and Chorley, 1982). An important point is that three variables which contribute towards defining heavy precipitation, namely depth of precipitation, duration of fall, and area covered, are all closely related. In general, storms of long duration tend to cover large areas, short-duration storms tend to have a greater proportion of precipitation concentrated in the first hours of the storm, and longer-duration storms have a more uniform temporal distribution of precipitation. Another important feature is the direction and rate of storm movement.

The mere occurrence of a heavy rainfall does not in itself ensure that a flood will follow. Much of the precipitation may be intercepted, or lost by evaporation, and much will depend on the soil moisture conditions before the fall and the infiltration capacity of the surface materials, which in turn will relate to geological and vegetation conditions, and to the time of year.

(ii) *Basin characteristics* Basin, drainage network, and channel characteristics are extremely important in determining the nature of the flood hydrograph, given a certain input of water (Fig. 6.18). Channel geometry is also important. As discharge is a function of width, depth, and velocity, and as velocity depends on hydraulic radius, channel slope, and bed roughness (see above), clearly the nature of these variables will influence the nature of discharge. For example, velocity of flow, all other things being equal, will be directly proportional to the square root of channel slope, and therefore channel slope will directly influence lag times.

Many other factors are also significant. For instance, peak discharge is related to the area of the drainage basin and is also reduced by basin storage (e.g. ponds, meander cutoffs, and the channels themselves). Perhaps it should be emphasized in this context that just as the basin characteristics help to explain differences of flood characteristics between drainage basins, they also influence flood characteristics at different stations in the same drainage basin.

(iii) *Land use* The importance of land-use characteristics can be illustrated by reference to the relations between urbanization and changes in the unit hydrograph (Leopold, 1968; Hollis, 1979). The unit hydrograph is the average time-distribution graph of discharge from a unit or standard storm, such as a storm which produced one centimetre of runoff. In Fig. 6.19 hypothetical unit hydrographs are shown for an area before and after urbanization. The building of the urban area normally has several consequences. Water runs off more quickly from the relatively impermeable surfaces of streets and roofs and flows through drains and sewers more efficiently than across 'natural' vegetated surfaces. This reduces lag time. As the time a given amount of water takes to run off shortens, so flood peak increases and the rising and falling limbs steepen. In general, the greater the area served by sewers and the greater the impervious area, the greater will be the ratio of peak discharge to discharge before urbanization, at least for smaller floods. This relationship is shown in Fig. 6.19B where discharge is defined as the mean annual flood (see above) for a drainage area of one square mile (2.59 km²), and the data are taken from various studies in the United States (Leopold, 1968). Similarly, the frequency of overbank discharge might be expected to increase as a result of urbanization. Fig. 6.19C shows the relationship between extent of urbanization and the ratio of number of floods to the number of floods preceding urbanization. This graph is also based on a drainage area of one square mile (2.59 km²), and employs data from various sources (Leopold, 1968). Thus for smaller floods, urbanization might be expected to increase the height and frequency of floods, to reduce lag time, and to increase rate of flood rise and fall; the effect is less clear for larger floods which may even be checked by urbanization.

Equally important, urbanization can alter the *location* of floods within drainage systems. For example, Fig. 6.20 shows the changing locations of major floods in Los Angeles County during the twentieth century, changes that reflect chiefly the spread of urbanization (with new settlements often

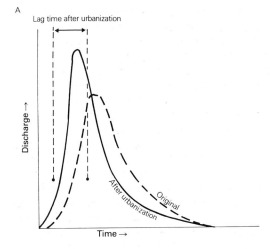

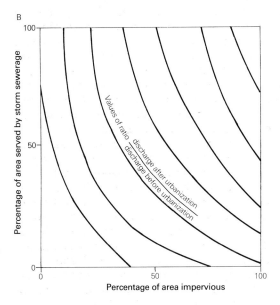

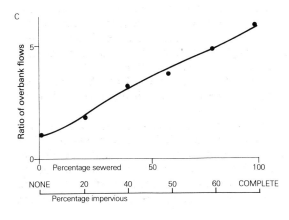

encountering unexpected flood hazards) and the progress of efforts to control the hazards in developed areas.

It is not just urbanization that influences flood hazard. Similar effects can follow from forest fires and devegetation, from changing agricultural land use, and from land-drainage schemes.

The previous discussion suggests that many flood characteristics should be precisely describable in terms of their numerous controlling variables. As the number of variables is so great, much effort has been directed towards identifying the most important variables that are relatively simply derived. Geomorphological variables figure prominently among the variables used. Amongst these, drainage area and shape, drainage density, and network features such as stream frequency and bifurcation ratios, channel slope, and storage area figure prominently. In ungauged catchments multiple regression analyses are commonly used to predict floods using a combination of climatic, land-use, and geomorphological variables, although such an approach is not without its problems (e.g. Richards, 1982). An example is the NERC (1975; Sutcliffe 1978) study of over 500 catchments in the UK, where it was found that:

$$\bar{Q} = 0.0201 \text{ AREA}^{0.94} \text{ STMFRQ}^{0.27} \text{ S1085}^{0.16}$$
$$\text{SOIL}^{1.23} \text{ RSMD}^{1.03} (1 + \text{LAKE})^{-0.85} \quad (6.39)$$

where

$\bar{Q}$ = mean annual flood,
AREA = drainage area (km²),
STMFRQ = stream frequency (junction per km²),
S1085 = stream slope (m/km),
SOIL = a soil index,
RSMD = 1 day rainfall of 5-year return period after subtraction of the effective mean soil moisture deficit (mm),
LAKE = a lake index fraction of catchment area drained through lakes and reservoirs.

FIG. 6.19 (A) Hypothetical unit hydrographs for an area before and after urbanization (after Leopold, 1968). (B) Effect of urbanization on mean annual flood for a one-square-mile drainage area (after Leopold, 1968). (C) Increase in number of flows per year equal to or exceeding channel capacity (for a one-square-mile drainage area), as ratio to number of overbank flows before urbanization (after Leopold, 1968)

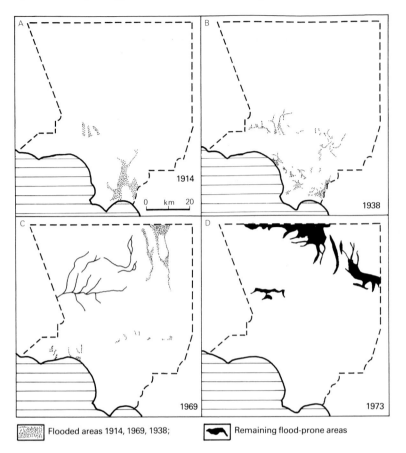

FIG. 6.20 Recorded areas of flooding in Los Angeles County (1914, 1938, and 1969) and the remaining flood-prone areas (1973) (after Cooke, 1984)

 Flooded areas 1914, 1969, 1938;   Remaining flood-prone areas

A simpler example was offered by Rodda (1967) for 26 small UK drainage basins:

$$Q_{ma} = 1.08 A^{(0.77)} \cdot R_{2.33}^{(2.92)} \cdot D^{(0.81)} \qquad (6.40)$$

where

A = drainage area (sq miles),
$R_{2.33}$ = mean annual daily maximum rainfall (in),
D = drainage density (miles/sq mile).

The value of the multiple correlation coefficient for this study was 0.90, and the factorial standard error of the estimate was 1.58.

Closely allied to the notions of flood description used in these studies and of flood probabilities discussed, is the shorter-term problem of flood forecasting (Chow, 1964). The importance of forecasting, of course, lies in the necessity of giving advance warning of a flood so that emergency action can be taken in potentially vulnerable areas. There are several prerequisites for an efficient forecasting system. Of great importance are adequate data on the history of rainfall and runoff in the basin. Much of this information is ideally described now in terms of real-time computer simulation models provided with telemetered data. But in many areas, graphs showing rainfall–runoff relations, unit hydrographs, routing methods (i.e. the ways of determining the timing and magnitude of the flood wave at successive points along a river), and stage–discharge relations are still used. Even more important is the creation of an information collection and dissemination network that can handle all the data relating to a particular flood event (Fig. 6.21). Satellite, telephone, telegraph, radio, or radar communications are often used, sometimes based on data provided by completely automatic recording stations. The early and poorest forecasts may be based on predictions of discharge from rainfall information; when a flood crest is initiated, better forecasts can then be related to routing methods,

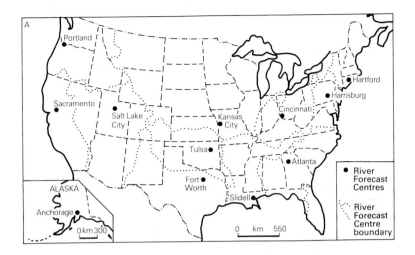

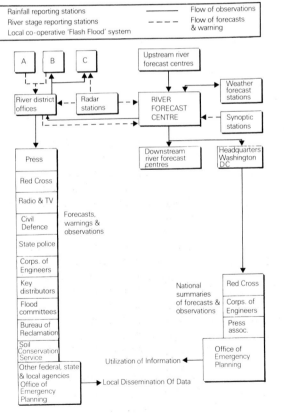

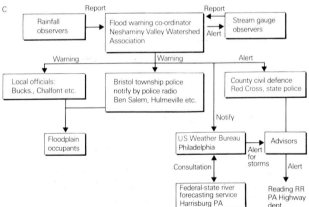

FIG. 6.21 United States river and flood forecasting service. (A) Location map. (B) Organizational flow chart. (C) Community-based flash-flood warning system—Neshaminy Valley Watershed Association, Pennsylvania organizational flow chart (after Ward, 1978)

using information on gauge heights, channel storage, and rates of travel.

(c) *Perception and responses*

How society responds to flooding depends, in the first instance, on how the hazard is perceived. The study of flood perception is fraught with difficulties and work on this subject suffers to some extent from the problem of deriving generalizations other than those that are immediately obvious. Nevertheless, it is of interest here to identify a few useful generalizations.

Firstly, perceptions will in part be related to some or all of the physical characteristics of the hazard itself. For example, Burton *et al.* (1979) showed that perception and consequent adjustment to flooding are directly related to flood frequency.

Secondly, perception of the hazard varies significantly between different cultural groups. For instance, the problem of adjustment to flooding is widely perceived as being important in the technically advanced societies of western Europe and the United States, but no such problem exists for technically primitive floodwater-farming groups in southern Arizona and northern Mexico. Here the difference is one of degree, for the Indians have successfully come to terms with flooding, whereas the technically advanced groups have only partially adjusted to the phenomenon. Equally, some groups may aspire to controlling floods, whereas others may see them as Acts of God to be endured.

Thirdly, within any particular cultural milieu there are often pertinent differences within and between groups, such as scientific personnel, resource users, and the general public. Scientists often differ in their perception of the flood hazard. In part such differences relate to inadequate data and the complexities of the problem, as the controversy over calculating flood insurance rates reveals (e.g. Kunreuther and Sheaffer, 1970). Similar variations occur between resource users. Differences might arise, for example, from variations in the nature and degree of personal experience and in the variety of ways that flooding affects different users of flood-hazard zones. For example, to one factory manager a few centimetres of flooding for a short time on rare occasions may present no problem; to another manager, faced with frequent, deeper flooding, there may be a serious problem. But if the second manager is new to his job he may fail to appreciate the problem,

especially if he has transferred from an area where there was no flooding hazard. Or again, the second manager might have serious labour-relations and production difficulties that distract his attention from the problem of flooding until the event occurs. And in any case, the managers' perception of flooding is likely to be quite different from that of neighbouring farmers or shopkeepers who are affected in different ways, and perceive the hazard in different economic and social contexts. There are also frequently conflicts between the attitudes of scientists, resource users, and the general public.

Individuals or groups can adjust to flood hazards in a number of ways. Research at Chicago by White and his students has led to the identification of a suite of possible responses (White, 1975*a*; Burton *et al.*, 1979). These are: bearing the loss; public relief; flood abatement and control; land elevation; emergency evacuation and rescheduling; structural adjustment; land-use change and regulation; flood insurance. Although this range of adjustments may theoretically be available to an individual or group, it by no means follows that any particular individual is free to select the most appropriate adjustments, for much will depend on the perception of the problem, and certain adjustments may be precluded for various technical, social, or economic reasons. The characteristics of the flood, the success with which it has been forecast, and the degree to which floods generally have been predicted are all important considerations here. Thus in examining flood responses, it is useful to bear in mind Table 6.4. The major responses to flooding are listed in Fig. 6.22.

(i) *Bearing the loss* Bearing the loss is more often an involuntary than a voluntary act, especially in areas where flood forecasting and drainage-basin management are poorly developed, or where such facilities fail to provide sufficient warning or prevent flooding. The ways in which the loss can be carried vary greatly, of course, according to the individual or group situation. A good example of a guide to accepting loss from flooding is the US Department of Agriculture's *First Aid for Flooded Homes and Farms* (1970).

(ii) *Public relief* In forms ranging from gifts from friends to government and international aid, public relief is a common response to a disaster. Such community help may ease immediate distress, but it may also encourage the persistence of inappropriate occupation of flood-hazard zones.

(iii) *Emergency action* The effectiveness of emer-

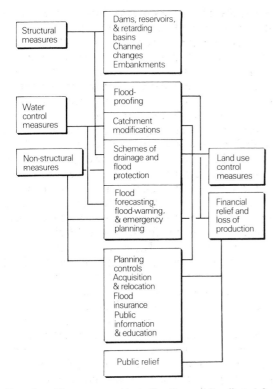

FIG. 6.22 Measures to mitigate flooding and its effects (after Victoria Water Resources Council, 1978)

gency action, which normally includes the removal of people and property, various flood-fighting techniques (e.g. sandbag defences, Plate 6.2), and rescheduling of activities, depends on advance warning of a flood and certain flood characteristics. Advance warning is related mainly to the availability of flood-forecasting techniques (e.g. Fig. 6.21) and to lag time (or flood-to-peak interval, is less than a day, only minimum emergency action is possible; if the interval is one to three days, responses will be governed by the effectiveness of pre-flooding planning by individuals and groups; if the interval is over three days, there should be sufficient time for most emergency adjustments. In addition, emergency action is generally most effective where flood duration is short, where velocity is low, and where frequency of flooding is high.

(iv) *Flood-proofing* The main forms of structural adjustments—sometimes referred to as flood-proofing—include control of seepage (for example, by sealing walls); sewer adjustment (by the use of valves, for instance); permanent closure of unnecessary openings, like windows; protection of openings (such as flood-control gates at entrances to underground stations); protective coverings to buildings, machinery, etc., to prevent such problems as rusting, and elevation of structures above flood level (e.g. Smith and Penning-Rowsell, 1982). These measures, and other similar ones, are designed to reduce damage to structures and goods within hazard zones. Some are permanent, some are contingent upon action being taken on receipt of a flood warning, and others are used only in emergencies. Some require structural changes. Many are restricted by hydrological limitations; neither depth of water nor hydrostatic pressure should be too high; velocity over two feet per second might begin to cause damage; duration should be relatively short; and generally flood-proofing is more applicable to pondage rather than floodway zones. The measures are therefore particularly beneficial in areas where flood-control measures are absent, where depth, velocity, and duration are short, and where forecasts are available.

(v) *Land use—planning controls* The regulation of land use in flood-hazard zones and the maintenance of adequate floodways are amongst the most successful and cheapest ways of limiting the damage produced by floods (e.g. Penning-Rowsell, 1981b). The strategies available include channel-encroachment statutes, flood-plain zoning ordinances, subdivision regulations, building codes, and various other devices such as permanent evacuation and the maintenance of wetlands as flood-storage areas. Critical to the successful adoption of such measures is the precise analysis of the physical geography of the floodplain and the appraisal of development potential in terms of such analysis. Inevitably there are wide differences of interpretation that are reflected, for instance, in methods of determining channel-encroachment lines (i.e. limits for development). Amongst several problems related to land-use regulation there are difficulties of estimating the damage and use potential of land, the danger of protection works giving a false sense of security and encouraging risky development, and confusion of terminology.

(vi) *Flood insurance* Flood insurance has been seen for a long time as a useful alternative in floodplain management, but its lack of availability and a heritage of bankrupt companies testify to the difficulties of creating realistic policies. Among the more intractable problems are lack of essential

PLATE 6.2 Sandbags form an embankment to a road that acts as a temporary flood channel in the Verdugo Hills, Los Angeles County, California

data such as those on frequency and magnitude, the formidable difficulties arising from creating policies that will attract floodplain subscribers and yet cover potential claims for a hazard that strikes erratically, and the possible encouragement to floodplain development by groups able to afford the premiums (e.g. Hoyt and Langbein, 1955; Kunreuther and Sheaffer, 1970). In the United States these difficulties have been met by a series of methods for determining flood insurance premiums. One was based on the applicant determining the amount of insurance cover he required in the light of information on the probability of floods reaching given heights. Another sought to allocate a single rate to a specific type of property within a major river basin regardless of location. More recently the National Flood Insurance Programme (NFIP) is based on the calculation of federally subsidized rates, in return for the agreement of communities to adopt and enforce floodplain regulations and flood-control measures (e.g. Cooke, 1984). Flood insurance is available and

adopted in the UK although its effectiveness is difficult to establish (e.g. Smith and Tobin, 1979).

(vii) *Abatement and control* Perhaps the most effective ways in which society intervenes to modify flooding in fluvial systems are through the alteration of land use within watersheds ('flood abatement') and through protective measures along flood channels ('flood control'). These two approaches are not mutually exclusive, although in some countries they have tended to be the responsibility of different authorities. In the United States different approaches and administrative rivalry have been reflected in the 'upstream–downstream' flood-control controversy (Leopold and Maddock, 1954). The principal protagonists were the Department of Agriculture and the Corps of Engineers, and a major issue was the relative merits of small 'headwater' dams (often associated with land-use management) and large, 'mainstream' dams—both types being designed to store floodwaters and reduce flood peaks. The large dams, some argued, flooded extensive areas of productive land and displaced people and settle-ments. On the other hand, it has been argued that numerous headwater dams—although they use less valuable land, affect no single community greatly, and are likely to cause relatively little damage if they fail—do not control larger, rarer floods downstream, and the cost per unit of storage increases rapidly as reservoir size declines. There are many other issues in the controversy, ranging from flood-routeing and sedimentation problems to the recreational and other uses of lakes.

Land-use practices, such as those discussed in the context of soil conservation in Chapter 4, may or may not be specifically designed to modify floods, but they often serve to reduce and delay surface runoff to river channels. Computations by the US Department of Agriculture show that, in general, land-treatment measures reduce peak flow from small storms giving a few centimetres of runoff, but have little effect on large floods (Leopold and Maddock, 1954).

Flood control by *channel change* has come to be focused on a relatively small range of engineering devices: structures to confine floods to the flood-way, by embankments, etc; modifications to natural channels to improve their efficiency and/or capacity by, for example, straightening, steepen-

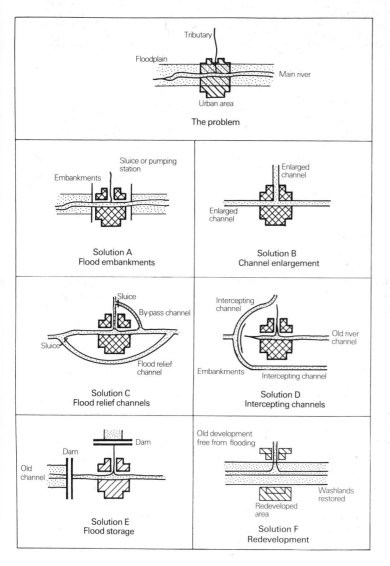

FIG. 6.23 The main methods of flood control in the United Kingdom (after Nixon, 1963)

ing, widening, deepening, or 'smoothing'; creating diversionary routes for use in times of flood; and construction of reservoirs to store floodwater and regulate the passage of water downstream. Many of these techniques are illustrated in Fig. 6.23 which shows various solutions commonly adopted in the United Kingdom to the problem of flooding in urban areas (Nixon, 1963). Solution A involves protective embankments, with sluice gates or a pumping station at the confluence of mainstream and tributary. The main problem with this solution is that it may increase flood levels upstream and downstream. An example is the Nottingham Flood Protection Scheme on the River Trent in England. Solution B simply requires enlargement of channels to accommodate larger discharges. One problem with such schemes is that as the enlarged channel is only rarely fully used by flowing water, weed growth may increase and riparian land may become overdrained. The River Don Improvement Scheme in Yorkshire adopted this solution in overcoming problems of land subsidence.

In Solution C a flood-relief channel removes excessive flows. This solution is often appropriate where it is difficult to modify the original channel, and it does not have the disadvantages of Solution B, but it tends to be rather expensive. An example is provided by the River Welland Improvement

Scheme in which the town of Spalding is bypassed by the Coronation Channel. Intercepting or cutoff channels are used in Solution D. These channels normally divert part of the flow away from the area subject to flooding, leaving some flow for riparian users in the protected area. The Great Ouse Protection Scheme in the Fenland is an example. In Solution E, flood-storage reservoirs are used. The arguments relating to reservoirs are admirably explored by Leopold and Maddock (1954); suffice it to say here that this solution is widely used especially as many reservoirs created for water-supply purposes may have a secondary flood-control role. An example of a multi-purpose reservoir project in Britain is the Bala Lake Scheme. The last solution, F, the removal of settlements threatened by flooding, is rarely used. Naturally, in many circumstances a combination of flood-control techniques is used to overcome a flooding hazard.

(viii) *A note on economic considerations* Almost all economic aspects of flooding are concerned in some way with determining flood losses and the costs and benefits associated with abatement and control efforts. Various methods have been used to put a value on these gains and losses, and most have been seriously criticized. Cost-benefit analysis, the most widely used technique, is an example.

A US Corps of Engineers' procedure for calcu-

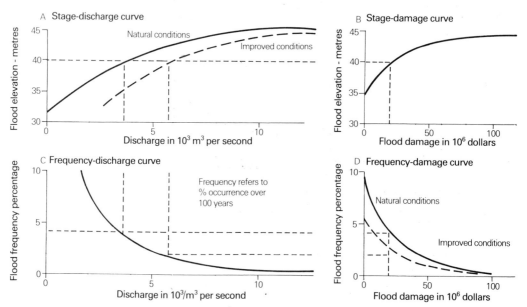

FIG. 6.24 Stages in the calculation of flood-damage frequency curves based on US Corps of Engineers procedure (after Kates, 1965)

lating damage-frequency curves is shown in Fig. 6.24. The flood characteristics used are flood depth (elevation), discharge, and frequency—other features such as sediment load are not directly involved. Calculations are made for both natural (initial) and improved (projected) conditions, with an assessment of frequency–discharge relations (c) first being combined with the description of stage–discharge relations (a) and then applied to the stage–damage curves (b), to give an average annual estimate of flood damage (e.g. Kates, 1965). (Specific examples of stage–damage curves for some establishments in the UK are shown in Fig. 6.25.) The average annual estimate can be

derived by measuring the area under the frequency–damage curve and dividing it by the number of years of record, or by multiplying the probability of each stage by the corresponding damage and summing the expected values. In this way, flood-control benefits are expressed as the difference between the 'natural' and 'improved' estimates.

There are, of course, numerous problems in making these calculations. Firstly, not all flood characteristics are considered. But it is certainly possible to take additional variables into account in calculating damages. Fig. 6.25B, for example, illustrates generalized relations of depth, duration, and velocity to urban flood damage for a represen-

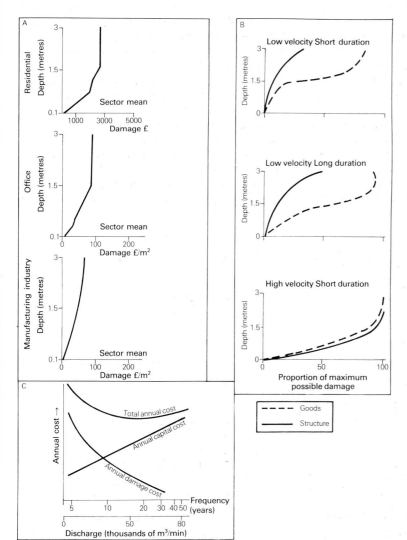

FIG. 6.25 (A) Relationship between flood depth and flood damage for different urban land uses in the UK for 1977 prices (after Penning-Rowsell and Chatterton, 1977). (B) Generalized relations between depth, duration, and velocity and urban flood damage for a representative commercial establishment suffering losses of goods and damage to structures (after White, 1964). (C) Hypothetical relations between discharge and frequency, annual damage costs and capital investment in a flood-protection project. The most economical structure is where the total annual cost curve is lowest

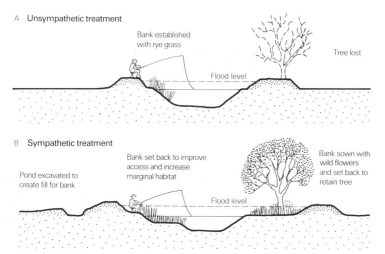

A   Unsympathetic treatment

Bank established
with rye grass

Tree lost

Flood level

B   Sympathetic treatment

Pond excavated to
create fill for bank

Bank set back to improve
access and increase
marginal habitat

Bank sown with
wild flowers
and set back to
retain tree

Flood level

FIG. 6.26 Sympathetic and
unsympathetic flood-bank engineering
(NCC, 1983)

tative commercial establishment. Secondly, while obvious features of structural damage may be easily costed, there is a host of incidental and intangible effects that may be difficult to identify and to cost. For example, what is the value of unanticipated leisure time, time that is possibly partly offset by sacrificing leisure at a later date; or how does one cost stress, worry, and damage to health (e.g. Green and Penning-Rowsell, 1985)? Further difficulties arise in projecting estimates into the future, when general economic conditions (interest rates, inflation, economic growth, etc.) and the period over which capital investment in flood-protection work is written off have to be considered.

Ecology and conservation are increasingly important considerations in river engineering. There is a growing and perceived need for river engineering and continuing management to be integrated with the sensitive conservation and enhancement of the environment, with special attention being given to habitat value and variety,

and to sites of special scientific interest (e.g. Nature Conservancy Council, 1983). Often the differences between sympathetic and unsympathetic flood control are quite subtle and inexpensive—for example, meanders do not *have* to be filled, pools and ripples *can* be encouraged, berms *can* be constructed to allow vegetative colonization, and flood banks *can* be shaped and positioned to encourage vegetation and recreation (Fig. 6.26).

In deciding the design of flood-control works, economics (as well as ecological considerations) have to be examined in the context of the hydrological situation. Usually it will be economically impossible to provide protection against the maximum probable flood, but if some smaller 'design flood' is selected, risks of damage will be greater. In short, damage costs must be offset against capital (construction and protection) costs, calculated on an annual basis, and the most economical structure is that where the total annual costs are least (Fig. 6.25C).

7 Drainage Basins and Sediment Transfer

7.1 Introduction

Water falling on the earth's surface tends to be organized within drainage basins. Such basins form the normal context for hillslope processes, including those of soil erosion (Chapter 4) and slope failure (Chapter 5) as well as forming the context of river channel behaviour including flooding (Chapter 6). Thus drainage basins form the natural unit within which most fluvial geomorphological processes operate.

In 1964, Strahler defined the drainage basin as the basic geomorphological unit on the grounds that it provided: (i) a limited, convenient, and usually clearly defined and unambiguous topographic unit, amenable to study at a variety of scales; and (ii) a physical process–response system, receiving inputs as thermal energy (from the sun), kinetic and potential energy (from precipitation), potential energy (form tectonic activity), and chemical energy (as a result of weathering processes). Outputs from the system include water, sediment, and dissolved material. Transport routes for the outputs are provided by the river channels, but their sources include the drainage-basin slopes, thus creating a complex but integrated system.

Drainage basins normally provide an outlet to the sea for both water and sediments. However, in some situations, mainly in arid lands or in tectonically active basins, rivers drain into inland depressions where lakes form and from which water is only lost by evaporation or human extraction. The largest of these internally draining basins include those in rift valleys (e.g. drainage to L. Baykal (and the Dead Sea)), and to inland depressions in deserts (e.g. Qattara Depression).

There is very little human activity, as expressed through planning policies, engineering projects, and land-use management, which does not have an impact upon drainage basins. There is clearly a direct connection in the case of reservoir construction or irrigation projects. There is also a direct link between the spread of fertilizers on drainage-basin slopes and the rising levels of nitrates and phosphates in rivers. Perhaps what is most often missing is an adequate level of awareness amongst environmental managers of how human activity affects the hydrology and water-quality aspects of a basin. Certainly drainage basin management is often inadequate or badly supported by legislation, except most notably in the United States.

Geomorphological accounts of drainage basins and river network systems are well described in *Water in Environmental Planning* (Dunne and Leopold, 1978); *The Fluvial System* (Schumm, 1977); *Rivers* (Richards, 1982); *Geomorphology* (Chorley *et al.*, 1984), and in *Earth's Changing Surface* (Selby, 1985). However, it is still worth turning to Strahler (1964) whose summary of drainage-basin morphometric studies has never been surpassed, and to the text by Gregory and Walling (1973), *Drainage Basin Form and Process*.

7.2 Drainage-Basin Systems

The critical feature about drainage basins is the extent to which they can be considered as integrated systems (Fig. 6.1). There is frequently an intimate link between cause and effect, or between different components of the system, and any human interference generally results in a response by the system. Well-known examples of drainage-basin responses include those where deforestation has increased surface runoff and soil erosion, or where the building of a dam has interrupted the downstream passage of sediment causing an increase in erosion downstream of the dam, or on a larger scale causing coastal erosion because the balance has been changed between sediment arriving down-river and its removal (e.g. along delta margins) by coastal processes. For example, the creation of dams across the River Nile is thought to have been the cause of erosion along the coast of the Nile Delta.

There are four principal aspects of drainage systems that are particularly important in applied geomorphology and which are considered below: stream network ordering and basin classifications; basin erosion–transport–deposition models; concepts of equilibrium; and basin responses to external influences. In the management context they also need to be examined in terms of their value in understanding sediment budgets and yield (Sect. 7.3).

(a) *Stream network classification*

Stream-ordering methods involve a numbering notation based on the branching network of river channels (Fig. 7.1), and the drainage basin is given the order number of the highest order present. Such ordering methods have been used as a basis for the *morphometric analysis* of both the network and the basin within which the network is contained.

Morphometric analysis has mainly been concerned with the statistical relationships that exist between network characteristics such as those listed in Table 7.1. Work of this kind includes that by Horton (1945) which provided the initial interest in such studies, and later by Melton (1958), Strahler (1964), Scheidegger (1965), Shreve (1966), and Gregory (1979c).

The early work demonstrated several general tendencies in the data analysed. These included the observation that stream order is proportional to the number of streams of a given order, the stream length, the stream slope, and the drainage-basin area.

Some of these relationships are not peculiarly geomorphological, since they apply to most natural branching systems. Greater applied value would certainly derive from any predictable relationships that can be found between these morphometric parameters (which tend to be easy to measure) and process parameters, such as river discharge or sediment load (which are more difficult and expensive to measure). A number of such relationships have been established that are generally applicable. River discharge (Q) and basin area (A) tend to be related by:

$$Q \propto A^k \tag{7.1}$$

(where k is an exponent that varies in value from place to place) but the relationship found in one place tends not to hold good elsewhere, reducing

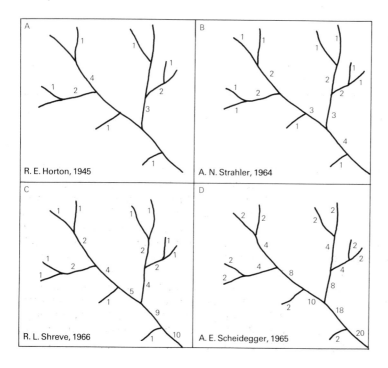

Fig. 7.1 Alternative stream-ordering systems (after Gregory and Walling, 1973)

TABLE 7.1 Morphometric properties of drainage systems and drainage basins

Variable	Symbol	Dimensions
Drainage network		
Stream order (used as subscript)	u	0
Number of streams of order u	N_u	0
Total number of streams within basin order u	$(\Sigma N)_u$	0
Bifurcation ratio	$R_b = N_u / N_{u+1}$	0
Total length of streams of order u	L_u	L
Mean length[a] of streams of order u	$\bar{L}_u = L_u / N_u$	L
Total stream length within basin of order u	$(\Sigma L)_u = L_1 + L_2 \ldots + L_u$	L
Stream length ratio	$R_l = L_u / L_{u-1}$	0
Channel volume	V	L^3
Basin geometry		
Area of basin	A_u	L^2
Length of basin	L_b	L
Width of basin	B_r	L
Basin perimeter	P	L
Basin circularity	$R_c = A_u /$ area of circle having same P	0
Basin elongation	$R_c =$ diameter of circle having same P/L_b	0
Measures of intensity of dissection		
Drainage density	$D_u = (\Sigma L)_u / A_u$	L^{-1}
Constant of channel maintenance	$C = 1/S_u$	L
Stream frequency	$F_u = N_u / A_u$	L^{-2}
Texture ratio	$T_u = N_u / P_u$	L^{-1}
Measures involving heights		
Stream channel slope	θ_c	0
Valley-side slope	θ_g	0
Maximum valley-side slope	θ_{max}	0
Height of basin mouth	z	L
Height of highest point on watershed	Z	L
Total basin relief	$H = Z - z$	L
Local relative relief of valley side	h	L
Relief ratio	$R_h = H / L_b$	0
Ruggedness number	$R_n = D \times H / 5280$	0
Network power	V/H	0

[a] The superscript bar indicates (here and throughout) a mean value.
Source: Doornkamp and King (1971). For further definitions see Chorley *et al.* (1984).

its predictive value. The model appears to improve if drainage density (a measure of channel network, D_d) is included:

$$Q_b \propto A \, D_d^2 \qquad (7.2)$$

where Q_b = bankfull discharge. The chief reasons for this is that measures such as basin area do not change as rainfall inputs change (within one basin), and discharge responds to such variations in rainfall. Thus while larger basins in general provide greater discharges, no empirical relationship based on these two parameters alone (i.e. independent of rainfall input) can possibly be expected to predict *particular* discharges. Wolman and Gerson (1978) were able to show, however, that different basins in different climatic belts provide different general relationships between basin area and discharge (mean annual runoff) (Fig. 7.2). But, basin morphometry does strongly influence several hydrographic characteristics, including the time to peak discharge and the concentration of flood waters within the basin (Chapter 6), and flood peaks increase significantly with drainage density (Fig. 7.3). The drainage network itself may be a sensitive indicator of drainage basin conditions. Drainage density is most commonly used to characterize the network, but because it only integrates length and area, the concept of *network power* (based on stream network *volume* and basin *relief*) has been introduced (Gregory, 1979c).

(b) *Basin erosion–transport–deposition models*

Drainage basins transport water and sediment from high ground to low ground. The classic sediment-transport model is epitomized by the concept of eroding headwaters feeding sediment to depositional lowlands and floodplains (Fig. 7.4). Although an extremely simple model of reality, Fig. 7.4 does provide an initial basis for placing a specific project within its dynamic context.

In general rivers seek to attain an equilibrium profile which tends to be concave upwards, but many fail to achieve this because of periodic interruptions by natural events or by engineering projects. In practice, the attainment of an equilibrium profile is on a time-scale far longer than is relevant to environmental management. In addition, attention in management is frequently

focused on small portions of a basin where more local variations in erosion–transport–sedimentation are important rather than on the basin as a whole, as exemplified in the discussion on soil erosion (Chapter 4).

(c) *Concepts of equilibrium*

Equilibrium is a difficult concept to identify in geomorphological systems, chiefly because the time-scale over which it can be recognized is longer than that over which measurements are made. In particular, short-term fluctuations may conceal long-term equilibrium (Chapter 3).

Equally important is the need to recognize those portions of a drainage basin, whether a hillside or a portion of the river channel, where there is a balance or an imbalance between the import and export of water and sediment. Those portions in balance are acting solely as units across which transport is taking place, without net loss or gain. Such sections are in *steady state* (Schumm and Lichty, 1965), and any interference with them by man must lead to an excess of either erosion or

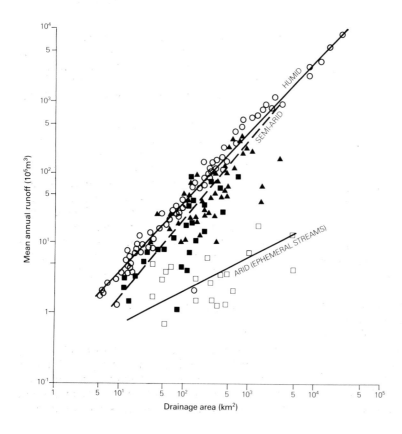

FIG. 7.2 The relationship between mean annual runoff and drainage-basin area for different climatic zones (after Wolman and Gerson, 1978)

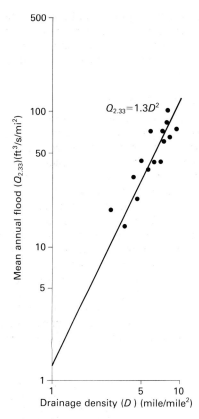

FIG. 7.3 The relations between drainage density and mean annual flood, for 15 locations in the eastern US (after Carlston, 1963)

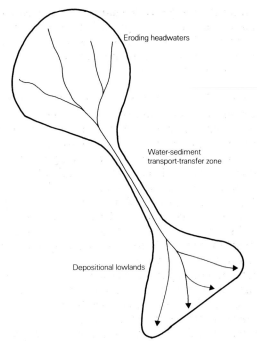

FIG. 7.4 A simple drainage-basin model

deposition, for the balance of forces will have been disturbed.

(d) *Basin responses to external influences*

Drainage basins respond to the external influences of climatic change and tectonics. Also 'external' to the drainage-basin system are the influences of human activity.

Climatic change has to be distinguished from seasonal and annual fluctuations in rainfall. Short-term fluctuations will affect discharges of water and sediment, but may not alter the gross morphology of either the drainage basin or the stream network which it contains. Short-term fluctuations can, however, have a distinct effect upon engineering and agricultural projects. The latter, for example, may be dependent upon the regularity of seasonal rainfall, or the adequacy of a storage reservoir to supply a sufficient amount of irrigation water. If there is a shortfall in this supply, because of inadequate rains, agriculture may fail. On the other hand, a significant increase in rain may lead to floods in excess of those for which culverts or bridges have been designed, and following the floods channels may even adopt new positions.

Longer-term climatic change, such as the sequence of climatic phases during and since the Pleistocene, which have left an imprint on many parts of the globe, has a quite different bearing on environmental management. Different climates have different styles of weathering, and these produce different soils. The results of these differences may still be present and have to be disentangled from present-day conditions if environmental management is to be effective. Tectonic influences are dealt with in Chapter 13.

7.3 Sediment Budgets and Yield

(a) *Introduction*

The movement of sediment, as suspended load, solution load, or bedload, through a drainage system is of fundamental importance in environmental management. In Chapter 4 the nature of soil erosion was examined on slopes in general, and on agricultural land in particular. But once the soil and other weathering products are removed from slopes, they are usually injected into the channel system, to be transported through it. The transport will probably be *intermittent*, so that at any one time some sediment will be *stored* in the system while only a proportion is removed. Fig. 7.5 shows a general model of sediment flow through a drainage basin. Within any one basin, the details of movement are in fact often extremely complex, as Fig. 7.6 suggests. As a result of such complexity, the precise routes of sediment through a basin are unknown and the sediment budget of the system is usually uncalibrated. Thus, in basins with sediment-related planning and engineering problems, estimates and predictions of sediment movement are commonly based on rather general, but nevertheless useful concepts and techniques. Three general concepts are particularly important (e.g. Chow, 1964; Vanoni, 1975; Jansson, 1982).

(i) *Gross erosion:* the total erosion within a basin arising from sheet, rill, gully, and channel erosion, expressed either as weight per unit time, or amount of ground-surface lowering per unit time;

(ii) *Sediment yield:* 'the total sediment outflow from a watershed or drainage basin, measurable at a cross-section of reference and in a specified period of time' (Vanoni, 1975, p. 438). It is usually expressed in weight per unit time per unit area (e.g. tonnes/km²/y) and is commonly based on measurement of suspended sediment load (*suspended sediment yield*).

(iii) *Sediment-delivery ratio:* the ratio between sediment yield and gross erosion in a drainage area, normally expressed as a percentage. By revealing the proportion of gross erosion being moved past a point, it also describes the proportion of *sediment storage* within the basin. These three terms and their relationships can be summarized as follows:

$$\text{Sediment-delivery ratio } (D) = \frac{Y}{T} \qquad (7.3)$$

where

Y = sediment yield,
T = gross erosion

and, therefore,

$$\text{Sediment yield } (Y) = TD. \qquad (7.4)$$

The importance of sediment movement to management can be illustrated by two quotations. The first relates to the Damodar River in India (Sundborg, 1983, p. 16) and emphasizes the close links between sedimentation processes in all parts of a river basin.

The Damodar river—a tributary of the Hooghly river in India—was developed by a multi-purpose project around 1950. Seven dams were built on the river and its tributaries, and a series of reservoirs were formed. Hydropower plants were built, and the irrigation systems in eastern Bihar and West Bengal were improved. But soil erosion is severe in the upper river basin, and huge amounts of sediment were transported downstream. In some regions arable land was sanded over and the new reservoirs were silted in at much faster rates than expected. The reservoirs' lives, it appeared, would be much shorter than predicted. The total water in-flow from the Damodar system to the lower Hooghly river decreased, which aggravated the sedimentation problems in the Hooghly part of the Ganga delta. The river reach between Calcutta and the Bay of Bengal was also choked with silt. The dredging operation to keep the bustling port of Calcutta navigable became an endless and costly undertaking. Although actions were taken to cope with the threatening development, conditions are still grave.

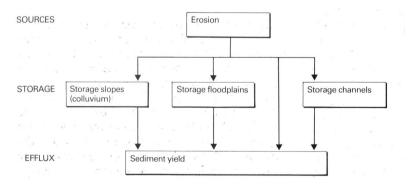

FIG. 7.5 General model of sediment movement in a drainage basin

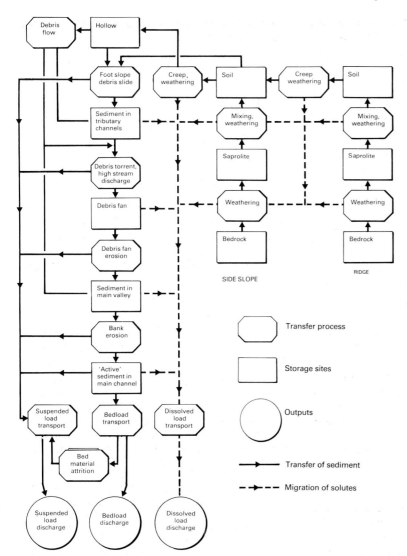

FIG. 7.6 Sediment-budget model for Rock Creek basin, central coastal Oregon, USA (after Dietrich *et al.*, 1982)

This example implies both ignorance and error, and unquestionably underscores the need for geomorphological appraisal. The second example leads to the same conclusions while also emphasizing the controversy generated by complex fluvial systems. It comes from Judge Kroninger's ruling over forest logging and related sediment problems in California (quoted in Wolman, 1977, p. 30):

While numerous expert witnesses in the field of geology, forestry, engineering and biology were presented, their conclusions and *the opinions they derived from them are hopelessly irreconcilable on such critical questions as how much and how far solid particles will be moved by any given flow of surface water. They were able to agree only that sediment will not be transported upstream.* [Our italics.]

Sediment transfer within river systems is important in several applied contexts. Firstly, sediment movement influences the character of the channel network and changes can alter the nature and loci of erosion and deposition, and channel geometry. Such changes may affect, in turn, channel navigability, flooding, property boundaries, and the stability of bridges, embankments, and other engineering structures (Chapter 6). Secondly, the turbidity of flows influences water quality and any increase of sediment concentration may damage fish and other biota in the system, engineering projects such as irrigation schemes and hydro-electric power plants, and the quality of water used for domestic and industrial purposes. Thirdly, reservoirs in a drainage system act as sediment stores, and the life of reservoirs is in part determined by the rate at which sediment accumulates (Plate 7.1). Thus in planning and designing dams and reservoirs, it is essential to be able to predict rates of sedimentation and to build them into calculations of cost and design life. Fourthly, it is often important to be able to predict changes in sediment movement and their consequent effects on reservoir sedimentation and pollution especially when changes of land use (for example, strip-mining, afforestation, urbanization) or engineering structures in channels are the likely cause. Finally, it has recently been recognized that sediment movement is fundamental to the transport of contaminants (such as pesticides, radio-nuclides, and heavy metals) and nutrients that are absorbed on to sediment particles.

Given such a wide range of economically significant problems associated with sediment movement, it is scarcely surprising that much effort has been directed towards understanding the geomorphology behind them. The relevant published literature is now very extensive. In recent years it has included several books, such as Vanoni's (1975) *Sedimentation Engineering*; Jansson's (1982) *Land Erosion by Water in Different Climates*; Laronne and Mosley's (1982) readings on *Erosion and Sediment Yield*, and Hadley and Walling's (1984) edited volume on *Erosion and Sediment Yield*. Several volumes of conference proceedings also address the sediment theme, notably those of the *Federal Inter-Agency Sedimentation Conference* (USDA 1963, 1976), the *International Association of Hydrological Sciences* (1982, 1986), and the US Department of Agriculture (Swanson *et al.*, 1982). In addition, research findings appear regularly in such journals as the *Journal of the Hydraulics Division of the American Society of Civil Engineers*, *Water Resources Research*, the *Journal of Soil and Water Conservation*, and *Earth Surface Processes and Landforms*.

PLATE 7.1 Debris basin, in the process of sediment clean-out, San Gabriel Mountains, Los Angeles, California

Some of the relevant literature is associated with that on soil erosion by water (Chapter 4).

(b) *Determinants of sediment movement rates*

Many of the fundamental controls on sediment movement in drainage systems are similar to those influencing soil erosion: the geometry of the slopes, channels, and basins; the nature of soils, bedrock, and vegetation; the magnitude, frequency, and duration of precipitation. Indeed, *sediment supply* is controlled largely by slope erosion processs and channel erosion. *Sediment transport* is, in turn, influenced by the nature of flow conditions, the characteristics of the channel network and the basin as a whole, and the characteristics of the sediment in motion (e.g. Bennett, 1974).

In the context of sediment yield and delivery ratios, three groups of variables are normally regarded as fundamental: the geometry of the basin, the climate, and the land use and vegetation. Multiple regression equations have been used extensively to describe and predict sediment movement in terms of these groups of variables. Jansson (1982) reviewed over 50 such equations, of which that by Jansen and Painter (1974) for temperate climates is fairly typical:

Log sediment yield = 12.133
$$\begin{aligned} &-0.340 \log Q + 1.590 \log H \\ &+ 3.704 \log P + 0.936 \log T \\ &- 3.495 \log C \end{aligned} \tag{7.5}$$

where

Q = annual discharge in 10^3 m³/km²,
H = altitude (m, asl),
P = annual rainfall, mm,
T = average annual temperature, °C,
C = natural vegetation index.

From the geomorphological perspective, many geometric properties of drainage basins are potentially valuable predictors. For example, there is much evidence to suggest that both sediment yield and sediment-delivery ratio *decrease* with size of drainage basin (Fig. 7.7). This is chiefly because the availability of storage for sediment increases with basin size (as there are, for example, more lower slopes and more extensive floodplains).

Lustig (1965) illustrated the potential of morphometric variables for predicting sediment yield. In a study designed to predict sediment yield for the purpose of dam design in one southern Californian mountain catchment without any sediment data, he developed a regression model for several basins with similar vegetation and climate in which known sediment yield was related to six morphometric indices. By calculating the indices for the catchment without sediment data he was able to predict the yield graphically on the regression graphs. The morphometric variables, and the rationale behind them were as follows:

(i) *Relief ratio* (H/L), which is a measure of overall watershed slope and therefore of energy input.
(ii) *Sediment area factor* (S_A), a measure of watershed surface area, defined as planimetric area divided by mean ground slope (average of 100 sample points).
(iii) *Sediment movement factor* (S_M), based on the belief that the steeper the slopes the greater the sediment movement, is defined as

$$S_M = S_A \times \overline{\sin \theta \text{g}} \tag{7.6}$$

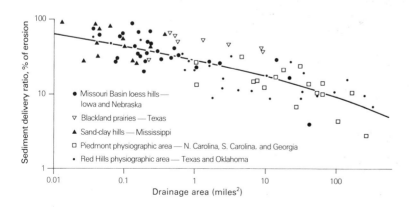

FIG. 7.7 Sediment delivery ratio as it relates to drainage area (after Boyce, 1975)

Sediment delivery ratio, % of erosion

- • Missouri Basin loess hills — Iowa and Nebraska
- ▽ Blackland prairies — Texas
- ▲ Sand-clay hills — Mississippi
- ▢ Piedmont physiographic area — N. Carolina, S. Carolina, and Georgia
- • Red Hills physiographic area — Texas and Oklahoma

Drainage area (miles²)

where $\overline{\sin \theta_g}$ = mean sine slope angle, which reflects the ratio of gravitational to shearing stress.

(iv–vi) *Transport efficiency factors* (*T*), which are functions of stream lengths, numbers, and gradients, and represent the ability of the channel system to remove sediment:

$$(iv) \quad T_1 = \bar{R}_b \times \Sigma L \qquad (7.7)$$

where $\bar{R}_b$ = bifurcation ratio and L = channel length.

$$(v) \quad T_2 = \Sigma N \times \bar{R}_c \qquad (7.8)$$

where N = channel number; $\bar{R}_c$ = mean stream channel slope ratio

and

$$(vi) \quad T_3 = (N_1 + N_2)(\bar{R}_{c1/2}) + (N_2 + N_3)(\bar{R}_{c2/3}) + \text{etc.} \qquad (7.9)$$

where R_c = mean stream slope ratio per order.

When applied separately, these indices predicted a yield *per unit-area* of between 114–979 m³/km²/y; a multiple regression equation predicted a yield of 764 m³/km²/y.

It is also clear that rates of sediment yield vary with climate (e.g. Fig. 7.8; Table 7.2). In an early study, Langbein and Schumm (1958) showed that for about 265 drainage basins in the United States, sediment yield is related to mean annual effective precipitation (the amount of precipitation required to produce a known amount of runoff under specified temperature conditions). The peak

of their curve occurs in semi-arid areas (i.e. *c*.300 mm precipitation and a mean annual temperature of 10 °C), and rates decline both towards areas of more and areas of less effective precipitation. They explained this variation by the operation of two variables. Firstly, the erosive influence of precipitation increases with its amount; secondly, and opposing this influence, is the protective effect of vegetation which also increases with precipitation. These factors can be summarized in an equation which, when solved empirically and converted into metric units (Douglas, 1967), states:

$$E = \frac{1.631 \, (0.03937P)^{2.3}}{1 + 0.0007 \, (0.03937P)^{3.3}} \qquad (7.10)$$

where E = suspended sediment yield in m³/km²/y, P = effective precipitation (mm), and where the numerator represents the erosive influence and the denominator, the vegetation-protection factor.

This model is not universally applicable. For example, Douglas used data from a selection of Asian rivers to demonstrate both a generally lower yield, and a second peak where mean annual runoff rises over 600 mm (Fig. 7.8). Wilson (1973), in a study of 1500 basins mainly in the United States based on data standardized for a drainage area of 259 km² also identified two peaks in sediment yield. The second (tropical monsoon) peak and a possible intermediate peak in Mediterranean climates share with some of the semi-arid

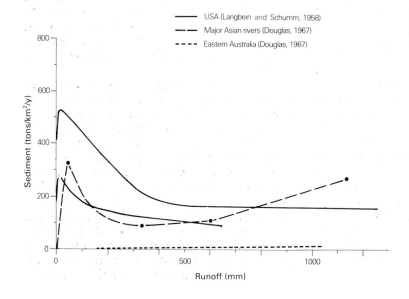

FIG. 7.8 Relationships between sediment yield and annual runoff (after sources shown)

— USA (Langbein and Schumm, 1958)
— — Major Asian rivers (Douglas, 1967)
- - - - Eastern Australia (Douglas, 1967)

TABLE 7.2 Catchment areas, water discharges, and sediment loads of some selected large rivers

River	Country	Catchment area (km²)	Mean water discharge (m³/s)	Annual sediment load (million tonnes/y)	Sediment load (tonnes/km²/y)
Rhine	Netherlands	160 000	2200	2.8	17
Po	Italy	54 300	1550	15	280
Wisla	Poland	193 900	950	1.4	7
Danube	Romania	816 000	6200	65	80
Don	USSR	378 000	830	4.2	11
Ob	USSR	2 430 000	12 200	15	6
Niger	Nigeria	1 081 000	4900	21	19
Congo	Zaire	4 014 000	39 600	72	18
Mississippi	United States	3 269 000	24 000	300	91
Amazon	Brazil	6 100 000	172 000	850	139
Paraná	Argentina	2 305 100	—	90	38
Indus	Pakistan	969 000	5500	435	450
Ganga	India/Bangladesh	955 000	11 800	1450	1500
Brahmaputra	India/Bangladesh	666 000	12 200	730	1100
Irrawaddy	Burma	430 000	13 500	300	700
Red	Socialist Republic of Viet Nam	120 000	3900	130	1100
Pearl	China	355 000	8000	70	260
Yangtze	China	1 807 000	29 200	480	280
Yellow	China	752 000	1370	1640	2480

Source: Sundborg (1983).

climates intense seasonal precipitation which effectively erodes the soil. This is made worse when the intervening periods of drought prevent the growth of a dense protective vegetation cover.

The curves such as those shown in Fig. 7.8 ignore the major influences of drainage-basin characteristics and land use/vegetation on sediment yield and are thus of only limited predictive value. Indeed Wilson (1973) suggested that land use is probably the most important variable controlling sediment yield in many areas, perhaps causing sediment load to rise many times beyond that which would have been produced under 'natural conditions'.

This view was confirmed by Dunne (1979) who showed the influence of different land uses upon sediment yield for a given runoff, within 61 Kenyan catchments (Fig. 7.9). It is likely that the data in Fig. 7.8 are affected by accelerated erosion associated with human activity (with the possible exception of some Asian rivers). Certainly *urbanization* can affect sediment yield. A common experience in many areas is for sediment production to rise quickly during construction, followed

by a decline after construction is completed to yield a lower value than existed prior to urbanization (e.g. Livesey, 1975). These sequences of change are so well established that urban development ought to require account to be taken of their effects. In particular sediment traps and the preparation of siltation reservoirs (or debris basins) should be a requirement in areas sensitive to soil erosion or high sediment concentrations in streams.

Other land-use changes, such as surface deforestation, overgrazing, and fire, have equally serious effects on sediment movement. The work of Trimble provides two examples. In a detailed study of the history of Coon Creek, Wisconsin, USA, Trimble and Lund (1982) analysed both the land-use trends since 1850 (Fig. 7.10) and related changes in sediment movement (Fig. 7.11). This study showed, *inter alia*, that whereas erosion in the period since 1938 (when conservation practices were widespread and farming was declining) was considerably lower than in the period from 1853 to 1938, sediment yield actually changed very little between the two periods, indicating that most

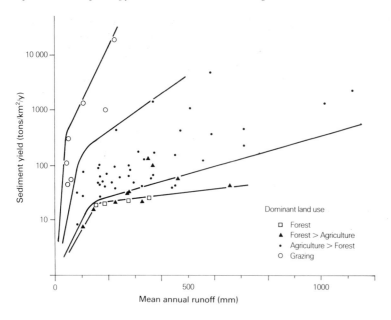

FIG. 7.9 Mean annual sediment yield and mean annual runoff for 61 catchments with different land uses in Kenya (after Dunne, 1979)

change was accommodated through storage within the system. Clearly, therefore, in this as in most examples, sediment yield is a poor predictor of gross erosion and should not be used for this purpose. In another study in south-eastern USA, Trimble (1977) showed that sediment-delivery ratio was only 6 per cent; as a result, he suggested, sediment yield may correlate poorly with factors controlling erosion, such as lithology, climate, vegetation, and land use, but may rather reflect such variables as channel efficiency, soil texture, basin morphology, and channel erosion.

(c) *Approaches to estimating and predicting sediment yield*

All approaches to estimating and predicting sediment yield are normally based on data collected from within drainage basins and on applying to them the principles of sedimentation. 'The accuracy of the predictions varies with the basic data

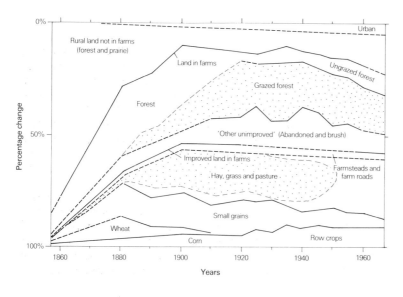

FIG. 7.10 Percentage change of land use in Coon Creek, Wisconsin, 1850–1975 (after Trimble, 1983). The basin covers an area of 360 km²

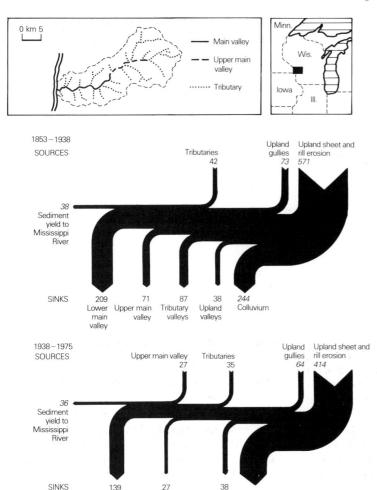

FIG. 7.11 Sediment budgets for Coon Creek Wisconsin, 1853–1938 and 1938–75. The numbers are annual averages for the period in 10³ Mg/y and account for 3.6 × 10⁷ Mg of sediment generated between 1853 and 1975. The upper main valley is a sink in the first period, and a partial source in the second (after Trimble, 1983; © and reprinted by permission of the American Association for the Advancement of Science)

available; knowledge of the processes of erosion, entrainment, transportation, and deposition of sediment; and the predictability of changes in the watershed that will alter sediment yield' (Holeman, 1975, p. 5). Most approaches are still firmly empirical, but mathematical modelling is becoming more important as understanding of drainage systems improves (e.g. Bennett, 1974).

Numerous methods have been adopted, and the choice is wide, but the method used in a particular situation is usually constrained by the nature of the data available (if any), the nature of the problem, and the time and funds available for solving the problem. Some problems require only a general assessment, some require detailed estimates, and others require both. For example, in

designing a reservoir, an initial general assessment is usually essential, short-term monitoring prior to construction is often desirable, and long-term monitoring following construction may be sensible (e.g. Fleming and Kadhimi, 1982).

Most methods rest ultimately on sediment data derived from erosion-plot studies (Chapter 4), suspended sediment sampling (Vanoni, 1975; Ward, 1984), and/or the survey of sediment accumulation in reservoirs (see Chapter 3). When such data are not available, various alternative strategies are possible. Some agencies responsible for estimating sediment yield adopt several methods. For example, in the United States, the US Department of Agriculture (the Bureau of Reclamation, the Soil Conservation Service, and

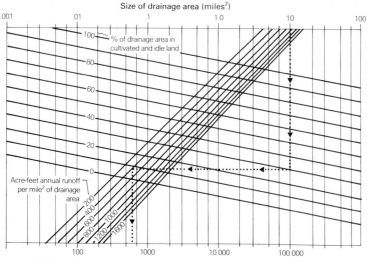

FIG. 7.12 Nomogram for computing sediment yield (for the Upper Mississippi Region). The example is for a 10 mile² basin—follow the arrow on drainage area, to percentage of cultivated land, to runoff and then to sediment yield (after Brune, 1951). Original units have been retained

the Forest Service) and the US Army Corps of Engineers use methods that include the analysis of reservoir sedimentation rates, suspended sediment load, gross erosion, sediment delivery, and attributes of the watershed such as slope, drainage area, vegetation, and runoff. Some of the techniques used by these and other agencies are reviewed below.

The empirical relations between sediment yield, drainage area, percentage cultivated land, and runoff were summarized many years ago for particular areas of the USA by Brune (1951). The use of Brune's nomogram (Fig. 7.12) for calculating the long-term rate of sediment production in the Upper Mississippi region is illustrated for a small basin by arrows on the diagram. Fournier's (1960) attempt to describe the global pattern of suspended sediment yield is also based on using relationships between sediment discharge, precipitation, and drainage-basin relief, and catchment area:

$$\log E = 2.65 \log (p^2/P) + 0.46 \log \bar{H} \cdot \tan \phi - 1.56 \quad (7.11)$$

where

E = suspended sediment yield (tons/km²/y),
p = rainfall in the month with greatest precipitation (mm),
P = mean annual precipitation (mm),
$\bar{H}$ = mean height of basin (m),
ϕ = mean slope in a basin,

and p^2/P is used as a measure of precipitation seasonality, or the incidence of rainfall concentration, to which sediment yield is often positively correlated.

The Fournier equation (7.11) requires the calculation of both the mean height and the mean slope of a drainage basin, it is tedious and time-consuming to use, and the results are at times unreliable. On a global scale it provides the general picture shown in Fig. 7.13. An alternative approach is first to classify drainage basins according to relief and climate into four categories:

Ia Low relief, temperate climate;
Ib Low relief, tropical, subtropical, semi-arid climate;
II High relief, humid climate;
III High relief, semi-arid climate.

then, only p^2/P is needed in order to predict sediment yield using empirically derived regression equations:

$$\text{Ia } Y = 6.14X - 49.78 \quad (7.12)$$
$$\text{Ib } Y = 27.12X - 475.4 \quad (7.13)$$
$$\text{II } Y = 52.49X - 513.21 \quad (7.14)$$
$$\text{III } Y = 91.78X - 737.62 \quad (7.15)$$

where Y = sediment yield (tons/km²/y) and $X = p^2/P$.

Fig. 7.13 demonstrates the contrast between low sediment yield in deserts and the high yields in

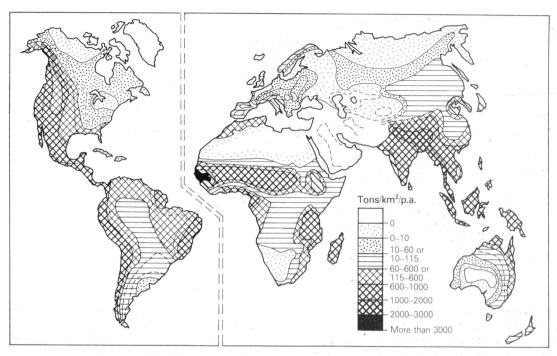

FIG. 7.13 World distribution of suspended sediment yield (after Fournier, 1960)

tropical areas with seasonally concentrated preci-pitation. On the global scale Fig. 7.13 provides a general context for more local studies. The Four-nier equation has also been used both at the regional and the local scale to predict specific values of sediment yield. This may be unwise since the regression equations generalize data with unknown variance and of unknown quality.

A second approach is to convert predictions of (gross) erosion into predictions of sediment yield through the use of an empirically derived sedi-ment-delivery ratio (e.g. McCaig, 1983). The method requires estimates of erosion based nor-mally on the resolution of the Universal Soil Loss Equation (Chapter 4) at standard field sites. It also requires prediction of the delivery ratio. This is possible because, for example, it is inversely related *inter alia* to the 0.2 power of basin area (e.g. Mitchell and Bubenzer, 1980). This method has been widely used in the USA but it may not be readily applicable in different climatic zones or areas with land use different from that at the test-plot sites.

This approach is one of several that attempt to predict sediment yield in areas *where no sediment data are available*. For example, in the Kotmale Reservoir catchment, Sri Lanka, Russell (1981) first applied Fournier's equation to 26 sub-catchments in the basin and from the results created a contour map of suspended sediment yield (Fig. 7.14A). This provided only a very approximate picture of reality because it ignores infiltration/runoff characteristics. Thus a second estimate was based on field measurements of infiltration capacity in each major soil zone (Fig. 7.14B) using the model defined by Kirkby (1974). In this model the daily rainfall totals likely to supply sufficient water to exceed critical infiltra-tion rates need to be estimated. From this, overland flow per annum (OF) is defined by:

$$OF = R \cdot e^{-(r_c - h)/r_0} \qquad (7.16)$$

where

R = annual rainfall (mm),
e = evapotranspiration (mm),
$(r_c - h)$ = soil water storage capacity,
r_0 = mean rain per rain day (mm).

By holding OF at 0 in the equation, a value of r_c can be calculated for each soil group and its water storage capacity estimated from the infiltration

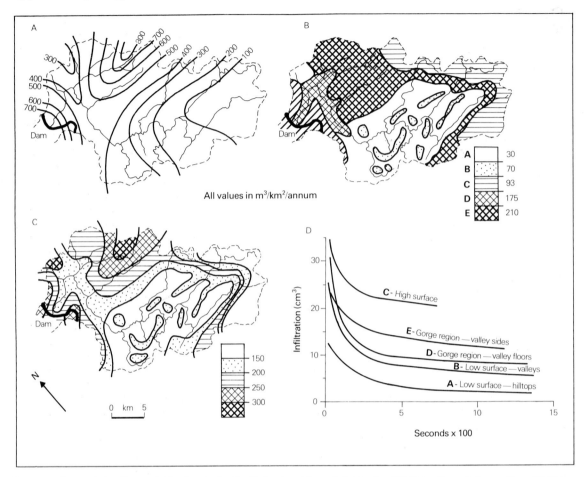

FIG. 7.14 Suspended sediment yield estimates for the Kotmale catchment, Sri Lanka. (A) Using Fournier's equation. (B) Using Kirkby's model. (C) Possible sediment yield based on both Fournier's and Kirkby's models. (D) Soil infiltration rates within the Kotmale catchment (after Russell, 1981)

data (but taking no account of antecedent moisture conditions). The values of overland flow are then put into the sediment yield equation:

$$SY = 170 \, (qOF)^2 \cdot \tan B \, 10^5 \qquad (7.17)$$

where

$$SY = \text{sediment yield (m}^3/\text{km}^2/\text{y)},$$
$$qOF = \text{overland flow discharge (m}^3/\text{km}^2/\text{y)},$$
$$\text{(derived from Equation 7.16)},$$
$$B = \text{mean slope angle.}$$

The results of the analyses for the five soil zones are shown in Fig. 7.14B. The next stage in the analysis was to combine the two independent estimates of sediment yield (Fig. 7.14C) to produce an average estimate. The novelty of this approach is based on

its rather risky application of Fournier's model to a specific catchment, the use of Kirkby's model which requires only infiltration data and basic climatic information, and its combination of the two predictions into an average estimate.

Another approach to estimating sediment yield in an area without sediment data involves extrapolation within areas of similar climate, soil, and topography. This was done, for example by Knott (1980) in southern California, in order to assess the effect of off-road vehicle use on sediment yield in one drainage basin. To do this he used a regression equation developed in a different but similar region by Scott and Williams (1978):

$$\text{Log SY} = a + b \log A + c \log ER + d \log SF$$
$$+ e \log FF + f \log K \qquad (7.18)$$

where

$\quad$ SY $=$ sediment yield,

$\quad\quad$ A $=$ drainage area,

$\quad$ ER $=$ elongation ratio (diameter of a circle with an area equal to that of the watershed divided by the maximum watershed length),

$\quad$ SF $=$ area of slope failures,

$\quad$ FF $=$ a fire factor (percentage of watershed burned, multiplied by percentage of non-recovery of vegetative cover),

$\quad\quad$ K $=$ storm-precipitation factor (the 50-year 10-day total precipitation, multiplied by the square of the 50-year 24-hour precipitation); a, b, c, d, e, and f are constants.

The effect of off-road vehicles was determined by calculating the volume of alluvium removed in gully networks produced by the vehicles and comparing predicted and actual rates in damaged and undamaged areas.

A third approach in the same region was developed by the Los Angeles County Flood Control District (e.g. Cooke, 1984) to predict sediment yield in small mountain catchments in order to provide data for check-dam and channel stabilization programmes. From an empirical analysis of many climatic, geomorphological, hydrological, and vegetational variables, the most powerful were selected and used in the predictive equation (in the original non-metric units).

$$F = \frac{35600\ Q^{1.67} R_r^{0.72}}{(5 + \text{VI})^{2.67}} \qquad (7.19)$$

where

$\quad$ $F =$ debris production rate (yd^3/mile^2),

$\quad$ $Q =$ peak runoff in ft^3/mile^2 resulting from the maximum 24-h rainfall of a given storm,

$\quad$ $R_r =$ relief ratio (as defined above),

$\quad$ VI $=$ vegetation index based on type, cover, and fire history.

Comparison of this model's *predictions* with monitored rates of debris production showed differences of only 2 per cent.

The major approaches to predicting sediment yield involving the monitoring of sedimentary accumulations or sediment movement are: reservoir sediment-deposition surveys; sediment-rating curve/flow-duration analyses; the sediment-delivery ratio method (see above); and the bedload function methods (Glymph, 1975).

The *reservoir sediment-deposition method* involves determining the volume of sediment deposited in a reservoir at known intervals. These volumes may be derived in several ways, including surveying (e.g. Rausch and Heinemann, 1984), the use of electronic echo-sounding, satellite remote-sensing mapping techniques, numerical modelling based on a knowledge of at least input discharge and reservoir geometry, and an empirical method that relates water depth, reservoir total depth, sediment-deposition depth, and sediment volume to different reservoir shapes (Jolly, 1982). Once the sediment volume is established it should be converted to an equivalent dry weight of sediment, on the basis of the density of the deposits. And the estimates also require adjustment for *trap efficiency*—a measure of the extent to which a reservoir traps the sediment supplied to it, which is dependent on the nature of the sediment, the rate of flow through the reservoir, and the outlet characteristics (e.g. Chow, 1964). In general, it is the finest material (which has the lowest settling velocity) which is most likely to be removed. Trap efficiency is not easily predicted, but predictive curves for storage-type reservoirs that relate efficiency to the ratio of capacity to inflow have been developed (Fig. 7.15). The method is often relatively cheaper, quicker, and more accurate than the rating curve method (below), and it has the advantage of requiring measurements to be made only occasionally (e.g. Rausch and Heinemann, 1984). Reservoir data can be used not only to determine yields for a specific basin, but also for the preparation of a regional or even national pattern (e.g. Fig. 7.16).

The sediment-rating curve/flow-duration method is probably the most useful means of predicting sediment yield because it is possible to extrapolate a short period of sediment records to much longer periods. But it requires the concurrent field measurement of both stream flow and sediment data, usually at gauging stations. The method used by the US Bureau of Reclamation and the US Army Corps of Engineers (e.g. Strand, 1975) depends on the creation of a *daily sediment-discharge rating curve* for a site (Fig. 7.17), sometimes disaggregated for different rainfall seasons, and *flow-duration curves* (Fig. 7.18).

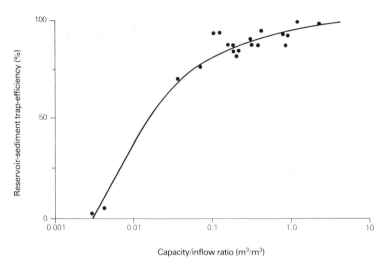

FIG. 7.15 Curves for predicting agricultural reservoir sediment trap efficiency (after Rausch and Heinemann, 1984)

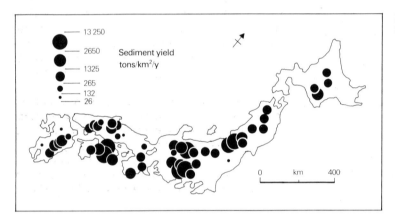

FIG. 7.16 Sediment yield in Japan (after Kadomura, in Jansson, 1982)

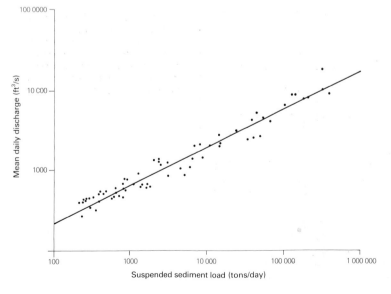

FIG. 7.17 Sediment-rating curve for the Elkhorn River, Waterloo, Nebraska, USA (after Livesey, 1975)

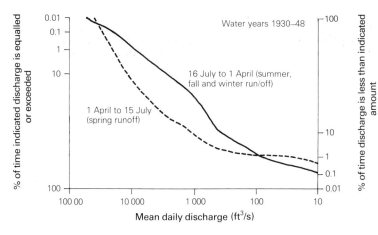

FIG. 7.18 Seasonal flow-duration curves for San Juan River at Bluff, Utah, USA (after Strand, 1975)

The method consists of a determination of suspended-sediment load values from the rating curve for corresponding increments of discharge from a flow duration curve. Multiplication of the suspended-sediment load and discharge increments by the time percentage interval gives a daily occurrence value. Totalling these daily average values produces the mean daily discharge and suspended-sediment load for the year. Further, multiplication of these mean daily values by the number of days in the year gives average annual rates' (Livesey, 1975, pp. 18–19).

Suspended-sediment load estimates must be adjusted for *bedload* and *solution load*. As neither of these is usually measured directly, empirically based correction tables are often used.

Suspended-sediment data, appropriately converted, can provide predictions at specific locations and regional descriptions. For example, Fig. 7.19A shows the pattern of sediment yield in Yugoslavia based on suspended-sediment measurements at 16 gauging stations. For comparison, the pattern that is obtained by using the Fournier equation (e.g. 7.11) is also shown (Fig. 7.19B).

The bedload function method uses equations designed to predict sediment movement in channels. Its application normally requires data on such variables as particle size, channel geometry, and flow duration. The method is as good as the estimating equations, of which those by Einstein and Toffaleti are the most widely used (e.g. Chow, 1964).

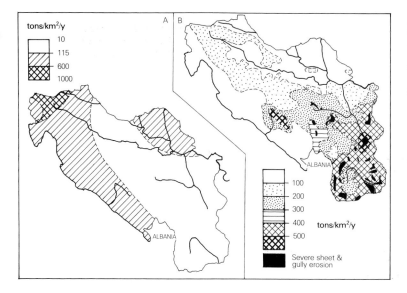

FIG. 7.19 Sediment yield estimates for Yugoslavia. (A) Based on Fournier's (1960) global prediction (from Jansson, 1982). (B) Based on suspended sediment measurements at 16 gauging stations (after Jovanovic and Vukcevic, 1957)

In recent years there has been a growing interest in the use of *dynamic simulation models*, in which sediment yield is simulated in order to produce a semi-continuous record, and the simulation is based on theoretical, dynamic equations that describe and link runoff, soil erosion, and sediment yield (Hadley *et al.*, 1985; see also Chapter 4). These models, which are especially useful for predicting short-term yields, fluctuations in sediment concentration, and sediment movement, use various empirical or theoretical functions to simulate the processes of erosion and sediment routeing.

7.4 Conclusion: Drainage-Basin (Watershed) Management

An understanding of the ways in which sediment passes through a drainage system and the prediction of sediment movement rates is fundamental to many aspects of river management. Sediment can affect navigability, flooding, property boundaries, and the ecology of the system. It can threaten engineering structures, such as bridges and irrigation systems. It is essential to know how quickly it may fill up reservoirs. It can affect water quality. Its rate of movement can be strongly influenced by land-use changes in the catchment. To predict sediment yield in drainage systems where there are no sediment data is difficult and inevitably approximate, but it can be done if some local data such as precipitation and soil infiltration rates are to hand; if sediment data are available in comparable environments elsewhere, extrapolation may be possible using, for example, regression techniques. Where sediment data are available together with water-discharge information, the most widely used techniques are based on bedload function equations, sediment-delivery ratio methods that commonly use USLE information, reservoir sediment-deposition surveys, and the sediment-rating curve/flow-duration method.

The discussion on sediment yield from drainage basins illustrates well the complexity of fluvial systems. Equally complex are the topics of groundwater conditions, water quality, hillslope hydrology, and soil erosivity, and yet so few of these are globally understood by specialists. The integration between them is even less-well understood. Yet management must take account of such integrated relationships.

If, as claimed above, drainage basins consist of organized dynamic natural process–response systems (Fig. 7.1), they clearly should be managed as a whole. Successful attempts are rather unusual. An exception is the Otago Catchment Board in New Zealand which manages 34 000 km² of drainage basin inland of Dunedin (Fig. 7.20). Responsibilities include appropriate management to deal with flood and river control, community drainage schemes, soil conservation, and the monitoring and management of water resources. Interestingly the Board also administers by law and its consent is needed for: gravel and sand removal from rivers; the construction of bridges and culverts: building construction next to watercourses; alterations to the natural drainage pattern. By this means the Board can apply its awareness of the whole catchment system to the proposals for a specific (local) action. They are in a position to assess the feedback effects of such action upon the system as a whole.

The organization of Water Authorities in England and Wales is different in that their responsibilities are dominated by water supply, sewage disposal, and flood protection. They do not carry responsibility for managing soil erosion (nor for providing planning consent although they may be asked for advice) in relation to specific (local) construction proposals.

The contrasts between these two examples is typical of the variety of ways in which drainage-basin management is carried out in different places. Universally, however, the need is for both data collection (Table 7.3) and continual monitoring (Table 7.4) to be part of the management function.

The call for a practical focus on drainage-basin management by Richter *et al.* (1985) includes a list of priorities for more research effort, especially additional monitoring of both the drainage-basin system and the effects of specific management actions upon it. In the humid tropical catchments, for example, they identify the need for *inventory surveys* to establish a priority for dealing with problems. However, in many parts of the world the existing management structure is inadequate and the political will to do something positive is lacking.

In many cases no time is left for research. The problems are here, now. A fire-fighting approach may be needed in the absence of adequate planning. For example, it might be useful to

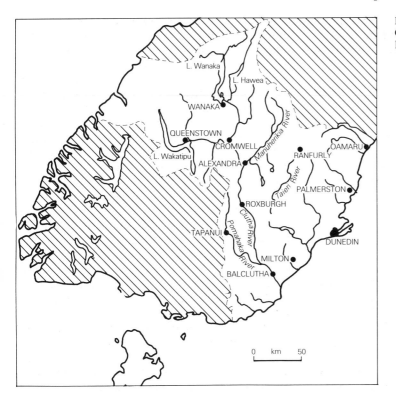

TABLE 7.3 Drainage-basin inventory data required for effective management

Geology	Lithologies: outcrop map physical properties chemical properties Structure: map of folds, faults, dips
Geohydrology	Subsurface water behaviour including position and depth of water-table Outcrop and structure of acquifers and acquicludes
Geomorphology	Relevant characteristics of surface form, materials, and process. Morphometric parameters (see Table 7.1)
Soils	Classification of soil types Map of soil distribution Physical and chemical properties Erodibility
Vegetation	Classification map of distribution
Hydrology	Drainage network

TABLE 7.4 Drainage-basin parameters to be monitored as part of an effective management programme

Hydrology	River discharges (water and sediment) Water quality Extension of drainage network (changing channel patterns)
Geomorphology	Erosion on selected slopes Channel bed and bank erosion Sedimentation within channels, deltas, lakes, and reservoirs Slope stability
Soils	Rates of erosion Changes in physical structure Changes in chemical properties
Vegetation	Depletion Changes in character
Climate	Rainfall—amount, duration Rainfall erosivity Temperature

recognize, where monitoring is absent, those geomorphological indicators of threshold states in which environmental degradation is indicated. Such indicators include signs of increasing erosion, the build-up of sediment, increasing landslide activity, more wind-blown dust in the atmosphere, increasing soil salinity, or waterlogging.

Even a recognition of these symptoms does not always bring the will to do something about it. Just as management needs to catch up with environment, so politics and administration need to catch up with operational (management) needs. Nowhere is this better displayed than in the environmental management of drainage basins.

8 Glacial and Periglacial Geomorphology

8.1 High-Latitude and High-Altitude Problems

(a) *The development context*

The settlement of high-latitude and high-altitude areas is not new, as the history of hunting, trapping, and transhumance testifies; what is relatively new is the penetration of technically advanced groups into these areas in order to exploit their animal, vegetable, and mineral resources and, perhaps as important, to secure the defence of remote frontiers (e.g. Harris, 1986). The groups mainly involved are those from the northern hemisphere circumpolar nations (the USSR, the USA, Canada, and the Scandinavian countries), and those with substantial mountainous areas (including China, the USA, Andean states, New Zealand, and Switzerland and its neighbours). The problems of high-latitude development have been longest recognized and most extensively managed in the Soviet Union, where experience stretches at least from the construction of the Trans-Siberian Railway at the turn of the last century to the building of modern towns such as Norilsk in areas of permanently frozen ground. Since 1938 a thorough survey has been mandatory before any structure can be erected on permanently frozen ground in the USSR. Soviet literature on the problems of high-latitude development is extensive (e.g. Kudryavtsev, 1978).

North American experience, rudely initiated by the Klondike Gold Rush in 1896 and slowly accumulated in the early decades of this century, was rapidly extended during the Second World War, with the construction of airfields and of transportation routes such as the Alcan Highway. Since the war, with changing political and military alliances and the discovery of great mineral wealth in the Northlands, the impetus for defence and development has increased and studies of environmental problems have burgeoned (e.g. Williams, 1979). In particular, the discovery of oil at Prudhoe Bay, Alaska, in 1968 has promoted enormous interest in these problems (Mackay, 1972).

High-altitude problems in mountain areas are in many ways similar to those in high latitudes, often featuring frozen ground ice and snow, as well as fluvial processes, but they often differ in having a context of steep slopes and high relief that can make more serious such gravity-related problems as avalanching, runoff, and debris movement (e.g. Slaymaker, 1981). They have been of concern for centuries in Europe, China, and the Andes, for instance, and only more recently in North America.

The essential problem is that 'modern' development of these areas, as with most other environments marginal to temperate lands, has tended to apply mid-latitude technology to conditions for which it is often unsuitable. Successful development requires, as ever, a sympathetic understanding of the environment. As Muller, in one of the few reviews in English on high-latitude engineering problems to consider the extensive Russian literature observed:

Costly experience of Russian engineers has shown it to be a losing battle to fight the forces of frozen ground simply by using stronger materials or by resorting to more rigid designs. On the other hand, this same experience has demonstrated that satisfactory results can be achieved if the dynamic stresses of frozen ground are carefully analysed and are allowed for in the design in such a manner that they appreciably minimize or completely neutralize and eliminate the destructive effect of frost action. Mastery of this working principle, however, can be achieved only if the natural phenomena of frozen ground are thoroughly understood and their forces are correctly evaluated (Muller, 1947).

(b) *Publications*

Academic and applied studies of high-altitude and high-latitude problems have flourished for over half a century, and they have recently become closely allied. The relevant literature is burgeoning. To give one example, a bibliography of permafrost publications between 1978 and 1982 alone includes over 4400 entries (Brennan, 1983). Fortunately, the literature is distilled in a number of books, and the most recent research is to be found in a number of conference proceedings and journals. Modern books on glacial matters include those on *Glacial Geomorphology* (Coates, 1974), *Glacial Systems* (Andrews, 1975), *Glacial Geomorphology* (Embleton and King, 1975a), *The Physics of Glaciers* (Paterson, 1975), *Glacial Till* (Legget, 1976), *Research in Glacial, Glaciofluvial and Glaciolacustrine Systems* (Davidson-Arnott et al., 1982), *Glacial Geology* (Eyles, 1983), and *Glaciers and Hazards* (Tufnell, 1984). Books on periglacial phenomena include *Geomorphology of Cold Environments* (Tricart, 1970), *Periglacial Geomorphology* (Embleton and King, 1975b), *Problems of the Periglacial Zone* (Jahn, 1975), *Periglacial Processes* (King, 1976), *The Periglacial Environment* (French, 1976), *General Permafrost Science—Geocryology* (Kudryavtsev, 1978, in Russian), Washburn's classic (1979) *Geocryology*, the Chinese work *Permafrost* (Natural Research Council of Canada, 1981), *Periglacial Mass Wasting* (Harris, 1981), *Periglacial Geomorphology and Climate in Glacier-Free Cold Regions* (Weise, 1983), and *The Periglacial Environment* (Harris, 1986). There is also a range of more specialized books, that include *Permafrost in Canada* (Brown, 1970), *The Mechanics of Frozen Ground* (Tsytovich, 1975), *Pipelines and Permafrost* (Williams, 1979), *Permafrost Engineering, Design and Construction* (Johnston, 1981), *Soil and Permafrost Surveys in the Arctic* (Linell and Tedrow, 1981), and *Frost Action and its Control* (Berg and Wright, 1984).

Conference proceedings are an indispensable part of the applied geomorphological literature on cold climates. Particularly important are the proceedings of the international permafrost conferences that provide bench-marks in recent research progress: the first at Purdue University, (National Academy of Sciences, 1966), the second at Yakutsk, USSR (National Academy of Sciences, 1973, 1978), the third at Edmonton, Canada (National Research Council of Canada,

1978, 1978), and the fourth at Fairbanks, Alaska (National Academy of Sciences, 1983). More parochial proceedings include the Canadian (Brown, 1969; Legget and MacFarlane, 1972; National Research Council of Canada, 1982) and Chinese permafrost conferences (e.g. Brown and Yen, 1982). In glaciology, too, conference proceedings yield a rich harvest, such as those on snow in motion (Glen et al., 1980), on processes of glacial erosion and deposition held at Geilo, Norway in 1980 (*Annals of Glaciology*, 2), or the engineering behaviour of glacial material (Midlands Soil Mechanics and Foundation Engineering Society, 1975).

Much relevant research is published in generally available journals, the most useful of which include the *Canadian Journal of Earth Sciences*, the *Canadian Geotechnical Journal, Arctic and Alpine Research, Quaternary Research*, the *Journal of Glaciology*, the *Journal of Glaciology and Cryopedology* (in Chinese), *Geografiska Annaler* (A), and *Biuletyn Peryglacjalny* (Lódź, Poland). Very usefully, developments in glacial and periglacial geomorphology are reviewed annually in *Progress in Physical Geography*. Unfortunately, some relevant work is not readily accessible, being printed in limited circulation reports produced by government and other public research agencies such as the Cold Regions Research and Laboratory (CRREL) of the US Army, the National Research Council of Canada's Division of Building Research, Moscow's V. A. Obruchev Institute of Permafrost Studies, the Permafrost Institute of Yakutsk, the Lanchou Institute of Glaciology, Geocryology and Desert Research in China, and the Swiss Commission for Snow and Avalanche Research.

Faced with such a large library of recent research, this review will be very selective. It will concentrate on the following themes: glacial hazards; problems of permafrost terrain including frozen-ground phenomena, frost action, and solifluction; and hazard appraisal in areas of present and past glaciation and periglaciation. In emphasizing these themes, it should be noted that other geomorphological processes are significant in the environmental management of glacial and periglacial areas: of these, fluvial and coastal processes are the most important, although, with some exceptions, they do not pose serious problems different from the same processes in other environments (e.g. Chapters 6, 7, and 10).

8.2 Glacial Hazards

(a) *Environmental context*

Glacial hazards are those associated with present-day ice-sheets and glaciers, and with areas that have been glaciated in the past. Fig. 8.1 summarizes, for the northern hemisphere, the location of contemporary ice sheets and the distribution of ice-sheets at the time of maximum glaciation in the Pleistocene. At present ice covers only some 10 per cent of the total land area ($c.15$m km²) especially, of course, in the polar regions; but over 30 per cent was formerly covered by ice ($c.40$m km²). Fig. 8.1 does not show the location of present-day glaciers, which actually occur, in a wide variety of climatic conditions where the requirements of mass balance between accumulating snow and the nature, amount, and use of incoming solar energy are met (Andrews, 1975). Glacial areas at present include parts of Iceland, Scandinavia, Alaska, the European Alps, the New Zealand Alps, the Rockies, and Himalayas, as well as individual mountains within the tropics (e.g. Ruwenzori in Uganda, Kilimanjaro in Tanzania).

Glaciers and ice-sheets provide resources and pose hazards, mainly around their margins (Table 8.1). Amongst the resources, water supply is particularly important and glacial hydrology is a distinct field of inquiry.

While the potential hazards of glaciers are substantial, glacier impact on human activity is relatively small because only about 0.1 per cent of the world's glaciers are associated with inhabited areas. For this reason, the hazards have received only minor attention in the huge literature of glaciology and glacial geomorphology (e.g. Grove, 1987). Only three of the hazards seriously threaten settlements and have strong geomorphological dimensions: glacier fluctuations, glacial flooding (jökulhaups, débâcles), and avalanches. Attention in this brief review will be focused on these.

Glaciated areas, those formerly covered by ice, pose quite different geomorphologically related possibilities and problems for environmental management. Positively, they may offer a complex range of sand and gravel resources, the distribution of which is best understood by a geomorpho-

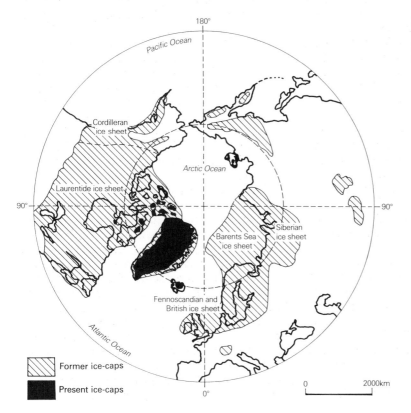

FIG. 8.1 The present distribution of ice, and its former extent in a glacial period, in the northern hemisphere (after Andrews, 1975)

TABLE 8.1 Some examples of glacial hazards and resources

	Ice	Meltwater	Snow
Positive utilization	Icebergs for water	Electric-power generation	Water storage in drainage basins
	Ice as a refrigerant	Irrigation	Skiing and tourism
	Waste disposal in glaciers		Frost protection
	Artificial ice islands		
	Drilling from ice shelves		
Hazards	Advancing/surging glaciers	Floods on meltwater rivers	Avalanches
	Glacier falls (ice avalanches)	Bursting of glacially and sub-glacially impounded lakes	Snow cover of transport lines, airfields urban areas
	Crevassing and calving		Snow loading on structures
	Icebergs as transport hazards	Damage by sediment load (e.g. in turbines)	Snow accretion on electric wires
	Sea-ice hazards: Harbour blocking, structural damage		
	River/lake freezing, ice-flow damage		
	Ice accretion (on ships, aircraft, etc.)		
	Loading limits on ice surfaces, on floating ice, etc.		
	Hazards of subglacial mining, glacier excavation		
Environmental control	On climate—global, local		
	On world sea-level		

Source: Embleton (1982).

logical appraisal (Chapter 11). Negatively, the heterogeneity of deposits, reflecting the intimate juxtaposition in many areas of glacial, fluvial, lacustrine, and marine sediments, can cause distinctive problems, such as overconsolidated tills, and rapidly ranging permeabilities and shear strengths (e.g. Derbyshire and Love, 1986). In addition, many glaciated areas are, or have been periglaciated, adding further complications to the geomorphology and the task of appropriate management.

(b) *Glacier fluctuations*

Tufnell (1984) and others have established that glaciers have advanced and retreated in historical times in response to natural climatic change (Fig. 8.2). There is considerable evidence, for example,

of a period of glacier advance between about 1550 and 1860 (the 'Little Ice Age') in Europe and elsewhere. Such changes affected settlement and land use in, for instance, the Alps, Iceland, Norway, the Caucasus, Alaska, and the Karakoram Mountains (Tufnell, 1984; Grove, 1987). The impact of glacier movement is well illustrated by the Chamonix Valley, France (Fig. 8.3, Table 8.2), where villages have been destroyed, and there have been floods, debris and ice falls, and avalanches. Most changes directly associated with glacier movement are normally slow, and an established trend gives ample warning. Whether the inhabitants heed the warning is, however, a different matter. But a major surge of a glacier can be dangerous and difficult to predict, as occurred when the Solda Glacier (in the Ortler Massif,

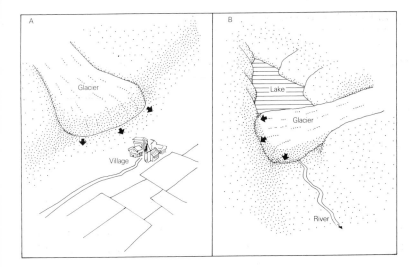

FIG. 8.2 Glacier advance leading to (A) the overrunning of a settlement and (B) the blocking of drainage to form a lake (after Tufnell, 1984; © 1984, and reproduced by permission of The Longman Group Ltd.)

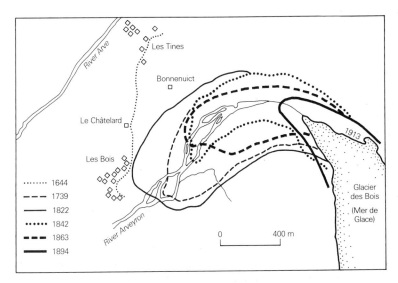

FIG. 8.3 Fluctuations of the Mer de Glace (Chamonix, France) between 1644 and 1919 (after Rabot, 1920 in Tufnell, 1984)

Switzerland) advanced 1200 m in 1817–18 (Tufnell, 1984). The hazards of changing glacier position are rarely catastrophic, but they may influence water supply, land use, forestry, routes, and tourism to the extent that their prediction is desirable.

(c) Glacier floods (jökulhaups, débâcles)

In contrast, the sudden release of water impounded in, on, under, or beside a glacier can pose a major threat to downstream settlements, land use, and lines of communication. Tufnell (1984) described several circumstances in which water may be impounded in association with a glacier—a tributary glacier may extend to block an unglaciated valley (Fig. 8.2B); ice from a 'hanging' tributary glacier may fall into a valley and block it; lakes may be formed between glaciers or between a glacier and a valley side; lakes may form on glacier surfaces during retreat and ablation; proglacial lakes may be created between a glacier and a terminal moraine; local melting of ice by fire or volcanic activity may cause thawing and ponding; and mass debris movements may block runoff.

The release of stored water, quickly and often without warning, can cause serious damage. For example, it is reported that a flood from a

TABLE 8.2 Glacier hazards and settlements in the Chamonix Valley

Dates	Events
About 1600	Advancing glaciers destroy 7 houses in the Argentière–La Rosière area, 2 at La Bonneville, 12 at Le Châtelard, and the entire hamlet of Bonnenuict. The Mer de Glace came so close to Les Bois that the village was damaged and had to be abandoned. It was also near Les Tines
1610	Water from the Argentière glacier destroys 8 houses and 5 barns. Torrents from the Bossons glacier severely damage Le Fouilly. 3 houses, 7 barns, and 1 mill destroyed at La Bonneville. Mer de Glace still close to Les Bois and causing damage
1613 or 1614	Glacial meltwater completes the destruction of La Bonneville
1616	Argentière glacier adjoining La Rosière. About 6 houses remaining at Le Châtelard, though only 2 inhabited. Glacier very close. At some time between 1642 and 1700 the village was finally abandoned and has never been rebuilt. Its inhabitants are thought to have settled at Les Tines
1628–30	Falls of snow and glaciers in the Chamonix valley. Flooding of the Arve due to glacial meltwater
1640s	Glaciers came close to Le Tour, Argentière, La Rosière, Les Tines, Les Bois, Les Praz and Les Bossons. 1641: Les Rosières (*not* La Rosière, but a village near Chamonix) destroyed by a flood from the Mer de Glace. 1642: avalanche of snow and ice destroyed 2 homes at Le Tour, and killed 4 cows and 8 sheep. 1641–3: property flooded and ruined by torrents from the Bossons glacier
1714	Several villages still threatened by glaciers
About 1730	Mer de Glace less than 400 m from the nearest houses at Les Bois
1818–20	Glaciers again almost at Le Tour, Montquart and Les Bois (the Mer de Glace was only 20 m from this last village). The Argentière glacier was little more than 300 m from the old centre of Argentière village
1826	Mer de Glace showering debris on to the chalets below
1835	*Séracs* from the Mer de Glace threaten to fall on Les Bois
1850	Mer de Glace about 50 m from Les Bois and causing blocks of ice to fall towards Les Tines
1852	Several glacier avalanches in the Chamonix Valley due to warm winds and heavy rains
1878	*Débâcle* from the Mer de Glace. Houses evacuated as a safety measure; fields flooded. Similar outbursts had occurred in 1610 and 1716
1920	*Débâcle* from the Mer de Glace floods the cellars and ground floors of many buildings in Chamonix. Much land inundated
1949	Avalanche from the Glacier du Tour kills 6 people. The worst ice avalanche in the French Alps since the Glacier de Tête Rousse disaster of 1892
1977	Le Tour threatened by glacier avalanche

Source: Tufnell (1984).

proglacial lake killed thousands in Tibet in 1953; a similar flood in the Cordillera Blanca of Peru in 1941 killed over 6000 people, and several less severe disasters are recorded in the Alps, Karakoram, Argentina, Iceland, Norway, and Alaska (Tufnell, 1984).

Jökulhaups (outbursts of subglacial water) are relatively common and serious in Iceland. The floods often carry huge quantities of debris that thoroughly modify the morphology of proglacial plains and can substantially damage downstream activity. Nye (1976) attempted to model mathematically the *jökulhaups* which have recently occurred every 5 or 6 years from subglacial lake Grímsvötn beneath the Vatnajökull ice-cap. He argued that the subglacial lake grows as a result of both surface runoff and geothermal melting until it is large enough to lift the superincumbent ice helped by a hydrostatic cantilever effect. Flow from the lake is in conduits that are subjected to melting by the frictional heat of flow and to closure by the plastic deformation of ice. The flood, Nye suggested, reflects the relationship between these forces when frictional heat is great; the flood may end, perhaps abruptly, when the lake level falls to a critical level and/or when the conduits are closed by plastic deformation.

The hazardousness of such floods is particularly high for several reasons. First, the lakes may have different causes, and sometimes they may develop beneath ice so that their growth cannot be seen or easily monitored. Secondly, the causes of release vary, and are not easily predictable. Thirdly, because it may be difficult to locate the areas of water accumulation and to predict the location and timing of water release, the floods may be unexpected. And finally they can be massive flows, against which engineering defences are unlikely to be effective. But there is some predictable periodicity to some floods: for example, *débâcles* in the Alps occur mainly in June, when floods are due to variations in glacier length, and in August, when rupture of water-pockets in ice is dominant. (Tufnell, 1984) and Embleton (1982) believed that their prediction is now well within the capability of glaciologists.

(d) Avalanches

The rapid movement of detached masses of ice and snow under the influence of gravity as avalanches is a particularly serious hazard in many mountain areas. Avalanches, especially when they are charged with debris, can cause considerable damage to buildings, roads, and other obstacles in their paths and they often cause loss of life (Fig. 8.4). As a result, avalanche research and forecasting is well advanced. For example, the Swiss Commission for Snow and Avalanche Research was founded in 1931.

There is a useful distinction to be drawn between ice, snow, and slush avalanches. *Ice avalanches*— which usually occur at a glacier snout—arise from the instability of ice on a steep slope, in which ice separates by calving and free fall or by detachment and sliding. The cause of breakaway may be associated with crevasse formation and melting, or there may be a trigger, such as an earthquake. Tufnell (1984) cited the example of the Huascarán, Peru, disaster in 1962, when 3m m³ of ice became incorporated into a 13m m³ debris-flow that achieved a speed of over 100 km/h and caused over 4000 casualties; in 1970, a landslide in the same area caused over 15 000 deaths (Chapter 5). Accounts of these events have frequently concentrated more on the landslide aspects (see Chapter 5), and less on the dislodgements of ice that caused the first movements.

Snow avalanches are of several different types (Fig. 8.5), the details of which are considered in Embleton and King (1975*a*), and Embleton and Thornes (1979). The nature and conditions of each type are distinctive, but in general snow avalanches occur when the weight of a snow mass exceeds the frictional resistance of the surface on which it lies. The main causes of these avalanches relate to meteorological conditions, snow structure and mechanics, and terrain characteristics (Fig. 8.6), and in particular to overloading with

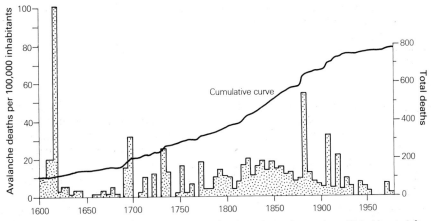

FIG. 8.4 Deaths caused by avalanches in Iceland relative to the number of inhabitants (after Bjornsson, 1980; reproduced by courtesy of the International Glaciological Society)

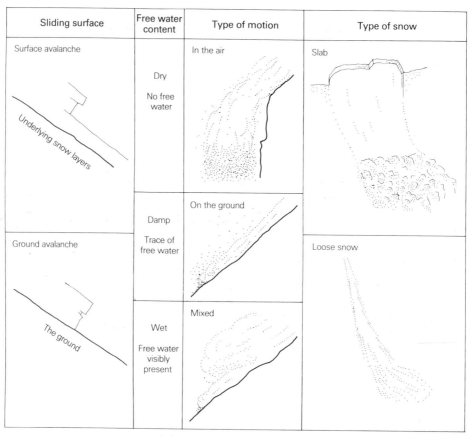

Sliding surface	Free water content	Type of motion	Type of snow
Surface avalanche	Dry No free water	In the air	Slab
Ground avalanche	Damp Trace of free water	On the ground	Loose snow
	Wet Free water visibly present	Mixed	

Fig. 8.5 Classification of avalanches (after Embleton and Thornes, 1979; © 1979 Edward Arnold (Publishers) Ltd.)

snow, structural weaknesses in snow masses, and melting (e.g. Embleton and Thornes, 1979). Geomorphologically, the important terrain conditions are slope inclination, length, orientation, and roughness. These characteristics are of predictive value. For example, there is a predictable relationship between slope angle and probability of avalanche release (US Dept. Agriculture, 1968) and the thickness of snow required to initiate an avalanche. In addition, Lied and Bakkehøi (1980) used empirical field evidence to derive an equation for predicting the 'run-out' distance for snow avalanches based on topographical parameters:

$$\alpha = (6.2 \times 10^{-1} - 2.8 \times 10^{-1} Hy'')\beta$$
$$+ (1.9 \times 10^{1} Hy'' - 2.3)^{\circ} + 1.2 \times 10^{-1}\theta \qquad (8.1)$$

where

    α = run-out distance (average gradient of avalanche path);

    y'' = second derivative of avalanche slope described by a second-degree function;

    β = average gradient of avalanche track (between rupture zone and run-out zone);

    H = total displacement of the avalanche;

    θ = gradient of the rupture zone.

Slush avalanches, a water-saturated form of rapid mass snow movement, cause damage to people, buildings, and communications in several mountainous high-altitude, and high-latitude areas (e.g. Hestnes, 1985; Onesti, 1985). Unlike snow avalanches, they do not require steep slopes in the areas of initiation, and the critical angle of repose may only be about 15°. In Norway, the hazard was found by Hestnes (1985) to be associated mainly with weak, cohesionless coarse-grained snowpacks; hard layers or crusts of ice in snow or on ground; and intense rain falling on

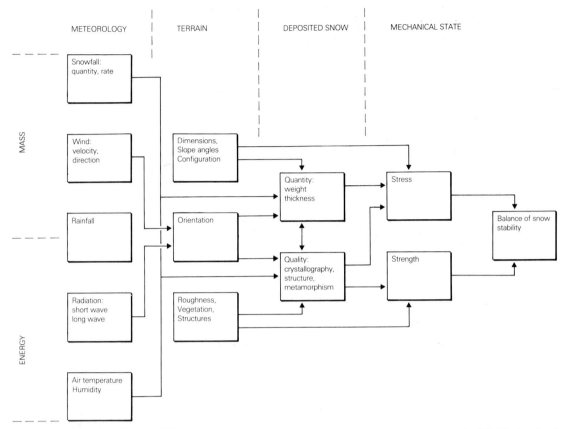

FIG. 8.6 The Skōda diagram illustrating the chain of causation in avalanche formation (based on the work of M. Skōda, after La Chappelle, 1980; reproduced by courtesy of the International Glaciological Society)

cohesionless new snow on these layers. Hestnes also found that the avalanches began mainly in drainage channels, sloping bogs, depressions, and open fields.

To a considerable extent, avalanche *location* can be usefully predicted on the basis of historical occurrence and landform data. Thus *avalanche hazard zoning* maps are produced in several countries, including Switzerland, France, Canada, the USA, and Norway. In Switzerland, for example, there is now a legal framework for avalanche zoning at federal, cantonal, and communal levels designed to prevent improper land use in potentially hazardous locations (Frutiger, 1980). Temporal forecasting of avalanches within topographically controlled contexts is usually focused on the variable characteristics of weather and snow (e.g. La Chapelle, 1980) and the monitoring of snow conditions. And a further response, also in the context of spatial analyses of the hazard, is the design and construction of avalanche defences such as fences and walls.

(e) *Hazards and resources in glaciated terrain*

Areas formerly covered by ice often leave distinctive, complex, and varied assemblages of landforms and deposits. These are described in detail in most geomorphological textbooks (e.g. Embleton and King, 1975a) and will not be repeated here. The common complexity of glaciated terrain and deposits is often a major problem for engineers because it can lead to enormous spatial variability in the *geotechnical properties* of materials, such as grain size, compressibility, consolidation, shear strength, void ratio, and Atterberg limits. The complexity reflects the diverse nature of sediments generated directly (chiefly tills and moraines) or indirectly (eskers, kames, etc.), by the ice and its drainage and the fact that such sediments are often intermingled with sediments produced in fluvial, lacustrine, aeolian, marine, and periglacial environments.

Fig. 8.7 illustrates some of these features in a glaciated valley and on a proglacial fluvial plain. Very often, the landforms are associated with characteristic deposits. For example, an esker typically is built of stratified sands and gravels. Given such relationships, a geomorphological classification of landforms in glaciated terrain is often a useful preliminary step in classifying the types of superficial sediments. What is especially important, however, is the fact that *geotechnical properties* of the sediments are also to some extent distinctively related to their origin, so that an understanding of glacial and postglacial history and geomorphology of a region should be useful in predicting the engineering properties of deposits laid down in glacial periods. Thus, geomorphology can assist site investigation in glaciated terrain (e.g. Derbyshire 1975; Fookes *et al.*, 1975; Boulton and Paul, 1976).

Boulton and Paul (1976) developed this argument in the context of glacial tills. They showed that some important properties of soils and geotechnical parameters are predictably related to transportational, depositional, and post-depositional processes that influence tills (Fig. 8.8). The geotechnical behaviour of tills is chiefly influenced by 'grain size distribution, state of consolidation, jointing, and the nature of the sequence in which they lie. Of these, grain-size distribution is determined partially by glacial erosion and transport, and partially by the mode of till deposition, whereas the others are determined almost entirely by depositional and post-depositional processes' (Boulton, 1975, p. 58). Thus, for example, lodgement till is commonly largely unsorted, contains a high proportion of rock flour, shows a preferred orientation of coarser particles and sets of fissures (which can influence shear strength), and is overconsolidated (which affects compressibility and void ratios); whereas ablation tills, which are also unsorted, are often coarser than lodgement tills, more disoriented, more permeable, and only normally consolidated (Fookes *et al.*, 1975). Table 8.3 compares qualitatively the engineering behaviour and properties of lodgement and ablation tills and fluvio-glacial sediments. It should be emphasized that while such generalizations are valuable, the specific values of geotechnical properties can be extremely variable even within one distinctive deposit (such as a till), so that only general

prediction is possible at present, and it is only a preliminary to, rather than a substitute for, laboratory analyses of samples from sites where development is proposed.

8.3 Periglacial Hazards: Environmental Context

In contrast to glacial hazards, periglacial hazards are more extensive, more serious, and have stronger geomorphological associations. The problems are mainly associated with areas of *permafrost* (Fig. 8.9). Permafrost is perennially frozen ground, that is, ground frozen continuously for two or more years. It is defined on the basis of temperature (i.e. ground below 0 °C) rather than on the presence or absence of ice. *Dry permafrost* occurs where there is no water present, as in certain bedrock areas of the Brooks Range, Alaska. Normally, however, ice is present. The upper limit of perennially frozen ground is a surface of some importance because it is relatively impermeable; it is called the *permafrost table*. Permafrost is often divided into at least two categories. *Continuous permafrost* normally has a mean annual temperature at 10–15 m depth of less than − 5 °C, and the permafrost table is usually no more than 0.61 m below the surface (although it may be up to 1.8 m deep in granular material). *Discontinuous permafrost* is thinner, is broken by thawed areas, has a lower permafrost table, and its mean annual temperature at 10–15 m depth is between − 5 °C and − 1.5 °C.

The *active layer* comprises the ground above the permafrost table. At least part, if not all of this zone is subjected to intermittent freezing and thawing, where freezing can occur from the top down and the bottom up, often on a seasonal basis: this highest zone, the zone of greatest temperature fluctuation, is called the *frost zone*, and its uneven lower surface is called the *frost table*. If the frost zone and the active zone do not coincide, there may be a residual zone of thawed ground between the frost table and the permafrost table: this ground is called *talik*, a term also applied to any thawed area within and beneath the permafrost. Talik may occasionally act as a viscous liquid and flow. The lower limit of the active layer may vary from year to year.

In areas beyond the limits of permafrost, where freezing occurs from the top down and thaw occurs from the top and bottom of frozen ground, freezing

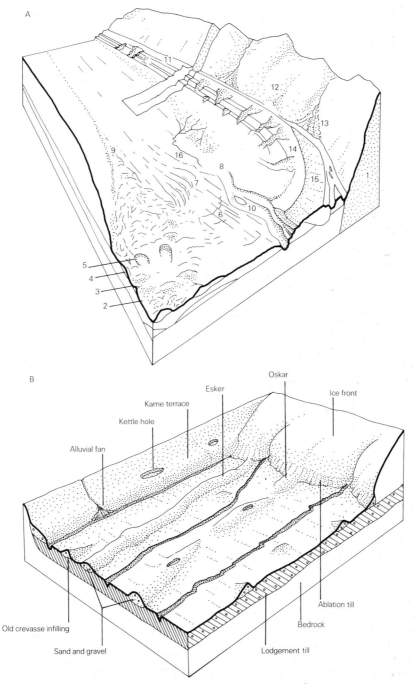

FIG. 8.7 (A) Landforms and sediments associated with the retreat of a valley glacier (after Boulton and Eyles (1979)). Key: 1, bedrock; 2, lodgement till with fluted and drumlinized surface; 3, ice cores; 4, supraglacial morainic till; 5, ice-cored and kettled supraglacial morainic till; 6, bouldery veneer of supraglacial morainic till with deposited crevasse fills; 7, dump moraine ridges showing internal deformed scree-bedding having formed as ice-contact screes; 8, supraglacial lateral moraine and till flows into meltwater streams resulting in till complexes; 9, supraglacial medial moraine; 10, proximal meltwater streams; 11, kame terrace; 12, truncated scree; 13, fan; 14, gullied lateral terrace; 15, lateral-terminal dump moraine; 16, fines washed from supraglacial moraine till into crevasses and moulins. (B) Landforms and sediments associated with a proglacial fluvial plain (after Fookes *et al.*, 1975)

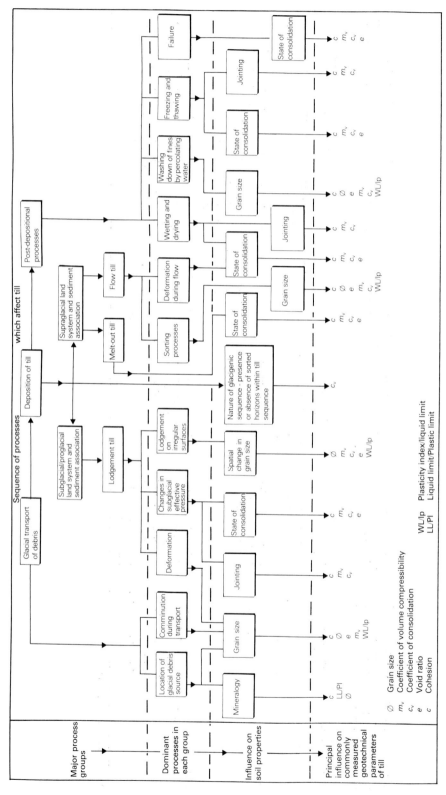

FIG. 8.8 The relations between the processes of till formation and soil properties and related geotechnical parameters (from Boulton and Paul, 1976; reproduced by permission of the Geological Society of London)

TABLE 8.3 Summary comparative table of the engineering properties and behaviour of the three basic categories of glacial materials

Land system and category of glacial materials	Bearing Capacity	Settlement	Slope stability	Excavation	Use as fill material	Construction materials
Lodgement till	Usually good, but liable to contain soft patches. Silt lenses liable to frost heave.	Usually small. Long term settlement. Differential settlement may be anticipated.	Generally stable at quite steep slopes. Water bearing sand/silt layers can cause instability.	Lodgement tills can be very tough and excavation hindered by high plasticity and boulders.	Good impermeable fill but sensitive to moisture changes. May be wet of optimum.	Generally unsuitable due to variability.
Ablation till	Generally good but may be variable.	Slighter higher, but more rapid settlement than lodgement tills.	As lodgement tills but with lower cohesion. When low fines content stand near angle of rest.	Excavation may be hindered by boulders and high silt content can hamper excavation in wet conditions.	Good fill material. Can be used as impermeable fill. Silt content causes sensitivity to moisture content changes.	Unsuitable without much screening and washing.
Fluvio-glacial cohesionless deposits	Generally good but lenses of till or openwork gravel can occur and relative density is variable (especially in kames).	Settlement largely during construction but differential settlement due to long term consolidation of clay or till or lacustrine clay lenses.	Materials generally all stand at angle of rest. Locally clay/silt layers may be unstable.	Usually easy. Face shovel preferred to mix sand and gravel. Sometimes hindered by wet silt/clay layers or by boulders.	Good granular, generally free draining, fill. Some selective digging may be required, particularly in kames.	May provide good sources of coarse and fine aggregate, but selective digging and washing often required.

Source: Fookes *et al.* (1975).

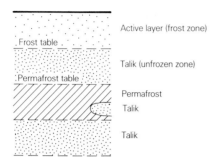

FIG. 8.9 Terminology of permafrost and some of its associated features

and thawing may be *seasonal* or *sporadic*. A phrase commonly allied to permafrost is *periglacial environment*. This may be taken to mean, broadly, the environment where frost processes predominate (or, more strictly, where permafrost occurs). In recent years *geocryology* is a term that has been widely adopted to describe the study of earth materials at temperatures below 0 °C and especially the study of permafrost, but excluding the study of glaciers (e.g. Washburn, 1979).

Approximately a fifth of the world's land surface is underlain by frozen ground of one kind or another. Descriptions of the distributions of frozen ground and related features are being refined continuously. Fig. 8.10 is based on several recent studies. Because there is more land in high latitudes of the northern hemisphere than the southern hemisphere, the permafrost area in the north is greater than in the south, being 22.4m km^2 compared with 13.1m km^2. Over 80 per cent of Alaska, 50 per cent of Canada, 47 per cent of the Soviet Union, and 22 per cent of China are underlain by permafrost. In general, permafrost extends further south on the eastern and more continental land areas. Beyond the permafrost limits, most of the land north of 30 °N is affected by seasonal or sporadic freezing and thawing (Fig. 8.10).

The vertical distribution of frozen ground and thawed ground also varies spatially. In Fig. 8.11 longitudinal cross-sections through Eurasia and North America show variations in the thicknesses of permafrost and the active layer. Permafrost reaches its maximum known thickness, 1500 m, in

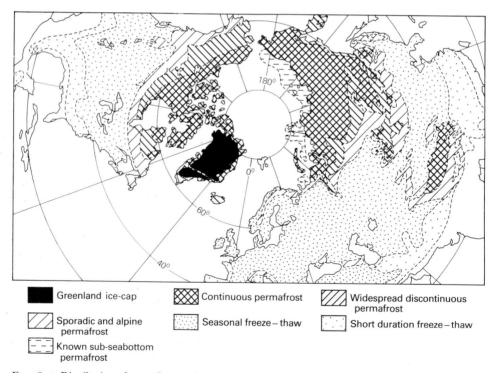

FIG. 8.10 Distribution of permafrost and related phenomena in the northern hemisphere (after Mackay, 1972; Heginbottom, 1984; and others)

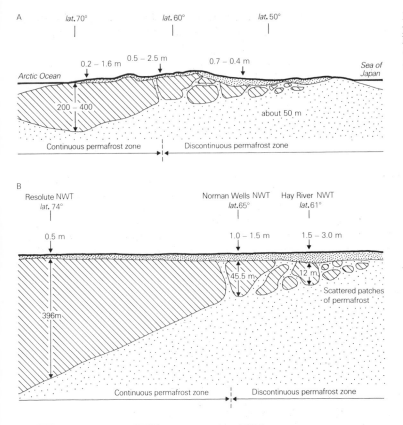

FIG. 8.11 Vertical distribution of permafrost and active zones in longitudinal cross-sections through (A) Eurasia and (B) North America (after Muller, 1947, and Brown, 1970)

Siberia, but it is not normally thicker than 600 m. In Eurasia the average thickness is 305–460 m, and in North America it is 245–365 m. The permafrost layer thins from the zone of maximum thickness northwards beneath the Arctic Ocean and southwards into warmer latitudes. Discontinuous permafrost is normally less than 60 m thick. The active layer may vary in thickness from a few centimetres over continuous permafrost to 4 m and more on discontinuous permafrost.

Formerly periglaciated areas, which generally lie equatorwards of the present periglacial zones and often extend beyond the limits of glaciation, may pose distinctive problems of management. Amongst these are the fact that the engineer may encounter unexpectedly variable ground conditions, materials, and features; and, in particular, potentially unstable low-angle solifluction slope deposits may be accidentally reactivated (see Chapter 5 and Sect. 8.7).

Thickness of different zones associated with

frozen ground, often an important consideration in engineering studies, can be determined in a variety of ways. Drilling is perhaps the most widely used method, but various geophysical methods, such as electrical resistivity and shallow seismic techniques, have been employed.

Numerous devices are available for monitoring thermal conditions above and within frozen soil, including simple glass thermometers, thermocouple resistance thermometers, thermistors, diodes, and transistors.

Fig. 8.12A shows a typical temperature profile through the active and permafrost layers, and Fig. 8.12B shows a specific example. This diagram shows several important features. Firstly, temperatures fluctuate most widely at the surface and in the active zone, largely in response to daily and seasonal changes in atmospheric temperature. Secondly, fluctuations decrease in depth until the level of *zero* (or *minimal*) *annual amplitude* is reached. Thirdly, below the level of zero annual

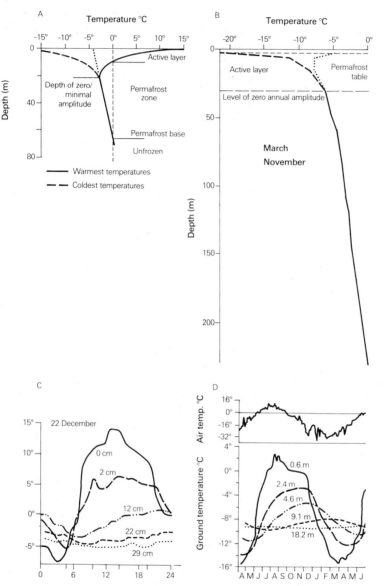

FIG. 8.12 (A) Typical temperature envelope for an area with permafrost (after Harris, 1986). (B) Temperature of permafrost in a shaft at Yakutsk, Siberia (after Muller, 1947). (C) Daily variations of air and soil temperature in Wright Valley, Antarctica (after Ugolini, 1966). (D) Seasonal fluctuations of air and ground temperature at Barrow, Alaska (after Price, 1972)

amplitude, temperature rises with depth; where temperature rises above 0 °C, permafrost ends.

Daily and seasonal fluctuations of air and ground temperature are exemplified in Fig. 8.12C and D. The daily march of soil temperature in Wright Valley, Antarctica, shows that daily fluctuations tend to decrease with depth and that maximum soil temperatures occur later at lower depths. Seasonal fluctuations (Fig. 8.12D) show similar trends. During the spring thaw, tempera-

tures rise relatively uniformly throughout the soil; but in the autumn freeze the change is not uniform owing in large measure to the fact that the latent heat of fusion is released when water freezes and soil temperature is thus maintained around freezing-point for some time. This so-called *zero curtain* is well shown by the 0.6-m depth curve for October and November in Fig. 8.12D.

The world-wide distribution of frozen ground reflects a kind of thermal equilibrium arising from relations between climatic conditions, the flow of heat from the earth's interior, the properties of earth materials, and the availability of water.

Climatic conditions are of overriding importance. According to Muller (1947), the following climatic conditions are most favourable for permafrost—cold, long winters with little snow; short, dry, and relatively cool summers; and low precipitation during all seasons. Seen in continental perspective, the occurrence of permafrost is broadly related to climatic conditions. In Canada, for example, Brown (1965, 1966) showed that there is a broad but not very close correlation between mean annual air temperature and permafrost distribution. Brown (1965) indicated that south of the $-1.1\,°C$ mean annual isotherm permafrost is rare; between the $-1.1\,°C$ and $-3.9\,°C$ mean annual isotherms permafrost near the ground surface is restricted mostly to peatlands; between $-3.9\,°C$ and $-6.7\,°C$ discontinuous permafrost is widespread; and north of the $-6.7\,°C$ isotherm permafrost is mostly continuous.

At a more practical level, several attempts have been made to predict the depth of freezing and thawing by means of climatic indices. One index, the *freezing index*, is based on the cumulative totals of degree-days below $0\,°C$. The freezing index is the number of degree-days between the highest and lowest points on the cumulative curve constructed from degree-days data plotted against time for one freezing season. There is a good correlation between the freezing index and frost penetration of the ground when dry unit weight and moisture content of the materials are taken into account.

Within the overall control exerted by climate, the distribution of permafrost is influenced by several other variables. Of these, vegetation and ground cover, hydrology and snowfall, terrain, and surface materials are fundamental. Most permafrost occurs either beneath the northern *boreal forest* or, north of the tree line, beneath *tundra* vegetation dominated by low-growing sedges, grasses, mosses, and lichens. The boundary between continuous and discontinuous permafrost in some areas appoximately coincides with the southern limit of the tundra. The importance of vegetation is mainly that its thermal properties determine the movement of heat into and out of the ground (Brown, 1966), although the precise effect of vegetation on permafrost characteristics is

often difficult to disentangle from the effects of related environmental variables. But it is clear, for example, that depth of thaw increases if vegetation cover is removed, and in the zone of sporadic permafrost, ice may occur preferentially beneath a blanket of insulating peats (see Sect. 8.4 below).

Surface-water bodies also influence the occurrence of ground ice (Ferrians *et al.*, 1969). Beneath lakes and rivers that do not fully freeze in winter, permafrost is thinner or may be absent, especially in the zone of discontinuous permafrost. Similarly, the permafrost table tends to reflect the shape of the ground, rising beneath hills and falling beneath valleys, and responding to insolation variations due to ground aspect.

Finally, the surface materials themselves and their water content relate significantly to frozen ground. Some of the important properties of the materials are their thermal conductivity, specific heat, volumetric heat capacity, and thermal diffusivity, and these, in turn, relate to fundamental characteristics of the materials such as texture, mineral composition, and packing of particles. Thermal conductivity, for example, is about four times greater for ice than for water, and therefore frozen ground has a much higher thermal conductivity than unfrozen ground. Equally, thermal conductivity varies with the kinds of material. Fig. 8.13A shows some relations between the heat conductivity of sand, clay, and water at different temperatures. This figure shows, for instance, that the thermal conductivity of sand is approximately half that of clay. Similarly, the susceptibility of soils to freezing and heaving as the result of ice segregation varies according to the composition of material. Many engineers use the percentage of grains smaller than 0.02 mm as a rule-of-thumb criterion. For example, fairly uniform sandy soils must contain at least 10 per cent of grains smaller than 0.02 mm for ice layers to form, and in soils with less than one per cent of such particles, no ice layers are likely to develop. Fig. 8.13B shows the limits of frost-heaving susceptibility according to grain size, taken from the work of various authors.

Table 8.4 shows the US Army Corps of Engineers' frost design soil classification based on particle size in which F1 is least susceptible to frost heave, and F4 is most susceptible.

The thermal pattern of frozen ground today reflects a quasi-equilibrium state that is related primarily to the refrigerating effects of climate and the loss of heat from the earth's interior, and is

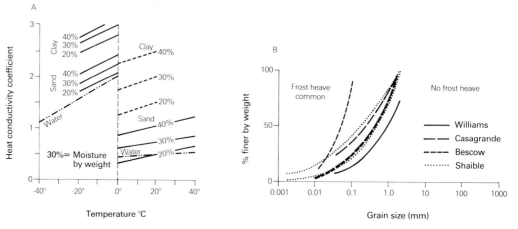

FIG. 8.13 (A) Relations between the thermal conductivity of sand, clay, and water at different temperatures (after Muller, 1947). (B) Limits of frost-heave susceptibility with respect to grain size (after Corte, 1969, and Embleton and King, 1975*b*)

locally conditioned by the nature of vegetation, the availability of water, and the properties of sediments. It is only in a shallow surface zone that temperatures fluctuate markedly, daily or seasonally.

While the thermal equilibrium may appear to arise from present conditions, it is clear that most permafrost has developed over millenniums, and some would argue that much continuous and sporadic permafrost is in fact a relic from colder climates. Certainly the well-known evidence of deep-frozen extinct mammals points to the antiquity of permafrost. Equally, permafrost is said to be actively forming in some areas. But it is

contemporary changes to frozen ground that are of the greatest interest to the resource manager.

8.4 Contemporary Aggradation, Degradation, and the Disturbance of Frozen Ground

(a) *Disruption of equilibrium*

The thermal equilibrium of frozen ground may have taken millenniums to become established but, even without human interference, permafrost conditions may still adjust to changing environmental conditions. It is thus useful to classify

TABLE 8.4 US Corps of Engineers frost design soil classification

Frost group	Soil type	Percentage finer than 0.02 mm, by weight
F1	Gravelly soils	3 to 10
F2	(a) Gravelly soils	10 to 20
	(b) Sands	3 to 15
F3	(a) Gravelly soils	> 20
	(b) Sands, except very fine silty sands	> 15
	(c) Clays, PI > 12	—
F4	(a) All silts	—
	(b) Very fine silty sands	> 15
	(c) Clays, PI < 12	—
	(d) Varved clays and other fine-grained, banded sediments	—

Source: Harris (1986).

permafrost in terms of its *natural thermal stability*. Harris (1986), for example, classified stability in terms of permafrost temperature: stable (colder than −5 °C), metastable (−2 to −5 °C), and unstable (warmer than −2 °C) (Fig. 8.14A). This analysis showed, *inter alia*, that unstable permafrost accounts for at least 40 per cent of the permafrost areas of Asia and North America. An alternative approach is to classify permafrost areas according to their sensitivity to the removal of soil and vegetation (Grave, 1983, 1984; Fig. 8.14B). This approach reveals, for instance, that where soils are relatively thin and on bedrock the susceptibility to human disruption is less. From the practical point of view, it is of the greatest significance that the thermal equilibrium can be

rapidly altered in some locations, and that permafrost may be reduced (degraded) or increased (aggraded) as a result of the modifications to existing conditions by human activity.

People modify existing conditions in a great variety of ways; many of their deliberate changes have unforeseen and undesirable consequences. The main changes are as follows: (i) removal of vegetation, perhaps deliberately for timber or to create cleared land for agriculture, or perhaps accidentally, as in the case of a forest fire, or the stripping of vegetation by tracked vehicles; (ii) modification of drainage conditions by, for instance, draining bogs, diverting rivers, or creating reservoirs; and (iii) construction of buildings, roads, pipelines, and associated ground pre-

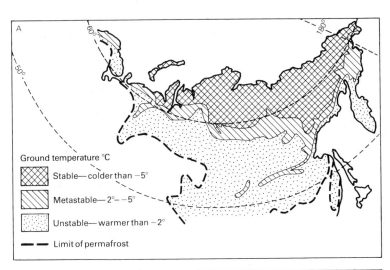

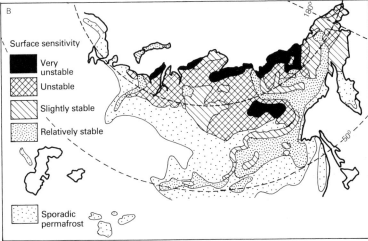

FIG. 8.14 (A) Thermal stability of permafrost in Asia (after Harris, 1986). (B) Sensitivity of ground in Asia to the removal of soil and vegetation cover (after Grave, 1983)

paration without appropriate precautions. In all these cases, the effect is usually permafrost degradation, and the change may be irreversible. The tendency is towards the establishment of a new thermal equilibrium in the ground. With small structures, the relaxation time may be rather short, and the new equilibrium might be established in a year or two; with large structures or extensive land-use modifications it may take decades to re-establish the equilibrium.

The nature and rate of change will vary according to initial ground and climatic conditions and the nature of the modification. Susceptibility to disruption is usually greatest in the extreme north where the active zone is shallow, ice content of the near-surface soil is great, and the insulating organic layer is thin; in general, susceptibility declines southwards.

Three examples will illustrate more precisely the influence of human activities on permafrost destruction: permafrost change with land use in Alaska; the influence of two building types on the permafrost table; and the predicted effects of an oil pipeline and permafrost.

Fig. 8.15 shows the degradation of permafrost under different land uses on silt over a 26-year period as observed by the US Corps of Engineers near Fairbanks, Alaska. Beneath the natural area of trees, brush, moss, and grass the permafrost

table was maintained at its original level of 1.1 m. Where trees and brush were removed, the table was lowered to 3 m in 10 years and to 4.7 m in 26 years. And where all vegetation was stripped and its insulating effect was therefore absent, the table was lowered to 3.8 m in 10 years and to 6.7 m in 26 years. In addition, the maximum depths of seasonal freezing were also lowered.

Structures, like houses, can be seriously damaged by both freezing and thawing. With freezing, if the surface material is bonded to the structure, heaving due to ice-segregation may deform the building (Harris, 1986). Two kinds of change to permafrost caused by small buildings constructed without significant precautions are shown in Fig. 8.16. If the building is unheated, the permafrost table may rise in the shadow, as it were, of the surface insulation provided by the building. If the building is heated, the permafrost table may fall perhaps for several years. In both cases, the depth of permafrost has changed and so has the thickness of the active zone.

If an oil pipeline is constructed in a permafrost area, the warm oil passing through the pipe will thaw a zone around it. Fig. 8.17 shows the *predicted* growth of the thawing zone around an oil pipeline, 1.2 m in diameter, with its axis buried 2.4 m below the surface, passing through 'medium silt' permafrost and carrying oil that maintains its

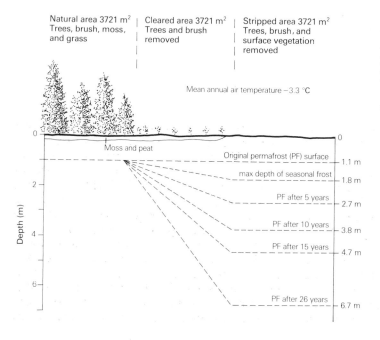

FIG. 8.15 Permafrost degradation under different surface treatments over a 26-year period. Mean annual permafrost temperature ranges from about −0.5 °C at a depth of 10 m under natural forested areas to about −0.2 °C where permafrost is degrading (after Linell, 1973)

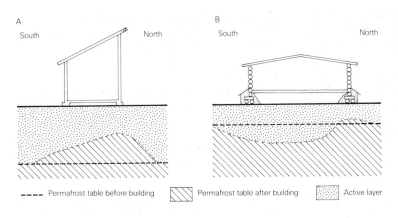

FIG. 8.16 Alterations to permafrost caused by (A) an unheated shed and (B) a heated cabin (after Tsytovich and Sumgin; from Muller, 1947)

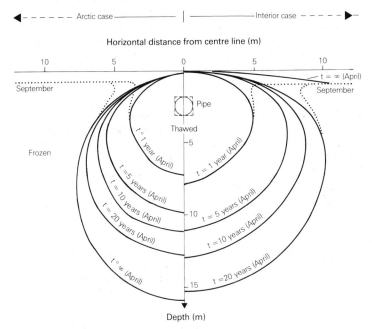

FIG. 8.17 Predicted growth of the thawed cylinder around a 1.21 m diameter oil pipeline with its axis 2.42 m beneath the surface and maintained at a temperature of 80 °C, near the Arctic coast of Alaska (left) and (right) near the southern limits of permafrost (after Lachenbruch, 1970)

temperature at 80 °C, for Arctic coast conditions (left) and (right) near to the southern limits of permafrost (Lachenbruch, 1970). These calculations were made in anticipation of an underground oil pipeline—the Alyeska pipeline—being constructed south from Prudhoe Bay, Alaska, and their implications are important. In the first place the thawed cylinder would continue to expand for several decades. Secondly, in certain circumstances the thawed zones might seriously affect the stability of the structure. For example, if excess water generated by the growing thawed cylinder cannot be removed as quickly as it is formed, part of the cylinder may behave as a fluid and the pipe

might settle or be broken, and the rate of thawing may be increased. Such implications were crucial to the design of the Alyeska pipeline, which was planned to cross both continuous and discontinuous permafrost. As a result of field experiments, it was finally decided to build much of the pipeline above ground on piles so that the heat was dissipated and the piles could respond to movements associated with temperature changes (e.g. Williams, 1979); where the pipeline was buried, thermokarst features (see below) have developed (e.g. Thomas and Ferrell, 1983). More recently, the problems of warm oil pipelines have been matched by those associated with chilled gas pipelines. The gas pipelines need to be chilled to ease movement, and the pipelines need to be buried for safety. In discontinuous permafrost, 'frost bulbs' can grow around such pipes and cause surface heave (e.g. Williams, 1979; French, 1980).

The consequences of changing thermal equilibria are extensive. Waterlogged channels may be created, erosion may extend along tracks, foundations may be disrupted, and the nature of frost action and solifluction may be altered. Properties of the ground surface, such as its bearing strength, may be changed. Almost all types of human activity are likely to be affected, including water-supply and waste-disposal systems; drainage; runways, roads, railways, telephone lines; buildings; mines, dams, bridges, and reservoirs.

Although human activity causes change by disrupting the established frozen-ground conditions, it is important to emphasize that such changes generally only influence human activities through processes that are in any case operating in the periglacial environment. These processes are briefly introduced below.

(b) *Ground ice and geomorphological processes*

The most destructive processes of concern to people in permafrost areas are related to the formation and thawing of ground ice, either within the permafrost zone or in the active zone above it. The two principal groups of processes are *frost action* (processes arising from ice formation in soils) and *solifluction* (processes of soil flow in the active zone). Frost action includes *frost shattering*, *frost heaving*, and *frost cracking*.

Critical to these processes are the nature and mode of origin of ground ice. There have been many attempts to classify ground ice and the classification adopted here is by Mackay (1972). This classification is based on the origin of water prior to freezing and on the processes by which water is transferred to the freezing plane. In all, ten ice forms are recognized (Fig. 8.18). From the practical point of view, the most important ice forms are those arising from thermal contraction of frozen ground, and those associated with segregated and intrusive ice.

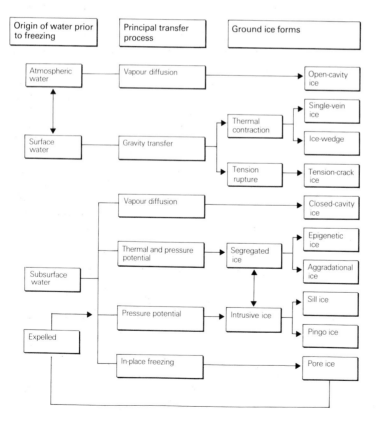

FIG. 8.18 Classification of underground ice forms (after Mackay, 1972)

Briefly, the characteristics of the main ice forms are as follows. *Open-cavity ice* is formed by sublimation of ice crystals directly from atmospheric water vapour. *Single-vein ice* forms when an open crack penetrating permafrost becomes filled with water and freezes. *Ice wedges*, like vein ice, arise from thermally induced cracking of frozen ground. *Tension-crack ice* grows in cracks produced by the mechanical rupture of the ground (usually by growth of segregated or intrusive ice). *Closed-cavity ice* is relatively unimportant, and forms by vapour diffusion into enclosed cavities in permafrost. *Segregated ice* includes *epigenetic ice* that grows as lenses in material predating the lenses and may grow into massive bodies. The second form of segregated ice is *aggradational ice* which develops when the permafrost table gradually rises and may incorporate epigenetic ice lenses. *Intrusive ice* is formed by intrusion of water under pressure, and its freezing, causing uplift of the ground above it. One type, *sill ice* is formed when water is intruded between horizontal sheets, as along bedding planes. *Pingo ice* occurs where intrusive ice domes the overlying surface. Finally, *pore ice* is that which holds soil grains together. The amount of ice in frozen ground may be more than available pore space (supersaturation), about the same as pore space (saturation), less than pore space (undersaturation), or the sediment may be cemented by ice that does not form veins or granules. Ground may be hard frozen (grains are embedded in an ice cement), plastic frozen (with some water remaining unfrozen), or granular frozen (in which grains are in contact and excess ice is absent).

8.5 Frost Action and Solifluction: Processes and Control

(a) *Frost heave*

Frost heave associated with segregated ice in supersaturated material involves two separate processes. The first causes displacement of the ground surface and arises from growth of ground ice. The surface disruption may be in the order of many centimetres, and the consequences for buildings, lines of communication, and other engineering structures may be serious (Berg and Wright, 1984). This process produces several distinctive landforms. The second causes the vertical sorting of particles, in which large stones

in mixed sediments migrate upwards to the surface and poles, pilings, fences, etc. are forced out of the ground. The most common natural features reflecting this process are the stone pavement, and stripes, nets, circles, and other patterns of surface stones.

(i) *Displacement of ground surface* Vertical displacement of the ground surface on a small scale due to the growth of ice crystals arises to a slight extent from the 9 per cent increase in volume of water when it is converted to ice, but largely by the force of crystallization of ice. The continuation of the latter requires the transfer of water to the ice crystal or growing ice lens, and there has been much discussion of the process since it was first recognized by Taber (1929 and 1930). Differential swelling of the ground may arise for three main reasons: unequal distribution of load, areal differences in the texture of the ground and ground cover, and differences in the amount of water available for ice growth (Taber, 1929; Muller, 1947). Two of the factors that make soils particularly susceptible to frost heave have already been mentioned: particle size and water availability. Also important are pore space, capillary properties of soils, rate and frequency of freezing and thawing, and depth of frost penetration. Examples of the effects of heave on structures are shown in Fig. 8.19.

Ground hummocks (known by various names such as *palsas* and *mima mounds*) are rather small features often attributed to ground heaving. They are usually in the order of 0.5–2 m high (and perhaps as high as 10 m) and up to about 10 m in diameter, and are commonly associated with poorly drained areas and peaty material (Lundquist, 1969). The origin of these features in periglacial and some mid-latitude regions has long been a matter of controversy. Their dimensions and composition are varied, and undoubtedly no single explanation is satisfactory in all areas. Some argue that hummocks were elevated by frost heaving following segregation of ice lenses; others have suggested that expanding patches of frozen ground squeezed earth between them into mounds; there are those who consider that degradation of permafrost (by removal of vegetation for instance) might cause ice wedges to thaw, causing subsidence and material to slump into the hollows, leaving mounds between them.

A large feature, associated with intrusive ice, is the ice-cored mound called a *pingo*, which may be

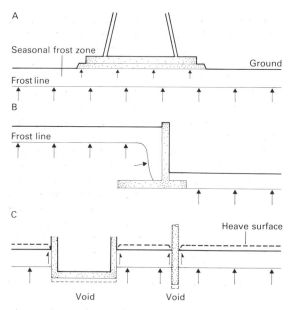

FIG. 8.19 The effects of frost heave: (A) Heaving of soil in seasonal frost zone causing direct upward thrust on overlying structural elements. (B) Freezing of frost-susceptible soil behind walls causing thrust perpendicular to freeze front. (C) Force at base of freezing interface tends to lift entire frozen slab, applying jacking forces to lateral surfaces of embedded structures, creating voids underneath. Structures may not return to original position on thawing (after Linell and Tedrow, 1981)

up to about 100 m high and 600 m wide (Flemal, 1976). Pingos occur in zones of continuous and discontinuous permafrost. Two main types have been distinguished—the *closed-system type* (Mackenzie type) and the *open-system type* (East Greenland type). The closed-system type (Mackay, 1966) usually develops as follows. If a lake that is not normally fully frozen and is underlain by unfrozen saturated sands becomes fully frozen (as the result, for example, of draining or sedimentation), permafrost will encroach upon the zone beneath the lake and create a closed system of unfrozen material. Growth of permafrost around this core increases water pressure in it, and the surface layer is heaved to relieve this pressure. Ultimately, the closed system is entirely frozen and the excess water forms an ice core beneath the dome.

The open-system pingo usually occurs on sloping ground in discontinuous permafrost where hydraulic pressure in taliks causes water to approach the surface and freeze. A continual supply of water builds up the ice mass and this domes the surface. Occasionally tension cracks develop on the crests of pingos. And the phenomenon of 'drunken forest' is often associated with them—raising of the ground causes the original trees to be inclined away from the vertical.

(ii) *Vertical migration* Vertical upwards migration of large objects in the soil or other unconsolidated material has been investigated under laboratory conditions, and Washburn (1979) recognized two types of mechanism. The *frost-pull mechanism* depends on the ground expanding during freezing and carrying the large objects (e.g. stones) upwards; on thawing, the fine material moves down and beneath the large objects so that they do not return to their original position. The *frost-push mechanism* depends on the fact that the thermal conductivity of stones is greater than that of the soil. As a result, stones heat and cool more quickly than the soil and ice would form first beneath them, forcing them upwards. During thawing, fine material would move beneath the stones and prevent them returning to their original positions. Such processes can upturn stones so that their long axes are vertical, a distinctive feature of *stone stripes* and *polygons* (Harris, 1986). More importantly, they can lead to the ejection and damage of pipes, culverts, piles, and foundations.

(iii) *Control of frost heave* The control of frost heave depends initially on a knowledge of climatic, material, and water conditions, and particularly on the nature of freezing and thawing. As discussed previously, it is useful to predict the depth of freeze–thaw in terms of freezing indices, for example, so that appropriate design measures can be adopted.

Frost heaving can be reduced in three main ways (Corte, 1969): by chemical treatment of the soil, by filling of soil voids, and by structural design precautions. Many different techniques have been tried, and the examples below illustrate some of the more successful methods in these three categories. The injection of calcium chloride (CaCl) has been shown to reduce effectively freezing temperature of water in the soil and to reduce the loss of strength of materials arising from freeze–thaw cycles. A neutral solution of waste sulphite liquor from the paper industry in Canada has been used to reduce soil capillarity (and thus reduce water flow to growing ice lenses) and to slow the rate of ice growth. Voids in the soil can be filled with a variety of often rather expensive cements, such as polymers, resins, portland cement, and other fine material. Chemicals have also been used to aggregate or disperse fines, and efforts have been made to wash out fine material from the soil. Design methods include increasing the thickness of pavement and sub-grade materials to prevent frost penetration, and insulation of the surface with natural materials such as straw, moss, or peat, or artificial materials such as cellular glass blocks. In some circumstances it may not be necessary fully to control frost heave; simpler measures may be satisfactory. For example, it may be adequate to create a pavement surface for a road that has sufficient bearing capacity during the thawing season when the soil is at its weakest.

(b) *Frost cracking, ice wedges, and patterned ground*

Permafrost areas are very frequently characterized by polygonal ground patterns formed by cracks that are the location of ice wedges. The importance of these features is that they provide evidence of ground-surface conditions (such as the type of material, and nature and amount of ground ice) and, if the ice wedges are caused to melt for any reason, they can become the loci of subsidence and seriously disrupt superincumbent structures (Black, 1976). The cracks may be up to 10 m deep, and the diameters of the polygons may exceed 100 m. The patterns of polygons are usually dominated by orthogonal junctions between cracks, but non-orthogonal patterns are commonly found.

Lachenbruch (1966) has reviewed the mech-

anics of frost-wedge formation. Fig. 8.20 illustrates four stages in the evolution of an ice wedge according to the thermal contraction theory. During the winter, because the thermal coefficient of contraction of frozen ground is rather high, tensile stress in the material is likely to exceed its tensile strength and cracking through both the active and the permafrost layers (A). By the following autumn (B), water has penetrated the cracks and a vein of ice has been formed. Compression caused by re-expansion of the melting permafrost in summer deforms the near-surface strata. In the following winter, the crack is reopened, first in the top of the ice wedge beneath the active layer. The next spring thaw fills the new crack with water, and so on. Analysing the stress in permafrost material, Lachenbruch suggested that stress depends mainly on the amount and rate of cooling, and on the temperature at the time of cracking.

The centres of polygons between ice wedges may be higher or lower than their edges. Where growth of the ice wedges causes upheaval of material adjacent to them, as for instance in areas where wedges are actively growing, the edges of the polygons will be higher than their centres. But where ice wedges are thawing and may be acting as loci of drainage concentration, the centres of the polygons may be higher. This contrast therefore reflects the condition of the ground, for actively growing ice wedges are largely restricted to zones of continuous permafrost, and they are less active or melting in the zones of discontinuous permafrost. There are numerous other patterned-ground

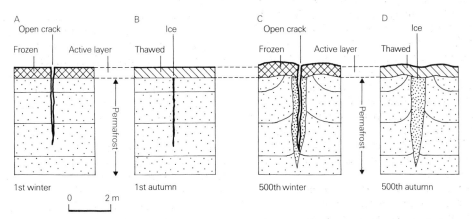

F IG. 8.20 Schematic representation of ice-wedge evolution according to the contraction-crack theory (after Lachenbruch, 1966)

phenomena in periglacial areas (see, for example, Washburn, 1979) and most but not all of them are associated with frost heaving and frost cracking.

(c) *Thawing, settling, and solifluction (gelifluction)*

The properties of thawed ground are usually markedly different from those of frozen ground. For instance, the mechanical properties of frozen ground tend to approach those of ice, having higher shear strength and thermal conductivity than similar but unfrozen material. When sediments in the active zone thaw, therefore, they behave differently from the frozen ground. Frequently, and especially beneath a superincumbent load such as that of a building, they *settle* or subside, and behave like a viscous fluid. Certain materials are particularly susceptible to settling: for example, gravel embedded in large quantities of ice is inherently unstable when thawed, and silty frozen soil with a high porportion of ice generally becomes waterlogged and spongy when thawed. If the ground surface is inclined, the thawed material may flow, even on very low-angle slopes. This process is called *solifluction* (or *gelifluction*, which is solifluction associated with frozen ground), and it is the most important of several periglacial mass-wasting processes (Harris, 1981).

Features arising from the degradation of ground ice in permafrost areas are collectively called *thermokarst landforms*. They include certain kinds of pits, dry valleys and lakes, and ground-ice slumps (e.g. French, 1974). All these subsidence features are formed by thawing of supersaturated icy soils at the top of permafrost (Mackay, 1970), and they are most pronounced where melting affects abundant ground ice in the upper permafrost. Clearly the cause of thawing is important. Climatic changes may cause regional thaw. Kachurin (1962), for example, has pointed out that thermokarst is most extensive in the southern permafrost areas of the Soviet Union, where regional climatic change may be responsible. Seasonal thaw can also lead to thermokarst development. Local thawing is often initiated by people through, for example, vegetation clearance, compaction of peat, fire, bulldozer tracks, ploughing, ditch-digging, and construction (e.g. French, 1975). Often the thawing pattern reflects the location of near-surface ice, perhaps following the line of an ice wedge or coinciding with thawed pingos. Degradation of permafrost may proceed

laterally, usually as a result of horizontal erosion by water. The extension of certain lakes (*thaw lakes*) is attributed to this cause. Vertical degradation is more common and may be illustrated by the development in Siberia of *alases*, large, flat-floored, vegetated basins (Czudek and Demek, 1970). These are initiated by the melting of ice wedges that become the loci of lakes. As the lakes deepen and coalesce, they may become too deep to freeze entirely in winter and the rate of permafrost degradation beneath them is increased. Eventually the large lakes may be filled or drained, vegetation may cover the floors, and the formation of alases is complete.

Solifluction is important because it can severely disrupt surface structures. Solifluction comprises two main processes, namely the flow of water-soaked debris, and the creep of surface material by freeze–thaw action (Benedict, 1976). The first involves the thawing of ground in the active zone or the addition of water to the thawed surface by snow-melt. The presence of permafrost is not essential, but if it exists its relative impermeability helps to maintain water saturation in the near-surface material. As the ground thaws or water is added to it the weight of material increases, and its shear strength is reduced because water reduces both internal friction and cohesion. The nature of flow that results depends mainly on the characteristics of the material (such as its texture and viscosity), water availability, depth of thaw, slope of the ground, and vegetation cover. Solifluction is especially facilitated by a high silt content and an abundant supply of water. It can occur on slopes as gentle as $2°$. Vegetation insulates the ground beneath and thus probably reduces the depth of the thawed zone and restricts the rate of *surface flow*. The second process, creep induced by freeze–thaw, is caused by the heaving of particles on a slope upwards normal to the ground surface and the lowering of them vertically upon thawing under the influence of gravity. This process, like frost heaving, depends on the frost-susceptibility of the materials, the availability of water, vegetation cover, and the nature of freeze–thaw activity.

Rates of solifluction have been measured in several areas, mainly by recording the translocation and vertical deformation of linear objects, such as pipes and cables, inserted through the mobile layer. Most movement usually occurs during the spring thaw, and field measurements show that rates may range from less than a centimetre to over 30 cm a year. Solifluction produces several distinctive landforms. The most common are lobes and terraces of material, of which some are terminated by stone banks and some are covered with vegetation. These features normally occur on slopes over 5°, their treads may by several tens of metres wide, and their risers may be a few metres high.

Various other forms of mass movement occur in periglacial areas. These are dealt with in Chapter 5; see also McRoberts and Morgenstern (1974).

8.6 Solving Permafrost Problems

(a) *Terrain evaluation*

Evaluation of permafrost conditions and the terrain associated with them normally involves the approaches and techniques described in Chapter 2. Because most permafrost terrain is remote and inaccessible, reconnaissance surveys are usually undertaken in the first instance by an analysis of available remote sensing imagery, notably panchromatic aerial photographs and satellite imagery (e.g. Frost, 1961; Linell and Tedrow, 1981; Gavrilov *et al.*, 1983; Gaydos and Witmer, 1983; Maüsbachar, 1983). Excellent examples of the valuable evidence provided by such imagery in Canadian permafrost terrain are to be found in Mollard's (no date) *Landforms and Surface Materials of Canada*. The purpose of most such surveys is to determine the nature of terrain, material, and ground-ice conditions and sources of construction materials, to plan access routes and provisional layouts for structures, and to select sites for detailed studies on the ground. Naturally these purposes will vary accoring to the specific project. For example, Mollard and Pihlainen (1966) identified five objectives for an air-photo interpretation exercise related to the building of an arctic road: to classify terrain, to locate the most likely route, to map and evaluate terrain conditions, to map granular materials, and to comment on terrain conditions and their engineering implications at previously selected sites. Often air-photo surveys will be carried out first on smaller-scale photographs or satellite images (e.g. 1:40 000–1:200 000) and then pursued in greater detail on larger-scale imagery (e.g. 1:20 000). A comparison of the utility of different types of imagery in terrain appraisal in permafrost areas is shown in Table 8.5.

It is in such reconnaissance surveys that techniques of land-system mapping, geomorphological mapping, and site classification are especially valuable. The methods described in Chapter 2 have been more or less explicitly employed in many environmental surveys of permafrost areas (e.g. Hughes, 1972: Mollard, 1972; Johnston, 1981; Maüsbacher, 1983). In preparing reconnaissance maps an understanding of landforms and drainage patterns, together with a knowledge of ecological conditions, is indispensable. In particular, as the discussion in previous sections has shown repeatedly, features such as pingos, ice wedges, high-centred and low-centred polygons, thermokarst phenomena, and solifluction lobes and terraces all provide valuable information on fundamentally important considerations of ground-ice and permafrost conditions, and the nature of superficial materials (see *Quaternary Research*, 6, 1976). Table 8.6 summarizes ground conditions commonly associated with different periglacial landforms, and Table 8.7 links landforms with their implications for site and route studies. Fig. 8.21 shows the normal relationships between permafrost landscape features and permafrost zones. These predictively valuable relationships should only be used with caution, however, because they are not perfect (see Ferrians *et al.*, 1969). For example, in better-drained soils or in arid areas, the diagnostic landforms may not develop (e.g. Harris, 1986).

The distribution of water bodies is also a valuable guide to the same variables and to such things as the possibilities of water supply. At the same time, an understanding of periglacial ecology also helps terrain evaluation. Plant communities are often sensitive indicators of soil-moisture conditions and communities differ in the insulation they provide to the ground; easily recognized peat bogs (known as *muskeg* in Canada) pose special engineering problems; and active surface disturbance is often revealed by, for instance, 'drunken forests' and the curvature of tree trunks. Plant associations can also be used to predict soil conditions (Table 8.8).

The field studies that invariably follow reconnaissance surveys are normally directed towards validating the air-photograph predictions and analysing in detail the conditions at specific sites. More information is generally required on the precise nature and distribution of permafrost, talik, the active zone, vegetation, surface

TABLE 8.5 Comparisons of imagery

Type of imagery and common scales	Applicability and relative usefulness for site and route studies in permafrost areas
High altitude black and white panchromatic photography. Common scales: 1:60 000 to 1:80 000	Very useful. Essential for almost all important site and route investigations. Many permafrost terrain features are recognizable or can be inferred reliably. Permits wide stereoscopic view.
Medium level to low level panchromatic photography. Common scales: 1:10 000 to 1:40 000	Very useful. Usefulness usually increased by prior stereoscopic study of high-level airphotos. Microrelief features identifiable. Provides a cross correlation of many features of the landscape.
Black and white infrared photography. Common scales: 1:30 000 to 1:60 000	Greatest assistance over the southern fringe of permafrost in mixed wood (coniferous and deciduous) forests. Shows much detail in drainage and high water-table; commonly reveals the relationship of vegetation to local drainage conditions. This film is best for identification of tree species (coniferous vs deciduous) and is relatively inexpensive.
Colour infrared photography. Common scales: 1:30 000 to 1:60 000	Applicable mainly to detecting healthy or stressed vegetation, one of several stresses being a thin active layer. This layer often increases in thickness after fires, slides, or clearing; and this may be revealed on colour infrared photos. Also shows thicker active layer (pinkish) around water bodies, where thawing is taking place or has taken place.
Radar. Scales: variable	Limited. Provides overall visualization of broad physiographic features (mountains, plains, lakes, etc.). Many terrain details are obscure and fuzzy. Almost all-weather capability. Can be taken day or night, but is of limited use.
Thermal infrared imagery. Scales: variable	Limited. Mainly, with some exceptions, confirms terrain relationships already suspected and better observed on suitably scaled panchromatic photographs. Requires a knowledge of the many factors that influence dark and light grey tones on imagery, some of which cannot be evaluated if details are not known at time of image-taking. Two-dimensional image analysis restricts use. Also, the change in surface temperature in response to vegetational-climatic effects (e.g. evapo-transpiration) and recent rainfall may cause tonal anomalies and give erroneous or misleading results.
Satellite imagery (ERTS now LANDSAT MSS) at 1:1 000 000 contact scale; enlargements to 1:250 000 scale	Limited. Reveals general hydrographic and vegetation patterns and effects of fire. Shows major geomorphic features. Most terrain details are obscure and fuzzy. Provides overview for initial appreciation of terrain along transportation routes.

Source: Johnston (1981).

TABLE 8.6 Common topographic features that indicate ground conditions in arctic and near-arctic regions

Feature and description	Associated ground conditions
Polygonal ground (ice wedges)—Usually indicates the presence of a network of ice wedges. Wedge networks are also common in wet tundra where no surface expression occurs. (Subject to extreme differential settlement when surface is disturbed.)	Typically indicates relatively fine-grained unconsolidated sediments with permafrost table near the ground surface; also known from coarser sediments and gravels where wedge ice is less extensive.
Stone nets, garlands, and stripes—Frost heaving in granular soils produces netlike concentrations of the coarser rocks present. If the area is gently sloped, the net is distorted into garlands by down slope movement. If the slope is steep, the coarse rocks lie in stripes that point downhill.	Strong frost action in moderate well-drained granular sediments that vary from silty fine gravel to boulders. Superficial material commonly susceptible to flowage.
Solifluction sheets and lobes—Sheets or lobe-shaped masses of unconsolidated sediment that range from less than a foot to hundreds of feet in width that may cover entire valley walls; found on slopes that vary from steep to less than 3°.	An unstable mantle of poorly drained, often saturated sediment that is moving down-slope largely by seasonal frost heaving. On steeper slopes they often indicate bedrock near the surface, and on gentle slopes, a shallow permafrost table.
Thaw lakes and thaw pits—Surface depressions form when local melting of permafrost decreases the volume of ice-rich sediments. Water accumulates in the depressions and may accelerate melting of the permafrost. Often form impassable bogs.	Poorly drained, fine-grained unconsolidated sediments (fine sand to clay) with permafrost table near the surface.
Beaded drainage—Short, often straight minor streams that join pools or small lakes. Streams follow the tops of melted ice wedges, and pools develop where melting of permafrost has been more extensive.	A permafrost area with silt-rich sediments or peat overlying buried ice wedges.
Pingos—Small ice-cored circular or elliptical hills that occur in tundra and forested parts of the continuous and discontinuous permafrost areas. They often lie at the junction of south- and south-east-facing slopes and valley floors, and in former lakebeds.	Silty sediments derived from the slope or valley, and groundwater with some hydraulic head that is confined between the seasonal frost and permafrost table or is flowing in a thawed zone within the permafrost. Those in former lake-beds indicate saturated fine-grained sediments.

Source: Ferrians *et al.* (1969).

materials, drainage, water supply, atmospheric and ground temperature and moisture conditions, and the performance of pre-existing buildings. The techniques used for these field investigations are numerous, and commonly include geophysical prospecting, drilling and other methods, test-pit studies, geological and geomorphological mapping, and meteorological observations.

Forearmed with the knowledge provided by such surveys, it has been possible to limit the impact of human activity in permafrost areas by planning regulations. In Alaska, for example, construction on tundra is prohibited in summer months, and north of 60 °N in Canada, land-use permits are required for any land development at any time of the year.

(b) *Engineering responses to permafrost problems*

There are four main engineering responses to the presence of frozen ground. It can be neglected, eliminated, preserved, or structures can be designed to take expected movements into account (Fig. 8.22). Permafrost can often be safely ignored where there is good drainage and ground materials are not susceptible to serious frost action and solifluction. Where permafrost is thin, sporadic, or discontinuous, and where thawed ground has a satisfactory bearing strength, the possibilities

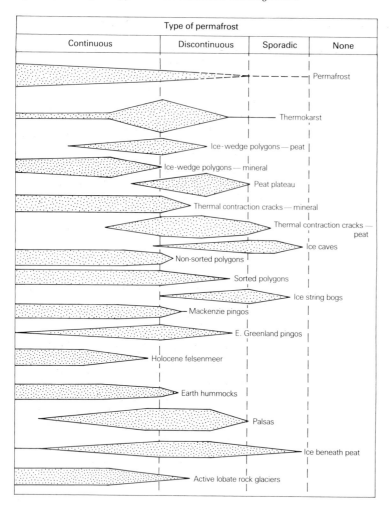

FIG. 8.21 Relationship between permafrost landforms and permafrost zones (after Harris, 1986; © 1986 S.A.H. Books Ltd.)

might be examined of removing the frozen ground by, for example, stripping the insulating cover of surface vegetation, by excavation, by treatment of the ground, or by thawing of the ice with steam. Material not susceptible to frost action may be placed on the surface. This approach is known as the *active method*. In regions of continuous permafrost, elimination of frozen ground is likely to be impracticable, and the *passive method* may be adopted, whereby efforts are made to preserve the permafrost by, for instance, insulating the surface with vegetation mats or gravel blankets, ventilating the undersides of heat-generating structures, preserving vegetation, or using piles, pads, or stable footings.

The techniques of design and construction in permafrost areas are numerous, and they have been extensively considered in other reviews especially in the context of buildings, roads and railways, airfields, oil and gas pipelines and related structures, mining, water supply and waste disposal (e.g. Muller, 1947; Price, 1972; Williams, 1979; Harris, 1986). Only a few important considerations will be mentioned here.

In the first place, it is axiomatic that successful responses will be based on an understanding of permafrost, and a local knowledge of permafrost conditions. Secondly, whether an active or a passive approach is adopted, efforts will normally be made to disturb as little as possible or to control the thermal regime established before construction begins. Thirdly, although there is inevitably likely to be some settling or heaving after a development has been completed, the engineering aim is usually

TABLE 8.7 Landscape features associated with permafrost. (Smaller features best observed on 1:10 000 photographs)

Permafrost zone(s)	Surface feature[a]	Implications for site and route studies
Continuous and discontinuous	Thermokarst (ponds, drained lakes, peat-filled depressions)	Ice-rich permafrost in waterlaid sediments; an indication of high contents of ground ice in hummocky terrain; less common in other materials; potential for bimodal flow, thaw settlement, and frost action; depth of thaw basins significant
Mainly continuous	Solifluction lobes, sheets, stripes, terraces, steps	Frost action and drainage are important; often indicates bedrock or compact till layer near ground surface; rate of solifluction creep is a function of slope; usually 1/2 to 5 cm/y and rarely more than 20 cm/y
Continuous and discontinuous	Vegetation: cottongrass tussocks	Poor overland trafficability. Frost-susceptible (fine-grained) materials—thus frost heave
Continuous and discontinuous	Earth hummocks	Poor overland trafficability. Frost-susceptible (fine-grained) materials—thus frost heave
Continuous and discontinuous	Cemetery mounds	Differential thaw subsidence, and slope failures in the active layer
Continuous and discontinuous	Active and fossil ice wedge polygons	Low centre polygons suggest ice-rich permafrost and poor surface drainage. High centre polygons in granular deposits suggest better surface drainage. Ice wedge polygons generally underlie but are not commonly evident on long smooth slopes, being obscured by mass-wasting. Distinct active ice-wedge polygons are limited at present to the continuous zone, whereas less conspicuous and commonly incomplete polygons with inactive ice wedges can sometimes be identified in the discontinuous zone, as well as fossil polygons in sand and gravel plains from which the ice wedges are melted out, with the sand and gravel unfrozen
Continuous and discontinuous	Beaded drainage	Stream erosion of edge ice at intersections of polygon fissures
Continuous (sporadic in discontinuous)	Pingos; breached and/or collapsed pingos	Closed-system pingos occur in fine-grained, frost-susceptible material in former lake-beds; open-system pingos form where groundwater moves through taliks beneath slopes
Continuous and adjoining discontinuous	Peat polygons	Indicator of southern margin of continuous permafrost; sometimes found on high plateaus well south of continuous permafrost zone
Discontinuous	Peat plateaus	Ice-rich sphagnum peat over other peat types; elevated and dry at ground surface
Discontinuous	Palsas	Ice-rich peat and organic silt, commonly prominent in vicinity of 25 °F (−4 °C) mean annual air isotherm but extending to 30 °F (−1 °C); usually occur in very wet settings, as at the intersection of 'strings' in patterned fens
Discontinuous	Collapse scars (thaw 'windows') and wide flat-bottomed drainage ways	Wet, thawed peat; in near-circular wet sedge meadows and marshes (thawed 'windows' in permafrost) and along drainageways

TABLE 8.7—*continued*

Permafrost zone(s)	Surface feature[a]	Implications for site and route studies
Continuous and discontinuous	Reticulated and ribbed fens (string 'bogs')	Thought to be polygenetic features, common in southern fringe of the discontinuous zone but occur rarely in the continuous zone as well; usually very wet and follow broad, low-gradient drainageways
Continuous and discontinuous	Skin flows and bimodal flows	Indication of local slope instability and a warning of a construction hazard
Continuous and discontinuous	Talus and rock glaciers	Indication of potentially unstable ground conditions
Continuous (mostly)	Subparallel (feather, horsetail) drainage	Often an indicator of massive ice bodies and icy sediments in the High Arctic

[a] Engineering problems associated with frost action in the active layer are commonly encountered in many of these features.

Source: Johnston (1981).

TABLE 8.8 Relations between soil conditions and plant associations in northern Alaska

Soil conditions	Plant associations
Xeric, shallow, rocky soils and bedrock with permafrost table about 1 m deep. Permafrost and seasonal frost are generally dry	Barren communities with lichens, mosses, and dwarf heaths. Where bedrock is exposed, crustose lichens and scattered herbs
Deep, mature, well-drained soils. Permafrost table about 1 m deep	Xerophytic mosses, lichens, dwarf heaths, and herbs
Gley (wet) soils (the most common condition). Permafrost table about 30 to 49 cm deep with ground ice	Cottongrass tussocks (*Eriophorum*) and dwarf heaths. Low areas colonized with lowland cottongrass (*Carex aquatilis*)
Organic (bog) soils. Permafrost table about 20 to 30 cm deep, but less where peaty material is well-drained	Lowland cottongrass (*Carex aquatilis*). Where there is shallow, standing water, *Dupontia* is generally present. Where peaty material is relatively dry, lichens and dwarf heaths colonize the substrate

Source: Linell and Tedrow (1981).

to minimize the effects of such changes either by modifying the process directly or by designing structures capable of withstanding them. Fourthly, lateral instability caused by solifluction is most commonly countered by avoiding affected sites or by firmly fixing structures in permafrost beneath the active zone.

In selecting the locations for buildings, certain conditions are preferable: a thin active layer, bedrock near to the surface, thawed material with adequate bearing properties, good drainage, lack of frost-susceptible materials, and a stable site away from possible areas of water seepage and icing. Because most building construction problems are associated with the active zone, foundations are frequently fixed in the underlying permafrost. A most important consideration here is the *adfreezing strength* of frozen ground. This is the resistance to the force required to pull apart the frozen ground from objects, such as foundations, to which it is frozen. In general, this strength is greater in sands than in clays.

Pilings are commonly used as foundations for buildings and the importance of adfreezing strength is usually recognized in their emplacement. In general, the aim is to freeze the pilings into the permafrost layer to a depth normally at least twice the thickness of the active zone, and to avoid adfreezing in the active zone by lubricating, insulating, or 'collaring' the piles in that zone. In

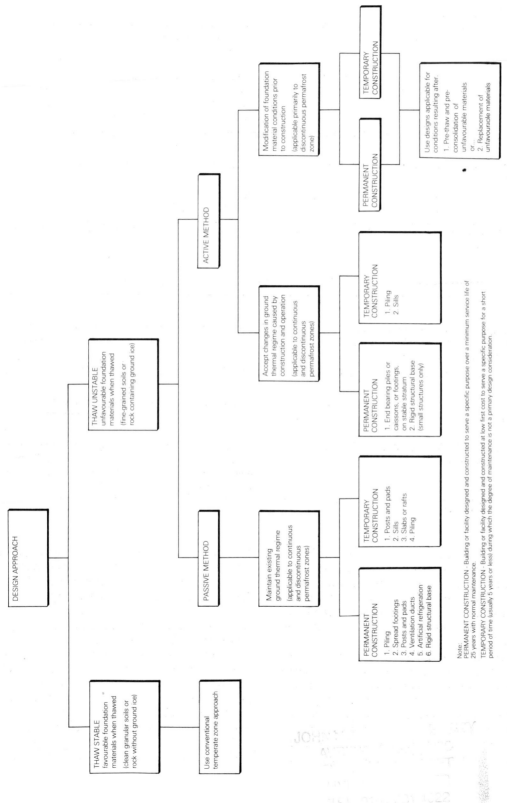

Fɪɢ. 8.22 Approaches to design and engineering in permafrost terrain (after Johnston, 1981; © 1981 and reproduced by permission of J. Wiley and Sons Ltd.)

this way, it is hoped that the downward-acting force in the permafrost zone (together with the load of the building and the weight of the pile) will be sufficient to overcome the upward-heaving force exerted on the protected piling in the active zone, thus preventing movement of the pile.

Other techniques frequently used in building construction include the creation of a trench around the building (possibly filled with material not susceptible to frost action) in order to eliminate possible horizontal stresses; the raising of the floor above the ground by 0.5–1 m to allow air circulation; the insulation of the floor to restrict temperature rise beneath it; and the provision of skirting insulation between ground and floor with air-vents that can be closed in the summer, and opened in winter to allow free air movement. Jacks may be used occasionally to adjust the building level if movement occurs. Gravel, wood, or concrete pads are often placed on natural vegetation to provide an insulating foundation for small and often temporary buildings.

The provision of utilities to communities also poses special problems in permafrost areas, problems that are mostly related to seasonal freezing and the presence of active and permafrost zones. A suitable water supply may be difficult to find—it usually comes from surface sources or from groundwater within, above, or beneath the permafrost. The disposal and decomposition of sewage may be a problem, leading to pollution and health hazards. Freezing and thawing of such services creates additional problems. One novel solution to some of these problems is for all linear utilities—water, sewerage, electricity, etc.—to be placed in a single insulated and heated conduit (known in North America as a *utilidor*) that may be placed above or in the ground.

Problems encountered in building roads, runways, and similar features are allied in many ways to those accompanying building construction (e.g. Harris, 1986). A passive approach demands the creation of adequate natural or artificial insulating layers, and an active approach often requires replacement of surface material with material that is not susceptible to frost. The aim is invariably to provide a stable track across what is frequently an inherently unstable surface. In addition, problems may arise because the route crosses varied permafrost terrain in which a variety of responses are necessary. Drainage is frequently difficult, especially where the route crosses drainage lines, and it is

desirable to stop as much water as possible from accumulating beneath the road, usually by providing ditches some distance away from it or preventing infiltration. Ditches may become eroded by flowing water. Road or railroad cuts may penetrate permafrost, lower the permafrost table, and cause landslides or solifluction in the newly created active zone—hence they are to be avoided as far as possible. In cuts and elsewhere, water may seep to the surface and freeze, producing treacherous *icings*. One way of avoiding this hazard is to transfer the icing away from the road into specially prepared ditches.

There are now numerous examples of the successful application of permafrost engineering principles to structures in permafrost areas (e.g. Williams, 1979; Thomson, 1980; Harris, 1986). One of the best is the new town of Inuvik, built in the 1950s as the regional capital of Canada's North West Territories, and replacing the initial hazardous settlement (Plates 8.1 and 8.2). The major problem in developing the town was undoubtedly the presence of permafrost and an active layer (Pihlainen, 1962).

The need to build Inuvik with as little disturbance as possible to ground conditions was recognized at the outset by several planning decisions. Firstly, the natural moss covering was to be left intact in order to retain its insulating benefits. Secondly, all permanent structures were to be placed on piles securely embedded in permafrost. Thirdly, road cuts and ditches were prohibited in order to prevent permafrost degradation, and culverts were to be installed in gravel-fill to accommodate surface runoff. Finally, all traffic routes and some temporary structures and storage areas were to be built up with gravel pads placed on top of the natural vegetation.

The use of piles posed several problems (Johnston, 1966). Each site had to be carefully prepared with a minimum of ground disturbance—the moss cover was maintained, vehicle movement was restricted, the piling equipment was placed on insulating gravel pads, and where vegetation had to be cleared, it was done by hand. Over 20 000 piles were used; most of them were of local spruce, but some larger wooden piles, and a few concrete and steel piles, were imported. As the upper part of timber piles would be in the active zone, they were treated with a preservative. Steam thawing was used to emplace most piles, except where the piles were to be closely spaced for carrying heavy loads,

PLATE 8.1 Subsidence of a house due to thawing of permafrost, North West Territories, Canada (photo courtesy of R. J. E. Brown)

PLATE 8.2 The new town of Inuvik, North West Territories, Canada. Note utilidors, foundation piles, and raised buildings (photo courtesy R. J. E. Brown)

when drilling was used to prevent too extensive thawing of permafrost.

A major problem was to decide how deep the piles should be buried. At Inuvik, as a general rule, piles were frozen into permafrost to a depth twice the maximum thickness of the active zone—that is to say, they were normally buried 4.5 m below the surface. Refreezing time for piles in permafrost is most important and depends on such variables as time of emplacement, ground temperature, and soil moisture. It was found that piles driven in the early spring could be loaded within two or three months, whereas those driven in autumn required about six months refreezing time. The fact that only a very small number of piles moved after they had been emplaced, and then only by a very small amount, is a tribute to the efficiency of the pile-construction techniques.

Other efforts were made to insulate the ground from the effects of new structures. For example, all major buildings had air-spaces beneath them of at least a metre. All heating, water-supply, and sewage-disposal lines were placed in an elevated and insulated utilidor. The heating supply also warmed the utilidor and thus prevented freezing of the other services. But the utilidor was expensive, costing $735 per m; utilidettes, leading from the utilidor to individual houses cost $440 per m (Yates and Stanley, 1966).

Water supply comes from the Mackenzie River when it only requires chlorination—it is pumped through the system to the lake behind the town. The lake also has its own catchment. Sewage is taken from the town to an artificial lagoon, where it is stored, treated, allowed to digest, and discharged into the river when there is no danger of contaminating water supplies.

The decision to relocate Aklavik at Inuvik was dramatically justified shortly after the new town was completed: in June 1961 an ice-jam on the Mackenzie River caused serious flooding of the old town.

8.7 Problems in Formerly Periglaciated Areas

Huge areas within and beyond the zone of maximum glaciation have suffered from periglacial processes in the past. Today many of these areas, including much of Britain (Fig. 8.23A),

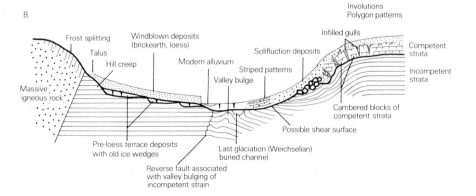

FIG. 8.23 (A) Generalized distribution of some important periglacial features in Britain. (B) Typical field relations of important periglacial features and deposits (reproduced by permission of the Geological Society from Higinbottom and Fookes, 1971)

preserve periglacial features that influence engineering development. In a systematic review, Higginbottom and Fookes (1971) classified these features (Fig. 8.23B) into *superficial structural disturbances*, *mass movements* and the *properties of periglacial deposits*. A summary of the engineering significance of these features is provided in Table 8.9. The main problem lies in their localized and variable nature, which means that rapid changes in foundation conditions can occur, perhaps unexpectedly. Higginbottom and Fookes (1971) pointed *inter alia* to the benefits of reconnaissance appraisal using aerial photography, supplementary test-boreholes, and *in situ* field-testing as responses to the problems of site evaluation. Within the range of engineering problems, the most serious are those relating to the reactivation, by inadvertent slope engineering, of slope failures and solifluction lobes. There are several examples

in the British Isles where such reactivation has required expensive engineering responses, as along the Sevenoaks bypass (e.g. Weeks, 1969; Chandler, 1970; Skempton and Weeks, 1976). Almost as serious are problems arising from *cambering and gull (tension-crack) formation* (the downward displacement of relatively strong, superficial strata across relatively weak strata in valley sides) and *valley bulge* (creation of 'anticlinal' structures along valley floors). Both features are developed under periglacial conditions and can lead to problems of route alignment, slope stabilization, and reservoir drainage (e.g. Horswill and Horton, 1976; Penn *et al.*, 1983). Within formerly glaciated areas periglacial processes commonly serve to obscure the relationships between glacial landforms and deposits, making the task of predicting geotechnical properties more difficult (Jones and Derbyshire, 1983).

TABLE 8.9 Periglacial phenomena and their engineering significance

Features	Engineering significance
A. Superficial structural disturbances	
Frost shattering	creates new fractures, increases deformability and permeability and reduces bulk density
Glacial shear	promotes surface instability
Hill creep	may require deeper foundations
Ice wedges and involutions	sudden, unexpected replacement of one material by another with different properties
Frost mounds, pingos	ditto, especially hazardous where peat deposits occur.
Chemical weathering (e.g. decalcification)	alteration of geotechnical properties of material (pipes and swallow-holes in Chalk may be related to periglacial processes; see Ch. 14)
B. Mass movements	
Cambering and bulging	gulls and bulge fractures create permeable zones of potential leakage requiring remedial measures
Landsliding	possible reactivation of fossil slides by inadvertent slope engineering
Mudflow (solifluction) activity	possible reactivation of low-angle flows by inadvertent slope engineering or by drainage changes. *Slip surfaces* may be present
C. Properties of periglacial deposits	
Sorted soils (loess)	wind-blown silt is characteristically metastable and can collapse when flooded
Unsorted soils (solifluction)	non-uniform soils, variable in nature and extent and erratic in their engineering behaviour. Not usually shown on geological maps

Source: Higginbottom and Fookes (1971).

8.8 Conclusion

The acceleration of human activity in the marginal lands of high altitudes and high latitudes, and the burgeoning of scientific research in these environments, have led to enormous progress in recent years. Much development in these areas is now planned, with prior evaluation of terrain conditions and sensitive responses to the complexities of the local conditions. While there is still much to be discovered, scientific exploration harnessed with engineering innovation has undoubtedly had substantial success. In this broad context, geomorphology has played a very significant role—in analysing the landforms and superficial processes of the areas, with outstanding contributions from geomorphologists such as J. R. Mackay, T. L. Péwé and A. L. Washburn; in appraising terrain conditions prior to development through mapping and monitoring; and in collaborating in the detailed process of site planning and development as revealed in the work of H. M. French and many others.

9 Aeolian Processes and Hazards

9.1 Introduction

In some circumstances, winds can erode, transport, and deposit particulate materials. Such aeolian activity can give rise to several hazards of importance in environmental management. The most serious problem is *soil erosion by wind*. It can damage crops and reduce productivity by removing seeds, exposing plant roots, and blasting leaves; and, by depleting surface humic soil horizons of organic matter and other fine particles, it can reduce the soil's water-retention capacity and productivity. Wind erosion is particularly serious in the context of agricultural and pastoral land use and has been widely recognized since it leapt into the public consciousness during the 'Dust Bowl' years in the Great Plains of United States between the two World Wars (Sect. 9.4). Consequent upon erosion, *dust storms* and *sand migration* can both cause serious hazards. Dust storms (in which visibility is reduced by dust below 1000 m) are associated with the transport of fine material in suspension, and can cause abrasion, transfer of weeds and pathogens, bronchial diseases, dangerous drifts on roads, and burial of canals, ditches, and fences (e.g. Péwé, 1981). Sand migration, which may be associated with active sand-dunes, can also create problems of burial and can menace land-use activities that lie in its path.

Because aeolian processes are most effective where the ground surface is relatively free of vegetation and dry, it is not surprising that aeolian hazards are most serious in arid and semi-arid lands. Dust storms and sand-dune migration are, of course, natural phenomena in many deserts (e.g. Goudie, 1983; Gerson *et al.*, 1985) and they commonly pose hazards for desert dwellers. But within drylands, the hazards are often most severe where inappropriate land-use practices, under conditions of variable and unreliable precipitation, leave the ground surface disturbed and

exposed. Such dry areas include 'pioneer' settlements in the plainlands of North America and the Soviet Union, and those drylands on desert margins where population pressures have caused widespread environmental disruption. As with soil erosion by water (Chapter 4), there is therefore a distinction to be drawn between 'natural' aeolian hazards, and those 'accelerated' or exacerbated by human activity. In some ways, the pattern of accelerated aeolian activity more closely reflects the pattern of desertification (Fig. 9.1) than that of aridity: the problems are most severe on desert margins where pressures on land and climatic uncertainties are greatest. Within these areas, undoubtedly some of the most hazardous areas are those dominated by fossil sand-dunes (Fig. 9.2)— sand-dunes that were once active but were subsequently vegetated and are now extremely vulnerable to destabilization by overgrazing and intensified farming. Most main areas of dust storms are also found within the desertified zones (Fig. 9.1).

Aeolian hazards are not confined to drylands. Elsewhere, they tend to be more localized, and to occur mainly on agricultural lands in humid areas that are windy, suffer occasional droughts, and have fine-grained or highly organic soils. In the UK, for example, wind erosion is occasionally a serious problem on the peaty soils of the Fens, the silty loams of the Vale of York, and the aeolian 'Cover Sands' of Lincolnshire. The loess areas of Europe (for example in Poland and Czechoslovakia) and China also suffer from aeolian hazards. And throughout the world, coastal sand-dunes can pose problems.

General assessments of the patterns of accelerated wind erosion have been made in several countries (Zachar, 1982). For example, Fig. 9.3 shows the areas affected by wind erosion in the USSR—some 200m ha, including 65m ha of drifting sand that is increasing by *c*.140 000 ha/y, and over 6.5m ha of cultivated land (Zachar, 1982). In

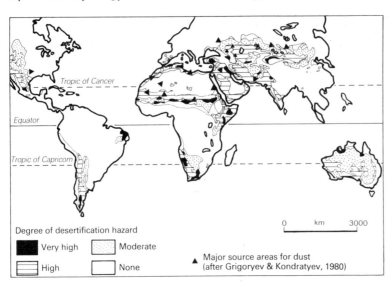

FIG. 9.1 Major areas of desertification (after UN, 1977), and major source areas of dust (after Grigoryev and Kondratyev, 1980)

Degree of desertification hazard

■ Very high   ▒ Moderate

≡ High   ☐ None

▲ Major source areas for dust (after Grigoryev & Kondratyev, 1980)

0   km   3000

the United States, over 170m ha of agricultural land are affected by wind erosion (Woodruff *et al.*, 1972), and on the Great Plains alone over 23.5 m ha were estimated to suffer wind erosion at a rate of over 11 tons/ha/y in 1980 (Lyles, 1981).

The foundations of modern scientific study of aeolian processes were laid in the 1930s by R. A. Bagnold, who published his outstanding treatise, *The Physics of Blown Sand and Desert Dunes* in 1941. Many of Bagnold's ideas and results provided the basis for subsequent research on aeolian hazards. In particular, the work of W. S. Chepil, A. S. Zingg, N. P. Woodruff, E. L. Skidmore, L. Lyles, and others, carried out largely under the auspices of the Wind Erosion Research Unit and the USDA's Agriculture Research Service and the Kansas Agricultural Experiment Station at Kansas State University in the aftermath of the 'Dust Bowl', built on Bagnold's foundations in the context of wind-erosion control of cultivated lands. This research attempted to identify and quantify the factors influencing the location and rates of soil erosion by wind, and to develop predictors of erosive conditions and soil loss based on a climatic index (e.g. Chepil and Woodruff, 1963). It led to the formulation in the 1960s of a wind-erosion equation that predicted potential soil loss from individual fields and facilitated the control of wind erosion by devices designed to manipulate factors affecting erosion so that potential erosion could be reduced to a 'tolerable level' (Sect. 9.2). This research continues today, concen-

trating on the tasks of refining the prediction equation and improving control techniques. Elsewhere, aeolian research has been strongly developed in the context of soil erosion in the Soviet Union and China, and of civil engineering in many countries of the Middle East (e.g. Cooke *et al.*, 1982). Progress has been reviewed in several books, including Cooke and Warren's (1973) *Geomorphology in Deserts*, Brookfield and Ahlbrandt's (1983) *Aeolian Sediments and Processes*, and Greeley and Ivesen's (1985) *Wind as a Geological Process*. Relevant articles are to be found in many journals, including especially the *Transactions of the American Society of Agricultural Engineers*, and the *Journal of Soil and Water Conservation*.

9.2 The Aeolian System

(a) *Variables in the system*

Fundamentally, the aeolian system comprises variables related to *erosivity* (surface winds and climate) and *erodibility* (surface materials and surface conditions). Wind erosion begins when air pressure on loose surface soil particles overcomes the force of gravity on the particles. Initially, the particles are moved through the air with a bouncing motion, known as *saltation*; their impact on other surface particles may promote further movement, by *saltation*, *surface creep*, or *suspension*. Chepil estimated that, for most of the soils he

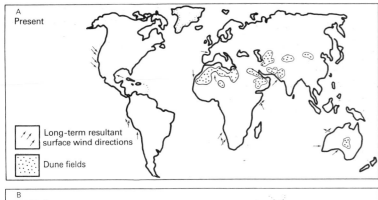

Fɪɢ. 9.2 Major sand seas of the world: (A) at present, (B) *c.*18 000 years ʙᴘ, and (C) *c.*6000 years ʙᴘ. The coastline in (B) represents a sea-level lowering of 120 m. Arrows indicate long-term resultant surface wind directions (after Sarntheim, 1978)

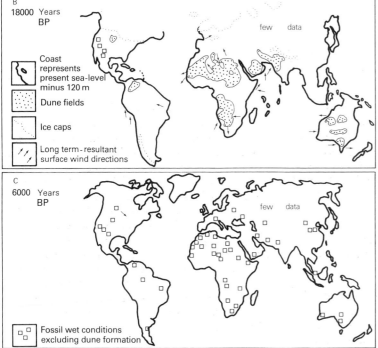

examined, 50–75 per cent of the weight of the eroded soil was carried in saltation, 3–40 per cent in suspension, and 5–25 per cent in surface creep. Clearly the process depends on the availability of particles which can be moved in saltation, and on the attainment of wind velocities at or above which particle movement can be initiated by saltation (fluid threshold) or by impact (impact threshold). The following is a brief summary; more extensive discussion can be found, for example, in Vanoni (1975), Wilson and Cooke (1980), and Zachar (1982).

(i) *Surface winds* Surface winds capable of initiating particle movement, are turbulent. The velocity of surface winds increases with height above the surface and the following equation is one description of such increases:

$$V_z = 5.75 V_* \log \frac{z}{\mathrm{k}} \qquad (9.1)$$

where V is mean velocity at any height z, V_* is drag velocity, and k is a roughness constant. If saltating material becomes incorporated into the wind, the momentum, and hence surface wind velocity, are reduced. Chepil and Milne (1941) demonstrated

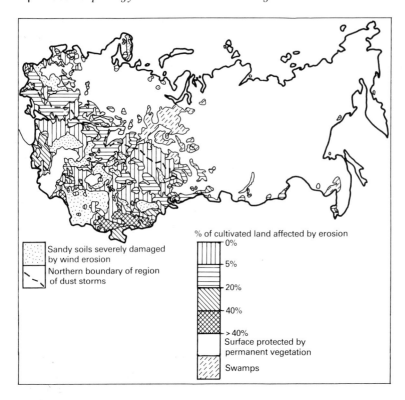

FIG. 9.3 General distribution of soil erosion by wind in the USSR. The map also shows areas of cultivated land affected by erosion (both wind and water) (after Zachar, 1982)

% of cultivated land affected by erosion
0%
5%
20%
40%
> 40%
Surface protected by permanent vegetation
Swamps

Sandy soils severely damaged by wind erosion

Northern boundary of region of dust storms

that the more erodible the soil, the greater the concentration of moving grains, and the greater the reduction of surface wind velocity.

Mean wind drag per unit horizontal area of ground surface, $\bar{\tau}$ (expressed, for example, in dynes/cm^2), is related to drag velocity and fluid density. In general the relations can be expressed by the equation:

$$\bar{\tau} = pV_*^2 \qquad (9.2)$$

where p is fluid density.

At the threshold of grain movement, three types of pressure are exerted on the surface soil particle: *impact or velocity pressure* (positive) on the windward area of the particle; *viscosity pressure* (negative) on the leeward area of the particle; and *static pressure*, a negative pressure on the top of the particle, caused by the so-called Bernoulli effect (which arises from pressure reduction where fluid (air) velocity is increased, as at the top of the particle). *Drag* on the top of the soil particle is due to the pressure difference against its windward and leeward sides, and *lift* is caused by a decrease of static pressure at the top of a particle compared with that at the bottom. The values of drag and lift

required to initiate movement are, of course, affected by the character of the grains.

Once the particle has been entrained, drag and lift change rapidly. The trajectories of saltating particles reflect the relations between the forward force of the wind, the mass of the grains, and the pull of gravity. Initially, grains usually rise almost vertically; as they enter more rapidly moving air their paths are flattened; and as the initial force of upward movement is dissipated, grains begin to fall under the influence of gravity. On striking the surface, grains may bounce back into the air and/or strike other particles, causing them to be pushed along or saltated. The height of the saltating grains' trajectory may be up to two metres, depending mainly on particle size and bed roughness. In general, saltation height is inversely related to particle size and directly related to bed roughness. The length of the trajectory is normally approximately ten times the height.

The critical characteristic of the wind is its velocity. Mean wind velocity is normally the only practicable field measurement of wind force and it is widely used. But in fact fundamentally important characteristics of turbulence and gustiness can vary greatly for the same mean velocity so that the mean value is not ideal (Wilson and Cooke, 1980). Other important climatic considerations concern the availability of water, which is in turn related to atmospheric, evaporation, and evapotranspiration conditions. In general, the drier the soil the more vulnerable it is to wind erosion. Thus wind erosion is usually most serious where wind velocities and evaporation rates are high, precipitation is low, and drought is common. Major phases of aeolian activity, including dust storms, are often associated with specific weather conditions, such as the passage of depressions, squall lines, and regional instability (e.g. Goudie, 1983).

(ii) *Surface materials and surface conditions* The principal characteristics of soil particles with respect to aeolian processes are size and density. Many soil particles are composed of quartz, which has a specific gravity of 2.65. The most erodible particles of 2.65 density are about 0.1–0.15 mm in diameter. Threshold wind velocities required to move grains larger than 0.1 mm are closely defined by the square-root law:

$$V_{*t} = A\left(\frac{\sigma - \rho}{\rho}\right)^{1/2} - gD \qquad (9.3)$$

where

V_{*t} = threshold drag velocity,
σ = specific gravity of the grain,
ρ = specific gravity of air,
D = diameter of the grain in cm,

A = a coefficient the value of which in air for particles above 0.1 mm in diameter was found to be 0.1 for the fluid threshold, and 0.084 for the impact threshold (Chepil, 1945). For smaller particles, threshold velocities do not conform to this law: the velocities increase with decrease in grain size, owing probably to cohesion between finer particles and to the fact that particles may be too small to protrude into the turbulent flow of air. The relations between particle size, specific gravity, and threshold and impact velocities are summarized in Fig. 9.4. Experimental studies suggest that particles less than 0.1 mm in diameter may be moved in suspension; those between 0.1 and 0.5 mm are commonly moved by saltation; and particles larger than 0.5 mm tend to be moved by creep.

The *equivalent diameter* of a particle with a density of 2.65 g/cm³ is defined as $peD/2.65$ where

pe is the bulk density of erodible soil particles, and D is their diameter. Chepil found that very few particles with equivalent diameters exceeding 0.5 mm (actual diameter = 0.84 mm) are eroded. In general, the potential erosion of material increases as the percentage of soil fractions greater than 0.84 mm in diameter declines (Woodruff and Siddoway, 1965).

So far this discussion of surface materials has only been concerned with individual loose particles. But such particles are frequently combined in various ways to produce soil structures resistant to erosion. Chepil and Woodruff (1963) distinguished four major types: primary (water-stable) aggregates; secondary aggregates, or clods; fine material among clods; and surface crusts. Primary aggregates are held together by water-soluble cements of clay and colloids. Clods are held together in a dry state by cements comprising mainly water-dispersible particles smaller than 0.02 mm in diameter. Cohesion between clods is provided largely by water-dispersible silt-and-clay-sized particles. Surface crusts arise from (a) raindrop impact which tends to reorientate clay particles parallel to the surface (Chapter 4) and (b) washing of the particles into near-surface pores.

The binding agents for these dry structures include chiefly silt, clay, and decomposing organic matter, and it is these agents that determine the mechanical stability of the structures. As Chepil and Woodruff (1963, p. 262) reported:

the relative effectiveness of silt and clay as binding agents depends somewhat on their relative proportions to each other and to the sand fraction. The first five per cent of silt or clay mixed with sand is about equally effective in creating cloddiness, but the quality of the clods is different. Those formed with clay and sand are harder and less subject to abrasion by windborne sand than those formed from silt and sand. For proportions greater than five per cent and up to 100 per cent the silt fraction creates more clods, but clods are softer and more readily abraded than those formed from clay and sand. The greatest proportion of non-erodible clods exhibiting a high degree of mechanical stability and low abradability is obtained in soils having 20 to 30 per cent of clay, 40 to 50 per cent of silt, and 20 to 40 per cent of sand.

The decomposition of organic matter on and in soils is also associated with the creation of temporary cementing substances. These substances are derived from the decomposition products of plant residues, the decomposer micro-

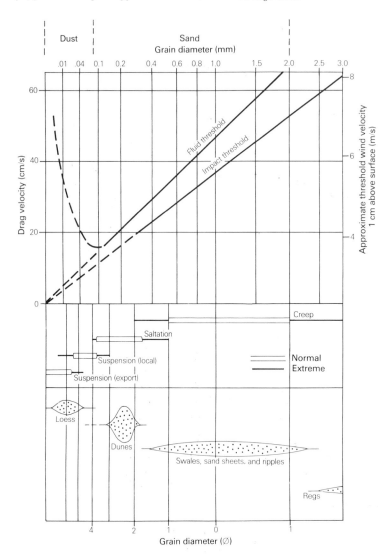

FIG. 9.4 Relationship between grain
size, fluid and impact threshold
velocities, characteristic modes of
aeolian transport, and resulting size-
grading of aeolian sand formations
(after Mabbutt, 1977)

organisms, and their secretory products, and they
serve to bind particles together and improve the
soil structure. Although the effect of decomposition products is temporary, it may last in places for
up to 5 years. In contrast, calcium carbonate tends
to decrease mechanical stability, and the erodibility of some calcareous soils may be high. Such is
the case in many arid and semi-arid lands.

The importance of structural units is that many
of them act as non-erodible or less-erodible
obstacles to wind erosion. For wind erosion to
proceed beyond the removal of existing loose
particles, the structural units must be broken
down by weathering forces, raindrop impact, or
wind abrasion. The structures vary in their
resistance to these forces. Their susceptibility to
wind abrasion (sometimes called 'abradability'),
for example, varies inversely with their mechanical
stability and thus with soil type (Fig. 9.5A).
Mechanical stability is a function of inter-particle
cohesion. Generally, in a dry state, primary
aggregates are most stable, and clods, crusts, and
the fine material among clods are less stable (in
that order). Wind abrasion may cause the progressive breakdown of soil structure as erosion continues.

Non-erodible particles (including stones) seriously restrict the progress of wind erosion, for the

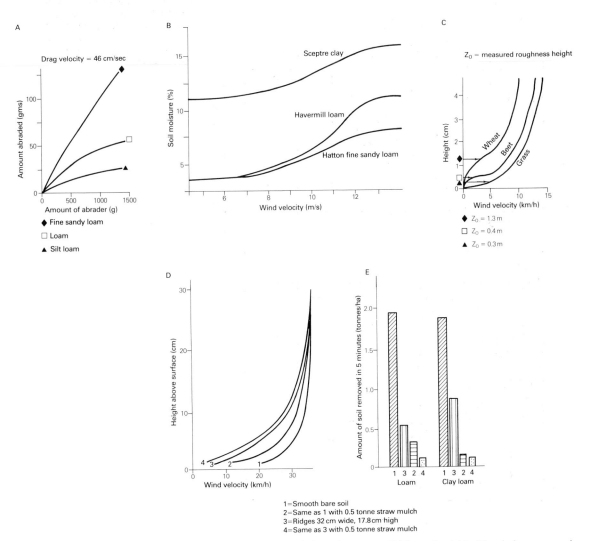

FIG. 9.5 Some empirical relationships between variables in the wind-erosion system. (A) Loss of weight of dry clods 25–50 mm in diameter of different soil types subjected to abrasion in a wind tunnel by drifting soil composed of particles less than 0.42 mm in diameter at a drag velocity of 46.0 cm/s (after Chepil, 1951). (B) Influence of soil moisture on the threshold velocity of soil movement by wind (after Bisal and Hsieh, 1966). Only the erosive fractions were used (<0.84 mm diameter) in these wind-tunnel experiments. The percentages of sand, silt and clay in these three soils were:

	Sand	Silt	Clay
Fine sand loam	75.8	15.0	9.2
Loam	33.4	48.2	18.4
Clay	25.5	32.3	42.2

(C) Wind velocity distribution over different types of vegetation and a snow-covered surface (after Frevert *et al.*, 1955). (D/E) Influence of surface treatment on wind velocity and (E) on soil erosion. 1, 2, 3, and 4 are the same in both (D) and (E) (after Frevert *et al.*, 1955)

amount of material removed is limited by the height, number, and distribution of the non-erodible particles. As erosion proceeds, the height and number per unit area of non-erodible particles increases until ultimately the non-erodible particles completely shelter erodible material from the wind, and a *windstable surface* is created. This final stage is defined by the *critical surface-barrier ratio* (the ratio of height of non-erodible surface projections to distance between projections which will barely prevent movement of erodible fractions by the wind).

Three further characteristics of the surface are important variables in the wind-erosion system: soil moisture, surface roughness, and surface length. Only dry or largely dry soil particles are readily erodible by the wind: *soil moisture* promotes particle cohesion and restricts erodibility (Fig. 9.5B). As described by Woodruff and Siddoway (1965), the rate of soil movement varies approximately inversely as the square of effective surface soil moisture. The soil moisture at any particular time, of course, is determined by the properties of the soil and the particular weather conditions. *Surface roughness* is composed of the physical elements comprising the surface configuration—clods, ridges, pieces of plant residue, etc. In general, a rough surface is more effective in reducing wind velocity than a smooth one and is thus less susceptible to erosion, provided the material contains non-erodible particles. Finally, the greater the length of the surface across which uninterrupted airflow occurs (the so-called 'fetch' of the wind), the more likely wind erosion is to reach its optimum efficiency.

Vegetation cover influences the nature of wind erosion in several ways. Firstly, the quantity of vegetation, as represented by the proportion of covered ground, governs the extent to which the surface is exposed to erosion. Secondly, vegetation tends to increase surface roughness, and hence reduces wind erosion. In general, the taller the crop, the finer the vegetative material, and the greater its surface area, the more wind velocity is reduced (Fig. 9.5C). Thirdly, plant residue is important in protecting the surface and in adding organic material to it (Fig. 9.5D).

Of the numerous variables in the wind-erosion system, some are permanent, others change. The characteristics of wind, structural units, organic residues, soil moisture, and vegetation may all change over short periods, and especially seasonally. In contrast, textural properties of surface material tend to be fairly constant unless they are progressively modified by weathering, erosion, and husbandry.

(b) *Initiation and progress of wind erosion*

Wind erosion may begin when the equilibrium of the system is disrupted by a change in one or more of the component variables to the extent that the saltation process starts. Changes that cause such disruption are numerous and include reduction of precipitation, increase of temperature, increase of wind velocity, destruction of soil aggregates, reduction of surface roughness, and reduction of vegetation cover.

For saltation to start, it is necessary that the fluid threshold velocity pertinent to the most easily erodible loose surface material be reached. If, as is often the case, there is a hard surface crust which has to be broken, then the initial fluid threshold velocity is higher than that required to move the most erodible particles. Impacts from saltating particles initiate movement of other particles if their impact threshold velocity is achieved. The fluid and impact threshold velocities are similar for most particles, but the impact threshold velocity becomes lower than the fluid threshold velocity for grains of increasingly greater size (Fig. 9.4). As erosion progresses across a surface, the quantity of debris in motion increases until it is at the maximum sustainable by the wind; the increase is known as avalanching. Avalanching is accompanied by increased surface abrasion—which tends to destroy soil structure and increase the supply of erodible particles—and by particle sorting. Abrasion by saltating particles is influenced by the abradability of soil particles and the character of the abrading particles (size, hardness, etc.), their velocity, and angle of impact (Hagen, 1984).

The rate of soil movement, q (in g/cm/s), may be described as follows:

$$q = a(D_e)^{1/2} \frac{\rho}{g} (V'_*)^3 \qquad (9.4)$$

where

V'_* = drag velocity above eroding surface,
D_e = average equivalent diameter of soil
    particles moved by the wind,
ρ/g = mass density of air.

Coefficient *a* varies with different conditions,

notably 'the size distribution of the erodible particles, the proportion of fine dust particles present in the mixture, the proportion and size of non-erodible fraction, position of the field, and the amount of moisture in the soil' (Chepil and Woodruff, 1963, p. 245). The quantity of material, X (in tons/acre) removable from a given area may be defined as follows:

$$X = a(V_*)^5 \tag{9.5}$$

In suitable circumstances, erosion will continue once it is initiated until a wind-stable surface is produced.

Eroded material is predominantly either sand-sized (0.08–2.0 mm diameter) or dust (less than 0.8 mm diameter). This distinction is fundamental in terms of the hazards of transport and deposition. *Dust*, which is often slightly less dense and somewhat more cohesive than sand, travels in suspension and may be raised to great heights by turbulence and involve huge clouds. *Sand* movement, in contrast, is chiefly by saltation and is concentrated near to the ground in the moving 'saltation curtain'. The force of bombardment of saltating grains rolls other, impacted grains (some

of which may be too massive to saltate) along the surface as creep load. Dust can be entrained by high-velocity winds alone, but in the presence of saltating sand, it may be entrained by ballistic impact. These different transport processes generally lead to the segregation of sand from dust, so that in deserts, depositional features arising from aeolian activity are usually composed either predominantly of sand or of dust (e.g. Cooke *et al.*, 1982).

Theoretically, a given airflow can continue to incorporate both sand and dust until it is saturated; at that point, no further removal can take place downwind. There is thus a significant distinction to be drawn in the context of sediment-transport monitoring between the passage of material across a surface that may be suffering no net loss, and the actual loss of material reflected in the depletion of a body of sediment. More broadly, it is convenient to recognize that the transfer of aeolian sediment is related to supply *sources*, and sediment *stores*, which may be *temporary* or *permanent*. Fig. 9.6 illustrates this concept for sand and dust movement at a site in Saudi Arabia (Jones *et al.*, 1986).

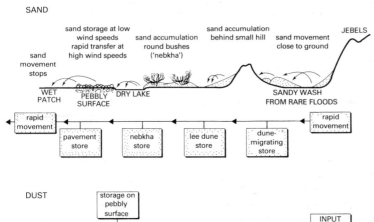

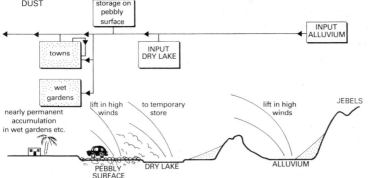

FIG. 9.6 Sources and stores for sand and dust on a desert surface in Saudi Arabia (Jones *et al.*, 1986; reproduced by permission of the Geological Society of London)

Chepil (1959*a*) attributed 'avalanching' to several causes. Firstly, there is a progressive increase in the number of grain impacts which results in the entrainment of more material by the impact mechanism. Secondly, because of the higher frequency of impact, abrasion increases and thus increases the supply of erodible material; in addition, the erodible material removed upwind supplements the erodible material downwind, making the soil generally more susceptible to erosion. Thirdly, particles dislodged from projections are trapped in depressions so that surface roughness is gradually reduced, which leads to an increase in shear velocity and hence in rate of transport. This process is usually called 'detrusion'. Whilst the saturated flow (q_s) for a given wind velocity was found to be independent of soil type and about the same for all soils, Chepil (1959*b*) noted that the distance from the point of initiation to q_s varied with soil erodibility. The distance was about 65 m for the most erodible soil and some 1900 m for the least erodible soil. These distances remained approximately the same for all levels of erosive winds.

As erosion progresses, the process of sorting due to differential rates of particle movement becomes more pronounced. Over the whole area erosion and deposition occur contemporaneously, but the net result at a particular site will be determined by the relations between the forces of erosion and deposition. At any given moment the surface grains reflect the associated airflow characteristics, but the ultimate depositional forms represent a sequence of events, probably taking their main characteristics from the peak period of activity (Wilson and Cooke, 1980).

(c) *The wind-erosion equation*

The major factors involved in wind erosion of agricultural land have been expressed by Chepil and others (e.g. Woodruff and Siddoway, 1965) in terms of a functional equation of predictive value that is similar to the universal soil-loss equation described in Chapter 4.

$$E = f(I', K', C', L', V') \tag{9.6}$$

where

E = erosion (per acre per annum, or metric equivalent),

I' = comprises the soil erodibility index (I)—the potential soil loss from a

wide *unsheltered, isolated* field with a bare, *smooth, non-crusted* surface; and the knoll erodibility index (I_s), or erodibility of windward slopes expressed as a percentage slope,

K' = soil ridge roughness factor, a measure of natural or artificial roughness other than that caused by clods or vegetation,

C' = local wind-erosion climatic factor, in which

$$C' = \frac{v^3}{(P-E)^2},$$

where

v = mean annual wind speed corrected to a standard height of (10 m),

and

$(P-E)^2$ = Thornthwaite's precipitation effectiveness index, a measure of effective *soil moisture*,

L' = field length (or equivalent) along prevailing wind-erosion direction, based on the *total distance* across a field measured along the prevailing wind-erosion direction (D_f) and the *sheltered distance* in the same direction (D_b),

V' = Equivalent quantity of vegetation cover based on R, the *surface residue* (washed, oven-dried residue × 1.2); S, the total *cross-sectional area* of vegetation cover (the finer the material, the greater its surface area); and K_o, a measure of vegetational *roughness*.

Because the relations between variables in the equation are complex, an estimate of wind erosion cannot simply be made by multiplying together different values of the variables. The equation can be solved graphically, using various charts and tables, or by means of a computer program (e.g. Skidmore *et al.*, 1970; Fisher and Skidmore, 1970). Like its water-erosion equivalent, the equation is not of universal applicability. It is designed to determine if a field is adequately protected and, if not, to calculate changes required to reduce erosion to within tolerable limits. Furthermore, many details are still lacking, and some variables are not easily transferred to areas beyond the Great Plains of the United States where the equation was developed.

The climatic index, C', illustrates some of the problems associated with the equation. It reflects the fact that the rate of soil movement varies approximately as the cube of wind velocity and inversely as the square of effective soil moisture. The base point for determining values of the index is the average annual value (2.9) at Garden City, Kansas. In fact, the value of C' may vary considerably through the year, and it may need to be calculated on a monthly or seasonal basis to allow short, highly erosive periods to be evaluated (e.g. Fig. 9.7; Skidmore and Woodruff, 1968).

Furthermore, the Thornthwaite index is only a very approximate index of soil moisture. And the index takes no account of climatic factors such as frost frequency (which has a role in breaking down clods) or the fact that certain winds may be associated with rainfall, which may be relevant in some areas (Wilson and Cooke, 1980).

9.3 Methods of Aeolian Hazard Assessment

As with water erosion of soils, scientists have adopted several approaches to the study of aeolian problems. Some have sought to *simulate* natural

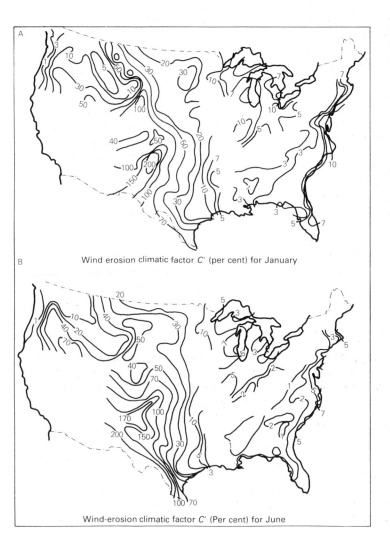

FIG. 9.7 Seasonal variation of the wind-erosion climatic factor, C', in the United States (after Skidmore and Woodruff, 1968)

A

Wind erosion climatic factor C' (per cent) for January

B

Wind-erosion climatic factor C' (Per cent) for June

conditions in the laboratory by carrying out wind-tunnel experiments. Others have devised special equipment for *monitoring* soil loss by wind erosion in the field, although this task is more difficult than that of monitoring water erosion. A third group, often drawing empirical evidence from field and laboratory studies, and meteorological data, has attempted to *predict* wind erosion through the use of several easily measured controlling variables.

(a) *Laboratory studies*

The advantage of using wind-tunnels to study aeolian movement is that it permits control and systematic variation of parameters governing aeolian activity, such as wind velocity and surface roughness. In order to use experimental measurements to supplement field data, the model and its environment should obviously simulate the field conditions as closely as possible. Similarity depends primarily on matching the appropriate Reynolds number (a measure of the balance between viscous and inertial forces resisting flow) and the geometrical boundary conditions. Reynolds numbers for full-scale structures are large and to obtain them in the laboratory the model must be relatively large; the size is, however, limited by its blockage effect on airflow which may distort velocity distributions round the model. Complete accounts of similarity requirements, together with assessments of the relative success of simulations have been given by several authors (e.g. Cermak, 1971).

Wind-tunnels have been used extensively since Bagnold's (1941) early work, and experimental results include prediction of soil erodibility and field-erosion rates (e.g. Chepil, 1950; Cole, 1984), calculation of threshold velocities (e.g. Greeley *et al.*, 1977), estimation of abradability (e.g. Whitney, 1978), and measurement of sand flow (e.g. Jones and Willetts, 1979). Portable wind-tunnels have been used in the field to determine, for example, the effect of different mulches on soil loss (e.g. Chepil *et al.*, 1963; Plate 9.1).

(b) *Field assessments*

Erodibility can be assessed in the field by using experimental plots (e.g. Butterfield, 1973), or by classifying soils and surface materials according to their erodibility as reflected in their surface condition, depth, structure, lithology, and horizonation. For example, Fig. 9.8 shows the relative potential of the land around a town in Saudi Arabia to generate sand; it is based on the qualitative interpretation of detailed geomorphological mapping, taking into account both natural conditions and the impact of human activity (Jones *et al.*, 1986). A number of authors have classified soil profiles according to the extent of their erosion by wind or burial by wind deposits (Zachar, 1982; Table 9.1) and used the classifications for mapping the patterns of damage.

A variety of instruments is available for monitoring the movement of soil, sand, and dust (e.g. Vanoni, 1975; Morales, 1979; De Ploey and Gabriels, 1980). Dust can be collected in settling-jars or similar devices, or by using electrostatic precipitation and other techniques; sand traps (Fig. 9.9) attempt to sample particles with as little

PLATE 9.1 Field monitoring: portable wind-tunnel for assessing soil erodibility by wind (photo courtesy US Department of Agriculture)

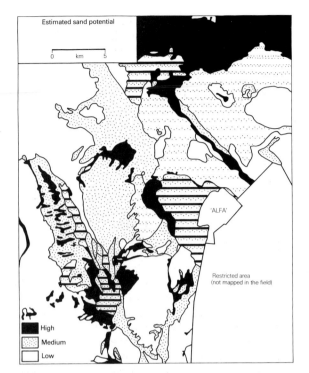

Estimated sand potential

0   km   5

'ALFA'

Restricted area
(not mapped in the field)

High
Medium
Low

FIG. 9.8 Map of estimated dust drift potential for an area of
desert surface surrounding a town in Saudi Arabia (Jones *et al.*,
1986; reproduced by permission of the Geological Society of
London)

disturbance as possible to airflow using either
vertically or horizontally oriented receptacles.
Dune movement can be monitored with reference
to marker stakes or by using detailed surveying
techniques. Closely related to sampling sediment
is the need to monitor associated airflow, for which
such devices as pitot tubes, anemometers, and
various flow-visualization techniques (e.g. bal-
loons and smoke-bombs) are available (e.g. Cooke
et al., 1982).

(c) *Meteorological data analysis*

Most studies of aeolian hazards will depend for the
assessment of wind erosivity on published meteorological data. Analysis of meteorological data
usually requires the construction of diagrammatic
generalizations, such as wind roses. The latter
indicate frequency and strength of all winds. But a
better approach is to calculate sand and dust roses
for effective (i.e. sand- and dust-transporting)
winds. Bagnold (1953) gave the following equation

for calculating sand- and (and dust-) movement
roses:

$$Q = \frac{1.0 \times 10^{-4}}{\log (100_z)^3} t(V-16)^3 \qquad (9.7)$$

where

Q = sand movement (tonnes/m),
t = the number of hours that the wind of
V km/h blows, and 16 km/h is the
threshold velocity,
z = height above ground in metres.

Q is summed for effective winds of all speeds for
each direction to give a sand-movement rose.

Skidmore and Woodruff (1968) assessed the
direction and relative magnitudes of wind-erosion
forces by computing wind-erosion force vectors, $\mathbf{r}_j$,
for each of 16 principal directions:

$$\mathbf{r}_j = \sum_{i=1}^{n} (\overline{V-V_t})_i^3 F_i$$

$$j = 0 \rightarrow 15 \qquad (9.8)$$

where

$(\overline{V-V_t})_i^3$ = the cubed mean windspeed
above threshold within the ith
speed group,
F_i = the percentage of total observations that occur in the ith speed
group and direction under consideration,
j = the direction (values 0 to 15
numbered in an anticlockwise
direction, starting with east).

The sum of the vector magnitudes gives the total
magnitude of wind-erosion forces for the location:

$$F_i = \sum_{j=0}^{15} \sum_{i=1}^{n} (\overline{V-V_t})_{ij}^3 F_{ij}. \qquad (9.9)$$

The relative erosion vector is given by $\mathbf{r}_j'$ such that

$$\sum_0^{15} \mathbf{r}_j' = 1$$

and

$$\mathbf{r}_j' = \frac{\sum_{i=1}^{n} (\overline{V-V_t})_i^3 F_i}{\sum_{j=0}^{15} \sum_{i-1}^{n} (\overline{V-V_t})_{ij}^3 F_{ij}}. \qquad (9.10)$$

The preponderance of wind-erosion forces can be
calculated. A preponderance of 1 indicates no

TABLE 9.1 Classification of soil eroded by wind and soils buried by wind deposits

Grade	Erodedness of soil	Wind removal of soil
1	Slight	Up to half of the A_1 horizon
2	Moderate	Up to the B_1 horizon
3	Severe	Up to half of the B_1 horizon
4	Very severe	Up to the C horizon

Grade	Soil burial	Thickness of aeolian deposit [cm]
1	Slight	<20
2	Moderate	20–40
3	Deep	>40

Source: Sobolev, in Zachar (1982).

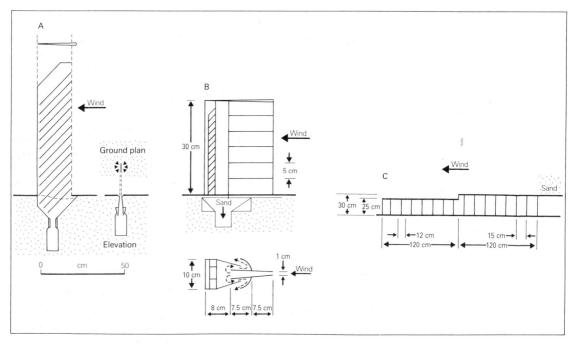

FIG. 9.9 Examples of sand collectors (after Bagnold, 1941 and Belly, 1964)

preferred wind-erosion direction, while a preponderance of 2 indicates erosion forces twice as great parallel with the direction line as normal to it: where preponderance is large, windbreaks oriented normal to the maximum erosion forces will perform efficiently. Thus, preponderance is useful in determining the correct configurations of windbreaks and fences to control erosion. The vector and its sum provide an excellent index of erosivity because they take account of the capacity

of the wind to erode, the prevailing wind direction, and the directional distribution of wind-erosion forces. Vectors indicate how the factors vary throughout the year, the time and place when erosion hazard is greatest, and thus, the time and place when the need for protection is greatest, and the correct orientation of barriers to reduce the problems.

A simpler way of using meteorological data to estimate the drift of aeolian material by month and direction, which is based on the observation that drift is approximately proportional to the rose of wind velocity was derived by Fryberger (1979) in which:

$$Q \propto v^2(V - V_t) \cdot t \qquad (9.11)$$

where

> Q = proportional amount of *sand* drift,
> V = average wind velocity at 10 m,
> V_t = impact threshold wind velocity for sand,
> t = duration of wind, from a given direction.

The data can be combined to produce a sand rose (or, using different thresholds, a dust rose) and further extended to show relative drift potential to a particular location in map form (Fig. 9.10).

Visibility measurements, in so far as they are reduced by dust, can be used to estimate dust concentration. Chepil and Woodruff (1957) established the following relationship (at 2 m height):

$$C = \frac{56.0}{V^{1.25}} \qquad (9.12)$$

where

> C = concentration of dust in mg/m^3,
> V = horizontal visibility in km.

Similarly, the climatic index C in the wind-erosion equation can be used to predict the potential severity of erosive conditions. Fig. 9.11 shows, for example, the relations between the frequency of dust storms and the mean value of C for the previous three years (C_3) at two locations on the Great Plains in the United States. The regression equation can be used to predict severity of erosion on the assumption that severity is equated with number of dust storms, although the relationship gives no indication of the amount or rate of erosion. The basis of the regression analysis is open to question. Firstly, as the authors pointed out: 'The reason for the relatively low incidence of dust storms during the period 1954–57 may be that farmers had learned more about how to control wind erosion and were in a much better economic position to control it' (Chepil *et al.*, 1963, p. 450). This suggests that the regression is based on different populations, each having their own statistical properties. A second criticism is that the spread of values does not allow accurate predictions of the number of dust-storm days to be made; it merely allows a general indication of the possible severity of conditions to be given.

It was suggested by Chepil *et al.* (1963) that severe erosion conditions occurred in the Garden City and Dodge City area when $C_3 > 25$ per cent and N (the number of dust storms predicted from the regression) exceeded 25, at which point special precautions were necessary in addition to those normally taken. This critical value of C_3 can be calculated for other locations, which are likely to have adapted to a similar probability of erosive

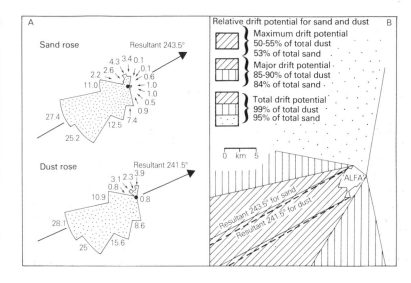

Fig. 9.10 (A) Sand and dust roses for an airport in Saudi Arabia using 1980 and 1981 data showing the percentage of mean annual drift potential of winds from 16 sectors. (B) Sand and dust drift potentials for a town in Saudi Arabia delimited by applying the rose data in (A) to the town perimeter (Jones *et al.*, 1986; reproduced by permission of the Geological Society of London)

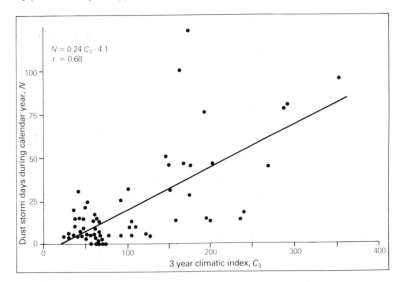

$N = 0.24\,C_3 - 4.1$
$r = 0.68$

FIG. 9.11 Regression of annual number of dust storms, N, beginning 1 January on the 3-year climatic index, C_3 ending 31 May of the preceding year (combined records for Dodge City and Garden City, Kansas) (after Chepil *et al.*, 1963; reproduced by permission of the Soil Science Society of America, Inc)

conditions occurring, using the equation

$$C_{3(crit)} = 125 \frac{C'}{100} \qquad (9.13)$$

in which C' is the mean value of the climatic index for that location.

Other equations exist that attempt to predict different aspects of aeolian hazards on the basis of climatic data (e.g. Cooke *et al.*, 1982; Zackar, 1982; Lyles, 1983).

(d) *Aerial photographs and satellite imagery*

These are not only useful as the basis for geomorphological mapping; they can also reveal much of value about the nature of aeolian activity. For example, successive images of the same location can reveal rates of dune movement (e.g. Hunting Surveys Ltd., 1977), and the form of *active* dunes can indicate the nature of wind regime (e.g. Petrov, 1976). Satellite imagery has been used to discern trends and corridors of sand drift not immediately obvious on the ground (e.g. Mainguet, 1984; Jones *et al.*, 1986); and to interpret the pattern of 'blowout distribution' and predict the locations of future wind erosion (e.g. Seevers *et al.*, 1975).

(e) *Maps of aeolian hazard*

The previous discussion has provided some examples of the prediction of the spatial variability of aeolian hazards. Geomorphological mapping and its subsequent analysis, based on field assessments and the interpretation of remote sensing imagery,

is clearly important; so, too, is the application of predictive equations, and factors within them, to specific locations. Different aspects of these approaches are occasionally brought together in specific studies. For example, aspects of several methods are integrated in a desert study by Jones *et al.* (1986) designed to predict the potential sand and dust hazard around a town in Saudi Arabia where a defensive 'green belt' was being planned (see Figs. 9.6, 9.7, 9.8, 9.10). Similarly, but in a humid environment, Briggs and France (1982) attempted to predict regional variability of the wind-erosion hazard in South Yorkshire, by applying the wind-erosion equation. Values for the major variables were estimated for each kilometre grid-square—wind velocity and Thornthwaite's index (from three meteorological stations), erodibility (based on analysis of soil samples from soil groups shown on the county soil map), surface roughness (0.75, a constant), field length (from 1:25 000 maps), and land use (from a county survey). 'Potential erosion' was predicted for each kilometre grid-square on the basis of climate, erodibility, and roughness; estimates of soil loss were based on these variables together with field length and vegetation factors.

9.4 Soil Erosion by Wind and its Control

(a) *The nature of the hazard*

The loss of soil by wind erosion can have physical and economic consequences that are sufficiently significant to justify a management response

(Table 9.2). In Britain, these consequences rarely create a crisis but there is some evidence that the hazard is increasing on light peaty soils growing a variety of market-garden crops, sugar-beet, grains, and potatoes that are especially prone to wind attack in dry periods during spring and early summer when fields are bare following recent cultivation and before a protective crop has grown. Reasons for the increased hazard, listed in Table 9.3, relate chiefly to changes in farm management and husbandry. But the precise contribution of these changes and their relative importance remains unclear. Indeed, little is known at present, other than by casual observation, of the nature and rates of wind erosion on cultivated land in Britain. The problems of studying the process, its physical damage and economic consequences are considerable. For example, monitoring wind erosion is time-consuming, expensive, and suitable equipment needs to be developed; and it is extremely difficult to disentan-

gle the economic consequences of wind erosion from the complex structure of the farm economy. Nevertheless, attempts are being made in the UK to cost the effects of wind erosion and preventive measures. For example, Rickard (1979) argued that the economic risk on easily eroded soils in the UK to sugar-beet is about £85 per hectare, and to onions up to £600 per hectare.

(i) *Lincolnshire* Lincolnshire provides a good illustration of the growing erosion hazard in the UK. Here, areas potentially susceptible to wind erosion include the sandy loams and alluvial soils of the Trent Valley, the regions of Cover Sands, the calcareous and sandy soils of the Lincolnshire Heights and Wolds, and the zones of fen peat and alluvium (Fig. 9.12). Much of the agricultural landscape of the county away from the fen peat and alluvium is recent, having been enclosed and 'improved' during the eighteenth and early nineteenth centuries when extensive 'wastes' were transformed with varying degrees of success from

TABLE 9.2 Some physical and economic effects of wind erosion

Physical effects	Economic consequences
Soil damage	*Soil damage*
(1) Fine material, including organic matter, may be removed by sorting, leaving a coarse lag.	(1, 2, 3) Long-term losses of fertility give lower returns per hectare.
(2) Soil structures may be degraded.	(3) Replacement costs of fertilizers and herbicides.
(3) Fertilizers and herbicides may be lost or redistributed.	
Crop damage	*Crop damage*
(1) The crop may be covered by deposited material.	(1–6) Yield losses give lower returns.
(2) Sandblasting may cut down plants or damage the foliage.	(1–3) Replacement costs, and yield losses due to lost growing season.
(3) Seeds and seedlings may be blown away and deposited in hedges or other fields.	(5) Increased herbicide costs.
(4) Fertilizer redistributed into large concentrations can be harmful.	
(5) Soil-borne disease may be spread to other fields.	
(6) Rabbits and other pests may inhabit dunes trapped in hedges and feed on the crops.	
Other damage	*Other damage*
(1) Soil is deposited in ditches, hedges, and on roads.	(1) Costs of removal and redistribution.
(2) Fine material is deposited in houses, on washing and cars, etc.	(2, 3) Cleaning costs.
(3) Farm machinery, windscreens, etc. may be abraded, and machinery 'clogged'.	(4) Loss of working hours and hence productivity declines.
(4) Farm work may be held up by the unpleasant conditions during a 'blow'.	

Source: Wilson and Cooke (1980).

TABLE 9.3 Increased wind erosion in Britain: some suggested causes and their consequences

1. An increase in arable land and correlative reduction in permanent pasture and the use and length of grass leys, which has reduced the protective effect of vegetation, reduced the time for soil to 'recuperate', and extended both the area susceptible to wind erosion and the periods of susceptibility.
2. A tendency towards monoculture, and a consequent reduction of the stability provided by crop rotations.
3. An increase in the practice of stubble and straw burning (and other crop residue removal practices) which reduces the protective effect of such material and reduces the provision of organic matter to the soil of value in maintaining soil structure.
4. An increase in the use of artificial fertilizers, some of which tend to disaggregate soil clods.
5. Improved weed control with herbicides, which reduces the protective effect of weeds.
6. The introduction and rapid extension of sugar beet, a crop requiring a loose, vulnerable seed bed and providing only limited surface protection early in its growth season.
7. An increase in the use of wide-spaced 'drill-to-stand' techniques which increase the area of ground vulnerable to wind erosion in a field.
8. A decrease in marling, a practice which can improve soil structure and reduce the erodibility of soil.
9. Continued removal of hedges and increase in field size, thus reducing the protective effect of field boundaries and increasing wind 'fetch'.

Source: Wilson and Cooke (1980).

heath and rough grazing into neatly geometric landscapes dominated by farms with mixed economies (Robinson, 1969). Rotations involving grass, root crops, and cereals have generally been used on the improved land, and such rotations included among their many advantages some degree of protection to the soil: the fields were bare for only part of the time and the rotation system helped to maintain soil structure. In addition, the hedgerows around fields afforded shelter.

In recent years, however, there have been significant changes in agricultural practices that may have tended to increase the erodibility of soil material. The widespread and continuous use of chemical manures and the burning of straw and stubble on arable land may have reduced the cohesive properties of the soils; the disc harrowing of light soils and the continuous cultivation of barley may have had similar effects. In addition, the percentage of grasslands has declined and the proportion of arable crops, especially cereals, has increased so that, for example, more land is now worked to a fine, vulnerable seed-bed in spring, and the ability of the soil to maintain its humus content and structure has been reduced by the reduction of grass. Finally, the demands of mechanical cultivation and the high value of land have led to the removal of trees and hedgerows, thus exposing more land to the attack of winds and increasing the fetch of eroding winds.

Under given climatic conditions, it seems probable that these changes lead to increased wind erosion, and when climatic conditions are most appropriate for wind erosion, soil blowing may be serious. Such was the case in March 1968. At that time, three climatic circumstances commingled to produce serious erosion. Firstly, precipitation in the first three months of the year, which is usually low in any case, was only 54 per cent of the standard average: the soils were abnormally dry. Secondly, the number of frosts in February was significantly higher than average in some areas: frost action may have helped to produce a finer soil tilth than usual. Thirdly, there was a period in mid-March of very strong westerly winds associated with a series of vigorous low-pressure troughs which followed the dry, frosty winter and preceded the growth of soil-protecting crops. The observations of soil blow on Fig. 9.12 are based on field observations and on reports made after the windy spell: the information is not comprehensive, but it probably reflects fairly accurately the distribution of wind erosion (Robinson, 1969).

The events of March 1968 had several serious consequences for the people of Lincolnshire. Many roads were partially blocked by wind-blown material: traffic was disrupted, and the cost of clearance was considerable. Clearing operations in Lindsey (the northern administrative unit) alone cost £4000 ($10 400). Ditches and drains were filled with sediment: it cost one drainage board £5000 ($13 000) to clear ten drains. And many farmers had to clear ditches on their own lands, at their own expense: the average cost was estimated to be approximately £5 ($13) per 22 yards (20.1 m), and in the Isle of Axholme alone the cost may have been £17 500 ($45 550). Also on farmland, productivity was reduced in places by uncovering or removal of seeds (e.g. barley, peas,

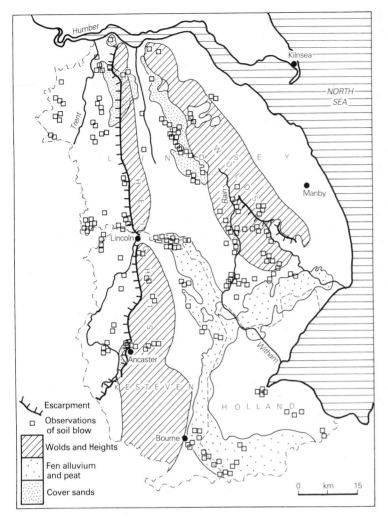

FIG. 9.12 Wind erosion in Lincolnshire, 1968 (after Robinson, 1969)

and beet), and by the 'scorching' of leaves and root-exposure of winter-wheat plants. Perhaps even more important than these short-term problems is the progressive loss of topsoil from the land and the decline of fertility associated with it. It is estimated that in Lindsey, some 2438 ha may be affected; in Kesteven, perhaps 6477 ha; and in Holland, where no figures are available, soil erosion is certainly a problem. The loss of soil is serious because it has implications for long-term productivity, and because it is not always indentified by farmers as a problem requiring remedial measures. The recent changes in agricultural practices which underlie the problem are not being significantly modified, and further erosion seems probable until solutions such as the construction of wind-breaks, marling, and the increased use of grass leys are adopted.

(ii) *The Great Plains of the United States* In Chapter 4 it was emphasized that successful control of soil erosion by water requires an understanding of *both* the physical system and the socio-economic system in which it is used. So it is with wind erosion. Unfortunately, such a broad understanding is commonly lacking. There are many reasons for this. In the first place, farmers fail either to recognize the phenomenon of wind erosion or to recognize it as a problem. Or, even if the problem is recognized, there may be no desire or motivation to solve it. These responses may be especially true in areas where there are numerous other farming difficulties, such as drought, insufficient capital, and uncertain market conditions. Secondly, although the wind-erosion system is now reasonably well understood by a small number of scientists, those responsible for wind-erosion control, such as farmers, may be unaware of the remedial measures or the physical principles underlying them. Thirdly, even if the desire to control wind erosion exists, the decision to act against the problem may be delayed by pressing financial circumstances or by the exigencies of the farming calendar. Often aid from insurance or governmental agencies is not available. And in any case the farmer *may* be unconcerned about the long-term future of his soil, especially if he is a pioneer, transient cultivator, or speculative land-purchaser. Finally, remarkably little is known about the ways in which farmers perceive and respond to the hazard of wind erosion.

Nowhere is the importance of understanding the socio-economic context of wind erosion more clearly demonstrated than in the 'Dust Bowl' region of the United States (Fig. 9.13) where, in the droughts of the 1930s, extreme wind erosion damaged over 9m ha, generated spectacular dust storms, decimated wheat harvests and livestock numbers, and brought about massive social hardship and upheaval (Plate 9.2). And nowhere is the social problem of wind erosion better explained than in Donald Worster's (1979) book, *Dust Bowl*. He stated (p. 13):

The story of the southern plains in the 1930s is essentially about dust storms, when the earth ran amok. And not once or twice, but over and over for the better part of a decade: day after day, year after year, of sand rattling against the window, of fine powder caking one's lips, of springtime turned to despair, of poverty eating into self-confidence.

Explaining why those storms occurred requires an excursion into the history of the plains and an understanding of the agriculture that evolved there. For the 'dirty thirties', as they were called, were primarily the work of man, not nature. Admittedly, nature had something to do with this disaster too. Without winds the soil would have stayed put, no matter how bare it was. Without drought, farmers would have had strong, healthy crops capable of checking the wind. But natural factors did not make the storms—they merely made them possible. The storms were mainly the result of stripping the landscape of its natural vegetation to such an extent that there was no defense against the dry winds, no sod to hold the sandy or powdery dirt. The sod had been destroyed to make farms to grow wheat to get cash.

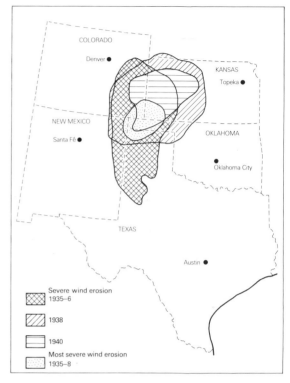

FIG. 9.13 The 'Dust Bowl' region in the United States in the 1930s (after Worster, 1979)

The question arises: why were the plains so extensively exploited with such disastrous consequences? There are many reasons. It became possible to transform the native grassland to productive arable land on a vast scale—'sodbusting' with tractors, harvesters, and other mechanical equipment. There was after the First World War a growing market for grain that promised huge rewards. Land exploitation was firmly based on attitudes at the heart of an expansionary, free-enterprise culture: the ethics of 'dominating nature' and the view that natural resources were inexhaustible; and other social attitudes were important;

that what is good for the individual is good for everybody, that an owner may do with his property what he likes, that markets will grow indefinitely, that 'the factory farm is generally desirable'—leading, in this last case . . . to irresponsible non-resident ownership, speculative commercialism, and land-abusing tenancy among the less successful operators. These attitudes were unmistakably associated with capitalism's evolution on the plains (Worster, 1979, p. 195).

PLATE 9.2 A powerful environmental message: dust storm in the 'Dust Bowl' of the Great Plains of the United States in the 1930s (photo courtesy US Department of Agriculture)

In addition, farmers consistently underrated, and still underrate the possibility of drought, and they minimize the risks. They are optimists of faulty perception:

When drought occurs, they insist that it cannot last long. Consequently, although they may become unhappy or upset by crop failures, they feel no need to seek out logical solutions or change their practices. They are prouder of their ability to tough it out than to analyze their situation rationally, because they expect nature to be good to them and make them prosper. It is an optimism at heart fatalistic—and potentially fatal in a landscape as volatile as that of the plains.

The source of that optimism is cultural: it is the ethos of an upwardly mobile society. When a people emphasize, as much as Americans do, the need to get ahead in the world, they must have a corresponding faith in the benignity of nature and the future. If they are farmers on the Western plains, they must believe that rain is on its way, that dust storms are a temporary aberration, and that one had better plant wheat again even if there is absolutely no moisture in the soil (Worster, 1979, p. 27).

Morover, Worster argued that the farmers and landowners often lacked a close communion with the soil under their feet—rather, they had their eyes firmly fixed on the urban world of markets, motorized mobility, fashion, cheap credit, and the pursuit of affluence. They continuously sought more from a soil that could not provide it. Thus:

The ultimate meaning of the dust storms in the 1930s was that America as a whole, not just the plains, was badly out of balance with its natural environment. Unbounded optimism about the future, careless disregard of nature's limits and uncertainties, uncritical faith in Providence, devotion to self-aggrandizement—all these were national as well as regional characteristics (Worster, 1979, p. 43).

The farmers were a part of the same culture that created the contemporaneous Great Depression—there was no great difference between 'Black Thursday' on Wall Street and the black days of the Dust Bowl (Worster, 1979, p. 44).

Responses to the wind erosion of the Dust Bowl sheltered under Roosevelt's 'New Deal'. Great strides were taken in alleviating hardship, removing unsuitable land from agriculture, and introducing soil-conservation practices. Many of the practices discussed below arose from the Dust Bowl experience.

But it is clear that even in this part of the United States, where more research has been undertaken and more aid supplied than anywhere else in the world, the problem has still to be completely solved. Despite the experience of the 1930s and the post-1945 years, the area of cultivated land is still increased when climatic, economic, and political circumstances are favourable in locations quite unsuitable for crops. And the US Dept. of Agriculture's Soil Conservation Service estimated in 1955 that there were still 5.6m ha in the Great Plains under cultivation that should be returned to grass. If the lessons have not yet been learned here, is it surprising that, in different cultural contexts, the same hazards still appear, for example on the Russian steppes and the plains of semi-arid Africa (e.g. Klimenko and Moskaleva, 1979)?

(b) *Management of wind erosion*

Control or prevention of soil erosion by wind requires the manipulation of key variables in the wind-erosion system. Fundamentally, the solutions lie in reducing the force of the wind and/or in improving ground-surface conditions so that particle movement is prevented and soil loss is reduced to a tolerable level. There are four general, constant principles of wind-erosion control: (i) establish and maintain vegetation and vegetative residues; (ii) produce, or bring to the soil surface non-erodible aggregates or clods; (iii) reduce field width along prevailing wind-erosion direction; and (iv) roughen the land surface (Lyles *et al.*,

1983). Fig. 9.14 summarizes the main variables and the methods commonly used to achieve management objectives.

Having identified the direction of the changes required in the variables, the question arises 'which is the *most efficient* way of altering the system in order to reduce wind erosion to an acceptable level?' Clearly, the desired result could be achieved in many different ways. The best answer will depend partly on the local environment, cost, and on the technical and financial resources of those making the adjustments.

The various control measures may be grouped conveniently under three headings: (a) vegetative methods, (b) ploughing practices, and (c) soil-conditioning methods.

(i) *Vegetative methods: Windbreaks* Perhaps the most familiar solution to the wind-erosion problem is to place a barrier across the path of the wind and thus reduce wind velocity at the ground surface both in front of and behind the feature, and at the same time reduce field length. Sometimes the barriers may be of materials such as netting, stakes or rows of palm fronds, but more commonly they are relatively permanent, growing vegetational structures.

A great deal of research has been carried out into the design, location, and effectiveness of wind-

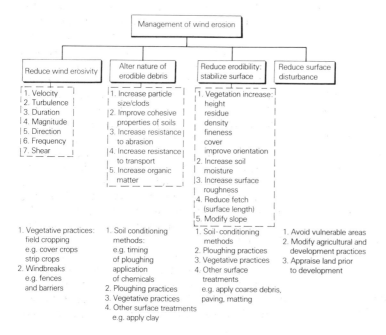

FIG. 9.14 Approaches to managing wind erosion of soil

breaks and the larger shelter-belts, and several important points arising from it should be emphasized. Firstly, because vegetational wind-breaks are relatively permanent their location must be carefully planned. It is essential, for example, that they should be set as nearly as possible at right angles to the most damaging winds, taking wind-erosion forces from other directions into account. Spacing is also important, and it should be related to the degree of shelter afforded by each obstacle. Secondly, the effectiveness of wind barriers depends mainly on wind characteristics, and the width, height, and porosity of the barriers. The protected distances either side of a barrier are usually measured in terms of barrier heights. For instance, percentage velocity reductions for average tree shelter-belts with wind approaching at right angles to the belts are as follows (FAO, 1960):

Percentage velocity reduction	Distance from barrier, downwind (in barrier heights)
60–80	0
20	20
0	30–40

On a percentage basis, these reductions are relatively constant regardless of open-wind velocities. In general, wind velocity is affected to about 5–10 times wind-break height on the windward side of the barrier and about 10–30 times on the leeward side. (It should be noted, however, that wind velocities might be *increased* at the ends of a barrier as a result of wind funnelling, and hence long barriers are generally preferable to short ones.) In terms of soil erosion, barrier effectiveness depends on the relations between open-wind velocities, sheltered area velocities, and fluid threshold velocities. For example, the fluid threshold velocity for many soils is between 19.3 and 24.1 k.p.h. A 50 per cent reduction of a 32-k.p.h. wind would provide complete control, but erosion might occur in the same area if open-wind velocity was 80 k.p.h. Clearly, the fully protected zone of any barrier is reduced as open-wind velocity increases.

A third point is that the most effective barrier is not completely impermeable because such barriers create diffusion and eddying effects on their lee sides. Semi-permeable barriers are most effective, for although they provide smaller velocity reductions their influences extend further downwind. Permeability can be controlled through vegetation density and width of the barrier. The shape of shelter-belts is also important. Experience suggests that a triangular cross-section is preferable to a streamlined shape or an abrupt vertical front. Shape is controlled by carefully selecting the plants to be grown and planning their distribution.

Finally, wind-breaks serve purposes other than wind-erosion control. Under suitable management they may provide such things as timber, refuge for wildlife, and protection for buildings.

The disadvantages of wind-breaks are that they require land, they are expensive, they take time to mature, the shelter they provide is limited, and they compete with other crops for moisture and soil nutrients in environments where both may be scarce. Wind-breaks also require a degree of forward planning.

Vegetative methods: Field-cropping practices
Vegetative methods related to field-cropping practices are often simpler, cheaper, and more effective than wind-breaks, and some of them have the additional advantage that they can be used as emergency measures.

One aim of wind-erosion control is to trap moving particles, especially saltating particles, and another is to protect the surface from attack. Various vegetative methods have been employed to achieve these ends. One is to plant 'cover' crops which grow to protect the surface when it would normally be exposed to erosion, notably in the period prior to spring planting. In some erosion-prone areas of Australia oats or barley are sown for this purpose. A second technique is to mix erosion-vulnerable and erosion-resistant crops in alternating strips normal to the prevailing winds (e.g. Chepil, 1975; Plate 9.3). The erosion-resistant crops trap particles, protect the surface beneath them and, in some cases, shelter the vulnerable crops. An example of this 'strip-cropping' technique (see also Chapter 4) is the alternation of wheat, sorghum, and fallow strips in fields in parts of the semi-arid United States. Other things being equal, width of strip increases as soil texture becomes finer (except for clays subject to granulation). Another variable is the angle of the wind to the strips (Table 9.4).

A general principle, of course, is to select, as far as possible, crops that provide the best surface protection: small-grain crops, legumes, and grasses are all fairly effective, once they are established. As wind erosion occurs on dry land,

PLATE 9.3 Sensible management: strip cropping in western Kansas, USA—winter wheat (dark) and sorghum stubble (light) (photo courtesy US Department of Agriculture)

TABLE 9.4 Strip dimensions for the control of wind erosion

Soil class	Width of strips		
	Wind at right angles (m)	Wind deviating 20° from a right angle (m)	Wind deviating 45° from a right angle (m)
Sand	6.09	5.48	4.26
Loamy sand	7.62	6.70	5.48
Granulated clay	24.39	22.86	16.46
Sand loam	30.49	28.04	21.34
Silty clay	45.73	42.68	33.53
Loam	76.21	71.64	51.82
Silt loam	85.36	79.26	57.92
Clay loam	106.70	99.08	76.21

Note: The table shows average width of strips required to control wind erosion equally on different soil classes and for different wind directions, for conditions of negligible surface roughness, average soil cloddiness, no crop residue, 1-foot high erosion-resistant stubble to windward, 64.4 km/h wind at 15.24 m height, and a tolerable maximum rate of soil flow of 203.2 kg/5 m width per hour.

Source: Chepil and Woodruff (1963) with metric conversion.

the choice of crop may be limited by the availability of water, although the choice may be extended if supplementary irrigation is available. The main advantage of strip cropping is that it restricts soil avalanching; disadvantages include possibilities of weed and insect infestation and the difficulties of grazing appropriate crops.

Closely allied to the cover-crop principle is that of stubble and crop residue management (see also Sect. 9.5). Stubble and other field-crop residues

protect the surface from erosion (Fig. 9.5D, E), trap moving particles, and provide organic matter for the soil. In some cases stubble may be left between strips of ploughed land normal to the prevailing wind. The two variables of stubble height and strip width can be manipulated according to wind intensity and soil erodibility. In general, for the same degree of protection, strip width decreases with stubble height, and thus the area of tilled land can be maximized by leaving tall stubble. The principles are that the strip should be wide enough to prevent saltating particles from jumping it and receptive enough to stop saltating particles that strike it. Stubbles vary in their effectiveness. In general, those of small-grain crops are more effective than those of large-grain crops. And some residues last longer than others: for instance, wheat and rye straw is more durable than legume residue. It is important that the stubble be disturbed as little as possible; thus, in areas where the ground has to be tilled, stubble cover may be retained by subsurface ploughing.

(ii) *Ploughing practices* Because soil erosion is approximately inversely proportional to surface roughness, it is clearly desirable to create a rough, cloddy surface when ploughing or creating a seed-bed. The effect of ridging cloddy soils by ploughing normal to the eroding wind is to reduce wind velocities and perhaps even reverse wind direction between ridges, thus promoting deposition of particles within hollows (Fig. 9.5D, E). If the spacing is correct with respect to saltation wave-length, then saltation may be stopped completely. If the ridging is done after rain, large clods may also be produced in suitable soil. But ridging is temporary, and the surface roughness may be reduced by wind erosion and other processes (such as frost action); the hope is that ridging will last through the period when the soil is vulnerable to wind erosion. Another problem of ridging is that it may promote drying of the soil, thus increasing the possibility of wind erosion. Emergency ploughing is usually undertaken when fields are most vulnerable to attack because plant cover is sparse. Some 2.4m ha are ploughed on an emergency basis to control wind erosion each year in the Great Plains of the USA (Lyles and Tatarko, 1982).

Considerable attention has been given to the design of agricultural equipment suitable for the cultivation of soils potentially vulnerable to wind erosion. One difficulty is that the conventional mould-board plough may break up the soil and bury plant residues, thus exacerbating the problem of wind erosion. Alternative tools have been employed, each with a particular purpose and a general concern about wind erosion in mind. For example, the lister or ridger creates a roughened surface and erosion-resistant ridges, and it is suitable for the emergency control of wind erosion in some areas. The disc harrow and similar equipment that tend to break down soil structure and leave a smooth surface are suitable for the

cultivation of light stubble but not for the cultivation of stubble-free soils. Various field and chisel cultivators—usually comprising points fitted to curved or straight shanks and pulled through the soil at depth—are often effective because they control weeds, crop residues are not greatly reduced at the surface, and a rough, ridged surface is created. These tools, and others, are described by the FAO (1960).

(iii) *Soil-conditioning methods* As wind erosion predominantly affects soil, it is usually desirable to conserve and maintain soil moisture. These aims can be achieved by reducing evaporation, reducing unnecessary plant growth (by weeding, etc.), and by reducing runoff. Surface mulching, for instance, conserves soil moisture and promotes infiltration. Rolling of a soil that is damp below the surface might in certain circumstances serve to moisten the surface; and occasional irrigation might serve the same purpose. In addition, infiltration of available rainfall may be increased by such practices as strip cropping and terracing.

Other methods of soil conditioning are directed towards the creation or maintenance of erosion-resistant clods. This aim may be achieved, for example, by timing ploughing so that it occurs soon after a rain, or by increasing soil organic matter. Finally, a soil-conservation practice that holds some promise for the future is the application by spraying of certain artificial compounds, such as water-based emulsions of resin, which protect the ground surface (Armbrust and Dickerson, 1971).

(iv) *Conclusion* Any fully informed farmer is likely to use a combination of techniques to limit the loss of his soil. The combination he chooses will depend on many considerations, but he will probably try as far as possible to create and maintain soil aggregates, roughen the surface, provide barriers to wind and to particle movement, and maintain soil moisture, vegetative cover, and plant residues at the soil surface.

9.5 Sand and Dust Problems

(a) *The hazards*

The movement of sand and dust by wind poses serious problems in drylands, and the problems are often unrelated to agricultural and pastoral land use. If buildings, cultivated areas, pipelines, and transportation networks in deserts are to avoid being attacked and perhaps buried by sand and dust, extensive control measures may be necessary. Urban areas in drylands may present obstacles to the natural and pre-existing patterns of aeolian debris movement. But, equally important, the nucleation of desert settlements dependent on localized water supply and limited suitable agricultural land causes pressure on land to be concentrated around settlements so that the desert ecosystem is likely to be most disturbed in the immediate neighbourhood of urban areas. As a result, vegetation may be destroyed and soil structure damaged, thus promoting sand and dust movement and leading to increases in the magnitude and extent of aeolian problems. In this way, the causes of sand and dust problems are often intimately related to the problems of desertification that have received much recent attention (e.g. Rapp *et al.*, 1976). Problems arising from sand and dust movement can be classified and examined in terms of deflation, transport, and deposition.

(i) *Deflation problems* Deflation, the removal of sand and dust by wind from desert surfaces, is primarily a problem because it leads to the depletion of some of the most important soil constituents—silt, clay, and organic matter—leaving beind coarser particles, lower levels of fertility, and a reduced ability to retain water. Other effects of deflation include scour and undermining of footings to telegraph-poles etc., and scouring beneath pipelines, railway sleepers, and even roads—effects that can lead to collapse of the structures. Deflation is a natural process, but it usually leads quite quickly to the establishment of wind-stable surfaces in deserts such as those comprising stone pavements, saline crusts, or soil aggregates. Only where stability cannot easily be achieved, as in mobile dune fields or on unpaved roads, is deflation a continuing natural problem. But deflation can be initiated where surface stability is disrupted by human interference. The Arizona Department of Transportation (1975) reported, for example, that the annual loss of deflation of silt and clay was in the order of 3–30 kg per vehicle km on unpaved roads.

(ii) *Problems of transport: abrasion problems* Although the saltation curtain reaches up to about a metre, the maximum abrasive effect of sand is felt at a height of 20–5 cm. Abrasion effectiveness is determined by numerous variables, such as particle shape, size, orientation, hardness, and surface texture of the material under attack; sizes, hard-

ness, mass, and sharpness of projectiles; and environmental factors such as topography, neighbouring particles, vegetation, and wind velocity (Cooke and Warren, 1973; Hagen, 1984).

Soil clods and other more or less indurated soil materials may be disintegrated or abraded as a result of saltating grain impact, thus promoting the impoverishment of soil structure, and rendering soil even more erodible. Abrasion can also be extremely injurious to certain plants, structures, and equipment. Abrasion heights may in fact be higher over hard, man-made surfaces, and the abrasion problem consequently more severe, because saltation heights and the rate of sand movement are increased over such surfaces as roads and runways. Sand abrasion can have an erosive effect on building materials: faces at angles greater than $55°$ to wind direction tend to be pitted, whereas more acutely inclined faces become fluted and grooved. Telephone- and telegraph-poles, and fences may be sand-blasted at their bases. Glass loses its transparency, first becoming pitted and then frosted: even the relatively high car windscreen may suffer (especially, of course, when the car is in motion). Paintwork is easily damaged. Maintenance and replacement costs for mechanical equipment are increased. Equipment such as generators and pumps may suffer from worn piston-rings, scored cylinders, damaged bearings, etc. Air filters require frequent changes.

Visibility and other problems. Dust storms are common desert phenomena, and they may vary in size from those covering as much as $1.5m \, km^2$ down to diminutive dust devils. Long-distance transport of dust has been reviewed by Goudie (1983). Problems associated with dust storms include the spread of disease through pathogen transport, suffocation of cattle, development of static electricity, interruption of radio, telephone, and telegraph services, disruption of transport, the damaging of property, and harm to human health (Idso, 1976; Morales, 1979). But the problem that has received most attention is that of visibility reduction, a problem of greatest importance to the transport industry. Flights into and out of airports affected by blowing dust may be delayed (and in places there may be a danger of runway skidding), as at Sharjah and Bahrain in the Arabian Gulf (e.g. Houseman, 1961). Sand and dust storms in deserts can halt road traffic (e.g. Péwé, 1981).

(iii) *Depositional problems: Dust* Problems associated with dust deposition include the burial and killing of young plants, the rendering of roads impassable, infiltration of dust into houses creating problems of sanitation and housekeeping, and contamination of food and drinking-water. For example, Clements *et al.* (1963) reported that dust between relay contacts and abrasion of switches in the Mojave Desert caused particular problems for a Californian telephone company.

Sand. The commonest depositional problems are associated with the bulk transport of sand *en masse* in bedforms. Dunes encroach upon and often completely bury obstacles of all types, including roads, railways, runways, pipelines, and cultivated gardens. Larger aeolian features than dunes have a similar effect but if they move, they move much more slowly.

A common outcome of sand encroachment is land abandonment, reduction in the intensity of land use, failure of communications, and depopulation. The town of In Salah, in the Algerian Sahara, provides a significant example: the inhabitants are fighting an endless war against encroaching sand that threatens to overwhelm their palmtrees, and which they attempt to ward off with palm-frond fences. Where control schemes have been implemented, their management is often poor or absent; sand fences are often wrongly constructed: they may be too low, or incorrectly spaced, or too close to the oasis. Ultimately, as elsewhere in this region of the Sahara on the fringes of the sand seas, the settlement may be abandoned as clearing costs become prohibitive.

In fabricating structures within urban areas the engineer and designer must not only consider wind loads for stability and safety of the buildings and inhabitants, but also the movement of sand and dust by winds. The debris may abrade the structures, and interactions between wind and buildings can produce local debris accumulations and debris infiltration into buildings. Areas of wind funnelling and vortices can result in unpleasant wind-abrasion conditions for people's comfort, vegetation, or traffic. Shelter in airflow separation zones will promote the growth of sand accumulations, blocking roads and passageways, burying vegetation, and encroaching on living areas.

Communication lines are vulnerable to sand movement and can be especially inconvenienced by migrating dunes: on roads, for instance, detours may be necessary around dunes, or costly sand-removal measures may be required; and heavy-duty four-wheel-drive vehicles are often essential. Some roads and railroads may be designed to be self-cleansing (e.g. Redding and Lord, 1982), but others, especially where they cross active sand fields, may be continuously in jeopardy. The road north of Al Kharga, in the New Valley, Egypt, was built on a dam about as high as the local *barchan* dunes, which led initially to sand storage behind the dam (and temporary relief to the road), but dunes soon began to climb over the drifts, advance on the road, and partially bury telephone-lines. Control measures, such as the dam, commonly fail because of a lack of understanding of the processes involved. One particularly common error is to assume that by flattening dunes the problems ensuing from their movement will be eliminated: this is rarely so, because the winds will soon recreate aerodynamically suitable surface forms similar to the original dunes.

Pipelines and similar features pose especially difficult problems, because deep burial by sand makes maintenance and inspection difficult; and unsupported pipes on active dunes can be left above the ground as dunes move on, putting them under torsion and possibly causing fracture. Such was the fate of part of the phosphate conveyor-belt at El Aaiun, Spanish Sahara (De Benito, 1974).

(b) *Management of sand and dust*

The hazards of sand and dust transport can be managed by reducing transport and/or by reducing sediment supply. For dust, little can be done to reduce flow because dust-storm winds cannot usually be significantly reduced by artificial means. Thus the major responses to dust transport involve avoiding activity in dust storms (e.g. by restricting vehicle movements) and excluding dust from locations where it is a problem (e.g. in houses). The ultimate solution to dust hazards is to reduce supply, as discussed above in the context of soil erosion; in deserts, reductions can also be achieved by preventing surface disruption caused by off-road vehicles and disruptive activities associated with engineering and settlements (e.g. Wilshire, 1980). But for sand movement, much more can be done, because motion is essentially near to the ground surface where management actions are more effective. In addition, any reduction of existing sand and dust movement carries with it complementary problems of deposition. Thus the problems of managing sand and dust transport and deposition are somewhat different from those of erosion control. Fig. 9.15 summarizes the major responses to sand movement and deposition. Recent reviews of this field are provided by Cooke *et al.* (1982) and Watson (1985).

(i) *Avoidance, removal, or control* A sound general rule in environmental management is to avoid hazard areas and surface disturbance as far as possible: avoidance is often more effective and cheaper than control both in the short and the long term. Similarly, if surface disturbance is inevitable, in general the less the disturbance, the better. Such rules are extremely pertinent to aeolian problems especially in the context of large and active dune fields, where the scale of natural forces is considerable and control measures, even expensive and extensive measures, are likely to be of only temporary value.

The best means of avoidance lies in sensible site selection, a procedure that requires prior knowledge of existing or potential problems and therefore requires antecedent environmental surveys. Even when the general location of development is determined by other factors, the site may be designed sensibly to avoid or to minimize the impact of aeolian problems. Al Ain in eastern Abu Dhabi, for example, has to be located for political and hydrological reasons near to the old oasis but the new city plan successfully seems to minimize (although not eliminate) the potentially serious sand-movement problems associated with a nearby, extensive, and active dune field.

Closely allied to the notion of avoidance in the eyes of some developers is that of removing the material by earth-moving equipment. An extreme and rarely practicable solution is to remove all the moving sand and thus eliminate the problem entirely—often an expensive solution, but possible if only small dunes are involved. Alternatively, where sand is continuously causing problems by advancing across roads, etc., clearance provides a temporary solution, but one that is generally both continuous and expensive in the long term. The solution of removing dunes by flattening them is unlikely to succeed because airflow working on the newly flattened surface is likely soon to reconstitute the mobile bedforms.

(ii) *Vegetational stabilization* Permanent stabilization of mobile sand and dust can often be achieved effectively only through the development of a vegetation cover. This solution usually requires the use of a combination of mechanical, chemical, and botanical methods at least until the

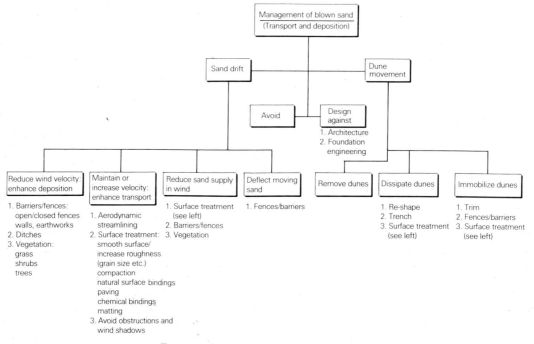

FIG. 9.15 Approaches to managing blown sand

vegetation has become firmly established (Tsuriell, 1974). Attempts at vegetational stabilization should consider the interrelationships between the following habitat factors—character of substrate, thickness of sand deposit, degree and nature of salinization; water-storage capacity, nutrients, and structure of the substrate or soil; quantity and quality of water available for the plants (such as precipitation regime, soil moisture, air humidity); depth of water-table and its chemistry; type of movement and rate of displacement of moving sand and dust, exposure to predominant wind direction, and solar radiation.

Natural recovery. Protection of an area (e.g. by enclosure) where relics of the original vegetation are still present, may result in the spontaneous redevelopment of the vegetation cover, although the speed and effectiveness of response will vary greatly with local conditions. Much depends on the stage of deterioration of vegetation and soil and also the respective sizes of degraded and undegraded areas adjacent to the recovery zone. In general, recovery is slower the more arid is the climate, the shallower the soil, and the more degraded the vegetation. Recovery is generally restricted to areas of shallower groundwater conditions where the problems of deflation are minor. *Artificial recovery* involves the transformation of the natural ecosystem by planting with species that may or may not belong to the native vegetation. Some of the measures used here are related to those designed to conserve soil from wind erosion, such as contour ploughing, cover cropping, strip cropping, and the use of fertilizers and mulches. In sand the nutrient content is commonly low, so that high initial amounts of fertilizer are often necessary to maintain a high plant growth rate—nothing is gained by sowing a complex seed mixture in a soil with insufficient nutrients; at the same time, the high permeability, low moisture capacity, and lower adsorptive power of sand makes the washing out of fertilizers a serious potential problem.

(iii) *Surface stabilization* In some circumstances, especially in extremely arid areas, vegetational stabilization of moving sands is impossible, so that sand fixation must depend on special surface coverings of man-made obstacles (e.g. Chepil *et al.*, 1963).

Water is only a good surface stabilizer if the surface is kept wet—if the surface dried out, as it quickly does in areas of high evaporation, the protection is lost. Permanent wetting requires frequent spraying, but minerals and sediment precipitated from irrigation water might form a protective cement in time. The method is expensive, but often an excellent temporary expedient. *Gravel, stones, and crushed rock* greater than 2 mm in diameter are stable under aeolian conditions and provide an excellent means of stabilizing a sand surface (e.g. Chepil *et al.* 1963). Even dune sand (where no traffic is involved) can be stabilized with fine, medium, or coarse gravel spread uniformly over the surface. It is often a problem, however, actually to spread material mechanically over unstable sand surfaces. Cost of transport of these bulky materials is an important factor in this method. *Oil*, despite its ugly appearance, has been used successfully to stabilize large areas at low cost. Where vegetation is ultimately to stabilize the surface, the oil must not restrict plant growth by toxic effects or prevent water penetration. Three types of oil are commonly used: low-gravity asphaltic oil (as used in road construction); high-gravity deep-penetrating waxy oil; and crude oil. *Chemical sprays* are widely used to stabilize surfaces. They usually require special equipment and trained personnel; amounts depend on soil structure, slope, spraying technique, and degree of stabilization required; sometimes seeds and fertilizers may be applied at the same time. In principle, the spray soaks a few millimetres into the surface where the water base evaporates and the particles remaining bind sand grains together. Germination and growth of plants are supported by water penetrating through the pores of the stabilized layer. Evaporative losses are often reduced by the stabilized layer. But germination may be retarded, especially where chemicals are highly concentrated. Local destruction of the chemical crust has to be prevented or the crust repaired by respraying, because rapid deflation may begin which can undermine and rapidly break up the crust. Many chemical stabilizers are only temporary, disintegrating after a year or so, and they are expensive, so they are normally used only in conjunction with other methods, especially vegetation stabilization. Costs and effectiveness of stabilizers are not easily evaluated because so much depends on local geomorphological and economic circumstances. For example, on level terrain it may be possible to use standard agricultural equipment and boom sprayers, whereas in dune country hand-spraying may be necessary or four-wheel-drive equipment required. For further evaluation see, for example, Armbrust and Dickerson (1971).

(iv) *Fences and wind-breaks* These have been reviewed in Sect. 9.4. In deserts, where vegetational wind-breaks may be difficult to sustain, a variety of diversion fences and impounding fences have been designed (e.g. Cooke *et al.*, 1982). Design is a key problem. It extends in deserts beyond the design of barriers to architectural features. By paying attention to layout, orientation, and geometry of buildings, much can be achieved in reducing the hazards of moving sand and dust, and controlling the loci of deposition (e.g. Gandemer, 1977; Cooke *et al.*, 1982).

(v) *Dealing with dunes* Finally, attempts may be made to *stabilize or destroy dunes. Flattening* dunes is not a successful solution, despite the fact that it is often attempted, because the flat surface is unstable and dunes redevelop. *Transposing* of dunes is also only a temporary solution, as is destruction by *trenching*. Stabilization is usually most effectively achieved by fencing or surface treatments. Dune stabilization by developing a vegetation cover, while it may be successful on coastal dunes in humid areas, is expensive and requires good management in deserts. For example, the supply of sand needs to be reduced to prevent plant burial and 'burning', and the planted surface needs to be stabilized.

(c) *Conclusion*

Sand and dust control in drylands differs fundamentally from that in humid regions because the areas involved are often more extensive, the problems more severe, the economics more precarious, and the scope for manipulation, especially through vegetation, is more limited. In general, several separate solutions, used in harmony, are likely to succeed better than a single panacea. Effective vegetation stabilization usually requires the combination of mechanical, chemical, and biological methods at least in the initial phase of vegetation growth. Once vegetation is established, other defences may be unnecessary. Successful control of sand and dust depends largely on adequate appraisal of the problem before development and then sound management practices including maintenance and surveillance.

9.6 Conclusion

Aeolian hazards are serious in many drylands, especially those affected by desertification and urban development; they also periodically affect more humid areas where climatic, soil, and land-use conditions lead to strong winds attacking dry, vegetation-free surfaces. There is evidence to suggest that, as with so many geomorphological hazards, they are becoming more serious, especially in drought-vulnerable semi-arid lands. This trend arises despite the fact that much is now known about the nature and dynamics of the aeolian system and there is very extensive experience of a wide variety of methods for controlling or ameliorating the hazards. In part the trend reflects increased pressures on the use of marginal land, and the continued application of inappropriate land-use practices. In part it also reflects the failure of scientists, who know how to manage the aeolian system, effectively to apply their solutions through those responsible for using the land. Equally fundamental, however, is the fact that many land-users fail to perceive the problems, or fail to repond to them even if they do perceive them. Such failure may reflect a callous attitude to the land; it is more likely to reflect the social, economic, and political constraints within which vulnerable land is used.

Future research on aeolian processes is likely to focus on the refinement of predictive equations, wind-erosion forecasting, and the more effective application of scientific principles to land-use practices within the context of local, specific circumstances. For the geomorphologist concerned with these problems, his major applied contribution is likely to focus on hazard appraisal and monitoring prior to the development of management plans.

10 Coastal Environments

10.1 Introduction

The coastlines of the world, over 440 000 km in length, represent one of the most dynamic of natural environments and one of the most important contexts in which human activity and geomorphological processes interact. Coastlines bring together a unique and extraordinarily varied group of processes; not only processes associated with the sea itself, but also, in certain locations, those arising from water and sediment transfer by rivers to the sea; from the subaerial degradation of cliffs and similar landforms above the water's edge; and from aeolian, glacial, and periglacial conditions. All are often also closely associated with distinct biological processes in different coastal ecosystems. In this dynamic, complex zone there are many human conflicts reflecting varied coastal resources and hazards. The coast is a locus for trade transfers and related port, industrial, and urban manifestations. Coastal resources include aggregates for construction (Chapter 11), often a bountiful source of inshore food (such as crustacea, fish, and seaweed), a major recreational environment, and areas suitable for land reclamation. The hazards include storm-generated floods, and an increasingly serious pollution problem; sediment transfer can interfere with coastal activities, and many engineering works, such as jetties, can deliberately or accidentally disrupt such sediment transfers (Plate 10.1).

The complexity of coastal zone environments and their management brings together management and research specialists from many disciplines, including engineers, planners, oceanographers, and ecologists, as well as geomorphologists. An example of the complexity of and conflicts within coastal management is shown in Fig. 10.1. The very extensive and rapidly growing literature on coastal problems reflects the convergence of cognate interests and the recent growth of interest in coastal problems. The biennial *Proceedings of the Coastal Engineering Conference* and the *Coastal Zone* conferences (e.g. *Coastal Zone 78*, 1978) both include papers from a range of disciplines, as do some edited books such as Davis's (1979) *Coastal Sedimentary Environments*, Hails and Carr's (1975) *Nearshore Sediment Dynamics and Sedimentation*, and Komar's (1983a) *CRC Handbook of Coastal Processes and Erosion*. Work in coastal geomorphology is represented in all the main geomorphology journals, and it is reviewed in *Progress in Physical Geography*. But geomorphological contributions are also to be found in such journals as the *Journal of Shoreline Management*, *Shore and Beach*, *Journal of the Institute of Civil Engineers*, *Coastal Engineering*, and the *Journal of the Waterway, Port, Coastal and Ocean Division of the American Society of Civil Engineers*. The publications of the Coastal Engineering Research Centre (CERC), such as its *Shore Protection Manual* (CERC, 1973) are also a major source of information. Amongst many recent books devoted exclusively to coastal geomorphology and its management are those by Komar (1976), Davis and Ethington (1976), Dyer (1979, 1986), Bird (1984, 1985), and Pethick (1984). In addition, several texts on coastal engineering are of geomorphological value (e.g. Wood and Fleming, 1981; Thorn and Roberts, 1981).

One of the features of coastal management problems is that they commonly transgress political boundaries. The drift of coastal sediment and pollution, for instance, rarely stops conveniently at administrative borders. Thus international collaboration is increasing, and management agencies have developed especially at national and regional levels. In western Europe, for example, coastal problems now come within the purview of the EEC, the Council of Europe, and UN agencies such as UNESCO and UNEP. Thus the European Spatial Charter, 1983 (under the aegis of the

PLATE 10.1 Groynes near Spurn Head, Yorkshire: the structures are left exposed as longshore sediment drift has depleted the beach

Council for Europe) recognizes the need for co-ordinated coastal development (Coccossis, 1985), and specific international initiatives include MEDPOL (the Mediterranean Pollution Monitoring and Research Programme, established in 1976 under the Barcelona Convention for the Protection of the Mediterranean Sea, in co-operation with UNEP) and the *Delta Plan* for the

control of flooding in the Scheldt, Rhine, and Meuse river estuaries, especially in The Netherlands.

In the United Kingdom, coastal research is helped by scientists from Hydraulic Research Ltd. in Wallingford, which has extensive laboratory facilities (Plate 3.1), and the Institute of Oceanographic Sciences; whereas coastal planning and

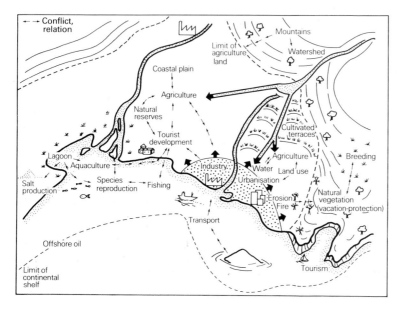

FIG. 10.1 Competition between different activities and functions in a coastal delta region (after Cocossis, 1985)

management is partly a matter for local authorities (including coastal protection authorities), for regional water authorities (under the Department of the Environment), and for the Ministry of Agriculture, Fisheries and Food, the latter bodies having responsibility for coastal erosion and flood protection (sea defence), effluent discharge, and water quality. In France, concern for coastal planning and management is more advanced and is reflected in the creation of *Le conservatoire de l'espace littoral et de rivages lacustres* (July 1975), which aims to safeguard coastal ecology, secure sensible economic development, and promote the attractiveness of coasts for tourists; and in the passing of *La directive d'aménagement national relative à la protection du littoral* (December 1979) and its successors (e.g. January 1986). Such laws are administered by the Ministère de l'Environnement, the *départements* and local *communes*. Coastal research in France is undertaken partly under the auspices of the *Comité d'Océanographie et d'Études des Côtes*. In the USA much of the national effort is focused on the Coastal Engineering Research Center (CERC) of the US Army Corps of Engineers, which has authority to undertake those measures considered necessary for the protection of public coasts and both field and experimental research. CERC assists the civil works programme of the Corps of Engineers in devising measures to prevent erosion, improve harbours and channels for navigation, and protect the coast from flood damage. CERC services are available to all federal agencies and, under certain conditions, to state and private agencies. National research institutes also contribute substantially to coastal studies, notably the Scripps Institute in California, the Virginia Marine Science Institute, and the Coastal Studies Institute of Louisiana State University at Baton Rouge. Sources of funds for coastal research include the Sea Grant College and the Geographical Branch of the Office of Naval Research.

The geomorphological contribution to coastal management is founded, as in other environments, on recording the nature of coastal conditions, and on monitoring change both for understanding historical development and for predicting future development. Within this context, the themes of major applied interest include the study of rates of erosion and deposition (from rivers into the sea, at the coast, and offshore); the nature of coastal equilibria and their changes, and longer-term

evolution; and the effects of coastal engineering and related sediment management on coastal dynamics (e.g. Hails, 1977*b*). In addition, of course, a major objective is to ensure that such knowledge is effectively integrated into management strategies.

10.2 Coastal Landforms and Dynamics

(a) *Classification*

The enormous diversity of coastal landforms reflects the infinitely varied resolutions of the conflicts between the processes operating on the coast, the rock lithologies and structures open to attack, and the complexities of coastal evolution that have usually included changes in level of the basic datum, sea-level. Given this variety, the task of classification and generalization is both difficult and controversial. As a backdrop to this discussion, Shepard's (1976) classification is shown in Table 10.1, despite its descriptive and genetic nature, because it is relatively comprehensive (Bird, 1984). Within the spectrum of coastal types shown in Table 10.1, two themes that figure prominently in coastal management are considered further below: cliffed coasts and shore platforms (including both land-erosion and wave-erosion coasts); and depositional coasts and beaches (including both land-deposition and wave-deposition coasts).

(b) *Coastal dynamics: the sea*

(i) *Sea-level* The action of the sea in the coastal zone relates in part to the level of the sea and in part to the nature of waves. The sea acts within only a narrow and often varying vertical zone, but the 'average' level of this zone may itself change over time. In the short term (daily, seasonally, annually) *tides* are the most significant of many factors. *Tidal range* is important because the higher the range the greater the vertical zone in which the sea can act (Fig. 10.2; Komar, 1976). In addition, *tidal currents* are generated in some areas, and they often have a velocity that can significantly influence the distribution of sediment beneath the sea surface, especially where there is abundant loose sediment and in areas of constricted flow such as estuaries. Fig. 10.2 shows the world-wide distribution of micro-, meso-, and macro-tidal localities, together with the observed variations of several depositional coastal features

TABLE 10.1 Classification of coasts

I. *Primary coasts.* Configuration resulting from non-marine processes.
  A. *Land erosion coasts.* Shaped by erosion of the land surface and subsequently drowned by sea-level rise, sinking of the land, or melting of the ice-caps.
    1. *Drowned river-cut valleys.* Relatively shallow estuaries, usually V-shaped cross-sections, deepening seaward, except at bay entrances. Example: ria coast of north-west Spain.
    2. *Drowned glacial erosion coasts.* Includes fiords and glacial troughs, has deep-water estuaries, U-shaped cross-sections, and greatest depths within the embayments. Example: Norwegian coast (fiords).
    3. *Drowned karst topography.* Embayments with oval depressions representing sink-holes, common in limestone areas. Example: Dalmatian coast of Yugoslavia.
  B. *Land deposition coasts.* Shaped by land-derived deposits that prograde the sea coasts.
    1. *River deposition coasts*
      a. *Deltaic coasts* form a lobe into the sea with distributary stream channels. Examples: birdfoot Mississippi Delta; cuspate, Tiber Delta; lobate, Rhone Delta; arcuate, Nile Delta.
      b. *Compound alluvial fan.* Coastal plain at base of mountains, usually straightened by wave erosion. Example: east coast of South Island, New Zealand.
      c. *Outwash plain.* Alluvial deposition formed along outer margin of large glaciers. Example: seaward margin of Malaspina Glacier, Alaska.
    2. *Glacial deposition coasts*
      a. *Partially submerged moraine.* Hummocky topography, usually straightened by marine deposition and/or erosion. Example: parts of Long Island, New York.
      b. *Partially submerged drumlins.* Oval hills, elongate in direction of ice movement. Example: Boston Harbour, Massachusetts.
    3. *Wind-deposition coast*
      a. *Dune prograded coast.* Rare. Example: part of south side of San Miguel Island, California.
      b. *Dune coast.* Where dunes are bordered by a beach. Examples: mouths of many river valleys in central California and Oregon.
    4. *Landslide coast.* Masses fallen from cliffed coasts, forming projections into the ocean. Example: Humbug Mountain, Oregon.
  C. *Coasts shaped by volcanic activity*
    1. *Lava-flow coasts.* Convex protrusions at volcano. Example: south coast of island of Hawaii.
    2. *Volcanic collapse or explosion coasts.* Concave bays on side of volcano. Example: Hanauma Bay east of Honolulu.
    3. *Tephra coast.* Where fragmental volcanic material has built out the coast. Example: San Benedicto Island, west of Mexico.
  D. *Coasts shaped by diastrophic movements*
    1. *Fault coasts.* Straight fault scarps continuing steep below sea-level. Example: north-west side of San Clemente Island, California.
    2. *Fold coasts.* Where coast has been recently folded.
    3. *Sedimentary intrusion coast*
      a. *Salt domes.* Example: small islands in south of Persian Gulf.
      b. *Mud lumps.* Small islands resulting from upthrust of mud. Short lived. Example: near mouths of Mississippi River distributaries.
  E. *Ice coasts.* Where glacier fronts extend into the sea. Example: along most of Antarctic coast.
II. *Secondary coasts.* Configuration resulting mainly from marine agencies or marine organisms.
  A. *Wave-erosion coasts*
    1. *Straightened by wave erosion.* Contrast with fault coasts in having gently sloping sea floors and wave-cut terraces. Example: seaward side of Cape Cod, Massachusetts.
    2. *Coasts made irregular by wave erosion.* Indentations differ from drowned river valleys in minor character of penetration. Examples: many irregularities along northern California coast.
  B. *Wave-deposition coasts*
    1. *Coastal flats or plains built seaward by waves.* Contrast with river-deposition coasts in lacking distributary streams and deltaic bulges. Examples: broad beaches on upcurrent side of jetties, and flats upcurrent of points.

TABLE 10.1—*continued*

2. *Barrier coasts.* Separated from mainland by lagoons or marshes. Example: south-eastern coast of United States.
3. *Cuspate forelands.* Large projecting sand points with or without lagoons on the inside. Example: Cape Hatteras.
4. *Mud-flats or salt marshes.* Where very gentle offshore slope stops waves from breaking near the coast. Example: Crystal River area north of Tarpon Springs, Florida.

C. *Coasts prograded by organisms*

1. *Coral-reef coasts.* Built out by corals and algae and later brought above sea-level by storm waves. Tropical areas. Examples: atolls of the Pacific.
2. *Serpulid-reef coasts.* Where beach sands have been cemented by serpulid worms. Small patches near sea-level, mostly tropical or subtropical.
3. *Oyster-reef coasts.* Built by oyster shells and thrown up by the waves. Examples: in many bays along the Gulf Coast of the United States.
4. *Mangrove coasts.* Mangrove trees often root in shallow water, mostly in bays, and sediments build around their roots to produce marshy land. Examples: south-west Florida.
5. *Marsh grass coasts.* Like mangroves, marsh grass may grow in shallow embayments and along gently sloping sea floors where wave energy is minimal.

III. *Man-made coasts.* Configuration resulting mainly from human activity. Example: reclaimed coasts of the Netherlands.

Source: Shepard (1976).

with tidal range. An example of tidal currents, and their daily changes are shown for the entrance to Teignmouth harbour, Devon, UK, in Fig. 10.3: here, the currents (both ebb and flow) can reach 6 knots (11.1 km/h), a velocity high enough to transport subsurface sediment in conjunction with wave processes.

On a longer time-scale, sea-level can change as a result of such processes as tectonic movements, isostatic recovery of land following deglaciation, or world-wide glacio-eustatic changes accompanying the melting of ice sheets (e.g. Bird, 1984; Vita-Finzi, 1986). Such changes lead to coasts dominated by the characteristics of *submergence* or *emergence*. And the changes are not merely a thing of the past: many coastlines are mobile today, creating problems for coastal management. For example, emergence can lead to the shallowing of harbours and the reduction of marine attack, and submergence can cause enhanced flooding hazard and coastal erosion. And a number of coastal scientists have argued that solutions to immediate coastal problems should take into account more than they do at present the fact that sea-level rise is continuing and inevitable, as is the consequential erosion (e.g. McCann, 1982).

(ii) *Waves and currents* With the exception of earthquake-generated waves (tsunamis), the most important waves are wind-generated. They are of two basic types: *sea*, those waves generated locally by contemporaneous winds; and *swell*, waves that are usually lower, longer and have travelled far from the area of their generation. Davies (1964) classified coasts in terms of wave types (Fig. 10.4). The *storm-wave environment* occurs in high latitudes where strong winds create steep, high, damaging waves that reach coasts before they are converted into swell; and where rocky cliffs and wave-cut platforms are typical in exposed locations. Lower latitudes are dominated by swell, with the higher *west-coast* swells, which have greater energy, differentiated from the weaker *east-coast* swells. In addition there are *low-energy environments* where waves are smaller.

The characteristics of waves have been studied intensively (e.g. CERC, 1973; Komar, 1976; Fig. 10.5). They are fundamentally generated by the wind through transference of energy from wind to water. In general, waves increase in size with wind strength and duration, and the *fetch* of open water. Empirical relationships have been established and graphs have been prepared that allow the prediction of the critical wave characteristics of *height* and *period* to be estimated from any specific combination of wind variables (e.g. CERC, 1973; Fig. 10.6). In fact, such predictions disguise the

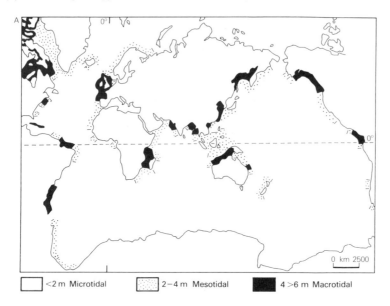

FIG. 10.2 (A) Classification of world coastlines according to tidal range (after Davies, 1964). (B) Variations in morphology of depositional shorelines with respect to variations in tidal range (after Hayes, 1975)

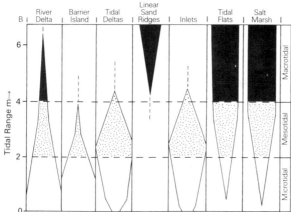

complexity of most wave trains that include a spectrum of waves of many heights and periods moving in a range of different directions.

In deep water, it is possible to predict wave characteristics from wind and wave data. For example, wave-form velocity, C depends on wave-length, L (for definitions, see Fig. 10.5) in the form

$$C = LT \qquad (10.1)$$

where T = wave-period (time taken for successive crests to pass a given point), and the relationship holds that

$$L = 1.56T^2 \qquad (10.2)$$

where L is in metres and T in seconds.

As waves approach shallow water, their length and velocity are reduced in the form:

$$C^2 = \frac{gL}{2\pi} \tan h \frac{2\pi d}{L} \qquad (10.3)$$

where

C and L are as above,
g = acceleration due to gravity,
d = depth of water,
$\tan h$ = the hyperbolic tangent.

In addition, the essentially *oscillatory waves* of the open sea are transformed into *shallow water waves* when the circular orbits of water particles encounter the sea-bed, become elliptical and are

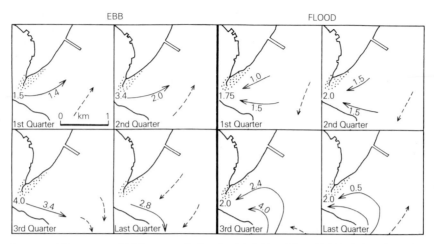

FIG. 10.3 The pattern of tidal currents during the ebb (left) and flow (right) within a single tide cycle in spring-tide conditions at Teignmouth, Devon, UK. Quarters refer to the tide phases within a cycle. Velocities (knots) relate to surface conditions (after Robinson, in Hails and Carr, 1975; © 1975, and reproduced by permission of J. Wiley & Sons Ltd.)

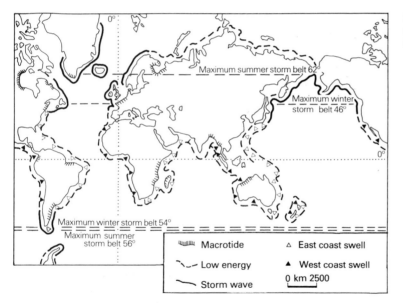

FIG. 10.4 Classification of coastlines according to the distribution of wave types (after Davies, 1964)

eventually broken, so that the waves break in the breaker zone (Fig. 10.5). Komar (1976) and others recognize a range of breaking waves of which the four main types (Fig. 10.7) are: *spilling* (the breaker spills down the shoreward wave front), *plunging* (the breaker becomes steeper and crashes forward), *surging* (when the base of the wave surges up the beach), and *collapsing* (intermediate between plunging and surging). Critical to the differentiation of these types is the gradient of the beach, as Fig. 10.7 shows, with surging character-

istic of steeper beaches and spilling commonest on gentle beaches. The type of breaker can be predicted, therefore, by the relationship

$$H_0/L_0 \tan^2 \beta \qquad (10.4)$$

where

β = beach slope,
H_0 = deep-water wave height,
L_0 = deep-water wavelength (Galvin, 1968).

Spilling breakers have values over 4.8, plunging

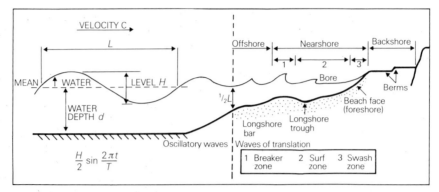

FIG. 10.5 Some terminology for the nearshore zone and beach profiles. Within the nearshore zone, the breaker zone is the area of wave breakage; the swash zone is characterized by swash and backswash between the upper limit of high-tide swash and the lower limit of low-tide backswash; and the surf zone lies between the breaker and swash zones, and is dominated by translation and edge waves (after Komar, 1976 and Holmes, 1975)

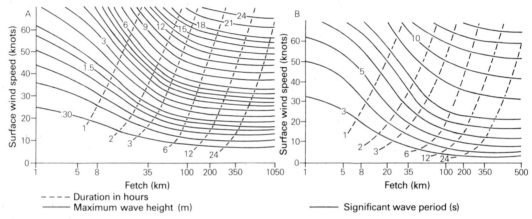

FIG. 10.6 Graphs relating wave heights (A) and wave period (B) to wind speed, wind duration, and fetch in oceanic waters (after Darbyshire and Draper, 1963)

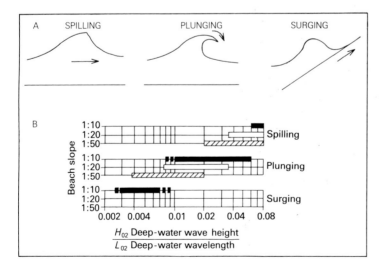

FIG. 10.7 (A) Schematic diagram of breaker types. (B) Probable relations between breaker type, beach slope and deep water wave steepness (after Huntley and Bowen, in Hails and Carr, 1975; © 1975 and reproduced by permission of J. Wiley & Sons Ltd.)

breakers 0.09–4.8, and surging waves less than 0.09. Shallow-water waves, on breaking, provide much of the energy for geomorphological activity in the surf and swash zones (Fig. 10.5) on coasts. *Edge waves* are oscillatory standing waves produced parallel to the shore with crests normal to the shoreline. There are several different types of edge waves; they are common on beaches steeper than 1 in 10, progress along the shore, and are increasingly recognized as strongly influencing nearshore depositional forms and the spacing of rip currents (e.g. Holman, 1983).

Whether waves are *constructive* or *destructive* (i.e. whether they add or remove sediment from the swash zones) depends crucially on their geometry and type of breaking. For example, plunging breakers may form little swash but create high backwash and are therefore likely to be destructive; and deep-water waves where the height–length ratio exceeds 0.025 are likely to be destructive whereas those with smaller ratios are constructive (Johnson, 1956; Bird, 1984).

The sea-floor has the additional effect of refracting wave crests so that they become more nearly parallel to submarine contours as they enter shallow water (Fig. 10.8). Refraction arises because wave-phase velocity is directly related to water depth, so that the part of a wave in shallower water is slowed down and is forced to become

parallel to the submarine contours. This process is especially important in waves that are long and therefore 'feel' the sea-floor in relatively deep water. Fig. 10.8 illustrates the process: orthogonals are at right angles to wave crests; if it is assumed that wave energy between any two orthogonals in deep water is preserved, then convergence should lead to an increase in wave height and divergence to a decrease (e.g. Holmes, 1975). Currents, if they are at an angle to wave direction, may also promote refraction.

Wave refraction diagrams can be constructed to assess the degree to which energy converges or diverges from the pattern of orthogonals. The wave refraction diagrams are an essential step in assessing the alongshore wave power, which also requires a knowledge of the wave dimensions and their angle of approach. The following empirical relationships have been established:

$$I_1 = K(EC_n)_b \sin a_b \cos a_b \qquad (10.5)$$

and

$$I_1 = K''(EC_n)_b \cos a_b \cdot (V_t)/U_m \qquad (10.6)$$

where

$I_1 =$ is the immersed load being transported along the coast,
$K = 0.17$, a numerical constant,

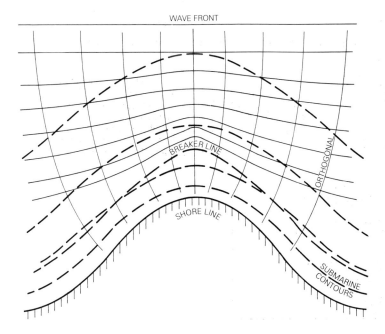

WAVE FRONT

BREAKER LINE

ORTHOGONAL

SHORE LINE

SUBMARINE CONTOURS

FIG. 10.8 The wave refraction process (after Holmes, in Hails and Carr, 1975; © 1975 and reproduced by permission of J. Wiley & Sons Ltd.)

$K'' = 0.28$, a numerical constant,

E = wave energy,

a_b = breaker approach angle,

C_n = wave-form velocity,

U_m = maximum horizontal velocity of the water particles in the wave orbit,

V_t = longshore current velocity.

Where waves break on the shore they not only move sediment, they also provide the energy for sediment transfer in the surf zone (Fig. 10.5). If the waves approach parallel to the shore a nearshore *circulation cell* is produced, mainly as a result of longshore variations in wave height, which includes *longshore currents* parallel to the coast on the shoreward side of the breakers that feed seaward-flowing *rip currents*. Rip currents appear to be located where breaker height is lowest, that is where edge waves and incoming waves are out of phase, or where refraction causes divergence. They can be a serious hazard to bathers and help to remove debris. Where waves approach the shore obliquely, a longshore current may be established

that increases in velocity as it flows parallel to the beach until it flows seawards as a rip current (Fig. 10.9). Commonly, both the cell circulation currents produced by longshore variation in wave height, and those produced by oblique wave approach are combined (Komar, 1976). The nature of sediment movement in the surf zone is determined by the balance between longshore current velocity and wave velocity, as tracer experiments have revealed (Fig. 10.10).

Longshore currents play an essential role in the transfer of sediment along a coast—and they are fundamental in influencing local patterns of erosion and deposition.

In the swash zone (Fig. 10.5), angle of wave attack is also important. If wave attack is oblique to the coast, the effect of longshore currents may be supplemented by *beach drifting*, in which breaking waves carry material up the beach in the swash zone in the direction of their approach, whereas the material moves backwards approximately down the maximum slope of the beach in backwash, thus creating a general movement along-

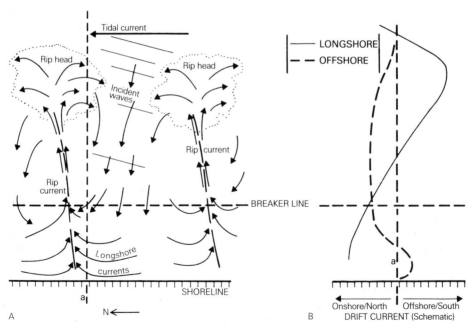

FIG. 10.9 (A) Schematic diagram of the proposed nearshore circulation system on the steep beach. The mean onshore drift under the incident waves is returned seaward in a series of regularly spaced narrow rip currents fed by longshore currents inside the surf zone. Some distance seaward of the breaker line each rip current broadens into a rip head and the seaward flow ceases. (B) The expected dependence of longshore and onshore drift currents on distance from the shore, along a line normal to the shoreline, just south of a rip current; (after Huntley and Bowen, in Hails and Carr, 1975; © 1975 and reproduced by permission of J. Wiley & Sons Ltd.)

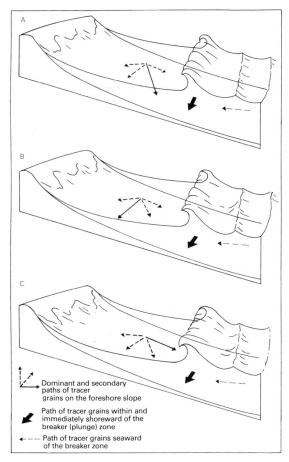

F IG . 10.10 The predominant paths of tracer-grain movement: (A) when longshore current velocity and wave motion exert equal influence, (B) when longshore current predominates, and (C) when wave motion dominates (after Davis, 1979)

shore. The direction and amount of movement depends crucially on the relations between swash and backwash velocity, beach gradient, and discharge.

On a broader scale, *circulation cells* may be identified within the coastal zone that are discrete sedimentation compartments within which the waves and currents transport material. In southern California, for example, Inman and Brush (1973) have identified five littoral cells (Fig. 10.11) each of which receives sediment from rivers at the coast, where it is transported by waves and currents until it reaches submarine canyons that direct the flow into the ocean floor.

Waves are a constantly varying force acting on the beach material. They vary through space and time, both because of changes in coastal aspect and offshore relief, and because of changes in wind direction and force over a very wide area of sea. This variability means that the beach is rarely in equilibrium with the waves, despite the fact that it may take only a few hours (relaxation time) for equilibrium to be reached under a new set of forces. This almost continuous state of disequilibrium is even more likely to occur where the tidal range is considerable, as the waves are always acting at a different level on the beach. It is, therefore, necessary to consider the range of variability within which the processes operate, and the resultant range of beach form. The time-scale must also be borne in mind. The day-to-day variability will usually be less than that between seasons, and still less than that between extreme storms and normal calm weather.

These variations are important from many practical points of view. For example, during periods of storm when destructive waves attack the shore, the beach is lowered, and under these conditions the backshore zone often suffers erosion. This may take the form of damage to coastal defences under severe erosional conditions. Under natural conditions the cliff-foot support may be removed, resulting in slumping or erosion by hydraulic action, in unconsolidated and hard rocks respectively. The foreshore foundation may also be eroded and lowered if the beach is completely removed. In this case even if the same quantity of beach material returns after the storm the profile will be permanently lowered and the coast made more vulnerable. In some areas where seasonal changes are marked, tidal or river inlets through barriers may be completely closed during the period of maximum accretion, especially if this coincides with low river flow. This may result in transport problems. The coastal barriers of parts of west Africa exemplify this situation. Thus in some areas inlet controls have to be built, which may have adverse repercussions along the adjacent shores.

(c) *Coastal materials*

Coastal landforms depend on the interaction of processes, such as those generated by waves, tides, and currents, and the materials available locally (e.g. from rivers, cliffs, offshore, perhaps the wind, biological deposition, and longshore drift). From the environmental management viewpoint, this relationship can provide evidence of both the

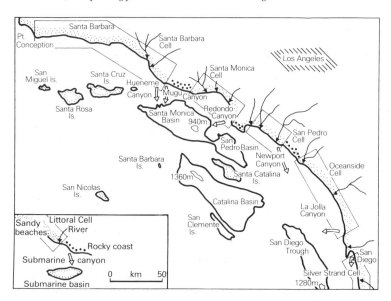

FIG. 10.11 Five littoral cells along the southern California coast, each of which contains a sedimentation cycle (after Inman and Brush, 1973; © by the AAAS)

influence of materials on process, and the effect of process on materials. The following comments exemplify this contention.

Storm waves will act differently according to the material on which they are operating. Steep storm waves breaking on solid rock may under certain conditions set up shock pressures that lead to disintegration of the rock, followed by attrition of the blocks so formed. Similar waves breaking on shingle will throw some of it to the backshore zone to create a storm-beach ridge, which may extend far above the high-tide level. For example, Chesil Beach, southern England, reaches 13.1 m above high-tide level at its south-eastern end. Storm waves may also drag much shingle down the foreshore to create a step at their break-point. On sand beaches storm waves are entirely destructive, creating a steep beach scarp at the limit of their action and carrying much sand offshore to their break-point where they create a submarine bar, while some sand is also carried to deeper water offshore. An exception occurs where waves can wash over a low sandy barrier island, carrying sand into the lagoon behind. This process usually takes place only when the water-level is raised abnormally by surge activity (Dolan, 1973).

Because of the different response of sand and shingle to waves, it is important to appreciate that dominant waves are those producing the most permanent forms. The nature of these forms will differ, however, with the type of beach material.

Storm waves will generate major backshore ridges if the beach is composed of shingle; while on sandy beaches storm waves can create a submarine bar at their break-point. This latter is especially the case under tideless conditions, as in the Mediterranean and Baltic seas, and in the Gulf of Mexico.

The type of material also influences the effect of tidal streams and wind action. The wind can only directly affect sand and finer material, as it cannot move pebbles. Wind is most important in relation to the building of coastal dunes where there is an abundance of sand available on the foreshore, a constructive wave regime, and a predominantly onshore wind direction. These conditions apply, for example, on the coasts of California, Oregon, and Washington in the USA.

Tidal streams are often fast enough to move fairly coarse sediment but the forms normally associated with tidal streams in the offshore zone or in tidal estuaries usually consist of sand and only occur where there is sufficient sand available. In quieter areas, where the tide ebbs and floods over large expanses of flat ground bringing muddy water periodically over the area, tidal mud-flats and salt marshes can be established, formed of fine sediment accreting vertically. In some circumstances mud can be deposited on the open foreshore, to the detriment of recreational activities on holiday beaches. Moderate- to fine-sized sand is the optimum for holiday beaches, because finer sand produces flatter beaches and hence a greater expanse of foreshore. Steep shingle beaches are the least desirable.

The different coastal processes operate in such a way that they leave their mark on the sediment-size distribution and character. Grain size is normally measured on a ϕ (phi) scale which has been so designed that the normal range of beach sediment sizes has positive values. The ϕ scale is a negative log. scale (to the base 2) but it can be directly related to grain sizes measured on a metric scale. Thus sand lies in the range -1ϕ to 4ϕ (i.e. 2 mm to 0.0625 mm), silt sizes are $+4\phi$ to $+8\phi$ (i.e. 0.0625 mm to 0.0039 mm), and clay has phi values greater than $+8\phi$ i.e. less than 0.0039 mm). It is possible to recognize various sedimentary processes from the nature of the size distribution curve plotted on logarithmic probability paper. The curve ideally should consist of three major log-normal sections (Fig. 10.12). The 'coarse' end represents the *traction load*, moved mainly by rolling along the bottom; the 'fine' end represents the *suspension load* and is usually limited in beach

foreshore sediments, but may be important in offshore, lagoon, and dune deposits. The main, central part of the curve represents the *saltation* load, which may be divided into two parts, the swash and backwash respectively, each of which is deposited under slightly different hydrodynamic conditions. The great fluctuation in velocity in this zone allows finer particles to become trapped amongst the generally coarser ones, so that this zone tends to show poor sorting. In general the zones of maximum energy expenditure are those where the material is coarsest, and where fine material is in short supply. This accounts for the negative skewness (King, 1972) on the ϕ scale due to a tail of coarse grains. Fine material cannot settle in these highly active zones except where it becomes trapped amongst coarser grains. Thus the breaker, surf, and swash zones are characterized by large mean size, poor sorting, and negative skewness on the ϕ scale. On the other hand, dune sand and some unidirectional current sediments are characterized by positive skewness, indicating a tail of fine sediment. For these purposes skewness may be defined as:

$$ \mathrm{Sk} = \frac{\phi_{16} + \phi_{84} - 2\phi_{50}}{2(\phi_{84} - \phi_{16})} + \frac{\phi_5 + \phi_{95} - 2\phi_{50}}{2(\phi_{95} - \phi_5)} \quad (10.7) $$

where ϕ_{84} is the ϕ value coinciding with the 84 per cent point on the size distribution curve and similarly for the other ϕ notations (Folk and Ward, 1957).

Sand is by far the most common material to be found on beaches, particularly in low latitudes. Much sand is formed of quartz material, because this mineral is particularly resistant to weathering. In some areas, however, sand consists of suitably sized basalt fragments, such as the beaches of Iceland and Hawaii. Other sands are organic, such as the foraminiferal sands of western Ireland and the coral sands of the tropical seas. Some beaches contain heavy mineral sands, including gold. The sorting action of the waves concentrates the heavy minerals in places forming rich reserves. These have been located on currently existing beaches and in both raised beaches (as in western New Zealand) and submerged beaches (as off Alaska), where they have been exploited.

Other more mundane, but more essential, resources in the form of sand and gravel for construction purposes are available on beaches and have been exploited. It is now realized,

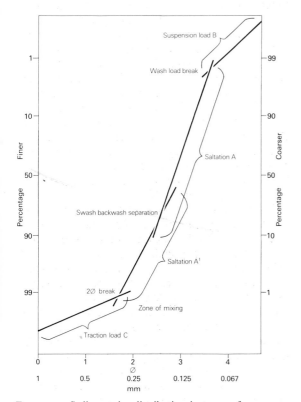

FIG. 10.12 Sediment-size distribution in terms of processes (after Visher, 1969)

however, that the beach sediments are one of the most vital, but decreasing, elements of coastal protection and conservation schemes. Instead of being removed, these materials are now frequently put on the beach (Sect. 10.4(c)). In seeking sand and gravel resources, contractors are now looking for the offshore zone to supply their growing needs. Material has been obtained from offshore for some decades, but now a systematic survey of resources has been undertaken in many areas, such as the eastern United States coast.

The shingle that forms such a conspicuous feature around the coast of Britain, including Chesil Beach and Dungeness, is more plentiful around the high-latitude coasts, probably because much of it was derived from reworked glacial drift deposits. Few rivers now bring shingle-sized material to the coast, but fluvio-glacial sediments and other drift deposits are widespread in the shallow marine zone in many parts of the coastal zone of north-west Europe. Eroding coasts in drift also supply mixed-size sediments, for example Holderness and north Yorkshire in eastern England and the cliffs of the northern part of Long Island, New York, USA.

10.3 Approaches to Studying Applied Coastal Geomorphology

In coastal environments, because change is often rapid and has an immediate impact on management, the geomorphological contributions of mapping, monitoring, and modelling are particularly closely linked, and much effort has been expended in developing them. This brief review examines the two main foci of attention: mapping and monitoring shoreline change, and laboratory and computer modelling.

(a) *Mapping and monitoring shoreline change*

Topographic maps, bathymetric charts, aerial photographs, and satellite imagery are useful in monitoring shoreline changes historically and at present. Charting shoreline changes to produce sequences of coastal evolution might appear to be straightforward, especially where historical sources are abundant. But in fact such sources can be misleading and intractable, and must be used with extreme caution (e.g. Carr, 1962). An outstanding example of an assessment of coastal change using historical sources, together with

PLATE 10.2 Cliff erosion on the Holderness Coast, Yorkshire: a hazard to cliff-top property

detailed shoreline surveys, is the US Army Corps of Engineers' *National Shoreline Study* (undertaken under the auspices of River and Harbor Act, 1968). This national survey revealed the varied pattern of erosion and accretion rates along 135 500 km of coast (Fig. 10.13): for example, over 33 000 km are 'seriously eroding' and the national average erosion rate of −0.8 m/y is also the average rate for the Atlantic coast and the Great Lakes' shores, whereas the Gulf Coast has a higher average rate (−1.8 m/y), and the Pacific Coast a lower average rate (−0.005 m/y); and within these zones, rates vary greatly, as might be expected, with coastal landform type (Table 10.2; Dolan *et al.*, 1983). Such historical data bases can be used to predict shoreline changes, assuming that historical trends will continue into the future (e.g. Dolan *et al.*, 1982).

Historical evidence has commonly been used to chronicle in detail the shoreline evolution of specific features and stretches of coast. Such studies are frequently undertaken because coastal changes are of direct relevance to coastal zone management. For example, the progress of coastal erosion poses a major hazard to farmland and property on the Holderness coast, Yorks., UK (Fig. 10.14; Plate 10.2); and coastal sediment accumulation is associated with channelization at Pangkah in the Solo Delta, Indonesia (Fig. 10.15).

Changing submarine topography is not only relevant to the interpretation of coastal forms, it is also of immense importance in offshore oil engineering. In the past, offshore topography was often gained from bathymetric charts for navigation, but recently geomorphologists have begun to map and monitor submarine topography using seismic and

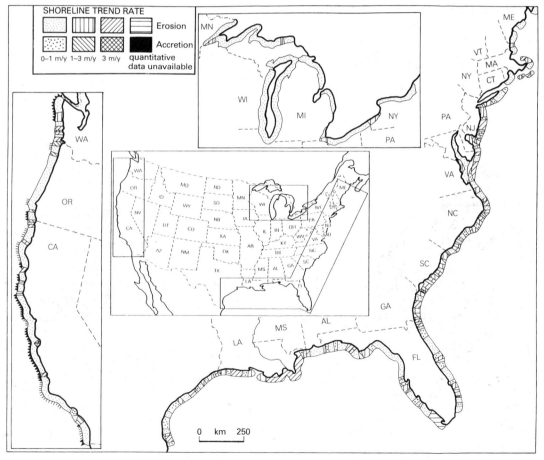

FIG. 10.13 Rates of erosion and accretion along the coastline of the United States (after Dolan *et al.*, 1983; reprinted with permission, © CRC Press Inc., Boca Raton, Fla.)

TABLE 10.2 Rate of change statistics for coastal landform types

Region	Mean $x(m/y)^a$	∂	Total range[a] Maximum	Minimum	N^b
Mud-flats					
Florida	−0.3	0.9	1.5	−1.5	9
La.–Tex.	−2.1	2.2	3.4	−8.1	84
All Gulf	−1.9	2.2	3.4	−8.1	93
Rock shorelines					
Atlantic	1.0	1.2	1.9	−4.5	36
Pacific	−0.5	—	−0.5	−0.5	7
Pocket beaches					
Atlantic	−0.5	—	−0.5	−0.5	9
Pacific	−0.2	1.1	5.0	−1.1	144
Sand beaches					
Maine–Mass.	−0.7	0.5	−0.5	−2.5	17
Mass.–NJ	−1.3	1.3	2.0	−4.5	22
Atlantic	−1.0	1.0	2.0	−4.5	39
Gulf	−0.4	1.6	8.8	−4.5	121
Pacific	−0.3	1.0	0.7	−4.2	19
Sand beaches with rock					
headland	0.3	1.9	10.0	−5.0	134
Deltas	−2.5	3.5	8.8	−15.3	155
Barrier Islands					
La.–Tex.	−0.8	1.2	0.8	−3.5	76
Fla.–La.	−0.5	1.7	8.8	−4.5	82
Gulf	−0.6	1.5	8.8	−4.3	158
Maine–NY	0.3	2.6	4.5	−1.5	12
NY–NC	−1.5	4.5	25.5	−24.6	153
NC–Fla.	−0.4	2.6	9.4	−17.7	256
Atlantic	−0.8	3.4	25.5	−24.6	421

[a] (−) values indicate erosion; (+) values indicate accretion.
[b] Total number of 3′ grid cells from which the statistics are calculated.

Source: Dolan *et al.* (1983).

side-scan sonar techniques, and this work has illuminated the nature and complexity of submarine slope failures (e.g. Prior and Coleman, 1979, 1980; see also Chapter 5).

The monitoring of contemporary erosional and depositional changes on coasts requires the use of techniques appropriate to local circumstances. In order to make effective predictions where process monitoring is difficult or absent, it may be possible to record data on independent variables (see Fig. 10.20 below for cliff erosion). Alternatively, the *consequences* of the interactions of these variables may be monitored in terms of the generally intermittent variables such as, in the case of cliff recession, cliff-toe erosion, mass movement, debris fall, and longshore transport (Fig. 10.16). Specific techniques of directly measuring cliff recession

include the micro-erosion meter (see Chapter 12) and the recording of change relative to marker pegs or nails driven into the rock (e.g. Sunamura, 1983).

On depositional coasts, contemporary erosional and depositional trends are often focused on sediment erosion/transport/deposition, their related drift directions, and their effects on coastal form. Chorley *et al.* (1984) recognized several approaches to the monitoring of such changes: (i) using the evidence of accumulation/erosion associated with barriers such as breakwaters; (ii) computations of sediment transfer based on, for example, wave energy estimates; (iii) the monitoring of tracers (such as painted, radioactive, or fluorescent particles, or dyes; e.g. Kidson and Carr, 1971); (iv) the use of sediment traps; and (v) the

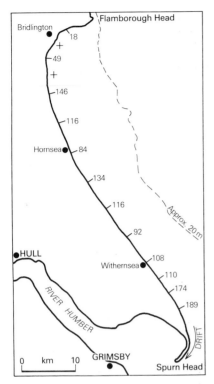

FIG. 10.14 Recession (in metres) of the Holderness coast, Yorkshire, UK, between 1852 and 1952 (simplified). The + indicates two sectors of aggradation in this period (after Valentin, 1954 and Bird, 1984)

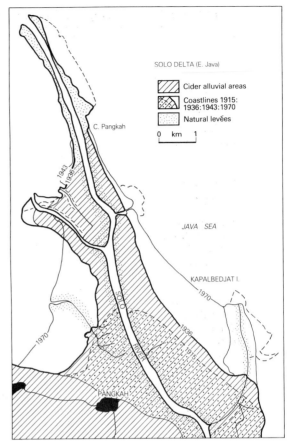

FIG. 10.15 Evolution of the Solo delta, Java, between 1915 and 1970 (after Verstappen, 1977*b*)

measurement of dilution rates of naturally occurring particles in a sediment from a known source. Equally important, current velocities and directions can be recorded with a variety of electrical flow meters or even simple floats or 'drifters'; and there are several different devices for recording the principal characteristics of waves. In many situations, a variety of techniques may be used, as in the construction of sediment budgets within circula-

tion cells or along coasts. An example is provided by Clayton's (1980) study of sand transport along the East Anglian, UK, coast. From observations made by some amateur observers over several years of beach volumes and cliff retreat, and of

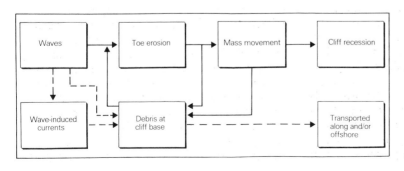

FIG. 10.16 Sea-cliff recession system. Toe erosion by waves is essential for continued cliff retreat (after Sunamura, 1983; reprinted with permission, © CRC Press Inc., Boca Raton, Fla.)

wave height and directions, Clayton produced regional average annual figures of sand transport (Fig. 10.17) of direct relevance to the evaluation of coastal-protection policy.

(b) *Laboratory and computer modelling*

Hardware modelling in the laboratory normally involves the scaled reconstruction of a particular coastal location in a wave-tank, and the simulation of coastal processes so that the consequences of various planned or possible management strategies can be evaluated. They are particularly valuable because they allow the contemporaneous manipulation of variables; the three-dimensional reproduction of flows through time; and relatively inexpensive opportunities to test ideas (e.g. Pethick, 1984; Silvester, 1974; see also Plate 3.1). Fig. 10.18 shows an example, in which the blockage of littoral drift by a proposed jetty in Cotonou, Dahomey, is predicted in a physical laboratory model. Laboratory hardware models based upon a narrow flume are also important in developing an understanding of relations within process–response systems, in developing theory, and in testing it (e.g. Dean and Maurmeyer, 1983).

Fox (1978) described a number of computer-based modelling approaches designed to illuminate coastal change. Of these, many are dynamic, and based on process–response systems relating coastal form to the forces affecting it, and chronicling the consequent changes over time. Amongst the most useful are mathematical or deterministic models, based on the conservation of mass and momentum (such as sediment volume; e.g. Dean and Maurmeyer, 1983); and probabilistic models where several simultaneously varying parameters can be evaluated (see Chapter 3, Fig. 3.5; and King and McCullagh, 1971). The applied problems to which computer-based modelling have been successfully applied include predicting equilibrium beach shapes/profiles; calculating beach nourish-

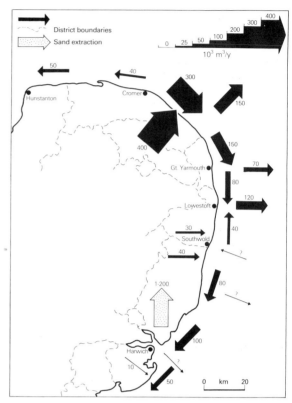

FIG. 10.17 Annual net sand transport along the East Anglian coast, and administrative boundaries: the mismatch between the two is clear (after Clayton, 1980)

ment requirements, and the effects of offshore dredging; and forecasting, shoreline changes arising from engineering construction (e.g. Willis and Price, 1975; Komar, 1983*a*). Where form/sediment changes are the focus of interest, as they often are, Komar (1983*c*) suggested that the approach normally involves (i) continuity equations appropriate to beach sediments, (ii) dividing the coast up into units (cells), and (iii) following the movement

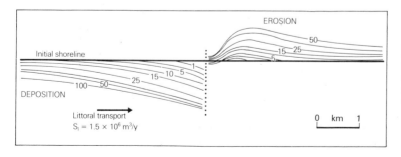

FIG. 10.18 Laboratory model prediction of the effect of a proposed jetty at Cotonou, Dahomey, on littoral drift. The model predicts shoreline changes (in years) and drift rate; 20 minutes of model operation is equivalent to a year of prototype conditions (after Sireyjol, 1965)

of sand between cells and consequent shoreline changes. It usually requires a definition of initial shoreline configuration, establishing the data base for sand sources and losses, modelling offshore wave conditions, and indicating the relationship between sand transport and wave factors, before the model is run (Komar, 1983c).

10.4 Management of Geomorphological Problems

(a) *The range of problems, resources, and responses*

The immense complexity of the physical and human activities in the coastal zone make the tasks of management difficult. Satisfactory solutions to problems require, according to Jolliffe and Patman (1985, p. 3) that certain basic and interrelated needs are satisfied: the need to integrate natural and human systems; the need for a sound environmental data base; the need to identify the respective roles of the public and private sectors; the need to match management organization with the scale of the problems; and the need to co-ordinate the coastal policies of different agencies. Even if these

prerequisites for successful management are met, sensible management does not necessarily follow: what is also required is a clear strategy for decision-making and monitoring (Fischer, 1985).

Geomorphology is relevant to coastal zone management in such contexts as ground conditions and foundation engineering; the conservation of marshes, estuaries, and beaches; flooding; the implications of coastal erosion and sedimentation, especially on harbour and estuary navigability and beach sustenance; the effects of hinterland changes on coastal erosion and deposition; and the consequences of aggregate exploitation. At any one location, there may be several such problems (e.g. Fig. 10.19).

The first and most important response to such problems, vested as they are within a wide range of other problems, is *planning*. Here, responses certainly reflect land-ownership patterns and the nature and location of management-agency responsibilities. In the United States, for example, federal, state, county, city, and private agencies may all have planning responsibilities (cf. Fig. 1.2): the US Army Corps of Engineers has responsibilities through the CERC for assembling data and undertaking major engineering works in collabor-

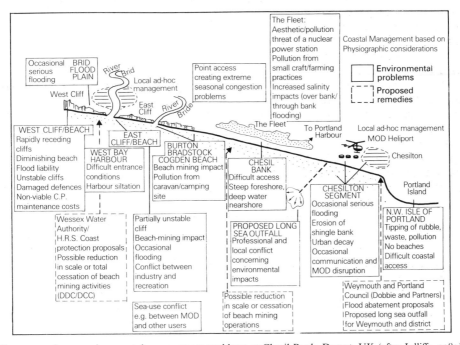

FIG. 10.19 Some environmental management problems at Chesil Bank, Dorset, UK (after Jolliffe, 1983)

ation with state agencies, and federal funds are available to assist beach erosion control; several states, such as California, exercise local planning control on coasts, and communities can often regulate development under state zoning laws (e.g. Howard and Remson, 1978). In the United Kingdom there is no national policy for coastal management and there is considerable potential for conflict between local authority, regional (e.g. water authority), and national interests in such areas as conservation, mineral extraction, and recreation; and effective integration may be difficult and time-consuming to achieve between those agencies responsible for coastal engineering, such as the Department of the Environment, the Ministry of Agriculture, Fisheries and Food, the Countryside Commission, and local authorities (Jolliffe and Patman, 1985).

Coastal geomorphology is enmeshed in such planning problems. One common problem arises because agency boundaries rarely coincide with geomorphological boundaries (see, for example, East Anglia, Fig. 10.17). Thus one agency's erosion may determine another's deposition: to stop the former may diminish the latter. Time-scales pose a second problem, because there is commonly a dissonance between the time-horizons of planning and the temporal dynamics of coastal processes and their adjustments to change. How long, for instance, will sea defences last and how long should they last? But understanding of coastal dynamics has advanced rapidly in recent years so that 'a coherent approach to the formulation of unifying criteria seems clearly within our grasp' (Inman and Brush, 1973).

The second major group of responses to problems of coastal geomorphology is *coastal engineering*. This is a well-developed field (e.g. Muir Wood and Fleming, 1981; Thorn and Roberts, 1981) that focuses on *protection* and *beach management*. The main responses, listed in Table 10.3, include both static and dynamic measures. Of the former, the most important are sea walls and revetments parallel to the coast, which restrict erosion, slope failure, and flooding; and groynes and breakwaters transverse to the coast which limit erosion and sediment transfer. From the environmental perspective, 'there are three fundamental steps necessary for the good design of coastal structures: (i) identification of the important processes operative in an environment, (ii) understanding of their relative importance and

their mutual interactions, and (iii) the correct analysis of their interaction with the contemplated design' (Inman and Brush, 1973, p. 30). Clearly all three steps require a geomorphological contribution. The dynamic responses involve chiefly beach nourishment (e.g. sand bypassing) and dredging, and the establishment of stabilizing vegetation. They, too, require geomorphological data for successful implementation. Some of these responses are considered further below, in the context of selected major geomorphological problems.

(b) *Cliffed coasts, shore platforms, and problems of erosion*

Erosion on many coasts can lead to the contemporaneous development of cliffs and shore platforms and, as a consequence, it can create a threat to local coastal defences, cause the loss of coastal land, and alter nearshore conditions. Probably the most serious hazard relates to property damage on or near cliff tops or along the cliffs (Plate 10.2). The form of cliffs and platforms is varied. Cliffs commonly reflect not only the lithology and structure of the rocks which compose them and the morphology and evolution of the coastal zone, but also the often complex relationships between marine and subaerial processes. Similarly, platforms vary in their width, gradient, and micromorphology. These are, in turn, partly a function of the depth of marine abrasion (e.g. *c*.10 m), the changes of sea-level (both long- and short-term), the availability of sediment, and the effectiveness of such processes as abrasion, potholing, weathering, and biological action by rock-boring and rock-browsing organisms (e.g. Sunamura, 1983). As cliffs retreat the bedrock surface at the cliff–platform junction is lowered (Zenkovich, 1967).

Fundamental to most cliff development is wave attack. Cliff erosion by waves can be summarized as follows:

$$X = \phi (f_w, f_r, t) \qquad (10.7)$$

in which

X = eroded distance,
t = time,
f_w = the assailing forces of waves,
f_r = the resisting forces of cliff material (Sunamura, 1983).

Erosion only proceeds when $f_w > f_r$. The factors involved in this system are summarized in Fig.

10.20. The key processes of cliff retreat are *hydraulic action* (including air compression and expansion, shearing, tension, and hydrostatic pressure) and mechanical action (such as abrasion and impact forces when waves are charged with debris): both are especially effective in storm-wave and major swell environments (Fig. 10.4).

Causes of coastal erosion (Table 10.4) may be either natural or induced by human agency, and each has its effects. Of particular importance in the context of management is the acceleration of coastal erosion as a largely accidental consequence of human activities. Coastal protection measures, such as groynes and sea walls, by protecting one part of a coast or keeping sediment in place, can lead to sediment depletion and thus to accelerated erosion elsewhere. This effect at Point Lonsdale in Victoria, Australia, has led to a continuous extension of sea walls (Fig. 10.21). In the case of sea walls, a common defence against cliff attack, if a high, broad beach is present waves may be prevented from breaking against the walls; but if the waves are reflected by the wall, scour may be accelerated and the beach eroded, as happened at Bournemouth (Fig. 10.22; Bird, 1979). Equally serious, the removal of beach materials for aggregates may leave a section of coast open to greater wave attack.

But cliff development is often not wholly dependent on the nature of wave attack. Commonly, where marine processes are intermittent and marine processes are counterbalanced by subaerial processes, cliffs may reflect a complex topography of slope failures (see also Chapter 5). In such circumstances, wave erosion leads to slope instability, and slope-failure debris comes to mask

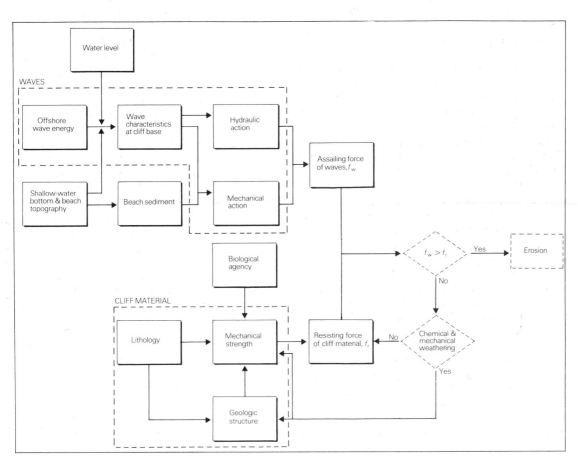

FIG. 10.20 Factors affecting wave-induced cliff erosion—the nature of erosion is determined by the relative intensity of assailing and resisting forces (after Sunamura, 1983; reprinted with permission, © CRC Press Inc., Boca Raton, Fla.)

TABLE 10.3 Static and dynamic measures of shore protection

1. Static measures

Form of protection	Function	Traditional method of construction	Traditional construction materials	Non-traditional method of construction	New construction material
Groynes	Barrier to movement of sand alongshore. Intended to reduce rate of longshore transport. Creates wider beach on updrift side and starves downdrift side.	Impermeable, extending perpendicular to shoreline from backshore into water normally beyond breaking waves.	Stone, riprap, concrete, wood, sheet pile, cribs.	Impermeable or very low, creating baffle for deposition or allowing some transport to downdrift beach.	Gabion mesh baskets filled with stone, Longard tubes (permeable polyethylene), acrylic or nylon bags filled with sand or grout, asphalt, nami rings, artificial vegetation, compressed solid waste, junk cars, ships or barges are possible but unsightly.
Bulkheads	Prevent undermining and slumping of backshore surface. Protect backshore from attack by swash and small waves. Stabilize shoreline position. Do not favour beach creation.	Impermeable, parallels shoreline at contact between beach and upland.	Same as groynes.	None	Same as above plus chemical soil solidification, synthetic nylon mat, plastic erosion control fabric, woven wire netting.
Sea-walls	Prevent attack of backshore by large waves and stabilize the shoreline position. Do not favour beach creation.	Same as bulkhead.	Usually, riprap or concrete but other materials as above.	None	None–few of the construction materials above offer sufficient strength.
Revetments	Dissipate wave or swash energies on sloping, immobile surface. Secondary function as sea-wall. Do not favour beach creation.	Variable.	Same as sea-wall.	None	See groynes and bulkheads.

	Purpose	Construction	Materials		
Breakwaters	Energy filter designed to dissipate wave energies and reduce erosive effects of waves. Energy shadow favours deposition from updrift sources and starves beach downdrift.	Offshore in depths which cause waves to be reflected without breaking or submerged to allow larger waves to break (not directly on the structure).	Usually stone riprap but can be other materials as indicated under groynes.	Construction with unconsolidated material. Floating breakwater, bubbling breakwaters, artificial seaweed.	Materials mentioned above, sand mounds, stretched polypropylene foam strands (artificial seaweed), plastic reeds attached to concrete base, floating tyres.
Foreshore obstructions (Perched Beach)	Low-cost, beach parallel structure designed to create swash zone deposition.	Development stage.	N/A	Located within swash zone (some permeable) creating perched beaches.	Concrete blocks, gabions, filled bags, Longard tubes.

2. Dynamic measures

	Purpose	Construction	Materials
Beach fill and bypassing	Increase protection afforded by beach and provide recreational space.	Hydraulic pipeline or trucking with bulldozing.	Sand.
Dunes	Barrier against flooding, reservoir of sand to replace beach losses.	Provide wind break (fences, vegetation), bulldozing.	Sand (plus structures) to interrupt air stream.
Offshore mound	Dampen wave energies, provide a reservoir of sand for eventual onshore migration.	Dumping from barges.	Sand (larger particles would be static).
Vegetation	Stabilize slopes, make unconsolidated sediments more resistant, dampen wave, swash, and wind energies, trap sand, improve habitat.	Planting.	Seedlings.

Source: Nordstrom and Allen (1980).

TABLE 10.4 Causes and effects of erosion in the coastal zone

No man-induced erosion		Natural erosion	
Causes	Effects	Causes	Effects
1. Beach mining for placer deposits (heavy minerals) such as zircon, rutile, ilmenite and monazite	1. Loss of sand from frontal dunes and beach ridges	1. Changes in location of river outlets after major floods and tectonic activity	1. Changes in pattern of sediment transport to and along the coast
2. Construction of groynes, breakwaters, jetties, and other structures	2. Downdrift erosion	2. Cyclones, hurricanes and storms	2. General shoreline recession resulting from the destruction of frontal dunes and beach ridges; formation of washover fans
3. Construction of off-shore breakwaters	3. Reduction in littoral drift	3. Fires	3. Destruction of vegetation, producing drift sand and transgressive dunes
4. Construction of retaining walls to maintain river entrances	4. Interruption of littoral drift resulting in downdrift erosion	4. Offshore islands	4. Reduction in littoral drift
5. Construction of sea-walls, revetments, etc.	5. Wave reflection and accelerated sediment movement	5. Promontories (headlands) and reefs	5. Downdrift erosion and interruption of littoral drift
6. Deforestation	6. Removal of sand by wind; sand drift	6. Tidal inlets/river mouths	6. Interruption of littoral drift
7. Fires	7. Migrating dunes and sand drift after destruction of vegetation	7. Variations in coastal configurations	7. Variations in sediment transport patterns
8. Grazing of sheep and cattle	8. Initiation of blowouts and transgressive dunes; sand drift	8. Vertical cliffs	8. Wave reflection and accelerated sediment movement
9. Off-road recreational vehicles (dune buggies; trail bikes, etc.)	9. Triggering mechanism for sand drift attendant upon removal of vegetative cover		
10. Reclamation schemes	10. Changes in coastal configuration; and interruption of natural processes, often causing new patterns in sediment transport		
11. Increased recreational needs in response to population pressures	11. Accelerated deterioration, and destruction, of vegetation on dunal areas, promoting erosion by wind and wave action		

Source: Hails (1977).

the wave-cut cliff foot, only to be removed when appropriate wave conditions return. For example, Prior and Renwick (1980) described clay-dominated cliffs in Denmark and France which are characterized by several *types* of slope failure that vary in their *location* on a cliff profile, *frequency*, *slope angle*, and *dependency on marine action* (Fig. 10.23). Fig. 10.23 illustrates an important generalization: that the complexity of the cliff system is such that cliff sections can be strikingly individual. As a result, cliff erosion and development can be very difficult to manage. Successful remedial

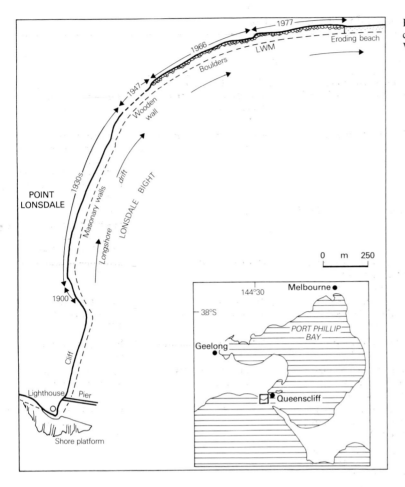

FIG. 10.21 Sequence of sea-wall construction at Point Lonsdale, Victoria, Australia (after Bird, 1979)

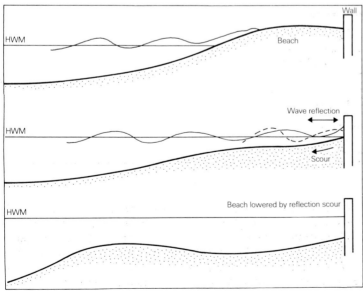

FIG. 10.22 Relations between sea walls and erosion. A broad, high beach prevents storm waves from breaking against a sea wall and will persist, or erode only slowly, but where waves are reflected by the wall, scour is accelerated, and the beach is quickly removed (after Bird, 1979)

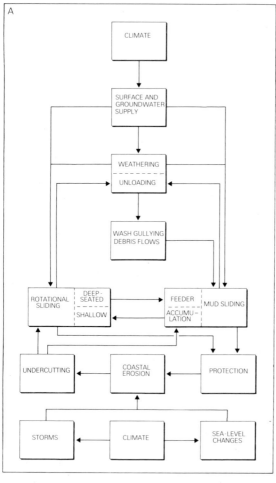

A

action requires both long- and short-term magnitude and frequency considerations to be taken into account, together with local appraisal of material properties, drainage, and slope stability characteristics; thus in the context of the clay cliffs described in Fig. 10.23, for example, control of a single variable such as wave attack is unlikely to be wholly effective as cliff-retreat may continue in response to other factors (Prior and Renwick, 1980).

Cliff erosion and platform development, whether at natural or accelerated rates, are not necessarily undesirable. Indeed actively eroding, steep cliffs may be a very desirable ingredient of attractive coastal scenery. But cliff development is certainly a serious hazard when it is associated with damage to adjacent buildings, roads, railways, and other valuable structures. The most sensible responses may often be to control development of buildings on cliffs through planning, or to remove the endangered structures (which may be much cheaper than building coastal defences). For example, it may be sensible to define a hazard zone in which development is severely restricted. Such a 'buffer zone' will need to be defined locally, particularly in terms of long-term and short-term rates of erosion. Kirk (1982) and others proposed the following general formula:

$$\text{Hazard zone width (m)} = (R \times 100) + S \qquad (10.8)$$

where

$$R = \text{long-term erosion rate (m/y),}$$

B

PROCESS	Rotational Sliding Subsidence	Translational Sliding	Transitional Sliding/Plug Flow/Viscous Flow		Shallow Slips	Marine Erosion
STRENGTH	Peak → Residual	Peak → Residual	Residual	Residual	Residual	
PORE-WATER PRESSURE	High → Low	High → Low	Low (<Hydrostatic)	High (> Hydrostatic → Geostatic)	High/Low	
SLOPE ANGLE	Reducing	Maintaining	Alternate Increase-Reduction	Reducing	Periodic Steepening	

1 Unloading Failures

2 'Weathering' Failures

3 Pore-Water Failures

4 Loading Failures

MOVEMENT TYPES

Movement

1. EPISODIC (Long term)

2. EPISODIC (Medium term)

3. ANNUAL CYCLES
Winter
Summer

4. EPISODIC (Short term)

15   30
Years

1   5
Years

1 2 3
Years

1 2 3
Years

FIG. 10.23 (A) General process/factor interrelationships for coastal areas subject to landsliding in Denmark and northern France. (B) Schematic representation of the interaction between factors affecting mudslide processes for clay-dominated coastal slopes in Denmark and France (after Prior and Renwick, 1980)

100 = assessment erosion rate (m/y),

S = the extent of short-term movements.

Such a formula is not universally applicable (for example, it should not be applied to situations of alternating aggradation and erosion); and many locations will have an inadequate historical record, but it does illustrate a sensible approach to planning solutions.

Alternatively, attempts to control cliff-retreat are likely to focus either on reducing assailing forces (f_w) or increasing resisting forces (f_r) (Equation 10.7) so that $f_w < f_r$. One approach is thus to remove or ameliorate the causes of erosion (Table 10.4). Given that the assailing forces are very difficult or impossible to modify, the commonest responses are associated with increasing resisting forces by, for instance, strengthening cliff bases with sea walls or by creating wide, high beaches. But, just as judicious planning can avoid engineering, so it is important to appreciate that engineering defences are not necessarily either sound geomorphologically or cost-effective. Clayton (1980), in a study of coastal protection in East Anglia (see Fig. 10.17), sounded an important note of warning:

Comparison of areas lost and gained by coastal erosion in the early years of this century with recent years shows that the building of coastal defences along parts of the coast was effective in reducing net land loss. At that time selection of relatively easy sites coupled with the continuing drift of sand from feeder bluffs showed the advantage in terms of net area to be gained by accepting the erosion of high cliffs and consolidating the progradation of neighbouring low-lying areas. Public pressure, willingly accepted by engineers, has brought extension of the defences to 60% of the coast, but no improvement in the erosional balance. Indeed the successful stabilization of some of the lower cliffs has removed local sand sources, so that the longer transport paths from the major feeder bluffs are now dominant. Current attempts to extend the defences to these active cliff systems seem unlikely to be successful in the long term although they may give a temporary reduction in basal removal by waves and thus in cliff retreat. *Our work suggests that more ambitious engineering schemes in such areas should be actively resisted, for the removal of these inputs of sand to the system would initiate a decline in beach volumes that could build up to catastrophic proportions.* The costs of sand feeds to maintain current transport volumes without the cliff input would be very expensive. Cost/benefit calculations on the retreat of cliffs in rural areas suggest that natural erosion levels are highly beneficial when the coast as a whole is considered

(Simmonds, 1977). *This is not an easy message to sell to the affected communities, but we must do our best to avert further expenditure on coastal defences where cliff retreat provides the sand on which our beaches depend for their very existence.* (Our italics.)

(c) *Depositional coasts: problems of sediment transfer*

If used with caution, coastal landforms can provide a valuable guide to coastal conditions for management purposes, because they reflect the integrated relations between the main marine processes (waves, tides, currents, etc.), the influence of the onshore and offshore topography, and the nature of coastal materials. This is particularly true of active depositional coasts. Fig. 10.24 shows the main types of loose sediment on coasts (mud, shingle, and sand) and some of the many wave-formed features associated with them.

Fine sediments can only be deposited in suitable environments, the resulting morphology depending on the environment. Mud can only accumulate in sheltered areas, such as salt marshes, lagoons, runnels landward of high ridges, the lower foreshore where shelter is provided by offshore banks, and the quieter shelf areas (such as the northern Celtic Sea off southern Ireland). Each environment differs from the other and each can be differentiated in terms of the local controls including climate, tidal range, type of sediment, and type of vegetation.

Sand can be deposited on a wide variety of forms in many different environments, including several types of coastal dune; beach features such as berms, ridges, cusps; and many types of bars and barriers, spits, and tombolos, as well as tidal features, offshore banks, sand waves, and sand ribbons (Fig. 10.24). Sand is the most widespread and versatile beach material, partly because it includes the sizes most easily picked up by moving air or water.

Shingle is less common than sand, but it does give rise to recognizable beach forms. The most important are the storm shingle ridges that are conspicuous in the backshore region. Such ridges can cover extensive areas of the backshore, for example at Dungeness and Orford Ness, UK. A careful study of such shingle structures, which are often clearly revealed on air-photos, provides valuable evidence for deciphering the development of the coastal zone through time, thus adding another dimension to the study.

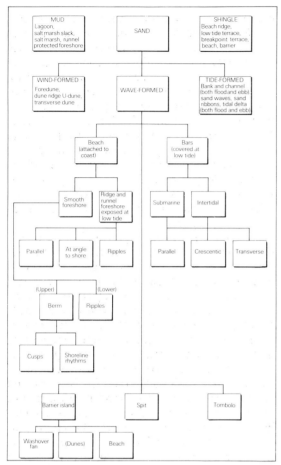

FIG. 10.24 Coastal sediments and related depositional coastal landforms

By studying the form of coastal features it is often possible to derive useful evidence concerning the processes that formed them. A good example of the relationship between process and form is the way in which a beach becomes aligned to the direction of approach of long swell waves that are refracted by the bottom morphology according to their length and direction of approach. This is a two-way feedback relationship, because the extent of wave refraction depends on the wave direction of approach and length as well as the bottom relief, so that both process and form are involved in explaining the final beach morphology, both in plan and profile. Once equilibrium has been established the longshore drift will be reduced to a minimum as the waves approach parallel to the shore at all points on the beach. A very important relationship between form and processes is that between wave direction, wavelength, and coastal outline, as these variables determine the amount and direction of longshore transport of material, and through this process the nature of many coastal forms. Thus zones of accretion and erosion can often be explained in terms of longshore transport, a process that exerts a greater effect on coastal morphology than does the on- and off-shore movement of material.

Each coastal feature can be used to provide evidence of the processes that formed it. Hurst Castle spit, near Southampton, UK, for example, can be interpreted in terms of the wave types that created it. Westerly waves supplied the material that storm waves built into the major ridge of this shingle structure. The lateral recurves were formed by waves approaching down the Solent from the east-north-east, while the proximity of the Isle of Wight to the south-east would account for the absence of long waves from this direction and hence the sharpness of the distal tip of the main spit. The essentials of the morphology are shown in Fig. 3.5; a simulation model of the system is discussed in Chapter 3.

Alternatively, it is also possible to monitor processes and relate them to changes in form (see Sect. 10.2(c)). Both observations of form and monitoring of change can help to answer many of the questions raised in planning coastal protection. In assessing a beach situation prior to developing protection works, for example, Muir Wood and Fleming (1981, p. 182) posed six questions that the geomorphologist can help to answer:

(1) What is the prevailing direction of littoral drift? Is there a seasonal variation? If so, what is the ratio of net to gross movement?

(2) Does the magnitude of littoral drift vary appreciably along the length of coastline adjacent to the site under study?

(3) What is the present rate of change of the line of the foreshore along the section under review? What does this represent as an annual volumetric loss (or gain) of material?

(4) What is the trend with time of this annual rate of change? Does it undergo cyclic variation?

(5) Must the shoreline be established at a particular position or may it be allowed to retreat? If so, at what rate and for how long?

(6) Are there additional factors that will affect the future trend?

In this brief review it is impossible to explore in detail the geomorphological problems associated with the major depositional coastal features such as barrier islands (e.g. Nummedal, 1983), deltas (e.g. Wright, 1978), coral reefs (e.g. Stoddart, 1971), marshes, and estuaries (e.g. Boothroyd, 1978; Frey and Basan, 1978). Attention will be focused here on one problem common to all of them, that of *sediment transfer*.

In managing depositional coasts and the transfer of sediment, it is preferable in most circumstances that the existing coastal features should be retained in or be allowed to develop into equilibrium. To achieve this aim requires both an understanding of what is the stable form and how best to retain it. For example, because the plan-form of many equilibrium beaches is log-spiral, it may be possible to place defences at intervals along a coast suitable to allow the equilibrium to develop. As indicated earlier (Table 10.3), responses to the lateral drift of coastal sediment involve both static and dynamic measures. Groynes are the best known of the static measures deliberately designed to impede drift, although drift may often be accidentally moderated by such features as breakwaters and sea walls (e.g. Figs. 10.21 and 10.22). For example, at Lagos Harbour, Nigeria, the construction of moles in 1907 and 1922 severely altered drift (Fig. 10.25; Usoroh, 1977).

Dynamic approaches to the management of sediment transfer are increasingly important (Table 10.3). Here, the common aim is to create a wide, high beach, a feature widely recognized as the most effective form of coastal defence. Artificial transfer, or beach replenishment, can involve dredging and bypassing operations (e.g. Fig. 10.26). The need for such transfers usually arises in order to protect sites under threat of erosion (as at the nuclear power station, Dungeness, UK), to preserve recreational beaches, and to redress imbalances created by coastal engineering works. For example, sand began to be pumped by suction from one side of South Lake Worth inlet, Florida, USA, to the other when groynes failed to prevent erosion induced by breakwater-construction. About 48 000 m³/y are moved, compared with an estimated longshore drift of about 225 000 m³/y. After five years, stability was reached over two kilometres down drift, although erosion restarted when pumping ceased.

Ideally, management responses should be compatible with both management and geomorphological requirements, often a difficult objective. Nordstrom and Allen (1980) provide a useful analysis of this objective at Sandy Hook spit, New Jersey, USA. There, the aim is to preserve wide recreation beaches and onshore facilities threatened by erosion. Conventional, static engineering measures are not necessarily the most appropriate: they represent an irreversible commitment of resources, are unsightly, pose safety hazards, and interrupt sand supply to downdrift beaches. But there are geomorphological alternatives compatible with management objectives. Beach fills, for example, when naturally reworked, lead to the establishment of equilibrium depositional forms: they can use local materials (if available), increase beach width thus increasing recreational potential and providing protection, and are more attractive than groynes. Other dynamic measures considered for use at Sandy Hook spit, and which have

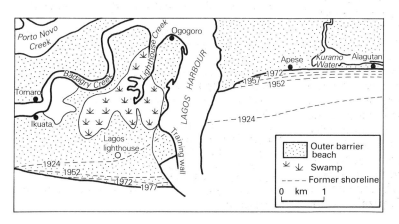

FIG. 10.25 Sand accretion and erosion around the entrance to Lagos harbour, Nigeria, 1924–77 (after Usoroh, in Faniran and Jeje, 1983)

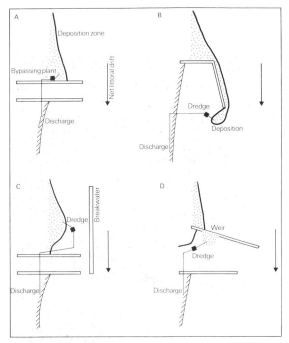

FIG. 10.26 Different types of bypassing systems used to transfer littoral drift past jetties and breakwaters (after Komar, 1983*b*; reprinted with permission, © CRC Press Inc., Boca Raton, Fla.)

potential for application elsewhere, are accelerated dune growth, vegetational stabilization, and offshore mounds (Table 10.3).

(d) *Coastal flooding hazard*

Problems of coastal erosion and sediment transfer are often associated with coastal flooding that may occur at times of unusually high sea-level generated by storms or exceptional tides. Behind the

major zone of wave attack on a beach, for instance, there may be a zone of surge damage and flooding (Fig. 10.27; Plate 10.3). The broad range of responses open to coastal managers faced with such hazards is summarized in Fig. 10.28.

Coastal flooding, the damage caused by hurricanes, and the human reaction to them along the shores of Megalopolis, the heavily populated north-eastern coast of the United States, have been studied by Burton *et al.* (1969). Barrier islands dominate this stretch of coast, which they have divided into four types according to their human adaptation: village, urban, summer, and empty shores. The environment is a fragile one, as the natural processes by which barrier islands migrate inland by wash-over flooding in storms cannot readily be combined with human occupation. Attempts at artificial control, as already mentioned, are not conducive to natural stability, which demands flexibility and not the rigidity imposed by man. The old village shores have become adjusted to flood hazards, but some of them, such as Chincoteague in Virginia are built less than 3 m above mean sea-level, and do suffer severe flooding under storm conditions. The flooding comes mainly from the bay side, as the dunes on the barrier protect the village from direct marine action. Local people are well aware of the danger during a storm, for their houses are built on piles, but even so evacuation follows storm warnings. The urban and summer shores are often protected by sea walls, as for example the summer shore at Sandy Hook, New Jersey, an area densely developed for recreational purposes. The artificial protection is necessary because the beach is only 60 m wide and dunes and their vegetation have been destroyed over much of the area, which is

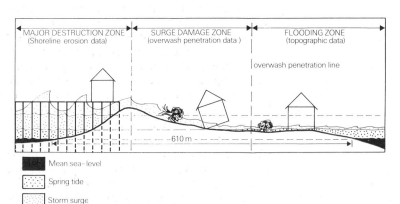

FIG. 10.27 Hazard zones normal to a coast: an example from the eastern seaboard, USA (after Dolan and Hayden, 1983; reprinted with permission, © CRC Press Inc., Boca Raton, Fla.)

PLATE 10.3 Flood damage in the surge zone behind Chesil Beach at Chesilton, S. England (cf. Fig. 10.19)

liable to heavy storm damage. Major floods reach 3 m elevation and 86 per cent of the area is below this level. The 1962 storm made 11 breaches in the Sandy Hook spit. There is a high degree of storm-hazard awareness in the zone and many private adjustments are made, while the major public adjustment is the 3-m high sea wall and a number of jetties to control sand movement. Sensible planning is essential if such coasts are to be used properly. In this area further intensification of land use is liable to take place, and this will require the maintenance and strengthening of coastal defences that are in many ways undesirable, since they prevent the natural processes operating that alone can maintain a dynamic stability in this delicate natural system. The ideal is to create zones in which the natural processes can operate unhindered, and this can be achieved by designation of national seashore zones, in which building and human interference are prohibited. The contrast between the barrier islands at Cape Look Out and Cape Hatteras, natural and man-controlled areas respectively, clearly reveals the dangers of interfer-ing with natural processes (Godfrey and Godfrey, 1973; and Dolan, 1973). Fig. 10.29 illustrates the response of both natural and man-controlled barrier islands to the effects of storms of varying intensity and direction.

10.5 Conclusion

Perhaps more than in any other geomorphological system, the complexities of process changes and management in coastal areas reveal how essential it is to understand geomorphological conditions before developing plans and investing in engin-eering structures. Fundamental to this understand-ing is the requirement that the interactions between forms, marine and subaerial processes, and mater-ials needs to be resolved at different time-scales, and that the geological time-scale (for example, as it relates to sea-level change) and historical changes have a significant role to play in planning and engineering strategies. Equally, an understanding of the geomorphological context of management

The Broad Options	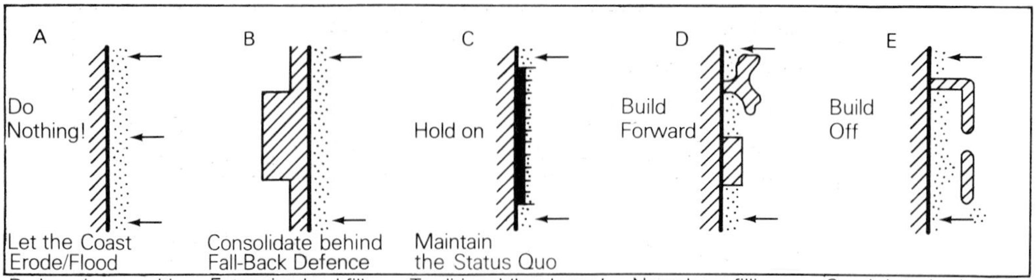 A Do Nothing! Let the Coast Erode/Flood	B Consolidate behind Fall-Back Defence	C Hold on Maintain the Status Quo	D Build Forward	E Build Off
Achieved by:	Re-housing; re-siting buildings/ highways; compensation etc.	Extensive land-fill; cliff-trimming; improved land drainage; clay embankments	Traditional 'hardware'- walling, groyning, revetments, etc.	Nearshore filling; beach nourishment; etc.	Semi-detached or detached breakwaters; nearshore/offshore recharge; etc.
Some possible physical benefits	Non-interference with shoreline sediment budget' does not interfere with coastal sediment flows	Preserves integrity of existing foreshore. Creates opportunities for environmental enhancement	Can often be applied 'comprehensively'; may locally increase beach dimensions	Increases coastal land; provides bigger beaches; can provide multiple benefits; need not be aesthetically intrusive and may create sheltered water areas	Increases coastal land; may lead to beach accretion; creates areas of low wave energy suitable for recreation
Some possible non-physical benefits	Cheap! Potentially cost-effective dependent on compensation. Minimises administration. Relative educational value	Relatively low cost but politically more acceptable than A. Visually stimulating. Potential for local resource reassessment and comprehensive development	Retention of access, property and coastal resources. Status quo likely to obtain public support as appears to be 'fair'. Funding arrangements well-established	Spreads benefits between user groups. Creates new coastal resources, particularly for recreation but also for education and science	Spreads benefits between user groups. Creates new coastal resources
Some possible physical dis-benefits	Loss of land; coastal serration if other parts have already been protected	Prevents natural landwards migration of untied beaches: does nothing for coast on either flank	Prevents natural landward migration of untied beaches; destroys integrity of foreshore/ backshore; end-groyne (terminal scour) problem; end wall/toe scour; visually intrusive?	Only provides 'local' protection; interferes with coastal sediment flows, nearshore current circulation patterns, etc.	Only provides 'local' protection; interferes with coastal sediment flows, nearshore current circulation patterns, etc.
Some possible non-physical disbenefits	General public dissent; invokes wrath among affected land-owners and tenants; inadequate politico-legal infrastructure for dealing with matters of compensation, etc.	Local land-use changes will probably create social conflict amongst affected property owners due to intensive transfers of resources and risks in favour of those behind the fall-back defence	High capital costs; possibly high recurrent (maintenance) costs. Ill-perceived resource and risk transfers may be substantial	High capital costs; Possible low recurrent (maintenance) costs; No well-established funding arrangements for build-forward programme	High capital costs; possibly high recurrent (maintenance) costs
Regional examples	NE coast of Isle of Man; parts of E. Anglian coast	The coast south of Aldeburgh, Suffolk; West Beach, Selsey	The coast at Whitstable and Reculver in Kent	The Metropolitan Toronto Shoreline; the Principality of Monaco	Parts of the coast of California
	A	B	C	D	E

FIG. 10.28 The broad options for coastal erosion management and flood abatement (after Jolliffe, 1983)

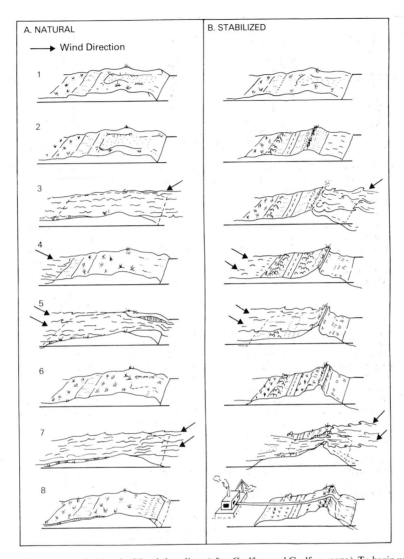

FIG. 10.29 Natural and man-controlled barrier island shorelines (after Godfrey and Godfrey, 1973). To begin with both the natural and the artificial barrier systems are alike (stage 1), thereafter they may differ as follows:

Stage	Natural	Controlled
2	No change	Stabilization of an artificial dune to protect a road
3	Storm has overwashed the natural barrier	Storm waves have eroded the artificial dune, foreshore, and nearshore zones, with some positive feedback
4	Overwash terrace is raised and recolonized	Badly eroded stabilized dunes and a rebuilding berm
5	Waves from the sound overwash the natural barrier	Sound storm tide erodes the land side of the dune, damaging the property and vegetation supposedly protected by it
6	Natural barrier undamaged by the storm	Further artificial rebuilding of the dune
7	Violent storm causes overwashing but no permanent damage to the natural barrier	Same violent storm breaches artificial dune with disastrous results to human installations
8	New salt marsh growing on over-wash sediment and stability of the natural barrier is increased	Artificial beach nourishment at great cost, attempts to repair the damage on the narrowed seaward slope and erosion of the marsh further narrows the barrier from the land side

problems may reveal that engineering solutions—especially if they are insensitive to geomorphological conditions—may not be cost-effective or successful. Recent work in coastal geomorphology has begun to reveal that 'geomorphologically compatible' solutions to management problems may be more acceptable environmentally and in terms of other management objectives; they may also be cheaper and more durable. Thus eroding cliffs may not only provide a source of sediment that ultimately preserves beaches; they may also be a significant environmental asset: it does not follow that because they are eroding they should be protected. Similarly, artificial beach replenishment may provide more attractive and effective protection than groynes in some circumstances. But general solutions do not necessarily solve all local problems: each situation must be examined in detail, *and* in the broader context of regional conditions. That examination requires, as in so many geomorphological environments, both mapping and monitoring.

11 Mineral Resources and Geomorphology

11.1 Introduction

This chapter explores the relations between geomorphology and earth materials as natural resources. Of particular importance in this context are aggregates for the construction industry, exploitable mineral deposits, and the agricultural soils of an area. Recent general studies of the aggregate industry include those by Harris *et al.* (1976) and Collis and Fox (1985) in the United Kingdom, and by Fookes and others in glaciated environments (1975) and desert areas (1980). In addition to such reviews, there are numerous specific reports of detailed aggregate surveys. Excellent public examples are provided by the Illinois State Geological Survey and the mineral assessment reports of the Institute of Geological Sciences (UK) (now the British Geological Survey). The interests of the aggregates industry are served by trade associates including the Sand and Gravel Association Ltd. (London), and the National Sand and Gravel Association (Washington); and by such journals as the *Quarry Managers' Journal*, *Cement Lime and Gravel*, *Mineral Planning*, and *Quarry Management and Products*.

Mineral resources at or near the earth's surface in relation to geomorphology are less fully reviewed, but the studies by Demek (1972), Piotrovski *et al.* (1972), and Verstappen (1983) provide useful guidelines.

Texts concerned with the relationship between soils and geomorphology include the early work by Jenny (1941), and several more recent works including those by Birkeland (1974) and Gerrard (1981).

11.2 Aggregates: Sand and Gravel

(a) *Industrial requirements and problems*
In terms of tonnage, sand and gravel are now the two most important materials extracted from the earth. Both are defined, geologically, on the basis of particle size. Sand is particulate material with diameters between 0.06 and 2 mm, and gravel particles have diameters between 2 and 64 mm. In practice, naturally occurring deposits of sand and gravel normally include particles having a wide range of sizes. But within aggregate deposits, size distribution (grading) is only one of several important and variable properties. Others, all of which need to be analysed, include mineral composition (including any potentially deleterious substances that may be present); particle shape, texture, and specific gravity; and absorption, shrinkage, abrasion, strength, and weathering characteristics (e.g. Fookes, 1980).

Sand and gravel are required mainly for use in the construction and building industry. Aggregates for concrete are in the greatest demand. Other requirements include gravel for road and railroad beds, macadam preparations and ballast, and sand for plastering, filters, and other uses.

One of the most important requirements in the sand and gravel industry, and one which the geomorphologist can help to meet, is for precise knowledge on the nature and distribution of available resources. There are several reasons why this requirement is so important. In the first place industrial demands for sand and gravel have been rising. Closely related to this rise is a trend towards more specific and rigorous requirements for particular purposes. For instance, the size, absorption, specific gravity, and composition of material for a civil engineering contract is normally specified very precisely, as are its resistance to abrasion, shrinkage, and weathering processes (e.g. Fookes, 1980). In short, the need is for greater quantities of higher-quality products. At the same time increased demand has led in many areas to depletion of the best and most suitably located reserves.

Sand and gravel are a bulky commodity of low intrinsic value which is extremely expensive to transport, so that proximity of supplies to markets is of fundamental importance. It has been estimated, for example, that the cost price of gravel at a point some 20-7 km from a pit may be double the cost price at the pit (e.g. Flawn, 1970). In the UK, over 80% of sand and gravel is transported by road over distances which are less than 32 km (D.o.E., 1976).

Sand and gravel reserves within and around major urban areas, where needs are greatest, are the most valuable. And yet it is precisely in these areas that competition for land resources is keenest and the conflict of interests is most pronounced. Urban development may itself preclude the use of sand and gravel. The building of Heathrow Airport, London, for example, resulted in the sterilization of about half of the 9311 ha of potentially workable sand and gravel in west Middlesex. In addition, many river-terrace sand and gravel deposits on urban fringes coincidentally provide optimum locations for those agricultural activities, such as market-gardening, which also benefit from proximity to urban markets. Thus, in order to help resolve such conflicts, the nature, extent, and quality of sand and gravel reserves must be known.

The extraction of sand and gravel near urban areas is seen by some to be objectionable on several environmental grounds. The plant is not only unsightly, but screening, washing, and crushing equipment is also often noisy. Heavy local traffic may be generated, and many hectares of land may be consumed. In the United Kingdom more than 1400 ha are used each year, with over half of them becoming water-filled pits. But dry or wet pits are not without their uses. Sites for the disposal of waste are scarce in most urban areas, and old sand and gravel workings may be suitable receptacles, especially for inert debris such as rubble and pulverized fuel ash. Once the pits are filled, they can be used in many ways, such as for forestry, recreation, building, or agriculture. Wet pits can often be transformed into recreation areas that provide facilities for water sports, as the Lee Valley Regional Park Scheme, London, demonstrates. Finally, the sand and gravel industry not only demands water, but it also may pollute rivers, and affect groundwater conditions (for example, by locally lowering the water-table).

These various characteristics of the industry pose many planning problems to which varied solutions have been devised. In the United States the designation of 'Sand and Gravel Zones', the issue of conditional use permits, and other planning devices, have been used for some years. In Ontario, Canada, the conflict between the need for aggregates in an urban market and the growing demands to protect the environment in the same areas as those in which the most desirable sources occur, has led to both legislation and policy formation (e.g. McLellan, 1975; Ministry of Natural Resources, Government of Ontario, 1977). In the United Kingdom, the working of minerals was first brought under general planning control in the Town and Country Planning Act 1947 and the general aims of mineral planning were set out in *The Control of Mineral Working* (Ministry of Housing and Local Government, 1960). In 1972 a Committee on Minerals Planning Control was established in the UK to examine planning control of mineral extraction, and an Advisory Committee on Aggregates was also formed in 1972 to advise the Secretary of State for the Environment (D.o.E., 1976). Responsibility for resolving problems of conflicting land use and ensuring future sand and gravel supplies rests largely with the planning authorities, although the task of exploitation remains in the hands of private operators.

As reserves have declined and as industrial requirements have become greater and more sophisticated, the sand and gravel industry has responded in a variety of ways. One response has been the increasing use of crushed rock as a substitute for sand and gravel in road-bed construction and similar activities. Indeed, while the demand for aggregate has increased from c.2m tonnes per annum in the early 1900s to over 300m tonnes per annum in the peak year 1973, the proportion of aggregates in this total from shallow water, land, and marine sand and gravel deposits has declined, from 55 per cent in 1965 to 45 per cent in 1973 (D.o.E., 1976; Collis and Fox, 1985). Equally important, mechanization and the development of such equipment as excavators have permitted a great increase in the rate or production from individual quarries. 'Beneficiation' of sand and gravel deposits by improved methods of crushing, screening, grading, and the elimination of unwanted material has helped to meet the specific requirements of industry and permitted the exploitation of 'poorer' deposits (Collis and

Fox, 1986). Unwanted elements in many natural deposits include clay and silt (which are not only useless but may harm concrete and asphalt mixes), porous particles (which are susceptible to weathering processes, Chapter 12), and other unsound particles. Where deleterious material has a specific gravity that differs significantly from that of the required sand and gravel, the Heavy Media Separation process can be used. Thus, as the technology of the industry has advanced, so poorer reserves have become usable (albeit, more expensively), dependence on specific deposits has relaxed, and the range of deposits suitable for a specific purpose has been extended.

But the fundamental requirement of knowing precisely the nature and distribution of potential reserves remains. And it is here that the geomorphologist has a contribution to make.

(b) *Sand, gravel, and the geomorphologist*

Sand and gravel supplies normally come from two major sources: stratified sedimentary rock formations and superficial sedimentary deposits. Only the latter are of major concern to the geomorphologist. Suitable sand and gravel resources often occur in four major geomorphological contexts: (i) the channels, floodplains, and terraces of rivers and alluvial fans; (ii) fluvio-glacial environments, where a wide variety of material may be laid down on, in, under, or in front of ice sheets or glaciers or carried in meltwater derived from them; (iii) marine environments, especially in the littoral zone; and (iv) on slopes, where suitable materials may be found in screes, etc. Thus in England and Wales approximately 10 per cent comes from Triassic deposits in the Midlands; about 60 per cent of sand and gravel supplies is recovered from river deposits; a further 25 per cent originates from fluvio-glacial deposits; and some 5 per cent is drawn from coastal localities (Dunstan, 1966).

Each of the principal sources of superficial sediments has distinctive characteristics. In the context of demands placed on such sediments by industry, several of these characteristics should be emphasized. Firstly, the better washed and sorted the deposit, by and large, the less processing will be required. Thus river deposits, in which natural sorting and washing have often been extensive, are usually preferable to glacial tills and other similar deposits. Particle-size distribution is a further important feature. For example, the presence of fine material (silt and clay) within and between

beds of sand and gravel is undesirable, and large boulders may make working more difficult. In addition, both sand and gravel are normally required in the construction industry and a mixed deposit is needed, the optimum proportion of gravel being normally between 40 and 60 per cent. Deviations from this proportion in a deposit may mean that some material is wasted. Closely allied to this point is the requirement that the quantity of material available for exploitation at one location should be adequate for sustained production for a period sufficient to justify the large capital investment needed in a modern operation. Also of importance are the shape, composition, and physical properties of the particles (e.g. Fookes, 1980). For example, coarse, 'sharp' (angular) sand—such as that often found in fluvio-glacial deposits—may be suitable for concrete, whereas 'soft' (rounded) sand may be preferable for mortar, and fine sand is required for plastering. Finally, the thickness of useless overburden on the deposit and the depth of the water-table affect the suitability of a reserve. In the case of the former, too much overburden may limit the economic viability of a deposit; and in the case of the latter, special extraction techniques may be necessary where the water-table is high.

The appraisal of superficial sediments with a view to their possible exploitation normally involves the following four phases:

(i) *Preliminary survey* of possible source areas from available geological and other maps and reports, and from aerial photographs and other remote sensing imagery (e.g. Cassinis, 1977; Beaumont, 1979). In this phase, both land-systems mapping and geomorphological mapping are often of great value (Chapter 2).

(ii) *Field survey* of deposits, including mapping of their areal extent, and determination of their depth and volume, and depth to water-table (e.g. Fig. 11.1). Here, field mapping may be supplemented by airborne and ground-based geophysical techniques (such as electrical resistivity and shallow seismic methods).

(iii) *Sampling* of deposits is commonly undertaken using exposures, boreholes, and trial pits; in this context, the sampling strategy may be assisted by geomorphological information provided in the preliminary phase.

(iv) *Laboratory analysis* of samples, to determine the physical and chemical properties of

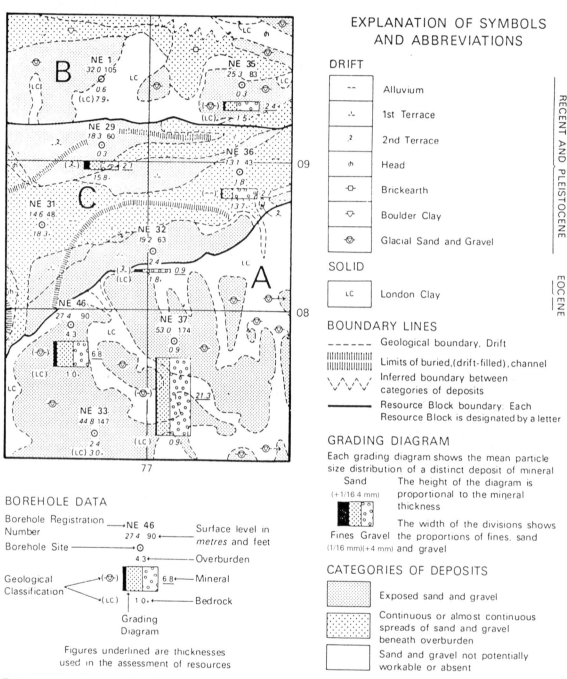

FIG. 11.1 Extract from a sand and gravel resource map, showing the contribution of potentially workable deposits in part of the Chelmsford area, Essex (after Thurrell, 1981)

aggregate deposits (grading, shrinkage, durability, etc.). Such analyses are often cast in terms of national standards for testing materials (e.g. Collis and Fox, 1985). Many countries have their own standards for aggregates. For example in the UK the British Standards Institute has prepared standards for sampling and testing aggregates, sands, and fillers (BS 812), for aggregates in concrete (BS 882, BS 3797), for building sand from natural sources (BS 1198, 1199), and for gravel aggregates in road surface treatments (BS 1984).

The relative cost-effectiveness and predictive accuracy of geomorphological contributions to the appraisal of aggregate resources is of great importance, but it has only rarely been evaluated. Chester (1980, 1982) provides a clear example from a study of fluvio-glacial sediments in Scotland (Table 11.1); he concluded *inter alia* that in

similar areas the use of soil maps and aerial photographs together can provide a reliable, cost-effective, and accurate preliminary assessment technique, which can be followed by assessment of resource volumes based on simple calculations of area and depth.

There are many examples of surveys of sand and gravel resources in which the techniques and expertise of the geomorphologist have been used successfully. It is also true to say that many research theses by geomorphologists in recent years, such as those on glaciated landscapes in Britain, contain much data pertinent to the interests of the sand and gravel industry, but unfortunately the authors have too often failed to pursue the applied potential of their studies. A brief example from the United Kingdom illustrates the contribution of geomorphologists to the assessment of sand and gravel resources.

S. W. Wooldridge was the principal geomor-

TABLE 11.1 Cost-effectiveness and predictive accuracy of aggregate assessment techniques used in Scotland

A. Cost-effectiveness of different data sources

	Area mapped/day km²	Cost of mapping[a] 1 km² (£ sterling)
1. Soils, topographical and geological maps	1250.0	0.04
2. Panchromatic vertical aerial photography	42.0	1.19
3. Electrical resistivity	0.24	208.3
4. Seismic logging	0.12	416.6
5. Field mapping	4.0	12.5

[a] Assuming consultancy rates of £50/day.
1, 2, 5 = estimates from pilot studies carried out in north-east Scotland; 3, 4 = estimates from G. Vann, (1965).

B. Predictive accuracy of different data sources

	Percentage explanation of ground survey map by different data sources
Aerial photographs	87
Soils maps	53
Geological map	10
Topographical map	10
Land-use maps	10
Literature: 'Sands and Gravels of Scotland'	8
All data sources	98
Soil maps and aerial photographs only	97

Source: Chester (1982).

phologist on the Ministry of Town and Country Planning's Advisory Committee on Sand and Gravel (Ministry of Town and Country Planning, 1948), and studies by Wooldridge and others (e.g. Wooldridge and Linton, 1955) on the geomorphological history of south-east England were of great value in assessing the resources of the London region and elsewhere. For many years the main sources of sand and gravel had been the river-terrace deposits, especially west of London, associated with the most recent major phase in the evolution of the river Thames and its tributaries. Studies of geomorphological evolution had shown, however, that the Thames formerly followed several different more northerly courses and that along these lines and elsewhere at elevations above the main terraces there were extensive fluvio-glacial and plateau deposits. Thus the report identified these deposits as the principal gravel reserves of the Metropolis and it pointed especially to the Vale of St Albans between Watford and Ware (Fig. 11.2). It was also recognized that these reserves were poorer in quality than those of the traditional sources west

of London. The main reason for this is that older deposits of the northern area are relatively poorly sorted and poorly bedded fluvio-glacial deposits, often containing silt and clay, and locally overlain by till, whereas the younger terrace deposits are often derived from the older material and, in the process, have become progressively washed and sorted. The older deposits clearly require greater beneficiation than the terrace sediments, but as the traditional sources have become exhausted or sterilized, so the exploitation of the less satisfactory reserves has increased. The importance of this work persists today. For example, it is not simply by coincidence that a major section of London's orbital motorway (the M25) closely follows the gravel deposits of a former course of the lower Thames in the north of London (Plate 11.1). Subsequent geomorphological work has not only elaborated the details of the evolution of the river Thames during the Quaternary, but it has also illuminated the complex nature of the sediments in ways that are relevant to future sand and gravel exploitation (e.g. Gibbard 1977, Green *et al.*, 1982; McGregor and Green, 1983).

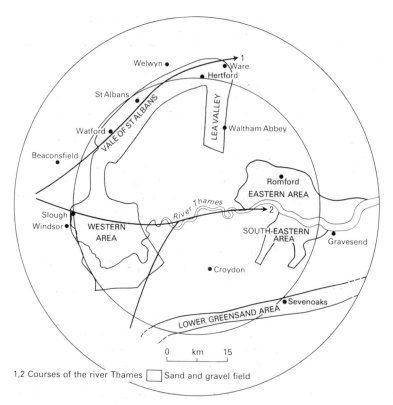

FIG. 11.2 The major sand- and gravel-producing districts in the London area, and the former courses of the River Thames (after, in part, Ministry of Town and Country Planning, 1948)

PLATE 11.1 London's orbital motorway (the M25) passes the major sand and gravel quarries exploiting a former course of the Thames near Watford, Herts.

11.3 Mineral Resources

The location of minerals is often closely related to the geomorphology of an area (Table 11.2). Perhaps the clearest example of this lies in the relationship between alluvial placer deposits and features of fluvial activity, such as river terraces, as exemplified by Hall *et al.* (1985) in the case of alluvial diamonds in Ghana. However, as is particularly shown by the work carried out in the USSR (Piotrovski *et al.*, 1972) geomorphological investigations have also played a leading part in the search for other types of minerals. The relationship of mineral deposits to relief is summarized in Table 11.2 and Fig. 11.3.

In searching for the association between minerals and relief it is important to establish as much as possible about the geomorphological history of an area. This is because some associations may arise not so much from the contemporary processes operating in the area but more from the recognition of former relief conditions. For example, some ore deposits are associated with weathering crusts, having been derived from the chemical breakdown of the original parent material. These crusts tend to be related to planation surfaces of low relative relief. The residual iron

TABLE 11.2 The relationship between major categories of minerals and relief

I. *Minerals directly related to relief*
(i) Placer deposits (e.g. gold, diamonds, tin (cassiterite), and other heavy minerals).
(ii) Weathering products (e.g. enriched copper, limonite, manganese, bauxite, cobalt, and kaolin).
(iii) Basin deposits (e.g. coal, iron and manganese ore, peats).

II. *Minerals indirectly related to relief*
Deposits associated with structures that are revealed in the relief, such as down-faulted blocks or areas of upwarping (e.g. gas, oil, salt, coal, mineralized magmatic bodies, and buried placer deposits).

III. *Minerals not related to relief*
(Geomorphological analysis of an area is only incidental)

ores of the Mayari and Moa Districts, Oriente Province, Cuba, were probably formed as part of a weathering residue of serpentine rock on an ancient land surface (Leith and Mead, 1911).

Where river dissection has followed planation the weathered crust will have supplied to the rivers

FIG. 11.3 Possible associations between geomorphological features and mineral deposits (after Demek, 1972)

TYPE OF MINERAL DEPOSITS / RELEVANT GEOMORPHOLOGICAL FEATURES	Placers (continental) (e.g. gold, diamonds)	Placers (littoral, marine and lacustrine)	Peat	Clays and sandy loams	Gravel sand	Brown coal	Black coal	Bauxite and refractory clays	Residual deposits of weathering (nickel, cobalt, iron, magnesite)	Infiltrated weathering deposits (sulphur, uranium, gypsum)	Deposits of translocated weathered crust (e.g. cobalt, iron)	Salt (e.g. potassium, sodium, gypsum)	Sedimentary, ferromanganese	Magmatic ore (metals, diamonds etc.)	Pegmatitic (e.g. muscovites, optical raw materials)	Carbonate (e.g. rare minerals)	Hydrothermal	Metamophic (e.g. iron, manganese)	Groundwater	Fissure water	Oil and natural gas
Tectonic-structural	*	+	+	+	+	*	*	*	*	*	+	+	*	*	+	*	+	+	*	+	*
Relief expression of deep intrusions	+							*	*		+			*	*	*	+ + *	* * *	+	+ + *	
Structures in sedimentary and metamorphic rocks	+	+			+		*	*	+	+	+	+	+	+	+	+	+	+	+	*	*
Valley pattern reflecting (a) folding	+	+			+	+	*	+	+	+	+	+	+	+	+	+	+	+	+	+	*
(b) dome structures	+	+			+	+	+	+	+	+	+	+	+	+	+	+	+	+	+	+	*
(c) faulted blocks	*	+			+	*	*	*	*	*	*	*	*	*	*	*	*	*	*	*	*
Old valley system changes induced by (a) tectonics	*	+	+	+	+	+	+				+						*		*	*	*
(b) exogenous factors (e.g. glaciation, sedimentation)	*	+	+	+	+		+										*		*	*	*
Longitudinal valley profiles	*										+										*
Slope steepness	*			*	+		+	+	*			+		+	+	+	+	+	+	+	+
River terraces } and	*			*	*				*	*											*
Floodplains } their	*	*	*	*	*				+												+
River beds } sediments	*	*			*																
Planation surfaces	*	*	+	*	+	+	+	*	*	*	*	+	+						*		
Eluvium and weathered crusts	*	*		*	+	+		*	*	*	*	+	+	+		+		+	+	+	*
Present-day karst	*	+	+	+				*	*	*	+	*			+				*	*	+
Fossil karst	*		+	+				*	*	*	*	*	*						*	*	*
Lake basins, lacustrine and marshy accumulation (a) humid regions	+	+	+	+	+	+			+		+	*	+						*	*	+
(b) arid and semi-arid regions			+	+	+	+		+			+	*	*						*	*	+
Marine and lacustrine littoral forms (primarily of wave and accumulation types)		*	+	+	*	+					+	+							*		
Glacial and fluvio-glacial forms of mountain glaciation	+			+	*				*										*		+
Glacial and fluvio-glacial forms of continental glaciation	+		*	*	*				*										*	+	
Cryogenic forms	+		*	*	+				*										*	*	*
Relief buried by poorly consolidated and young volcanic rocks	*	*			*				*	*	*						*		*		

+ Association possible—feature mapping useful
* Association probable—feature mapping indispensable

Highly modified from Denmark (1972)

minerals which may have become concentrated as placer deposits within river terraces. These may subsequently be buried (e.g. by hillslope colluvial material) or may themselves become the source for a colluvial feature. This hypothetical case illustrates some of the complexities that may occur in a landscape, and that need to be unravelled by geomorphological analysis during a search for likely sites of mineral deposits. Such an analysis could do much to eliminate from a field investigation those sites which are unlikely to yield minerals.

An example of studies linking geomorphology and the supergene alteration of minerals is the work of Clark *et al.* (1967) in the southern Atacama Desert, Chile. In this area, enriched copper deposits have been known for many years, and they have been essential for the economic progress of small-scale mining operations. Such enriched deposits, characterized by chalcocite and other copper sulphide minerals, are associated with chemical changes in the vicinity of present or former water-tables. They are normally found below a zone of oxidation or leaching, and above relatively unaltered hypogene mineral assemblages. As the shape and position of the water-tables at present and in the past are closely related to the form of the ground, an attempt was made to establish the relations between enrichment zones and topography, so that the latter could be used to predict where new enriched copper deposits might be found.

The identification of enriched, leached, oxidized, and hypogene mineral zones showed that there had been at least three major periods of supergene mineral alteration, and that the resulting zones are roughly horizonal and sheet-like in form. This evidence suggested that their development was controlled by horizontal or gently sloping surfaces, such as the major pediplains recognized in the region. But the relationship between enriched ores and pediplains was not simple. For instance, the lowest pediplain in places clearly truncates one of the enriched zones, exposing rich ores at the surface. It seemed probable that supergene enrichment was related to dissection phases between periods of uninterrupted pediplanation, but that the less dissected areas of the planation surfaces permitted both the development and preservation of enriched ores. Canyon development followed the formation of the pediplains and major enriched zones, causing water-tables to be lowered and oxidation of the upper parts of the enriched zones. Later gravel deposition in the canyons raised water-tables adjacent to valley floors and led to the formation of 'sooty' chalcocite in a zone above the present water-table. Subsequent incision of these gravels was accompanied by renewed oxidation.

This study led to the suggestion of guidelines for determining the location of new enriched deposits. For example, copper deposits within canyons could largely be ignored because they are likely to have had their enriched zones eroded or oxidized.

And undissected remnants of the lowest pediplain are the most promising localities for the discovery of enriched ores relatively near to the surface.

Another example occurs in the Urals where a series of Tertiary and Quaternary terrace and floodplain diamond placers has been formed following the weathering and erosion of fossil marine placers (Piotrovski *et al.*, 1972). The latter occur in Palaeozoic littoral sandstone and coarse gritstone beds, and the primary source of the diamonds has never been found. In South Africa, on the other hand, diamond placers occur within rivers draining from the kimberlite sources in the Kimberley area. The kimberlites rise as gentle domes above the general surface and can themselves be recognized by direct reference to the relief.

These examples show that a knowledge of the geomorphological history in each case can guide very materially the continued search for further diamond-bearing placers in these areas.

In mineral prospecting it is also important to know something of the reaction of the minerals themselves to geomorphological processes. For example, those with a high specific gravity (e.g. gold) tend to be concentrated in specific locations and are better sorted than those with a lower specific gravity (e.g. zircon and diamonds). Heavy minerals may also accumulate in placers nearer their sources than minerals of lower specific gravity. The resistance of minerals to weathering and abrasion is also important. Diamonds are very resistant, can be carried far, and even recycled, as in the example quoted above from the Urals region. Cassiterite (SnO_2) on the other hand quickly disintegrates during transport, and sites of alluvial and colluvial tin in south-west Uganda, for example, are close to the original source on the valley sides. In the valley shown in Fig. 11.4 the distance from source to payable alluvial deposit is less than 500 m. Beyond a distance of 800 m no deposits have been worked, and no one has yet thought it worth while to look for such deposits beneath the cover of lava and volcanic material which buries the lower part of the valley containing these cassiterite-bearing deposits.

On a larger scale whole ore bodies may be directly responsible for particular relief features because their resistance to denudation differs from that of the rocks around them. For example, the lead–zinc lode at Broken Hill, Australia, is marked by a conspicuous ridge. Conversely, calcite veins

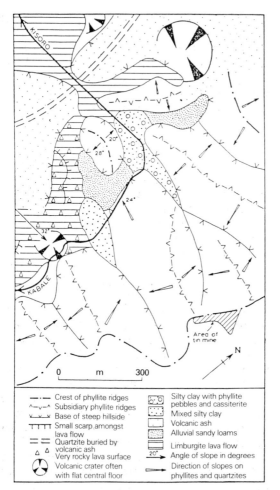

Fig. 11.4 An alluvial cassiterite site in south-west Uganda

at Oatman, Arizona, coincide with surface depressions (Thornbury, 1954). If the geomorphological expression of a specific ore body can be identified for any one areas then the recognition of similar ore bodies may become a comparatively simple task. In such cases remote sensing imagery becomes an invaluable aid in the recognition of sites worthy of detailed field examination.

A similar approach has been used in the USSR where recent folding and upwarping has been detected by geomorphological means. These have included the recognition and mapping of unwarped planation surfaces and depositional (downwarped) plains or basins. The upwarped structures provide locations favourable for the development of oil and natural gas. In some instances such upwarping needs to be inferred

from radial drainage patterns, or from the observation that rivers are skirting around an area that may prove to have been upwarped. The most recent tectonic movements may be reflected in knick points in the river thalweg (Piotrovski *et al.*, 1972).

A systematic approach to the search for minerals and ore bodies through landform analysis is provided by geomorphological mapping (Chapter 2). In the case of placer deposits this can be carried out within the context of the drainage basin likely to contain them (Chapter 7), though past changes in river patterns must be taken into account. The most valuable gold placer deposits of the Sierra Nevada are to be found predominantly within gravels associated with prevolcanic, early Tertiary, or Eocene drainage channels which do not coincide in position with the present river channels. As a result the problem of locating the richest placer deposits is one of reconstructing the early Tertiary drainage pattern.

11.4 Soils

(a) *Introduction*

This brief section seeks only to indicate some of the reasons why a study of soils cannot be undertaken without also making a critical assessment of the geomorphology of their site.

Soils form the most ubiquitous natural resource available for human activity. Soil is the essential element in all agriculture in that it forms the basis for plant growth. From the soil plant roots receive mechanical support, water, the minerals required for growth, and oxygen. The physical condition of the soil determines its ability to supply water to plants. This is largely a function of available pore spaces. Soil chemical properties, on the other hand, determine the capacity of soils to supply nutrients. Both physical and chemical properties determine the nature of plant root extension, adequate extension being essential to healthy plant growth. A good physical state usually indicates that a soil is also in a good chemical condition.

The main physical properties of the soil that matter to agriculture are texture, structure, depth of horizons, total depth of the soil, soil consistency, and temperature. Field assessment of physical properties such as texture, structure, and consistency may be made by reference to the *Soil Survey*

Manual (Soil Survey Staff, 1951) of the US Department of Agriculture. These physical properties determine water movement through the soil, weathering processes, and the translocation of colloids and minerals. A soil acquires these properties as a result of the weathering of parent materials and the denudational (i.e. erosional and depositional) processes that have affected it since its formation began. The influence of parent materials may also be seen in the mineral composition of the soil, though this is particularly vulnerable to weathering processes. Parent materials containing a large proportion of aluminosilicate minerals weather readily into clays. Parent materials rich in calcium, magnesium, or potassium (i.e. alkaline rocks) tend to weather more slowly than acid rocks. Intense chemical weathering may cause minerals to be leached out of the soil as weathering proceeds. Thus, although a parent material may have a high calcium content (to take but one example) this leaching process may leave a soil with a low calcium content. That is why liming is sometimes necessary for agricultural soils developed on calcareous bedrock. As time proceeds, however, the depth of weathering tends to increase and climate may begin to impose characteristics upon the soil that override the earlier dominant influence of parent materials. Similarly the influence of vegetation may increase with time. The direct influence of parent materials may also become obscured by the mixing of materials, as where a low site receives material from hillslopes above.

It is generally assumed that climate may be a dominant influence over parent materials in tropical areas where chemical weathering is intense and may produce a weathered profile tens of metres deep. Nevertheless, tropical soils may still be dependent upon site for some of their characteristics (e.g. Moss, 1965; McFarlane, 1971).

Applied geomorphology is concerned not only with landforms but also with earth materials and denudational and weathering processes. Soils thus provide a key element in geomorphological investigations. In addition, a geomorphologist's interests lie not only in the soil itself, but also in its environmental setting, its relationship to landform, parent materials, and the processes that are likely to affect its development. The interest also continues into an assessment of the soil's potential vulnerability to erosion by water (Chapter 4) or by wind (Chapter 9). The links between the geomorphologist's interest in soils and those of the agriculturalist are therefore clear. Through his work the geomorphologist has much to contribute to a proper understanding of the agricultural assets of a soil. An analysis of soil genesis and an evaluation of soil environment provide useful information to those whose task it is to manage agricultural land.

(b) *The geomorphological context of soils*

Soils represent the interaction between processes of weathering and rock materials. Their accumulation or removal is a function of both time and processes of denudation. They are therefore closely related to the geomorphological history of the site they occupy. For example, sandy-textured soils with a granular structure may be directly related to their situation on an alluvial terrace. Likewise the soils of floodplains, alluvial fans, screes, pediment surfaces, or deeply weathered etch-plains may be predictable in terms of site geomorphology. Sometimes soils have been shown to vary with the age of the land surface on which they occur (Ruhe, 1956; Mulcahy, 1960). Conversely, however, the soil can be thought of as an indicator of geomorphological history. Thus a thin skeletal soil or a truncated soil profile generally indicates that material has recently been removed from the site. On the other hand a buried soil profile shows that deposition has taken place. A soil which displays a full profile development is indicative of a relative geomorphological balance between the processes of weathering and denudation. To the extent, therefore, that soils are determined by geomorphological processes, the recognition and analysis of those processes has a bearing on soil investigations for resource management.

The influence of relief on soil-profile formation, and therefore soil type, can be direct in that steep slopes tend to develop thin soils, which also tend to be less weathered than those on gentler slopes in the same area. Relief also influences the movement of water. Sites which are freely drained develop eluvial soils, with the tendency for material to move out of the soil either in suspension or solution. Conversely where water tends to accumulate it also brings in material to form illuvial soils, or even an organic soil. The amount of water available to a soil usually increases down-slope (see the discussion in Chapter 4 on throughflow), providing a down-slope sequence of soils related to hydrological conditions of the slope. Valley-floor sites or lowland depressions may remain wet for most of the year, giving waterlogged conditions that strongly influence soil-profile development. Such hydrological sequences are particularly common in areas of hummocky glacial deposits, and in certain areas of stabilized sand-dunes (Fitzpatrick, 1971). Thus, a progressive change down-slope from a freely drained site at the top of a slope to a waterlogged site at the bottom can give rise to a soil sequence from podzol through gleys to peat. The podzol is identifiable because of its eluvial upper horizons, while the gley has illuvial characteristics and the peat arises from the accumulation of organic material.

(c) *Mapping soil resources*

In many areas landforms and soils are closely related, not only because the former may reflect processes that influence soil formation (Chapter 12) or soil erosion (Chapters 4 and 9), but also because they may both be related to the parent materials upon which the soils have developed.

In 1935 Milne reported for parts of East Africa a recurring and definable relationship between soil site and soil type. He suggested, perhaps somewhat optimistically, that a predictable relationship could be established, in any one area, between a soil and its position down a valley-side profile. Indeed, all similar valley-side positions might be expected to have the same soil type. Milne (1935) called this relationship a soil catena. This principle has been very important in terms of soil mapping chiefly because once a definite relationship between soil and site has been established it is much easier to delimit provisional soil boundaries by mapping sites than by digging soil pits. This principle is common to many reconnaissance soil maps derived from aerial photographs, and it is echoed in the principles of the land-systems approach described in Chapter 2.

Common examples of catenas are quoted by Young (1972). In temperate latitudes the down-slope sequence often to be found is brown earth→gleyed brown earth→gley. In upland Britain there is often a repeating sequence of hill peat→thin iron-pan podzol→iron podzol→shallow acid brown soil, becoming deeper downslope→gleyed acid brown soil→gley→basin peat. In some locations not all members of this catenary sequence are present.

The soil catena, therefore, is an expression of the direct relationship between geomorphological and soil-forming processes. The forces that produce the geomorphologist's 'materials' create the pedologist's 'soils'.

The most useful general-purpose model of site–process relationships in terms of predicting soil characteristics is that developed by Dalrymple *et al.* (1968). These authors recognized nine recurring landform units which have definable surface and subsurface processes. The model provides a ready framework into which areas may be subdivided and from which primary inferences may be made about the processes influencing soil development. All nine units of this model will not be present on every hillside, but it should be possible to classify any hillside in terms of these units. More complex hillsides may have one or more of these units repeated down their profile. Equally valuable is the mapping of each unit laterally along the slope as an initial estimate of the likely extent of associated soil types. Such a model is especially useful during soil mapping at the reconnaissance level.

Whenever a reconnaissance survey of soils is required, especially over large or remote areas, recourse normally has to be made to the interpretation of aerial photographs or remote sensing imagery (Chapter 2). Important elements of a landscape that can be directly interpreted from aerial photographs and which have a direct bearing on soil mapping include the following geomorphological properties (Buringh and Vink, 1965):

(i) elements having a positive and direct correlation with soils (e.g. waterlogged sites, patterned ground);
(ii) elements related to terrain morphology (e.g. land type, relief form, slope, drainage pattern, watershed pattern, rivers);
(iii) elements related to special aspects of terrain (e.g. gully dissection (form and pattern), lithology, tonal, or colour variations).

A common starting-point for soil mapping from aerial photographs is a classification of the relief of the area to provide a framework for the soil map legend. The units mapped may resemble those of a land-systems analysis (Chapter 2), and they may be adapted, through subsequent fieldwork and laboratory analysis, into soil-mapping units. Such a classification can also save much time and expense during subsequent fieldwork as it can form the basis of a stratified random sampling programme of soil-pit investigations. A limited number of soil pits are required as only the within-unit variability needs to be tested in order to confirm boundaries. It is also easier to check the reality of provisional soil boundaries than it is to make a primary definition of them during a field survey. This approach, however, frequently leads to a classification of soils by their site characteristics rather than on the basis of their own intrinsic properties.

11.5 Conclusion

Since a close relationship exists between site and sand and gravel deposits, mineral deposits, and soils, it is often possible to evaluate these resources through the geomorphology of their sites. This evaluation may be applied at a reconnaissance stage in order to discover the likelihood of a particular type of material resource being present in an area. It may also be undertaken, in the case of soils, in order to discern the physical controls on their continuing development. Deciphering the geomorphological history of a deposit or site may form an essential component of the evaluation process if predictions are required of the composition of the materials and of their likely internal variability.

12 Weathering of Rocks and Stone

12.1 Widespread Interest, Disseminated Literature

Rock is a material of fundamental importance to many aspects of human activity. It is a primary resource, as the previous chapter has shown; it is the parent material for soils upon which food production in part depends; it is widely used for construction of buildings, roads, and other engineering works; and it forms the natural foundations on which many engineering structures are built. All rocks may be subjected to attack by natural *weathering* processes, and most tend to be altered and destroyed by them, especially if they were formed under conditions different from the present environment. The nature and consequences of these processes are not only of interest to scientists but have a fundamental relevance to the work of engineers, architects, builders, and others who use rocks as raw materials in their work.

Several different groups have long-standing interests in the weathering of rocks. Geomorphologists have studied weathering processes theoretically, in the field, and in the laboratory, mainly with a view to interpreting weathering landforms and subsurface weathering profiles, and to understanding the supply of materials to rivers and other erosional systems. Recent substantive geomorphological surveys include Trudgill's *Weathering and Erosion* (1982), Ollier's *Weathering* (1984), Paton's *The Formation of Soil Material* (1978), and several general reviews (e.g. Brunsden, 1979; Chorley *et al.*, 1984; and Ollier, 1977).

There are close links between weathering processes, landforms, and soils, and these constitute an important focus of research for scientists who view soils in terms of their agricultural potential and role in land management (see also Chapters 4 and 9). Recent geomorphological contributions in this field include Birkeland's *Pedology, Weathering and Geomorphological Research* (1974), Gerrard's *Soils and Landforms* (1981), Jungerius's *Soils and Geomorphology* (1985), and Richards *et al.'s Geomorphology and Soils* (1985). Surface and near-surface weathering has traditionally been of fundamental importance to geochemists and pedologists (e.g. Loughnan, 1969; Soil Science Society of America, 1977); recently, however, it has attracted much more geomorphological attention, especially in the context of surface sediments and duricrusts (e.g. Goudie and Pye, 1983). Soil is such an important element in agriculture that many countries have specific agencies devoted to its study, such as the Soil Conservation Service of the US Department of Agriculture, the Soil Survey of England and Wales, and the CSIRO in Australia. Weathering is a problem of research interest to many in these agencies.

The nature and extent of weathering of surface and near-surface materials is of equally fundamental importance to construction engineers and engineering geologists concerned with site appraisal and foundation design (e.g. Hoskin and Tubey, 1969; Attewell and Farmer, 1976; Fookes and Vaughan, 1986). Here the interests of engineering are closely allied to those of soil mechanics and links between the two fields have been securely forged over the past 50 years or so (e.g. Scott and Schoustra, 1968). Work in this area owes much to Karl Terzaghi and his successors at Harvard University and elsewhere (e.g. Terzaghi and Peck, 1948). There are today many foundation engineering consultants, larger engineering firms usually have a geotechnical section, national agencies are common (e.g. Soil Engineering Group, CSIRO, South Africa; and the Venezuelan Society for Soil Mechanics and Foundation Engineering), and international conferences are flourishing (e.g. regional conferences for Africa on Soil Mechanics and Foundation Engineering).

The building industry has long been concerned with those weathering processes that tarnish or destroy natural and artificial building materials, with determining the durability of the materials they use, and with preventing their decay (e.g. Schaffer, 1932; Honeyborne, 1982). Recent contributions include Ashurst and Dimes's *Stone in Building* (1979), Winkler's *Stone* (1973), and Leary's assessment of *The Building Limestones of the British Isles* (1983). National research in this area is focused on such public agencies as the Building Research Establishment (UK), the Division of Building Research (Canada), the Division of Building Research (CSIRO, Australia), CNRS (Caen, France), and the United States Environmental Protection Agency. International groups include RILEM (Réunion Internationale des Laboratoires d'Essais et de Recherches sur les Matériaux et les Constructions), and ICCROM (International Centre for the Study of the Preservation and Restoration of Cultural Property). Interest is also reflected in such conferences as those on stone conservation (e.g. Bologna, 1975, 1981), weak rocks (e.g. Tokyo, 1981), and building-stone deterioration (e.g. Athens, 1976; Venice, 1979). Journals like *Monumentum* and *Lithoclastia* also promote the exchange of ideas.

Despite their common interests in rock weathering, the cross-fertilization of ideas *between* the different interested groups is surprisingly limited. Take the example of rock breakdown by salt weathering, where each discipline appears to have gone its own way. Geomorphologists have devised tests to simulate the effect of salt-weathering processes under different climatic conditions in the laboratory (e.g. Sperling and Cooke, 1985) and in the field (e.g. Johannessen and Feiereisen, 1982); and building-research workers employ a wide variety of tests using salt-crystallization procedures to determine stone durability and the 'soundness' of aggregates (e.g. BS1438 in the UK; ASTMS C88 in the USA; DIN 52111 in West Germany; AS-A77 in Australia; IS1126 in India; see, Ross, 1984). Such tests are descended from Brard's work in the 1820s, when salt was the experimental substitute for ice; they often leave much to be desired (e.g. Hudec, 1982/3). There is clearly scope for collaboration between the interested disciplines in order both to rationalize and improve the study of rock weathering.

Despite the efforts of journals and international conferences, the dissemination of weathering research amongst so many disparate groups and the proliferation of unpublished reports that are often difficult to obtain means that applied studies of rock weathering are difficult to review comprehensively. In this chapter, some of the scattered literature will be explored through four important themes: the nature of the weathering system in general (Sect. 12.2); approaches to the study of rock weathering (Sect. 12.3); the weathering of building stones (Sect. 12.4); and weathering and engineering problems (Sect. 12.5).

12.2 The Weathering System

(a) *Introduction*

From most practical points of view there are two main categories of weathering processes—*disintegration* processes and *decomposition* processes. The first group causes rock breakdown without significant mineral alteration; the second causes chemical changes to the rock. Table 12.1 lists examples of these two groups of processes. Much discussion in the literature relates to individual processes, often examined separately in controlled environments. Thus, it is essential to emphasize at the outset of this discussion that in most field situations, rock disintegration and decomposition occur together, and the results of weathering

TABLE 12.1 Classification of weathering processes

I. *Processes of Disintegration*
  a. *Crystallization processes*
    Salt weathering (crystal growth, hydration thermal expansion)
    Frost weathering
  b. *Temperature/pressure change processes*
    Insolation weathering
    Sheeting, unloading
  c. *Weathering by wetting and drying*
    Moisture swelling
    Alternate wetting and drying
    Water-layer weathering
  d. *Organic processes*
    Root wedging
    Colloidal plucking
    Lichen activity

II. *Processes of Decomposition*
  a. *Hydration and hydrolysis*
  b. *Oxidation and reduction*
  c. *Solution, carbonation, sulphation*
  d. *Chelation*
  e. *Biological chemical changes*
    Micro-organism decay, bacteria, lichens

---

normally reflect the combined effects of several different processes. It is often difficult or impossible to be certain in any one locality of the processes involved and of their relative importance. Indeed, the quantitative evaluation of the relative importance of different processes is a subject that has been grossly neglected, especially in the important context of stone conservation.

The nature and effectiveness of any weathering process, or group of processes depends largely on *three* sets of variables: the weathering environment (in which the climatic conditions are of great importance), the nature of the materials, and, where appropriate, the nature of biological conditions. The relations between these variables and weathering processes, and their consequences in terms of weathering products are summarized in Fig. 12.1. These general relationships can be

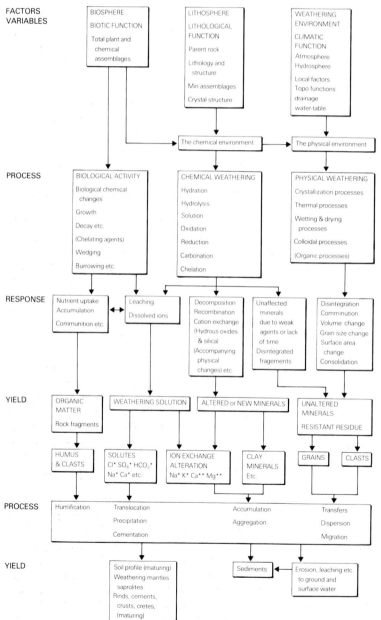

FIG. 12.1 General summary of the weathering system (after Brunsden, 1979; © Edward Arnold (publishers) Ltd.

simplified for specific applied problems. For example, Atkinson (1970) summarized the total deterioration (TD) of building materials in terms of the following expression:

$$TD = f[(D), (W), (E), (U)][(P)] \quad (12.1)$$

in which D = design factors, W = workmanship, E = environment, U = use, and P = intrinsic properties of materials.

(b) *Weathering processes and climate*

As most weathering processes are at least partly dependent on climatic conditions, there have been many attempts to map spatial variations of weathering processes at global and regional scales using climatic indices. At the global scale, this is an exceptionally difficult task for, as Trudgill (1976) has argued, it is first necessary to measure the relationships between weathering and process factors and their spatial variability. In fact, most world-wide classifications have not done this and have merely assumed a relationship between weathering processes and the primary controls of temperature and precipitation (e.g. Strakhov, 1967), basing the analysis of quantitative parameters derived from climatic statistics. An early example is that of Peltier (1950) who used mean annual temperature and mean annual precipitation as a means of describing the world-wide variability of chemical and mechanical weathering processes (Fig. 12.2). He assumed that chemical weathering increases with the availability of water (and hence with mean annual precipitation, in so far as the two are related) and with temperature. Thus, it is argued, chemical weathering is strongest in hot, wet climates. Mechanical weathering processes are equated to frost weathering (salt weathering is ignored), and mean annual temperature is used to define the intensity of frost action. Fig. 12.2D reinterprets Peltier's weathering zones in terms of concrete durability (after Fookes, 1980).

But such generalizations are dangerous, because the relations between temperature and precipitation are so complex. In considering temperature and frost action, for example, information is required not only on mean values, but also on absolute values, range over different time-periods, and the frequency, magnitude, and rates of temperature fluctuation about freezing-point. Indeed, it is not altogether clear which aspects of temperature best predict the effectiveness of, say, frost action. In the case of chemical weathering, the complexity of interrelationships and the conditions leading to rapid chemical weathering are shown in Fig. 12.3, where it can be seen that the climate is only one of many relevant factors. Table 12.2 provides a summary of the weathering characteristics of four major climatic zones based largely on empirical observations. Such generalizations are inevitably problematical at the world scale, not only because the complexity of the climatic and other controls prejudices the use of simple climatic parameters, but also because appropriate climatic data are limited, and knowledge of the processes and their relative importance is incomplete.

Slightly more meaningful and perhaps more useful regionalization can be obtained if the scale of inquiry is enlarged and more refined parameters are derived from climatic data. Two practical examples of studies at national scales will illustrate this point, one related to building-brick durability and disintegration processes, the other concerned with highway-foundation engineering and decomposition processes.

The American Society for Testing Materials (1971b) has related the effect of weathering on facing bricks in the United States to a *weathering index* that is the product of the average annual number of freezing-cycle days and the average annual winter rainfall in inches. A freezing-cycle day is any day during which the air temperature passes either above or below 0 °C, a crude measure of freeze–thaw frequency. Winter rainfall is the sum of the mean monthly rainfall occurring during the period between and including the normal date of the first killing frost in the autumn and the normal date of the last killing frost in the spring. In effect, by incorporating measures of frost frequency and water availability, this index records susceptibility to frost action. Its pattern in the United States, in terms of negligible (< 100), moderate (100–500), and severe (> 500) weathering, is shown in Fig. 12.4.

A contrasted example at the national scale relates to the engineering performance of road foundations in South Africa (Weinert, 1965). Field experience shows that weathered basic igneous rocks, notably the Karroo dolerites, are generally 'sound' in the west and 'unsound' in the east of South Africa (Fig. 12.5). In constructing Fig. 12.5, the unsound dolerites are taken to be those with a liquid limit greater than 30 and a plastic index

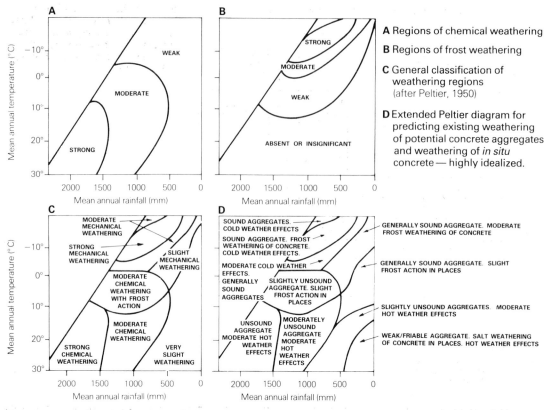

FIG. 12.2 (A) Regions of chemical weathering. (B) Regions of frost weathering. (C) General classification of weathering regions (after Peltier, 1950). (D) Highly idealized extension of Peltier's diagram (C) for predicting existing weathering of potential aggregates and weathering of *in situ* concrete (reproduced by permission of the Geological Society from Fookes, 1980)

greater than 10. 'Liquid limit' is defined in Chapter 5, and 'plastic index' is the difference between the liquid and plastic limit values. In terms of Fig. 12.5 disintegration prevails west of the 'weathering and performance boundary' defined by these values, and decomposition prevails to the east of it.

Weinert analysed climatic data in order to determine those climatic variables that could describe the weathering contrast, assuming it to be the result of *contemporary* climatic conditions. Only a few climatic parameters were needed in order to define the weathering boundary. These were potential evaporation and total precipitation during the warmest month, and annual precipitation. An appropriate climatic index expressing these variables was derived as follows. The potential evaporation and the precipitation during the warmest month (January) was expressed by the ratio:

$$R = \frac{E_{\jmath}}{P_{\jmath}} \qquad (12.2)$$

where

E = potential evaporation (Meyer formula),
P = precipitation,
$\jmath$ = January (the warmest month).

As temperature and evaporation tend to be higher in summer, and the seasonal distribution of precipitation varies from place to place, weathering will vary according to the availability of water, and a measure of annual rainfall is required. Such an expression is:

$$D = 12 \frac{P_{\jmath}}{P_{a}} \qquad (12.3)$$

where index $_{a}$ = annual. Summer rain occurs where

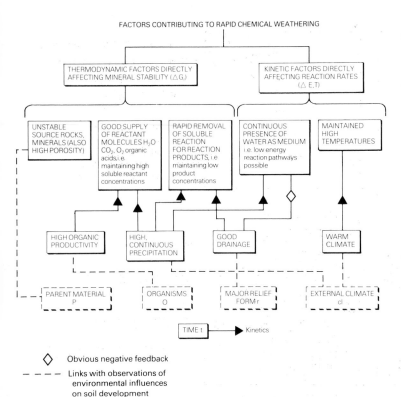

FACTORS CONTRIBUTING TO RAPID CHEMICAL WEATHERING

FIG. 12.3 Environmental factors influencing the nature, extent and rate of chemical weathering (after Curtis, 1976, in Derbyshire, 1976; © and reproduced by permission of J. Wiley & Sons Ltd.)

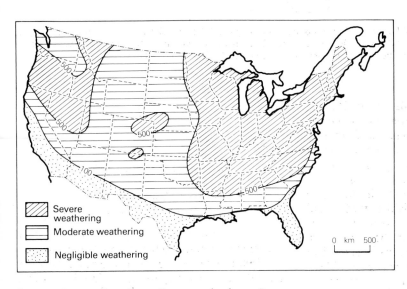

FIG. 12.4 Weathering regions in the United States based on the weathering index (after ASTM Standard C216, 1971*b*). For explanation, see text

$D > 1$, winter rain where $D < 1$, and where $D = 1$ rain is distributed equally throughout the year.

A numerical expression of the balance between the relevant climatic variables can be derived by multiplying R and D, to produce a precipitation ratio, N:

$$N = \frac{12E_\mathrm{j}}{P_\mathrm{a}} \qquad (12.4)$$

N-values are shown for South Africa on Fig. 12.5. $N = 5$ is the line that most closely follows the weathering boundary. Where $N > 5$, disintegration

TABLE 12.2 Generalized weathering characteristics in four climate zones

	Processes	Grading	Strength	Permeability	Compressibility and consolidation	Rates (mm year^{-1})
Glacial/ periglacial	Frost very important. Susceptibility to frost increases with increasing grain size. *Taiga*, fairly high soil leaching. Low rates organic matter decomposition. 1:1 clays can form. Iron mobility varies but generally low clay formation. *Tundra*—moist conditions, slow organic production and breakdown. May have slower chemical weathering. Algal, fungal, bacterial weathering may occur. Granular disintegration occurs. Hydrolytic action reduced on sandstone, quartzite, clay, calcareous shales, phyllites, dolerites. Hydration weathering common due to high moisture.	Hard rocks breakdown to well-graded and gap-graded material. Soft rocks change slowly. Chem. weathering increases plasticity and decreases grain size.	Peak shear strength reduced with weathering. Residual strength not obtained. Mode of failure changes from brittle to plastic.	Increases with production of coarse debris. Decreases with further disintegration.	Soil becomes compressible. Preconsolidation effects removed.	Narvik 0.001 Spitsbergen 0.02–0.2 Kärkevagge 0.32–0.5 Alaska 0.04
Temperate	Prec. evap. generally fluctuate. Both physical weathering and chemical weathering occur. Iron oxides leached and redeposited. Carbonate deposited in drier areas, leached in wetter areas. Increased precipitation, lower temperatures reduce evaporation. Organic content moderate-high, breakdown moderate. Silicate clays formed and altered. *Deciduous* forest areas—abundant bases, high nutrient status, biological activity moderate-high. *Coniferous* areas, acidic, low biological activity, leaching common.	Grading changes in steps particularly in late stages, e.g. threshold slopes. Plasticity increases in late stages by decomposition.	Marked loss of strength mainly occurs in stages but also in steps during course of weathering.	Generally as related to index properties of parent rock.	As expected from index properties. Preconsolidation effects removed.	Askrigg 0.5–1.6 Limestone 0.083 Austria 0.040 Austria 0.015
Tropical arid/semi-arid	Evaporation exceeds precipitation. Rainfall low. Temperature high. Seasonal. Organic content low. Physical weathering, salt weathering, granular disintegration, dominant in driest areas. Thermal effects possible. Low organic input relative to decomposition. Slight leaching produces 2:1 clays and $CaCO_3$ in soil. Sulphates and chlorides may accumulate in driest areas. Increased prec. and decreased evap. toward semi-arid areas and steppes yield thick organic layers, moderate leaching and $CaCO_3$ accumulation.	Coarse-grained but good sorting by transport agencies produces poorly graded deposits. Becomes finer-grained away from source. Evaporites cause aggregation. Salts, crusts, detrital sands, gravels, silts and calcareous fragments common.	Generally high, related to grading. Plasticity characteristics generally unknown.	Generally high for weathered soils near source. Decreases away from parent material. Local variations common.	As expected from index properties. Low near source. Loessic soils may be metastable.	Egypt 0.0001–0.0005 Australia 0.6–1
Tropical humid	High rainfall often seasonal. Long periods of high temperatures. Moisture availability high. Weathering products (a) removed or (b) accumulate to yield red and black clay soils, ferruginous and aluminous soils (lateritic), calcium-rich soils. Calcareous rocks generally heavily leached where silica content is high, soluble weathering products removed and parent silica in stable products are sandy. Where products remain, iron and aluminium oxides are common. Usually intense, deep weathering, iron and alumina oxides and hydroxides predominate. Clay minerals of 2:1, mixed and 1:1 lattice occur with increasing rain. Organic content high but decomposition high.	Generally well-graded in initial stages. Finer grained and poorly graded with time.	Published work indicates consistent, isotropic. Strength increases with increased grain size. Laterization and cementing increases strength.	Variable—often low, related to index properties rather than weathering process.	Effects of pre-consolidation removed. Some soils metastable after leaching. Coeff. of consolidation variable but high. Compressibility values average.	e.g. Florida 0.005

Source: Brunsden (1979).

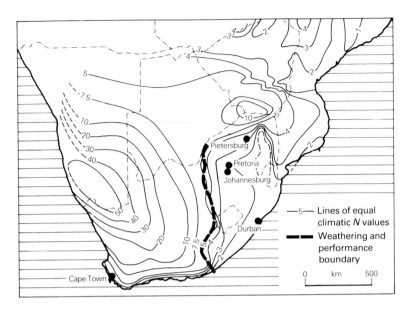

FIG. 12.5 *N*-values and the weathering and performance boundary in South Africa (after Weinert, 1965). For explanation, see text

Lines of equal climatic *N* values

Weathering and performance boundary

is more important than decomposition and the foundation materials are relatively sound. Where $N < 5$, the reverse is the case. The formation of hydrous mica is the most important product of chemical alteration where $N > 6$; where N is between 2 and 5, montmorillonite is the predominant clay mineral arising from the decomposition of basic igneous rocks, and kaolinite is produced in acid igneous rocks. It should be emphasized that the *N*-index was derived for, and is only of proven application in, southern Africa.

(c) *Rocks and their properties*

The susceptibility of rocks to weathering depends in part on their physical properties, and two groups of properties are particularly important: those properties that determine responses to weathering processes, and those properties that are significantly changed by weathering processes. Sometimes one property is important in both contexts.

The group of properties determining weathering responses, as it relates to the effectiveness of disintegration and decomposition processes, is summarized in Tables 12.3A–F. Rock type, and in particular mineral composition, is clearly fundamental. The tables give an indication of the relative proportions of major rock types and minerals at the surface of the earth, and indicate the relative susceptibility of major rock types and rock-forming minerals to various weathering processes.

The scales for rock decomposition reflect not only mineral susceptibility, but also crystallinity, particle size, and other properties.

The ability of water and solutions to penetrate a rock affects its resistance to most weathering processes and is determined primarily by its *porosity* (the proportion of voids in the rock) and its *permeability* (a measure of the ability of water to pass through it). Of particular importance to predicting building-stone durability are *porosity*, *microporosity* (normally defined as the proportion of pores < 0.005 mm), *water absorption* (the amount of water absorbed in a unit of time, usually 24 hours), and the *saturation coefficient* (a measure of the amount of water absorbed in a unit of time expressed as a fraction of total pore space). There are standard tests for these properties which include traditional methods of weighing, wetting, and drying (e.g. Schaffer, 1959), and more advanced methods of characterizing *pore structure* such as mercury porosimetry, and ultraviolet fluorescence microscopy (e.g. Ross, 1980, 1984). The relations between these properties are complex, but some empirical and predictively valuable relationships have been established. For example, limestones with high microporosities and saturation coefficients in general tend to be less durable than specimens of the same stones with lower saturation coefficients and microporosities (Honeyborne and Harris, 1958), and stones with saturation coefficients above 0.8 have been found to be more susceptible to frost action because expansion of water on freezing in pores cannot be adequately accommodated.

Several other properties are known to affect the durability of rock. For example, *coefficients of volumetric expansion* of both rocks and minerals play a significant role in determining the efficacy of insolation-weathering processes. The *thermal conductivity* and *diffusivity* is also relevant in this context and in others involving crystallization processes. And *tensile strength* is important in

determining the resistance of rock to the pressure of crystal growth in confined spaces within it. Resistance to decomposition processes, in addition to being related to porosity and pore structure, permeability, and mineral composition, involves many other variables such as the crystal structure and shape of minerals, grain texture, lines of weakness (cracks, bedding planes, etc.), and the nature of salts in the rock.

The second group of rock properties—those significantly affected by weathering—include

TABLE 12.3A Occurrence of rock types and minerals

Rock	% Porosity	Relative permeability	Percentage of land area	Mineral	Percentage of land area
Shale	18	5	52	Felspars	30
Sandstone	18	500	15	Quartz	28
Granite	1	1	15	Clay minerals	
Limestone	10	30	7	and micas	18
Basalt	1	1	3	Calcite and	
Others			8	dolomite	9
				Iron oxide	
				minerals	4
Sedimentary			75	Pyroxene and	
Igneous and				amphibole	1
metamorphic			25	Others	10

Source: Leopold et al. (1964).

TABLE 12.3B Chemical weathering sequences

1. Weathering sequence of very fine-grained minerals	2. Susceptibility to weathering of minerals in igneous rocks

most susceptible ↑

Primary minerals
 1. Gypsum
 2. Calcite
 3. Olivine-hornblende
 4. Biotite
 5. Albite

Secondary minerals
 6. Quartz
 7. Illite
 8. Hydrous mica intermediates
 9. Montmorillonite
10. Kaolinite
11. Gibbsite
12. Haematite
13. Anatase

 1. Olivine
 2. Ca–plagioclase felspar
 3. Ca–Na plagioclase
 4. Na–plagioclase
 5. Biotite
 6. Orthoclase felspar
 7. Muscovite
 8. Quartz

least susceptible

Source: Modified from Goldrich (1938) and Jackson et al. (1948).

TABLE 12.3C Resistance to weathering related to rock properties

Rock properties	Physical weathering (disintegration)		Chemical weathering (decomposition)	
	Resistant	Non-resistant	Resistant	Non-resistant
Mineral composition	High feldspar content Calcium plagioclase Low quartz content $CaCO_3$ Homogeneous composition	High quartz content Sodium plagioclase Heterogeneous composition	Uniform mineral composition High silica content (quartz, stable feldspars) Low metal ion content (Fe-Mg), low biotite High orthoclase, Na feldspars High aluminum ion content	Mixed/variable mineral composition High $CaCO_3$ content Low quartz content High calcic plagioclase High olivine Unstable primary igneous minerals
Texture	Fine-grained (general) Uniform texture Crystalline, tightly packed clastics Gneissic Fine-grained silicates	Coarse-grained (general) Variable textural features Schistose Coarse-grained silicates	Fine-grained dense rock Uniform texture Crystalline Clastics Gneissic	Coarse-grained igneous Variable textural features (porphyritic) Schistose
Porosity	Low porosity, free-draining Low internal surface area Large pore diameter permitting free draining after saturation	High porosity, poorly draining High internal surface area Small pore diameter hindering free-drainage after saturation	Large pore size, low permeability Free-draining Low internal surface area	Small pore size, high permeability Poorly draining High internal surface area
Bulk properties	Low absorption High strength with good elastic properties Fresh rock Hard	High absorption Low strength Partially weathered rock (grus, honeycombed) Soft	Low absorption High compressive and tensile strength Fresh rock Hard	High absorption Low strength Partially weathered rock (oxide rings, pitting) Soft
Structure	Minimal foliation Clastics Massive formations Thick-bedded sediments	Foliated Fractured, cracked Mixed soluble and insoluble mineral components Thin-bedded sediments	Strongly cemented, dense grain packing Siliceous cement Massive	Poorly cemented Calcareous cement Thin-bedded Fractured cracked Mixed soluble and insoluble mineral components
Representative rocks	Fine-grained granites Some limestones Diabases, gabbros, some coarse-grained granites, rhyolites Quartzite (metamorphic) Strongly cemented sandstone Slates Granitic gneiss	Coarse-grained granites Poorly cemented sandstone Many basalts Dolomites, marbles Soft sedimentary (poorly cemented) Schists	Igneous varieties (acidic) Metamorphic (other than marbles, etc.) varieties Crystalline rocks Rhyolite, granite, quartzite (metamorphic), gneisses Granitic gneiss	Calcareous sedimentary Poorly cemented standstone Limestone, basic igneous, clay-carbonates Slates Marble, dolomite Carbonates (other) Schists

Source: Lindsey *et al.*, 1982, table 4, from Chorley *et al.* (1984).

TABLE 12.3D Relative resistance of rocks to salt action

Goudie *et al.*, 1970[a]	Leary (1983)[b]
1. Cotswold Limestone	A. Fenacre stone, Devon
2. Arden Sandstone	B. Ketton stone, Lincs.
3. Red Sandstone	C. Bath stone, Combe Down
4. Chalk	D. Ancaster freestone
5. Black shale	E. Portland stone, Basebed, Easton
6. Diorite/Dolerite/Granite	F. Happylands stone, Broadway, Glos.

[a] Based on Na_2SO_4 test (1 most susceptible, 6 least susceptible).
[b] Selected from durability classes based on sulphate crystallization tests on British limestones (A most susceptible, F least susceptible).

TABLE 12.3E Relative rates of rock decomposition in Illinoian Hills (Birkeland, 1974)

1. Quartzite
2. Chert >
   Granite
3. Basalt >
   Sandstone
4. Siltstone >
   Dolomite
5. Limestone

TABLE 12.3F Relative resistance of rocks to frost action

Field studies		Experimental studies— 'Icelandic' conditions	
Ardennes	Dartmoor	Potts	Wiman (1963)
1. Phyllite—pure schist[a]	1. Metamorphosed sediments	1. Igneous rocks	1. Slate
2. Calcareous schist	2. Fine-grained granite	2. Sandstone	2. Gneiss
3. Phyllite—quartz schist	3. Diabase	3. Mudstone	3. Porphyritic granite
4. Limestone	4. Elvan	4. Shale	4. Mica-schist
5. 'Macigno'	5. Tourmalinized medium-grained granite		5. Quartzite
6. Grits/sandstones	6. Quartz-shorl		
7. Quartzite	7. Coarse-grained granite		
8. Conglomerate			

[a] 1 most susceptible, 8 least susceptible.

Sources: Ardennes, Dartmoor and Potts, in Potts (1970). The Ardennes data are based on the work of J. Alexandre, the Dartmoor data on a study by R. S. Waters.

many of those examined by the soil engineer that are relevant, for example, to the analysis of foundation conditions. They include tensile strength, Atterberg limits (liquid, plastic, and solid limits, and plasticity index, as described in Chapter 5), compressive and shear strengths (Chapter 5), colour, texture and particle size, permeability, infiltration capacity, and erodibility (Chapter 4), bulk density, consolidation, bearing capacity, and many others. There are standard methods for determining these properties (e.g. Attewell and Farmer, 1976; Whalley, 1976). In the context of weathering, the important point is that they can all change over time as a result of weathering, and the changes can in some circumstances lead to thresholds of failure being crossed.

(c) *Local variables*

In addition to the properties of climate and rock, several other variables may influence the effectiveness of weathering processes at particular locations. The weathering of rocks in the landscape is considerably affected by topographic position (including aspect and slope), drainage conditions, the nature of vegetation and animal life, and the age of the surface (e.g. Birkeland, 1974; Gerrard, 1981). Indeed, it is probably only at the local scale that it is possible to study precisely the unique relationships between the complex of variables that determine specific weathering phenomena; and it is only micro-scale features, such as stone tablets placed in particular locations, that are likely to respond quickly to short-term, contemporary conditions and processes (e.g. Trudgill, 1976).

Nowhere is the importance of local variables in weathering more clearly seen than in buildings, where aspect, exposure, microclimates associated with the building, and the availability of moisture with respect to features such as damp courses are of fundamental significance. Certain parts of buildings, for example, are more likely than others to become saturated with water, such as cornices, copings, string courses, plinths, sills, and steps, and weathering may be concentrated at these places. Where sculptured stonework is exposed, weathering is likely to increase as the ratio of surface area to stone volume increases, so that gargoyles and the like are especially vulnerable (Plate 12.1). Leary (1983) classified buildings into four exposure zones, ranging from the zones most vulnerable to weathering and requiring the use of

the most durable stone (1) to the least vulnerable areas of plain walling (4) (Fig. 12.6). Weathering in a building may be promoted because of faulty craftsmanship or errors in choice of materials (Schaffer, 1932). Craftsmanship faults include 'face bedding' (the placing of laminae in stone blocks vertically and parallel to the surface of the building), inadequate 'seasoning' of stones that require it, and damaging of stone by inappropriate quarrying or dressing methods. Errors of material choice include the use of easily weathered stone in places especially susceptible to weathering, the use of unsuitable materials (such as corrodable iron in stonework), and the juxtoposition of incompatible building stones.

PLATE 12.1 Weathering of St Andrew, carved in Portland stone, at St Paul's Cathedral, London

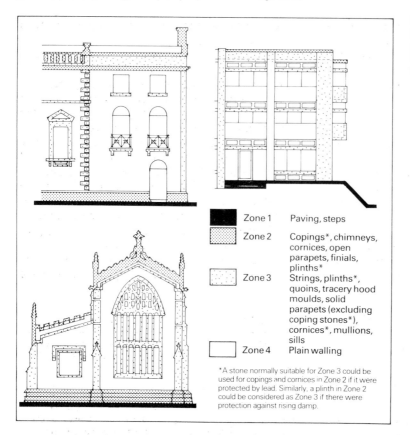

FIG. 12.6 Exposure zones of buildings: a guide to stone selection and susceptibility to weathering (after Leary, 1983; Crown copyright). For explanation, see text

	Zone 1	Paving, steps
	Zone 2	Copings*, chimneys, cornices, open parapets, finials, plinths*
	Zone 3	Strings, plinths*, quoins, tracery hood moulds, solid parapets (excluding coping stones*), cornices*, mullions, sills
	Zone 4	Plain walling

*A stone normally suitable for Zone 3 could be used for copings and cornices in Zone 2 if it were protected by lead. Similarly, a plinth in Zone 2 could be considered as Zone 3 if there were protection against rising damp.

12.3 Approaches to Studying Rock Weathering for Applied Purposes

(a) *Introduction*

A range of field and laboratory methods for studying rock weathering are suitable for recording *degrees and rates* of weathering, and *predicting rock durability*. The following brief review looks at each major approach and illustrates it with an applied example.

(b) *Recording degrees of weathering*

Degrees of weathering are commonly assessed by measuring properties of rock, stone, or weathering mantles. *Petrological approaches* often involve the microscopic analysis of thin sections and the evaluation of imagery from scanning-electron microscopy. For example, thin sections of Doulting stone emplaced at different times since the thirteenth century in Wells Cathedral, south-west England, reveal the near-surface accumulation of

gypsum and associated stone disintegration. Associated X-ray diffraction analysis and differential thermal analysis of materials may help to identify the minerals present in a weathered material, including the nature of clay minerals. *Chemical analysis* of weathering products, using such techniques as atomic absorption spectrophotometry and ion chromatography, can further illuminate the nature of weathering products in a sample. Such analyses normally express mineral content of weathered rock as a ratio of that in fresh rock. And data for many of the rock properties described in Sect. 12.2 may be used comparatively to reveal weathering changes. For example, microporosity may increase, and tensile strength decrease, over time.

In the field, the engineer is often less concerned with studying the processes and progress of weathering than he is with recording the properties of weathering mantles in space and depth. One aim is often to classify degrees of weathering using simple criteria. Soil scientists, geomorphologists (e.g. Ruxton, 1968; Ollier, 1984), and engineering geologists (e.g. Fookes and Horswill, 1969; Little, 1969; Fookes *et al.*, 1971) have all adopted qualitative scales for classifying weathering profiles and degrees of weathering. Tables 12.4A and 12.4B show two examples for engineering purposes, one a general scheme, and the other used by the Snowy Mountains Authority Australia (Ollier, 1984). A useful, slightly more quantitative method for determining weathering grades *in situ* involves the Schmidt hammer, an impact device originally designed for concrete testing which gives a rebound number (R) that is an index of rock 'hardness' or compressive strength (Day and Goudie, 1977). In a study of stone durability in Bahrain (Doornkamp *et al.*, 1980), for example, it was found that the mean rebound numbers of the

15 rock types ranged from 57.5 (the hardest) to only 12.8 (the softest).

Most such schemes have several common features. Firstly, they are based on criteria that are simple to use in the field. Secondly, they include a small number of arbitrarily defined categories (usually about 6). Thirdly, they describe the alteration of material down a profile to fresh bedrock, which usually forms the first class. The boundary between weathered and unweathered material in depth is called the *weathering front*. The form of the front is extremely varied, reflecting the relations between climatic conditions, groundwater and water-table, the nature of rock lithology and structure, and changes in weathering processes that have affected the area through time. In general, weathering is deepest where materials are most susceptible to weathering, where they are

TABLE 12.4A Engineering grade classification of weathered rock

Grade	Degree of decomposition	Field recognition (after Fookes & Horswill, 1969)		Engineering properties of rocks (after Little, 1969)
		Soils (i.e. soft rocks)	Rocks (i.e. hard rocks)	
VI	Soil	The original soil is completely changed to one of new structure and composition in harmony with existing ground-surface conditions.	The rock is discoloured and is completely changed to a soil in which the original fabric of the rock is completely destroyed. There is a large volume change.	Unsuitable for important foundations. Unstable on slopes when vegetation cover is destroyed, and may erode easily unless a hard cap present. Requires selection before use as fill.
V	Completely weathered	The soil is discoloured and altered with no trace of original structures	The rock is discoloured and is changed to a soil, but the original fabric is mainly preserved. The properties of the soil depend in part on the nature of the parent rock.	Can be excavated by hand or ripping without use of explosive. Unsuitable for foundations of concrete dams or large structures. May be suitable for foundations of earth dams and for fill. Unstable in high cuttings at steep angles. New joint patterns may have formed. Requires erosion protection.
IV	High weathered[a]	The soil is mainly altered with occasional small lithorelicts of original soil. Little or no trace of original structures.	The rock is discoloured; discontinuities may be open and have discoloured surfaces and the original fabric of the rock near the discontinuities is altered; alteration penetrates deeply inwards, but corestones are still present.	Similar to grade V. Unlikely to be suitable for foundations of concrete dams. Erratic presence of boulders makes it an unreliable foundation for large structures.

TABLE 124A—*continued*

III	Moderately weathered[a]	The soil is composed of large discoloured lithorelicts of original soil separated by altered material. Alteration penetrates inwards from the surfaces of discontinuities.	The rock is discoloured; discontinuities may be open and surfaces will have greater discolouration with the alteration penetrating inwards; the intact rock is noticeably weaker, as determined in the field, than the fresh rock.	Excavated with difficulty without use of explosives. Mostly crushes under bulldozer tracks. Suitable for foundations of small concrete structures and rockfill dams. May be suitable for semi-pervious fill. Stability in cuttings depends on structural features, especially joint attitudes.
II	Slightly weathered	The material is composed of angular blocks of fresh soil, which may or may not be discoloured. Some altered material starting to penetrate inwards from discontinuities separating blocks.	The rock may be slightly discoloured; particularly adjacent to discontinuities which may be open and have slightly discoloured surfaces; the intact rock is not noticeably weaker than the fresh rock.	Requires explosives for excavation. Suitable for concrete dam foundations. Highly permeable through open joints. Often more permeable than the zones above or below. Questionable as concrete aggregate.
I	Fresh rock	The parent soil shows no discoloration, loss of strength, or other effects due to weathering.	The parent rock shows no discoloration, loss of strength, or any other effects due to weathering.	Staining indicates water percolation along joints; individual pieces may be loosened by blasting or stress relief and support may be required in tunnels and shafts.

[a] The ratio of original soil or rock to altered material should be estimated where possible.

Source: Fookes *et al.* (1971).

porous and permeable, where surface erosion is limited, and where climatic conditions favour chemical decomposition. Finally, the criteria used in these schemes vary greatly from those based on qualitative impressions to those requiring precise measurement. Thus some schemes refer, for example, to 'degree of coloration', whereas others record colour in terms of hue, value, and chroma, as shown on Munsell Color Charts.

(c) *Recording rates of weathering*

One approach to monitoring weathering rates is to record rates of *debris production* from rock surfaces over time. Rockfalls from cliffs, for example, can pose a threat to beaches, and lines of communication at their bases. As a result, there may be adequate archival information to construct a picture of the frequency and causes of the falls. Thus, Hutchinson (1971) showed that 40 chalk-falls along the Kent coast, south-east England,

between 1810 and 1970 occurred only in the October–April period and corresponded with monthly average effective precipitation (total precipitation minus potential evapo-transpiration) and the frequency of air frost, the former probably being more influential earlier in the season, and frost being apparently more significant later (Fig. 12.7). The processes causing the falls are uncertain, but they probably relate to the development of transient water pressures in pores, cracks, etc., the action of ice, and the increase of weight of blocks due to increased saturation.

A second approach has been widely used by geomorphologists to estimate rates of limestone weathering in Karst regions (e.g. Corbel, 1959) and has recently been adopted in studies of weathering on buildings. It involves comparing the *chemical composition* of rainfall incident upon a drainage area, and the runoff derived from it in order to determine the nature and magnitude of

TABLE 12.4B Weathering classes used by the Snowy Mountains Authority, Australia

1. Fresh rock (Fr). Rock which exhibits no evidence of chemical weathering. Joint faces may be clean or coated with clay, calcite, chlorite or other minerals.
2. Fresh with limonite stained joints (Fr St). Joint faces coated or stained with limonite but the blocks between joints are unweathered.
3. Slightly weathered rock (SW). Rock which exhibits some evidence of chemical weathering, such as discoloration, but which has suffered little reduction in strength. Except for some inherently soft rocks, slightly weathered rock rings when struck with a hammer.
4. Moderately weathered rock (MW). Rock which exhibits considerable evidence of chemical weathering, such as discoloration and loss of strength but which has sufficient remaining strength to prevent dry pieces about the size of 50 mm diameter drill core (of inherently hard rock) being broken by hand across the rock fabric. Moderately weathered rock does not ring when struck with a hammer.
5. Highly weathered rock (HW). Rock which is weakened by chemical weathering to the extent that dry pieces about the size of 50 mm diameter drill core can be broken by hand across the rock fabric. Highly weathered rock does not readily disintegrate when immersed in water.
6. Completely weathered rock (CW). Rock which retains most of the original rock texture (fabric) but the bond between its mineral constituents is weakened by chemical weathering to the extent that the rock will disintegrate when immersed and gently shaken in water. In engineering usage this is a soil.

Source: Ollier (1984).

the chemical changes taking place. For example, Livingston *et al.* (1981) analysed the rainfall on, and the runoff from, the variable marble statue of Phoenicia on Bowling Green Custom House, Manhattan, New York, during eight rainfall events. They revealed an eightfold increase in calcium concentration in runoff, which represents the amount of stone lost from the statue during rain events; there was also a significant increase of sulphate in runoff, but only about half the amount required to balance the increase of calcium sulphate.

Direct monitoring of weathering changes to rock surfaces provides another approach. One method involves the micro-erosion meter (MEM) in which the height of the rock surface above an arbitrarily established datum on three studs securely fixed in the rock is measured using a sensitive micrometer gauge. Measurements can be repeated and changes in the rock surface monitored at the same location as often as required (High and Hanna, 1970). For example, measurements at six sites of Portland limestone on St Paul's Cathedral, London, over 5 years revealed an average rate of lowering of 0.062 mm/y.

On some buildings and other monuments there may be useful indices of stone-weathering rates. For examples, Sharp *et al.* (1982) showed that the surface of stone blocks on the balustrade at St Paul's Cathedral had been lowered by an average of 0.078 mm/y between 1771 and 1981 on the basis of the fact that lead plugs in the blocks, the surfaces of which were originally flush with the rock surface, now stand proud on the surface by 8–30 mm.

Tombstones provide another potentially useful weathering index because they are usually of known age, in fixed positions, and made of only a few, widely used rock types. Equally, they pose problems—sites may not be comparable, variability of exposure can influence rates, and stones are often moved. But several studies have successfully recorded change by measuring the legibility (or depth of incisions) of the inscriptions on the tombstone, or by comparing the decline in tombstone thickness. For example, Feddema and Meierding (1987) mapped damage, such as face recession and exfoliation, to marble tombstones in the Philadelphia region, USA, and demonstrated a close relationship between extent of damage and airborne pollutant concentrations. In Israel, Klein (1984) showed that limestone tombstones were weathering at a rate of 0.005 mm/y, and that an exponential curve best described the relationship between tombstone age and degree of weathering over time. Similar studies by Kupper and Pissart (1974) of limestone tombstones in the Liège region of Belgium gave a mean rate of 0.0025 mm/y, and a range from 0.0012 mm/y in a rural cemetery to 0.00365 mm/y for a cemetery in an industrial district.

An alternative to using dated features is to place prepared samples of known characteristics in weathering situations and monitor changes to them over time. The 'tray tests' of the Building Research Establishment (Honeyborne and Harris, 1958) represent one approach; another is to place

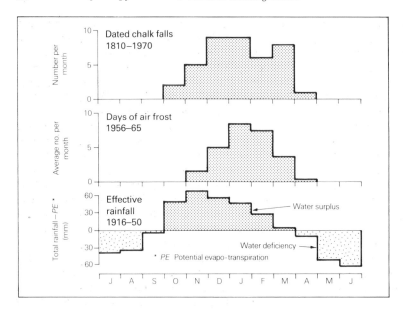

FIG. 12.7 The incidence of chalk falls, frost, and effective rainfall on the Kent coast (after Hutchinson, 1971)

tablets of rock in soils or aggressive weathering environments (e.g. Trudgill, 1976); and a third is to mount tablets on freely rotating carousels that overcome problems of fixed orientation and submergence in standing water (e.g. Plate 12.2; Jaynes and Cooke, 1987). Such tablets or blocks are usually monitored in terms of weight loss over time, but changes in surface micro-morphology or chemistry may also be recorded.

(d) *Predicting stone durability*

Broadly, stone durability is predicted by relating known performance of stone to climatic and stone properties, or by accelerated weathering tests in the laboratory. One field approach is to monitor climatic conditions and rock characteristics to determine, on the basis of known or assumed relationships in the weathering system, if weather-

PLATE 12.2 Exposed and sheltered carousels of rock tablets designed to monitor contemporary rates of stone weathering

ing is taking place. Roth's (1965) study of a boulder in the Mojave Desert, California, illustrates this approach. More commonly, the spatial variability of weathering is predicted on the basis of climatic data (see Figs. 12.4 and 12.5). A study by Honeyborne (personal communication, 1972) for the Building Research Establishment in the UK demonstrated the use of rock properties to predict the durability of Portland stone, a limestone widely used in British buildings (Fig. 12.8). The four classes (A = most durable; D = least durable) are defined by values of the saturation coefficient (S) and the microporosity (M), and each class is attributed to suitable weathering environments. Such predictions say nothing about the weathering processes involved, being based largely on observations of stone performance in buildings, but it seems likely that crystallization processes involving salt and frost are of major importance.

In the laboratory, accelerated weathering tests are widely used to predict durability. In these tests rock samples are subjected to cyclic weathering processes (e.g. freezing–thawing, wetting–drying) under controlled conditions at much higher frequencies than normally occur naturally. Such tests have the advantages of saving time and controlling environmental conditions, but results are open to question because of the substantial differences between laboratory and field conditions, and because it is impossible to ensure that by increasing cyclic frequency the effects of weathering are merely produced at an increased rate. The most widely used tests are 'standard tests'. Examples include the ASTM Standard C666 (1971*a*) *Standard method of test for resistance of concrete to rapid freezing and thawing* and the French test for resistance to freezing, NFB 10-513; a multitude of tests based on the use of salts, known as 'crystallization' tests, such as ASTM C88, *Standard method of test for soundness of aggregates by use of sodium sulphate or magnesium sulphate* and many others (see also Sect. 12.1; Ross, personal communication, 1985). Such tests clearly have the virtue of consistently providing comparable data; but their drawbacks include the fact that they relate to a 'standard' or even a wholly unreal climate and therefore ignore both climatic variability and the probability at any one location of durability being related to several processes working together. Thus, there have been attempts more accurately to simulate in the laboratory the environmental conditions in which weathering occurs using 'climatic cabinets'. For example, geomorphologists have attempted to predict the relative effectiveness of frost action in 'Icelandic' and 'Siberian' winters (e.g. Tricart, 1956; Wiman, 1963; Potts, 1970), and

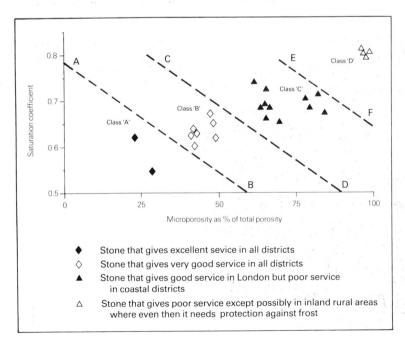

FIG. 12.8 Classification of Portland stone on the basis of microporosity and saturation coefficient data (after the Building Research Establishment, UK; Crown copyright)

◆ Stone that gives excellent sevice in all districts
◇ Stone that gives very good service in all districts
▲ Stone that gives good service in London but poor service in coastal districts
△ Stone that gives poor service except possibly in inland rural areas where even then it needs protection against frost

of salt crystallization and hydration in desert climates (e.g. Sperling and Cooke, 1985).

12.4 The Weathering of Building Stones

(a) *Contemporary concern*

Although stone weathering and conservation have been matters of concern since at least the beginning of the nineteenth century, they have recently become a major political and scientific issue in the context of the widespread debate on 'acid rain' (e.g. House of Commons, 1984). Contemporary concern arises mainly because there is a widespread feeling that many ancient monuments are threatened, the processes are accelerating especially in urban areas, and the options for conservation are becoming increasingly expensive. Examples of ancient monuments that are often cited as suffering severely from weathering include the Parthenon in Athens (e.g. Skoulikidis and Papakonstantinou-Ziotis, 1981), the Taj Mahal in Agra (e.g. Gauri and Holdren, 1981), buildings of marble in Venice (e.g. Fassina, 1978), numerous European cathedrals such as Cologne (e.g. Luckat, 1981), and Cleopatra's Needle in New York (Winkler, 1973).

(b) *The effects of weathering*

Most of the stone buildings that are suffering seriously from weathering are built of limestone, marble, or sandstones with calcareous cement. The effects of weathering are many and varied. At St Paul's Cathedral, London, for example, there are several weathering zones, each of which is related to distinctive weathering features. The occurrence of solution pits in drip zones, and gullies and lapiés in flow zones is reminiscent of many limestone terrains; in addition, in sheltered areas and in the lower part of some flow zones, there are thick, black cumuliform accumulations of sooty calcium sulphate. Elsewhere, there may be evidence of spalling and splitting, of granular disintegration, of case-hardening, and of surface decay by lichens and other micro-organisms.

One major research objective is to determine the nature and relative importance of weathering processes; a second is to determine their spatial variability; and a third, arising from the others, is to apply the knowledge of processes to methods of conservation. These three themes are considered below.

(c) *Weathering processes and building stones*

There is evidence that a range of processes may be responsible for the weathering of building stones. *Wet deposition*, either by '*rainout*' (in which pollution is involved in the development of condensation, and the stone is wetted by dew, cloud, or fog) or by '*washout*' (in which the stone is wetted by polluted precipitation), leads to solution of calcium carbonate, and features of solution weathering such as pitting and granular disintegration. Under 'normal' conditions, CO_2, transformed into carbonic acid, is the principal agent of attack. But if the pH of precipitation falls below 5.6, other gases, notably SO_2 and NO_x are involved. The following equations summarize the main reactions (Jaynes, 1985):

(a)
$$CO_{2(g)} \xrightleftharpoons{H_2O} CO_{2(aq)} \qquad (12.5)$$

$$CO_{2(aq)} + H_2O_{(l)} \rightleftharpoons H_2CO_{3(aq)} \qquad (12.6)$$

$$H_2CO_{3(aq)} \rightleftharpoons H^+ HCO_{3(aq)}^- \qquad (12.7)$$

$$CaCO_{3(s)} \xrightleftharpoons{H_2O} CaCO_{3(aq)} \qquad (12.8)$$

$$CaCO_{3(aq)} \rightleftharpoons Ca^{2+} CO_{3(aq)}^{2-} \qquad (12.9)$$

$$Ca^{2+} CO_{3(aq)}^{2-}) + H^+ HCO_3^- \rightleftharpoons$$
$$Ca^{2+} 2(HCO_3^-)_{(aq)} + H_2O_{(l)} + CO_{2(aq)} \qquad (12.10)$$

($CaCO_3$) passes into solution as hydrocarbonate which, when dried, revers to $CaCO_3$ and is seen, for example, as a thin white coating on limestone surfaces.)

(b)
$$SO_{2(g)} \xrightleftharpoons{H_2O} SO_{2(aq)} \qquad (12.11)$$

$$SO_{2(aq)} + H_2O_{(l)} + \tfrac{1}{2}O_{2(aq)} \rightarrow 2H^+ + SO_{4(aq)}^{2-}. \qquad (12.12)$$

$CaCO_3$ dissolves in water (12.8 above) and then

$$Ca^{2+} CO_{3(aq)}^{2-} + 2H^+ + SO_{4(aq)}^{2-} \rightarrow$$
$$Ca^{2} + SO_{4(aq)}^{2-} + CO_{2(g)} + H_2O_{(l)}. \qquad (12.13)$$

(The solution may pass out of the system or, on drying, become $CaSO_4$ or precipitate out as gypsum ($CaSO_4$, $2H_2O$). Similar reactions apply to SO_3, leading to the formation of sulphuric acid (H_2SO_4) rather than sulphurous acid (HSO_3).)

(c) There are several forms of nitrous oxides, such as NO and NO_2. NO_2 reacts as follows:

$$NO_{2(g)} \underset{H_2O}{\rightleftharpoons} NO_{2(aq)} \quad (12.14)$$

$$2NO_{2(aq)} + H_2O_{(1)} \rightarrow H^+NO^-_{2(aq)} + H^+NO^-_{3(aq)}. \quad (12.15)$$

Nitric and nitrous acids react with $CaCO_3$ dissolved in water, then:

$$2(H^+NO^-_3)_{(aq)} + Ca^{2+}CO^{2-}_{3(aq)} \rightarrow$$
$$Ca^{2+}2(NO_3)^2_{(aq)} + H_2O_{(1)} + CO_{2(g)} \quad (12.16).$$

$Ca(NO_3)_2$ is highly soluble and if produced by weathering is usually quickly removed, although traces have been found in stone.

Dry deposition of pollutants on stone surfaces can lead to the fundamental process of *sulphation*. This takes the form of a direct reaction between SO_2 gas deposited on the surface and the stone itself, which occurs in the presence of a catalyst (e.g. Fe_2O_3) and when the surface is wetted by precipitation or high relative humidity (e.g. Skoulikidis and Papakonstantinou-Ziotis, 1981):

$$SO_{2(g)} \underset{air, catalyst}{\rightleftharpoons} SO_{3(g)} \quad (12.17)$$

$$SO_{3(g)} + CaCO_{3(s)} + H_2O_{(g)} \rightarrow$$
$$CaSO_4.2H_2O_{(s)} + CO_{2(g)}. \quad (12.18)$$

The product, as before, is gypsum. There is no comparable process involving NO_x.

The chemical processes are often accompanied by mechanical processes. One of these is *frost weathering*, which tends to fracture stone into coarse angular fragments. The mechanism may partly be the result of the volume change that takes place when water freezes: water freezing at 0 °C in a closed system increases in volume by some 9 per cent, and a pressure of as much as 2100 kg/cm² is exerted at −22 °C. But there is also a substantial force exerted by the *growth* of ice crystals. It seems that when a saturated porous material freezes, the crystals begin to form in the larger pores and continue to grow by withdrawing water from smaller pores; mechanical disruption may accompany such crystal growth (e.g. Everett, 1961). This is one context in which microporosity is important, because the process depends on the juxtaposition of large and small pores.

Probably of greater importance than frost, however, is the activity of salts. Salts that may be responsible for stone disintegration fall into three categories (Schaffer, 1932): salts present in the material before its incorporation in a building (e.g. emplaced during sedimentary deposition, from sea spray in coastal quarries, or during the manufacture of bricks and terracotta tiles); salts derived from external sources (e.g. from jointing materials such as mortar, and backing materials, such as bricks and plaster, from the atmosphere and from groundwater); and, most importantly, salts, like gypsum, derived from prior decomposition by solution or sulphation.

Salts on a stone surface can create an unsightly *efflorescence*; crystallization of salts within pores and cracks, *cryptoflorescence*, may cause *crumbling*, *flaking*, *scaling*, and *blistering*; case-hardened skins may be formed as a result of acid attack, and these may reduce surface evaporation and lead to the deposition of soluble salts beneath them causing subsurface disintegration, and may themselves *exfoliate* and blister. Three major processes are associated with cryptoflorescence (e.g. Cooke and Warren, 1973): thermal expansion of crystallized salts (e.g. Johannessen and Feiereisen, 1982), the growth of salt crystals from solution in pores, a process similar to that described for ice above (Evans, 1970); and the hydration of hydratable salts such as sodium sulphate (e.g. Sperling and Cooke, 1985). The latter two forces are summarized as follows:

(a) *Crystal growth* (Correns, 1949):

$$P = \frac{RT}{V} \cdot \log n\left(\frac{C}{C_s}\right) \quad (12.19)$$

where

P = pressure on the crystal,
R = gas constant,
T = absolute temperature,
V = molar volume of crystalline salt,
Log n = natural logarithm,
C = concentration at saturation point without pressure effect,
C_s = concentration of a solution under external pressure P.

Correns (1949) concluded that salt crystals can continue to grow against a confining pressure when a film of solution is maintained at the salt–rock interface. The maintenance of this solution depends on the interfacial tensions at the salt–rock, salt–solution, and solution–rock interfaces: when the sum of the last two is smaller than the first, the solution can penetrate between the salt and the surrounding rock. Experiments suggest that this is the most important salt-weathering process, and it is most effective in hot, dry evaporative environments such as deserts and south-facing surfaces in buildings (e.g. Cooke, 1981).

(b) *Hydration* (Windler and Wilhelm, 1970):

$$P = \frac{(nRT)}{(V_h - V_a)} \, 2.3 \log\left(\frac{P_w}{P'_w}\right) \qquad (12.20)$$

where

P = hydration pressure in atmospheres,
n = number of moles of water gained during hydration to the next higher hydrate,
R = gas constant,
T = absolute temperature (degrees Kelvin),
V_h = volume of hydrate

$$\frac{\text{(molecular wt. hydrated salt)}}{\text{density (g/cm}^3)},$$

V_a = volume of original salt

$$\frac{\text{(molecular wt. original salt)}}{\text{density (g/cm}^3)},$$

P_w = vapour pressure of water in the atmosphere (mm mercury at given temperature),
P_w = vapour pressure of hydrated salt (mm mercury at given temperature).

The circumstances in which the hydration pressure can exceed the tensile strength of stone are difficult to define, but experimental studies suggest the process is most effective when the hydration is accomplished in 12 hours (so that it can be completed in a diurnal temperature cycle), and the hydrating salt cannot escape from the pores; temperatures and relative humidities are crucial, and specific to particular salts (Winkler and Wilhelm, 1970); and the process is normally less effective than crystal growth (Sperling and Cooke, 1985).

Salt-weathering processes are also moderated by the nature of the salts present and the properties of the stones (e.g. Goudie *et al.*, 1970). For

example, Na_2SO_4 and its hydrates, closely followed by $MgSO_4$, are the most effective. $NaCl$ and $CaSO_4$ are less aggressive but are particularly common in coastal and polluted areas. In general, the susceptibility of rocks to salt attack is directly related to their microporosity and water-absorption capacity.

(d) *Acceleration of weathering in polluted atmospheres*

In Chapter 4 a distinction was drawn between 'normal' and 'accelerated' rates of erosion. A similar distinction can be made between 'normal' and 'accelerated' rates of weathering. 'Normal' rates of weathering occur where the atmosphere has not been significantly polluted by people. In such circumstances, most limestone weathering is likely to be the result of carbonic acid attack with the pH of rainfall over 5.6. Weathering is principally accelerated by modifying the composition of the boundary layer of the atmosphere with SO_2 and NO_x gases, and thereby activating some of the processes described above, principally solution, sulphation, and salt weathering. Carbon dioxide, always known to be relevant in rural weathering, itself tends to be found in higher concentrations in urban areas and in certain circumstances it too may accelerate urban weathering (e.g. Winkler, 1966, 1970). The sources of the gaseous pollutants are well known. In the UK, for example, it is estimated that power stations emit 65 per cent of the SO_2 and 46 per cent of the NO_2 whereas vehicles emit 29 per cent of the NO_2. Since 1900, emissions have more than doubled, but they have declined recently mainly as a result of economic recession, changes in fuel use, and legislation (Table 12.5; Watt Committee on Energy, 1984). In terms of atmospheric concentrations, and take-up of pollutants in stone, there are similar trends over time, suggesting that consequent weathering may also be accelerating. Equally, important, the concentration of these pollutants varies spatially, with concentrations being characteristically highest in urban areas (e.g. Fig. 12.9) and declining rapidly away from them. This has led to the suggestion that there are corresponding weathering gradients from rural to urban areas.

Certainly there is evidence of such weathering gradients. For example, a study in south-east England by Honeyborne and Price (1977), which involved measuring the weight loss from blocks of Portland stone exposed for 10 years at a rural site

TABLE 12.5A European and UK emissions since 1900 (million tonnes per year)

	1900	1910	1920	1930	1940	1950	1960	1970
SO_2 Europe	16	22	22	25	25	25	32	52
SO_2 UK	2.8	3.2	3.2	3.2	3.6	4.6	5.6	6.0
NO_x UK	0.68	0.69	0.72	0.72	0.82	0.99	1.35	1.64

Note: NO_x expressed as equivalent NO_2.

TABLE 12.5B UK emissions of SO_2 and NO_x since 1971 (million tonnes per year)

	1971	1972	1973	1974	1975	1976	1977	1978	1979	1980	1981	1982
SO_2 Power stations	2.80	2.87	3.02	2.78	2.82	2.69	2.74	2.81	3.10	2.87	2.71	2.60
All sources	5.83	5.64	5.80	5.35	5.13	4.98	4.98	5.02	5.34	4.67	4.23	4.00
NO_2 Power stations	—	0.73	0.81	0.72	0.76	0.79	0.79	0.81	0.88	0.85	0.82	0.77
Vehicles[a]	—	0.42	0.45	0.44	0.47	0.45	0.46	0.48	0.49	0.49	0.48	0.49
All sources	—	1.73	1.85	1.72	1.70	1.74	1.77	1.80	1.89	1.79	1.71	1.67
NO All sources[a,b]	—	7.86	8.30	8.07	7.80	8.06	8.27	8.62	8.78	8.85	8.62	8.83
Total hydrocarbons methane equivalent weight	—	2.40	2.60	2.73	2.77	2.93	3.03	3.16	3.35	3.37	3.36	3.29

[a] These are the best estimates currently in a consistent format, but recent WSL work suggests that vehicles emit somewhat more NO, ($\leq 40\%$ of total) and significantly less CO.

[b] Principally vehicles.

NO_x represented as equivalent NO_2.

Source: Watt Committee on Energy (1984).

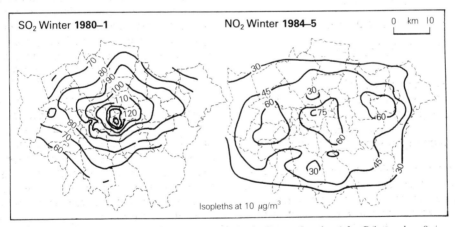

FIG. 12.9 Regional patterns of pollution concentration in Greater London (after Brice *et al.*, 1984/5; Culley *et al.*, 1984)

(Garston, near Watford) and at an urban site (Whitehall, central London). As Fig. 12.10 shows, the weight loss in central London was over double that in rural Garston. SO_2 levels at the same time, however, were over five times higher in Whitehall than in Garston, so that while the rate of weathering was higher in the more polluted atmosphere, the weathering gradient was not as steep as the pollution gradient, indicating that other factors need to be taken into account, such as frequency and duration of wetting. More recent work by Jaynes and Cooke (1987) confirms London's urban–rural weathering gradients and reveals that they are quite complex, reflecting in part the complexity of atmospheric pollution in a large metropolis.

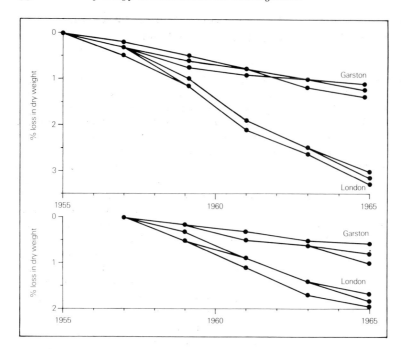

FIG. 12.10 Weight loss (%) from two sets of Portland stone samples exposed at Garston, near Watford, UK and Whitehall, London (after the Building Research Establishment, UK; Crown copyright)

(e) *Management of stone weathering*

Weathering damage to buildings has become so widespread and so unsightly that many methods have been devised for preventing weathering and removing its products. Unfortunately, such methods have often been introduced without an understanding of the responsible weathering processes. Recently there has been a concerted effort to match the treatment to the cause (e.g. Ashurst and Dimes, 1979; Gauri, 1978; Rossi-Manaresi and Vannucci, 1978; Winkler, 1973). Broadly, conservation measures fall into five main groups: *design, cleaning, surface treatment* and *consolidation, maintenance,* and *climatic protection.* The choice of methods should depend on the causes of damage, the types of stone and material to be removed, the condition of the surface, and the extent of the area to be treated. All too often, the choice is also determined by cost.

Design is particularly relevant in new buildings where access of water and degree of exposure can be controlled, and the materials can be selected according to the micro-environment in which they will be placed (e.g. Fig. 12.6). For example, sometimes the effects of frost can be prevented by omitting from the design of a building features liable to be frequently saturated, or by using materials in such features that have low-saturation coefficients and microporosities. The provision of a damp course, not as universal as might be imagined, is another example of a design measure which prevents penetration of groundwater and related saline solutions above its level (see Sect. 12.5).

Cleaning is perhaps the most widespread response to stone decay. Its purpose is normally to remove soluble salts, unsoluble and often unsightly encrustations, soot and other soiling particles, micro-organisms, parasitic vegetation, and bird and animal droppings—in short, to remove many of the causes of weathering and to improve appearance. Major approaches (e.g. Ashurst and Dimes, 1979) include: *washing* (e.g. using water from jets, as runoff, as steam, or in high-pressure lances); *abrasive blasting* (e.g. on harder stones, using compressed air or water jets containing abrasives); *mechanical cleaning* (e.g. with brushes, power-drills); and *chemical cleaning* (e.g. based on acid or alkaline solutions, absorbent clay packs or solvent-jelly coatings). There are advantages and disadvantages associated with all of these methods. For instance, washing can leave a brownish stain and cause undesirable water penetration into a building; abrasive cleaning may remove protective weathering rinds, exposing a fresh surface to decay, and it may damage severely weathered stone; and chemical cleaners may etch stone, and must be wholly removed if salts are involved.

The third approach is to apply surface treatments to buildings either to consolidate or to seal the surface and restrict the access of water and solutions. *Consolidation* processes are designed to improve the cohesion of well-weathered surfaces, and include both organic (e.g. synthetic resins) and inorganic products for small-scale work, and adhesives and fillers for larger gaps and cracks (e.g. Rossi-Manaresi and Vannucci, 1978). These methods are similar to *protective* methods designed mainly to waterproof the stone, and indeed one treatment may serve both purposes. The range of protective treatments is great, ranging from the use of polychromatic beeswax coatings on the fifteenth-century sculptures in Bologna (Rossi-Manaresi, 1972); to the limewash/ hot-lime poultice/sacrificial coating of lime, dust, and casein now used on several English cathedrals; to the new, deep-penetrating silane-based, irreversible Brethane preservative treatment developed by the Building Research Establishment in the UK (Price, 1981). Problems with many preservatives include the fact that they are not fully effective waterproofing agents and therefore permit continued subsurface weathering or, if they are effective, their depth of penetration is insufficient to prevent subsurface moisture fluctuations and related salt weathering beneath the treatment zone. Also, new stresses may be created by juxtaposing materials of very contrasted properties. Although Brethane may overcome some of these problems, studies by Clarke and Ashurst (1971) using eight other silicone-based preparations in an experiment on 24 buildings in the UK showed that none of them had any overall beneficial effect in retarding stone decay.

Three other responses to stone weathering deserve emphasis. *Maintenance*, including periodic inspection, is essential if preventive measures are to be taken as soon as they are needed and their effectiveness is to be adequately assessed. The *replacement* of damaged stone is often both essential and desirable, especially in buildings of historic importance. And *'climatic protection'* (e.g. Rossi-Manaresi and Vannucci, 1978) can involve the protection of especially valuable artefacts from the atmosphere; it can also include, of course, the improvement of air quality by controls on air pollution.

12.5 Weathering and Foundation Engineering Problems

(a) *Introduction*

Engineers have extensive interests in two aspects of rock weathering. Like those concerned with buildings, they require information on the durability of rocks in order to allow selection of appropriate materials for dams, breakwaters, and other engineering structures, and for concrete aggregates. These interests are reflected in many studies, such as Sayward's (1984) analysis of the effects of salt on concrete, and evaluation of the durability of rock in breakwaters (Poole *et al.*, 1983). In general, these interests are similar to those of the architect and builder described in Sect. 12.4.

Engineers, especially foundation engineers, are also concerned about the nature of the surface and subsurface materials and weathering conditions present at a site or in an area where development is planned. The first aspect of this concern, the recording and evaluation of the characteristics of weathering mantles, was discussed in Sect. 12.3(b) and illustrated in Table 12.4. The broad relations between rock weathering, major climatic zones, and engineering foundation features are summarized in Table 12.6.

The second aspect, examined here, relates to the appraisal of weathering conditions within ground scheduled for development. Nowhere is this problem more important than in arid lands.

(b) *Salt weathering and foundation engineering in arid lands*

In many arid lands, the groundwater and the capillary fringe above it are in places close to the ground surface (Fig. 12.11), and the groundwater may be made quite saline, especially by chloride and sulphate salts (e.g. Goudie and Cooke, 1984). This is particularly true in *sabkhas* (broad coastal flats arising from intertidal sedimentation) and *playas* (the floors of enclosed drainage basins; e.g. Cooke, 1981). Occasionally, the groundwater table and the capillary fringe (which can be over 3 m thick) may intersect its surface and give rise to distinctive salt efflorescence and patterned ground phenomena.

If the engineer builds structures that intercept either the capillary fringe or the ground water, saline waters may penetrate the structures and cause substantial salt-weathering problems (e.g. Cooke *et al.*, 1982). There is therefore an important need to define the limits of the salt-hazard

TABLE 12.6　A summary table on rock weathering, climate and related foundation engineering features

Climatic zone—predominant weathering process	Periglacial zone—disintegration	True temperate zone—decomposition and disintegration	Arid zone—disintegration	Humid tropical zone—decomposition
Classification characteristics	(a) Susceptibility to frost weathering and shattering generally increased with increased grain size of parent material (b) Hard rocks break down to produce well-graded material, often gap-graded (c) Soft rocks show little change during initial stages of weathering but chemical breakdown increases plasticity and decrease grain size	(c) Grading appears unaffected to any large degree except in late stages (b) Decomposition processes may alter plasticity especially in late stages	(a) Good sorting by transporting agents produces poorly-graded material; becomes more fine-grained with distance from parent material (b) Evaporites can cause aggregation	(a) Tests often dependant on pretreatment techniques; large scatter of results makes comparison difficult (b) Generally well-graded in initial stages of weathering, becoming finer-grained and poorly-graded
Strength	(a) Reduction in strength with weathering, but residual strength not obtained (b) Mode of failure changes from brittle to plastic	(a) Marked loss in strength does not occur until late stages of weathering	(a) Generally high and directly related to grading	(a) Published work indicates consistent and isotropic (b) Increases with increased grain size of parent material (c) Laterization increases strength owing to cementing
Permeability	(a) Increases in early stage but generally reduced with advanced disintegration	(a) Generally as expected from index properties	(a) Generally high for weathered soils near parent material (b) Away from parent material, lower permeability (c) Local variation frequent	(a) Variable, but generally as expected from index properties; often low
Compressibility and consolidation	(a) Soil becomes more compressible with increased weathering (b) Chemical weathering removes effects of preconsolidation	(a) Generally as expected from index properties (b) Chemical weathering removes effects of preconsolidation	(a) Generally as expected from index properties (b) Low for soils near parent material (c) Metastable structure of loess can cause dramatic consolidation (d) Sand dunes of variable density result in changing characteristics	(a) Chemical weathering removes all effects of preconsolidation (b) Some soils metastable from leaching processes (c) Coefficients of consolidation vary over wide range but are often extremely high (d) Compressibility values are average (e) Cementing (e.g. laterites) may reduce compressibility

Soil stabilization	(a) Normal procedures of stabilization satisfactory, but presence of organic matter can reduce the rate of gain of strength	(a) Small thicknesses of weathered material generally make removal more economical than stabilization	(a) Addition of coarse-grained material to produce a well-graded soil (b) the addition of lime or cement (c) presence of organic matter can reduce the rate of gain of strength	(a) Satisfactory results for most purposes obtained by few per cent of lime or cement or oil with additives (b) Firing to high temperatures; suitable for black clays (c) Presence of organic matter can reduce the rate of gain of strength
Engineering uses	(a) Scree slopes of disintegrated material, often suitable for road stones	(a) Limited, but provides suitable topsoil for plant growth, for stabilization, or for improving visual appearance	(a) Provides satisfactory borrow material for roads and runways; destroy metastable structure of aeolian deposits by wetting prior to rolling (b) Surface crusts of calcareous and saline soils provide good bearing layers (c) Vertical cuts in loess stand almost indefinitely (d) Generally good foundation materials; deep water level has little effect on most engineering constructions (e) Scree slopes of disintegrated material make fair to good road stones	(a) Generally makes satisfactory base courses for roads and runways; often requires stabilizing for wearing surfaces (b) Fired black clay can be used as low-grade aggregate (c) Satisfactory material for slopes; use near-vertical cuts, as flatter slopes erode badly (d) Some soils compact to very low density but make acceptable fill
Engineering problems encountered	(a) Weathering of soft rocks associated with decrease of strength and increase of both primary and secondary consolidation (b) Continual freeze–thaw cycles can result in graded bedding (c) Glacially induced rock fractures parallel to topography (d) Successive deformation results when foundations become water-logged during thaw, together with the formation of solifluction lobes	(a) Weathering of soft rocks associated with decrease in strength and increase of both primary and secondary consolidation	(a) Unless destroyed, the metastable structure of aeolian deposits can cause catastrophic collapse (b) Infilling of excavations by blown sand (c) Often the salts resulting from evaporation of mineral bearing waters necessitates the use of sulphate resistant cement	(a) When used as a wearing surface can corrugate owing to loss of fines (b) Compaction difficult to attain if high Proctor densities specified (c) Seasonal climatic variations can cause swelling and shrinking to great depths especially the montmorillonitic black soils (d) Weathered products are often minerals of unusual engineering behaviour (e.g., halloysite or montmorillonite) (e) Strength may decrease with depth particularly when laterization has occurred (f) Softening on contact with water necessitates satisfactory drainage

Source: Saunders and Fookes (1970).

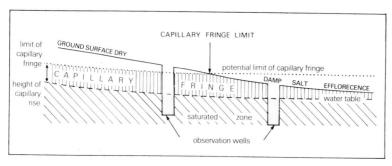

FIG. 12.11 Groundwater terminology: the capillary fringe and related phenomena

zone and the variability of hazard intensity within it. Three main approaches have been followed in attempting to meet these needs (Cooke, 1981). Firstly, to survey and classify the severity of salt damage to existing structures with a view to correlating damage with hazard intensity. This approach, unfortunately, encounters several difficulties of which the most important are that it requires a precise knowledge of the age of buildings and types of material involved, which is rarely available, and it demands a consistency of observation that is usually difficult. Secondly, it may be possible to monitor over time the degradation of blocks of material of known properties placed in the hazardous areas. This is a useful approach, but it often requires years of observations to produce results.

A third approach is to map the spatial variability of those variables in the salt-weathering system that are assumed to determine the intensity of the

hazard. In this approach, it can be postulated for example that the potential saline aggressiveness of an area is determined chiefly by the salinity and chemical composition of the groundwaters; by the position and stability of the water-table, the capillary fringe, and the soil climate; and by the relations between the capillary fringe and the ground surface. This approach has now been used widely in the Middle East (e.g. Doornkamp *et al.*, 1980; Cooke *et al.*, 1982). In a study of this problem in Bahrain, the hazard zone was initially defined in terms of the capillary fringe limit (Fig. 12.11); this was redefined in terms of the 10-m contour, a planning boundary which included between it and the coast all areas where foundations might penetrate the capillary fringe (Fig. 12.12). Within this zone it was assumed that the intensity of the hazard is directly proportional to the *shallowness* and the *electrical conductivity* (i.e. salinity) of groundwater. Mapping of these

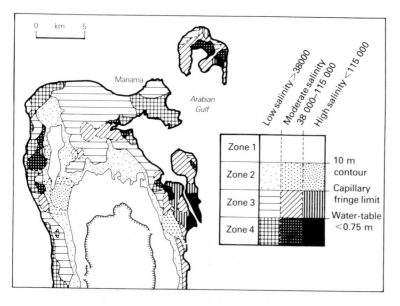

FIG. 12.12 Aggressive ground conditions in northern Bahrain and predicted hazard intensity (after Doornkamp *et al.*, 1980. Salinity in umhos/cm)

TABLE 12.7 Outline of potential physical and chemical problems encountered in road construction

	Bituminous Wearing Course/Basecourse		Unbound base e.g. wet or dry bound macadam, wet mix or all-in granular material	Unbound sub-base e.g. gravel crushed stone
	Thick/dense	Thin/porous		
Potential migration of water	*Very low*	*Moderate*	*High* (varies with aggregate etc.)	*High* (varies with aggregate etc.)
Potential migration of salt	*Very low*	*Moderate to high*	*Moderate* (varies with aggregate etc.)	*High* (varies with aggregate etc.)
Physical changes in presence of groundwater—permanent, intermittent or capillary	*Unlikely* in short or medium term	*Probable*	*Probable*	*Probable*
Distress	Possible long term aggregate disintegration, surface erosion and stripping	Short term aggregate disintegration, surface erosion scabbing, blisters, potholes	Short to medium term disintegration, stripping, settlement	Medium term disintegration settlement
Physical change in presence of transient water (rain, dew, etc.) depends on aggregate and salt content	*Unlikely* unless salt in aggregates high	*Unlikely* unless high salt in aggregates	*Unlikely* unless salt in aggregates	*Unlikely* unless salt in aggregate
Distress	Very slow aggregate disintegration	Slow aggregate disintegration	May be slight disintegration, settlement	May be slight disintegration settlement
Chemical changes in presence of groundwater—movement intermittent, capillary. Depending on salt levels and type	*Unlikely*	*Possible* bitumen and aggregate may decompose and disintegrate if salt high	*Possible* if salt content high	*Possible* if salt content high
Distress	*Unlikely*	Potholes, scabbing, stripping	Volume changes and loss of strength	Volume changes and loss of strength
Chemical changes in presence of transient water (rain, dew etc.)—depends on aggregate and salts present	*Unlikely*	*Unlikely* bitumen may decompose if salt high	*Possible* with some aggregate and some salts	*Possible* with some aggregate and some salts
Distress	*Unlikely*	*Unlikely*	Small volume change or loss of strength	Small volume change or loss of strength

Source: Fookes and French (1977).

variables led to the creation of a potential hazard intensity map (Fig. 12.12) that is a basic planning document. The durability of the various local stones in this hazardous environment was predicted by subjecting them to an accelerated salt-weathering test (Doornkamp *et al.*, 1980).

Within the hazard zone, each type of engineering activity faces distinctive problems. For example, *road construction* may encounter a range of problems (Table 12.7) most of which relate to the position of the road with respect to soil moisture zones (e.g. Fig. 12.13; Fookes and French, 1977). Such problems have been recognized in many arid lands, including Australia (Cole and Lewis, 1960), India (Mehra *et al.*, 1955),

and South Africa (e.g. Netterberg, 1979) as well as in the Middle East. The main weathering processes include chemical attack by soluble sulphates of cement in road bases, physical and chemical weathering of aggregates used in base or sub-base, and the concentration of salts by evaporation at the base of, or in the wearing course. The consequences of such attack include cracking, pot-holing, blistering, heaving, decomposition of bitumen and aggregates, and surface crumbling. Various strategies have been adopted to avoid these problems. These include relocation to avoid saline ground; careful selection of aggregates and limitation of the salt content in base and lower layers of the road; road design that recognizes the

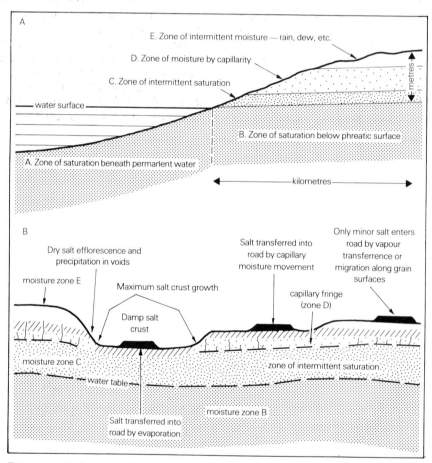

F IG . 12.13 (A) Schematic diagram to show the interrelationships in soil moisture zones. (B) Schematic section to show the relevance of soil moisture zones to road construction (after Fookes and French, 1977)

problems and includes, where appropriate, such features as impermeable membranes; and a high standard of maintenance.

12.6 Conclusion

Weathering studies in geomorphology are concerned mainly with the relationship between weathering processes and climate, the effect of weathering on rock and the soils which this produces, and the nature of landforms developed or modified by weathering processes. Such concerns for weathering processes, the environmental systems within which they operate, and the effects of the processes clearly overlap with the interests of soil scientists, engineers, and building-research workers.

Extending the study of natural weathering processes into a study of their effects on building stone is a natural step by the geomorphologist towards practical application. There are direct parallels, as has been shown, between sites of weathering on a building and weathering sites within the natural environment. Also, the use of laboratory techniques for simulating climatic conditions, originally developed for testing the weathering characteristics of rocks drawn from the natural environment, can be used equally well for testing the durability of building stone.

For the engineer, the geomorphological context of the products of weathering—the engineering soil—provides a basis for understanding how and why particular soil conditions exist. In fact, it is often possible to take the principles of weathering described above and predict the general nature of the soils that the engineer may encounter. This is of value particularly in those areas of the world where precise field data are absent. However, the geomorphologist's awareness of other influences (such as slope form) on likely soil conditions also plays an important part in such predictions. Other complications may also arise. For example, in many parts of the world soil conditions are inherited from weathering conditions under past climates. If the nature of such climates is known (e.g. from palynological evidence) then the nature of the weathering processes at that time can be revealed. More usually reliance has to be placed on someone recognizing a fossil soil. This requires extensive field understanding of soil development under different climates, and a basic knowledge of the related weathering processes.

13 Neotectonics, Earthquakes, and Volcanicity

13.1 Introduction

Previous discussions have focused mainly on problems in environmental management posed by specific geomorphological processes. There has been passing reference to the relevance of geological materials (both soils and bedrocks), as in Chapter 5 on landslides and Chapter 12 on weathering, but mention should be made of other geomorphological problems that arise out of the geological character of an area. In the main such problems occur either because of geological structure (especially in the case of earthquakes), or because of bedrock composition (especially lithology). The study of landforms affected by recent earth movements, *neotectonics*, is sometimes called *morphotectonics*. Problems relating to specific lithologies are considered in Chapter 14.

The importance of seismic activity to environmental management is increasing. Every year there are over one million earthquakes. While many of them do no human damage, because they occur in unoccupied areas, others may devastate whole cities. An average of 10 000 people every year are probably killed by earthquakes. (Costa and Baker, 1981). Although not the worst in history, the 1976 earthquake at Hopei in China alone took 650 000 lives. Even in countries such as Britain, not usually thought of as subject to earthquakes, there have been more than 100 tremors this century. This causes particular concern to those involved, for example, in locating nuclear power stations or in disposing of nuclear waste, for the sites of both ideally should be free of earthquake risk. Finding such sites may be difficult.

13.2 Morphotectonics and Earthquake Prediction

Within the existing level of our knowledge, geomorphological studies can help to identify *where an earthquake is likely to occur, but not precisely when*. Deciding 'where' is a matter first of identifying the nature of crustal movements up to the present time, and second, from this, of assessing where geological stresses continue to exist that will need to find release in the form of an earthquake. It may be possible to say that this is likely to be sooner rather than later. Yet, if it is to be of practical value the prediction of 'when' has to be on the time-scale of hours, or at the most days, and not years. People may respond to a warning that an earthquake is imminent, and take evasive action; they do not know what to do, and tend to do nothing, if the warning suggests that an earthquake will occur, for example, 'sometime over the next 5 years'. Greater precision over time predictions is more likely to come from seismologists than from geomorphologists although, as will be shown, the latter can make a singular contribution to the study of where ground stresses are at a maximum, and hence where an earthquake may occur. The subject of morphotectonics is still in its infancy: and has not yet received the attention it deserves. Up to the 1970s, a traditional approach was used in the study of morphotectonics in which recent earth movements were used to explain large landforms, especially in areas of active tectonics such as New Zealand (Cotton, 1948; and more recently Soons and Selby, 1982), parts of Africa (King 1951), and elsewhere (King, 1962). Many examples are given in Ollier (1981) and Coates (1981).

Much of this earlier work relied almost exclusively on assumptions and observations concerning the relationship between surface form and tectonic events. An obvious example is that of rift valleys, where the inward-facing scarps have always been seen as fault-line scarps directly resulting from a rifting process. This same philosophy of using land *form* as an indicator of neotecto-

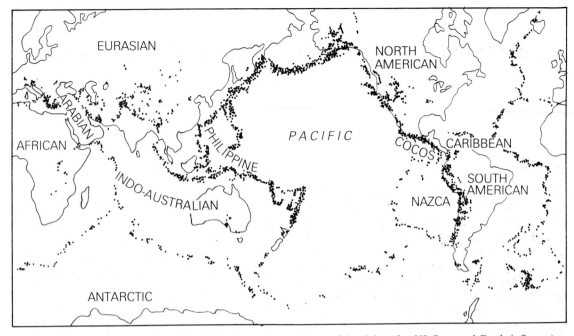

FIG. 13.1 Earthquake locations in relation to the main structural plates of the globe (after US Coast and Geodetic Survey)

nic activity is present in more recent work, but the emphasis has swung towards deciphering smaller, more detailed, and sometimes quite complex landforms and relationships. It is to this latter approach that most attention is given here.

Drainage is also a sensitive indicator of neotectonic events. Streams and rivers can either be displaced by such an event or have their gradients changed. In either case the response may be quite rapid. Complications exist, however, because causes other than tectonics can produce similar changes, especially as a response that could be mistaken for a tectonically induced change in gradient.

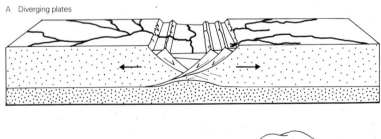

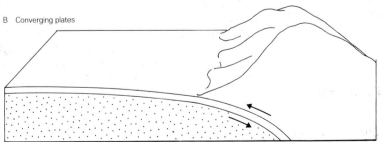

FIG. 13.2 Motion of diverging and converging structural plates

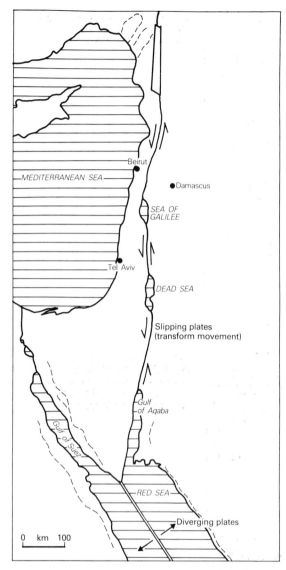

FIG. 13.3 Lateral motion of plates to produce the Dead Sea rift system, a site of many earthquakes

Recent reviews of the geomorphological contribution to the study of neotectonics include those by Keller and Rockwell (1984), Morisawa and Hack (1985), and Doornkamp (1986), while Vita-Finzi (1986) provided an introduction to neotectonics in *Recent Earth Movements*. Accounts of the geophysical processes involved, and of methods used in seismological investigations may be found in Badgley (1965), Lomnitz (1974), Wyss (1975), Bolt (1978), Eiby (1980), Kasahara (1981), Open University (1981), and Hsu (1982). The study *Earthquake Prediction* by Mogi (1985), of the Earthquake Research Institute at the University of Tokyo, has now been translated into English, and includes many examples of work done in Japan, including an account of earthquake monitoring.

Reviews of earthquake damage and risk are often produced as a reaction to a large earthquake. Some have been produced by national and international agencies. In 1978, UNESCO published *The Assessment and Mitigation of Earthquake Risk*; in the United States reports have been produced by the Office of Emergency Preparedness (1972), the Panel on the Public Policy Implications of Earthquake Prediction (1975), and the Panel on Earthquake Prediction (1976). A well-documented review of hazards from earthquakes is given by Bolt *et al.* (1975).

13.3 Causes of Earthquakes

It is no coincidence that earthquakes are concentrated around the margins of the plates that make up the earth's crust (Fig. 13.1). It is at these margins that there is most movement, either by one plate converging on another (Fig. 13.2A), by two plates moving apart (Fig. 13.2B) or by two plates sliding past each other, as they do through the Dead Sea (Fig. 13.3). Movement occurs as a result of stresses built up within the Earth's crust. When the resistances to movement are exceeded by these stresses, further movement occurs, often accompanied by earthquakes. Such earthquakes generally occur on faults and may be centred deep below the surface. The shock waves are transmitted outwards from this epicentre creating vibration and ground movement at the surface. In the case of a large earthquake the motion may be detected several hundred kilometres away from the fault or epicentre, and even at this distance great damage to buildings can be done.

13.4 Detecting Structural Stress

There are two main approaches to the detection of structural stress. The first is generally known as looking for the *seismic gap*. The second involves the recognition of *incomplete tectonic activity*. Sometimes the gaps and incomplete tectonic activity coincide.

A seismic gap is an area which, though it occurs

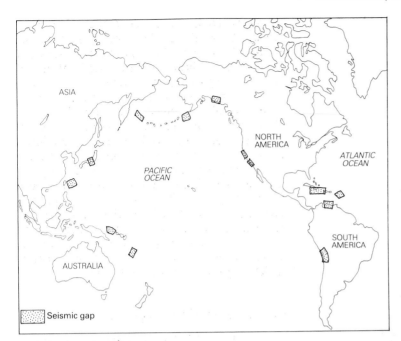

FIG. 13.4 Some major seismic gaps (after US Geological Survey, and Costa and Baker, 1981)

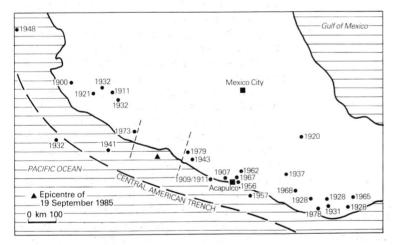

FIG. 13.5 Epicentres of larger twentieth-century Mexican earthquakes showing how the 1985 earthquake falls into the Michoacan seismic gap (after Degg, 1986)

in a belt of seismic activity, has not itself experienced an earthquake. Some major seismic gaps of the world are shown in Fig. 13.4. At a more detailed scale, Fig. 13.5 shows the location of twentieth-century earthquakes in Mexico, with the centre of the 1985 earthquake neatly fitting into the Michoacan seismic gap that existed prior to the earthquake.

Clearly seismic gaps can only be recognized if there is an adequate record of previous earthquake history. The record needs to include not only the date of the earthquake but also the position of its epicentre. Most of the world's earthquake epicentres occur within a short distance of a plate margin (Fig. 13.1) and along a specific fault. The most famous of these is the San Andreas fault in California (Fig. 13.6), along which many tremors have been recorded this century. Many of these have filled previously existing seismic gaps. Of course, the absence of a record does not of itself mean that there has been no earthquake, and so the concept of a truly 'seismic' gap is here open to considerable doubt.

In other studies an allied, although apparently

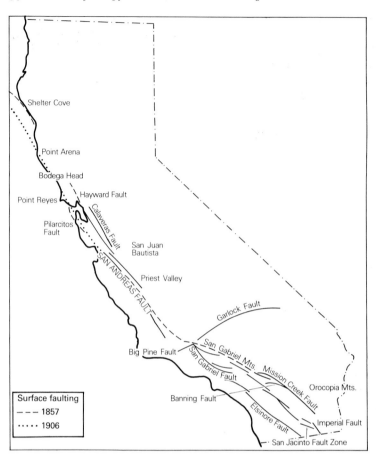

FIG. 13.6 The main active faults of California (after Tank, 1973)

Shelter Cove

Point Arena

Bodega Head

Point Reyes

Hayward Fault

Pilarcitos Fault

Calaveras Fault

SAN ANDREAS FAULT

San Juan Bautista

Priest Valley

Garlock Fault

Big Pine Fault

San Gabriel Mts.

San Gabriel Fault

Mission Creek Fault

Orocopia Mts.

Banning Fault

Elsinore Fault

Imperial Fault

San Jacinto Fault Zone

Surface faulting
- - - 1857
· · · · · 1906

opposite, concept is that of a *periodicity* in earthquake events. Evidence to support the idea that earthquakes occur in a region at a definable recurrence interval (i.e. with a periodicity of known length) is sometimes based on actual seismic records. But it may also be based on other historical records or on the morphotectonic evidence of landforms directly affected by earthquakes. An analysis of historical records of earthquakes in the Middle East by Degg (1987a) showed that by comparison with the previous 1900 years (Fig. 13.7) there has been little earthquake activity in Syria this century (Fig. 13.8). This is also true for the whole of the Dead Sea fault-system, with only moderate shocks recorded in Jordan, Israel, and Lebanon. Drawn as a graph (Fig. 13.8), the earthquake record for Syria shows a cycle of alternating active and quiet periods. The nineteenth century was active and the twentieth century has been quiet. Renewed activity from the twenty-first century onwards seems likely.

An example of the use of morphotectonic evidence to indicate periodicity in earthquake activity is provided by Adams (1980a). He found that the landforms along the central part of the Alpine Fault in New Zealand revealed fault scarps that affected a sequence of dated river terraces. This dating was based on ^{14}C methods applied to samples of wood found within the terraces. The magnitude of the earthquakes associated with each of the fault scarps was calculated from the geometry of the faults which showed that displacement of up to 90 m took place once every 500 years or so. Interestingly, further analysis of the altitude of an alluvial surface 8–12 000 years old showed that it had been warped and not faulted: folding can take place in a tectonically active zone as well as faulting.

To be of any real value the concept of earthquakes taking place within seismic gaps, and that of earthquakes occurring periodically needs a physical explanation. Present-day explanations

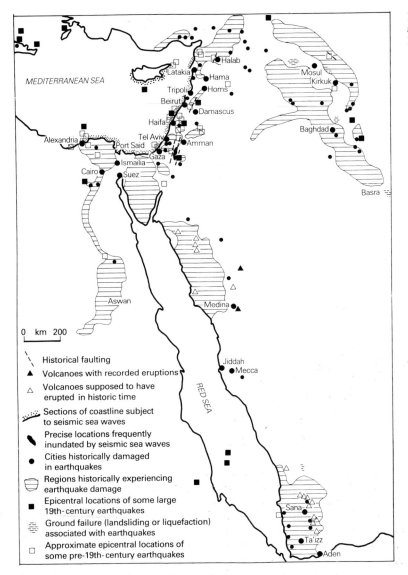

MEDITERRANEAN SEA

Halab
Latakia
Hama
Tripoli
Homs
Beirut
Damascus
Haifa
Tel Aviv
Amman
Alexandria
Port Said
Gaza
Ismailia
Cairo
Suez

Mosul
Kirkuk

Baghdad

Basra

Aswan
Medina

0 km 200

Jiddah
Mecca

RED SEA

Sana

Ta'izz
Aden

\ Historical faulting

▲ Volcanoes with recorded eruptions

△ Volcanoes supposed to have
   erupted in historic time

⋯ Sections of coastline subject
   to seismic sea waves

◣ Precise locations frequently
   inundated by seismic sea waves

● Cities historically damaged
   in earthquakes

◡ Regions historically experiencing
   earthquake damage

■ Epicentral locations of some large
   19th- century earthquakes

▤ Ground failure (landsliding or liquefaction)
   associated with earthquakes

□ Approximate epicentral locations of
   some pre-19th- century earthquakes

FIG. 13.7 Earthquakes and associated hazards in the Middle East, AD 1–1899 (after Degg, 1987a)

are usually based on plate tectonics, plate motions, the activity at plate margins, and structural stresses set up within the plates themselves. On a world-wide scale (Fig. 13.1) the pattern of earthquakes coincides very closely with that of active volcanoes, and together these reveal the pattern of the larger plates of which the earth's crust is composed. Movements at these plate boundaries produce earthquakes of varying magnitude. These magnitudes when associated with damaging effects are measured on the Modified Mercalli Scale (Table 13.1). The largest recorded earthquakes were one of $M = 8.6$ in Alaska in 1964, two of

$M = 8.5$ off the north-west coast of South America in 1960, and one off China in 1920. Such offshore and indeed similar coastal earthquakes can generate huge waves (tsunamis) which can cause extensive coastal flooding and damage.

Despite the general and close coincidence between earthquake and plate margins, there are many records of earthquakes occurring within the area of (so-called) stable plates. Indeed they include some of the most destructive, including the New Madrid, Missouri, earthquake of 1911. While the geophysical reasons for their existence are not entirely clear, many of the earthquakes may be in

TABLE 13.1 Modified Mercalli Scale of earthquake magnitude

Magnitude	Effects
I	None to barely perceptible
II	Suspended objects may swing
III	Vibration similar to that produced by passing truck
IV	Building fabric and fittings disturbed. Parked cars rocked
V	Most sleepers awakened, some windows broken, unstable objects overturned
VI	Frightening, heavy furniture moved. Slight structural damage
VII	Everyone runs outdoors, chimneys broken, further damage slight-moderate in well-built ordinary structures
VIII	Considerable structural damage, changes in well-water levels
IX	Partial collapse even in substantial buildings, ground cracked, underground pipes broken
X	Most buildings destroyed, ground badly cracked, rivers splash over their banks
XI	All buildings destroyed, broad fissures in ground putting all underground pipes out of service
XII	Damage total. Waves seen on ground surface

Source: US Geological Survey (simplified).

localities undergoing isostatic adjustments in response either to large-scale removal of material by erosion or as a consequence of stress release in the rocks since the melting of glacial ice.

One of the most important contributions of geomorphology, therefore, is to identify those morphotectonic features of the landscape that indicate past earthquake activity. Out of this can also come an awareness of locations which are in a state of structural stress and may experience earthquakes in the future. Above all, palaeoseismology (as the study of earthquake history is called) stands to benefit from any improvement in the seismic record and hence the analysis of periodicities and seismic gaps.

13.5 Morphotectonic Features

There now exists a considerable list of geomorphological indicators of earthquake (tectonic) activity (Table 13.2), some of which are illustrated in Fig. 13.9. An equally wide variety of methods can be used to analyse the evidence (Table 13.3). One country in which the indicators and methods are used extensively is China, where reliance is placed especially on river terraces (Fig. 13.10) and alluvial fans (Fig. 13.11) as morphological and sedimentological clues to neotectonic activity which may be accompanied by earthquakes (Doornkamp and Han, 1985).

In fact, through the use of such indicators and

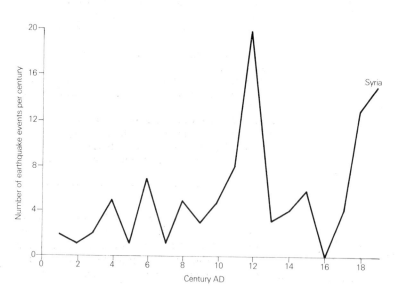

FIG. 13.8 Earthquakes in Syria up to the nineteenth century (after Degg, 1987a)

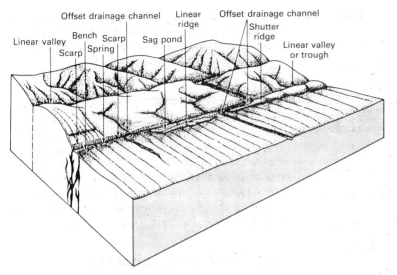

Offset drainage channel · Linear ridge · Offset drainage channel · Shutter ridge · Linear valley · Bench · Scarp · Sag pond · Linear valley or trough · Scarp · Spring

FIG. 13.9 Some morphotectonic landforms (after Wesson *et al.*, 1975)

TABLE 13.2 Geomorphological indicators of neotectonic activity

Direct
Emerged coral reefs (Taylor *et al.*, 1980; Neef and Veeh, 1977)
Displacement of dated beaches (Temple, 1964)
Deformed shorelines (Ota, 1975)
Offset in coastline configuration (Moore and Kennedy, 1975)
Distortion of river terraces (Popp, 1971)
Segmentation of alluvial fans (Hooke, 1972)
Deformation of alluvial fans (Bull, 1964a, b, 1977)
Displacement of dated terraces (Wellman, 1972)
Changes in lake depth (Clark and Persoage, 1970)
Offset glacial features (Richter and Matson, 1971)
Warping of planation surfaces (Doornkamp, 1972)
Displacement of synthetic structures (Rogers and Nason, 1971)
Fractured cave structures (Lange, 1970)
Fault scarps (Cotton, 1948)
Spur and facets (Thornbury, 1954)
Shutter ridges (Cotton, 1948)
Separation of river terraces (Lensen, 1968)
River reversal (Wayland, 1929)
Displacement of man-made structures (Rogers and Nason, 1971)

Indirect
Responses of stream channels (Cooke and Mortimer, 1971; Adams, 1975)
Downstream changes in river sinuosity (Adams, 1980b)
River capture (Biancotti, 1979)
Rates of sedimentation (Lofgren and Rubin, 1975)
Fluvioglacial gravel disposition (Sharma *et al.*, 1980)
Formation of lakes (Cotton, 1948)

TABLE 13.3 Geomorphological methods used in the study of neotectonic activity

Geodetic data and precise levelling (Thurm *et al.*, 1971; Naumov, 1975; Karcz and Kafri, 1975; Reilinger *et al.*, 1980; Vanicek and Nagy, 1980; Fourniquet *et al.*, 1981)

Analysis of tide gauge records (Balling, 1980)
Reconstructed planation surfaces (Sigov and Romashova, 1969)
Analysis of break points in planation surfaces (Capdeville *et al.*, 1978)
Variations in slope angles (Wallace, 1977)
Analysis of the morphology of scarp slopes (Palmer and Henyey, 1971)
Analysis of terrain types (Bull and McFadden, 1977)
Analysis of river patterns (Popp, 1971)
Reconstructed river networks (Sigov and Romashova, 1969)
Analysis of the longitudinal profile of rivers (Zhukovskiy, 1980)
Analysis of tilted lake levels (Wilson and Wood, 1980)
Measurement of water levels along aqueduct (Leary *et al.*, 1981)
Lichenometry (Nikonov and Shebalina, 1979)
Biostratigraphy (Baranova and Biske, 1971)
Vigour of plant growth (Babcock, 1971)
Soil morphology and mineral alteration (Douglas, 1980)

by comparison with past earthquake events, experience in China (Han, 1984) has shown that earthquakes tend to occur in the following specific locations: at the ends of faults (especially hinge faults); where faults make a sharp change in

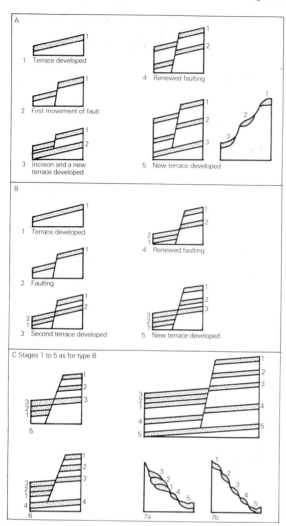

FIG. 13.10 Variations in river terrace displacement as a result of faulting (after Doornkamp and Han, 1985)

direction; at the intersection of active faults; and at the pivotal point of a compression-shearing pivotal fault. On the basis of these criteria and through associated geomorphological evidence, severe structural stress was identified in the Oujiang Valley. Later two earthquakes took place, one in 1965 (force $M = 5.2$) and one, at Tonghai, in 1970 (force $M = 7.7$). Detailed geomorphological studies have also provided evidence of recent (and possibly current) movement along faults which is direct evidence of a response to internal (geological) stress conditions. For example, along the San Andreas fault zone in the southern Indio Hills, Coachella Valley, California, Keller (*et al.*, 1982)

recorded beheaded streams, deflected and offset streams, sags, shutter ridges, pressure ridges, and fault scarps (Plate 13.1). Along one portion of the fault an offset of 9.7 km was recorded on an alluvial-fan/pediment complex. The estimated minimum slip rate along the San Andreas fault here is about 30 mm/y.

13.6 Earthquake Hazard Zoning

Research in California and in Mexico has shown that the response of the ground to an earthquake is closely related to the rocks and soils at or near the ground surface. Borcherdt (1975) (following Lawson, 1908) showed this to be the case for San Francisco for he found that during the 1906 earthquake ground-shaking was most intense in the areas underlain by bay muds and alluvium, and least in areas of bedrock. The same is true of Mexico City where earthquakes, such as that in 1985, cause devastation in that part of the city built on old lake-bed sediments and hardly affected properties built on the rocks of the ground above the old lake margin (Fig. 13.12).

Such observations lead naturally to the idea that areas can be classified, or zoned, according to their potential exposure to damage from an earthquake. There have been many attempts to do this. For example, Fig. 13.13 shows a seismic risk map for the United States which was issued by the US Coast and Geodetic Survey in 1969. It is based on the historical record of known earthquakes. It has found practical use, for both insurance rates and building codes are based upon it. However, as Costa and Baker (1981) showed, not only is this one of several such maps to have been published, but it is also restricted in its value because it does not consider the probable frequency of damaging earthquakes. This map indicates 'where'; it does not give an indication of 'when' or even 'how often'.

The fact that insurance and reinsurance companies find such zonation maps useful is demonstrated by the fact that they produce their own hazard maps. For example, the Swiss Reinsurance Company (1978) has produced an earthquake hazard atlas for the whole world, and the Reinsurance Offices Association (ROA), based in London, has produced maps for individual countries in a series of separate publications circulated amongst its members. The ROA has also funded research into earthquake risk analysis for areas in the

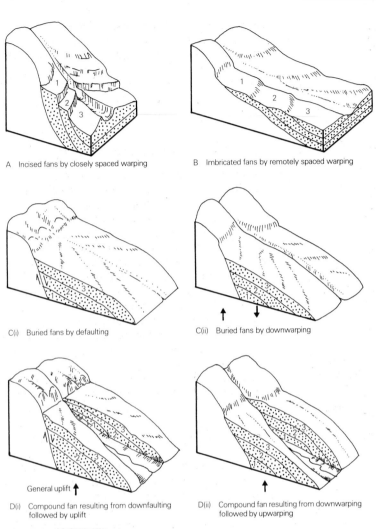

FIG. 13.11 Variations in alluvial-fan
morphology as a result of
neotectonics (after Han, 1984)

A Incised fans by closely spaced warping

B Imbricated fans by remotely spaced warping

C(i) Buried fans by defaulting

C(ii) Buried fans by downwarping

General uplift ↑

D(i) Compound fan resulting from downfaulting
followed by uplift

D(ii) Compound fan resulting from downwarping
followed by upwarping

Middle East which since the 1970s have seen much industrial and urban development but for which there is little or no history of earthquake damage claims. This research is based not only on historical records but also on the analysis of geomorphological and geological criteria. The approach helps to extend historical knowledge back into the Quaternary, providing a longer perspective on earthquake activity that can be very valuable in determining the relative (or absolute) state of activity of, for example, a fault. There has been considerable encouragement for such time-extended studies to be carried out elsewhere, particularly in the United States (Allen, 1975).

Cross *et al.* (1985) indicated how a hazard zoning map can be derived from historical data (Fig. 13.14) if an earthquake frequency map (the time dimension) can be combined with geology and soils (physical attributes of the ground), and a seismo-tectonic map (recorded earthquake activity). Even if the earthquake frequency element is missing (as it often is) the hazard zoning map can largely be derived from knowledge of ground conditions and the position of former earthquakes. For planning, engineering, and insurance purposes such hazard maps are highly relevant at the scale of the city or town.

13.7 Human-induced Earthquakes

Earthquakes can be generated, even outside the realms of war and explosions, by human activities.

PLATE 13.1 Fault scarps along the San Andreas fault zone in the Carrizo Plain, central California

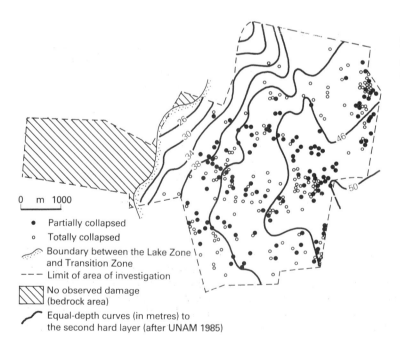

FIG. 13.12 Site of partly or completely collapsed buildings in the Mexico City (1985) earthquake in relation to depth of 'hard' deposits (after Degg, 1987*b*)

0   m   1000

● Partially collapsed

○ Totally collapsed

Boundary between the Lake Zone and Transition Zone

– – – Limit of area of investigation

No observed damage (bedrock area)

Equal-depth curves (in metres) to the second hard layer (after UNAM 1985)

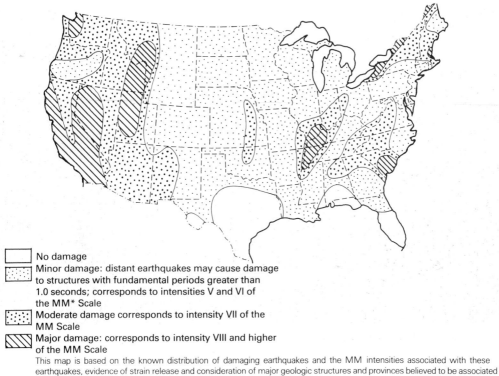

No damage

Minor damage: distant earthquakes may cause damage to structures with fundamental periods greater than 1.0 seconds; corresponds to intensities V and VI of the MM* Scale

Moderate damage corresponds to intensity VII of the MM Scale

Major damage: corresponds to intensity VIII and higher of the MM Scale

This map is based on the known distribution of damaging earthquakes and the MM intensities associated with these earthquakes, evidence of strain release and consideration of major geologic structures and provinces believed to be associated with earthquake activity. The probable frequency of occurrence of damaging earthquakes in each zone was not considered in assigning ratings to the various zones.
*Modified Mercalli Intensity Scale of 1931

FIG. 13.13 Seismic risk map of the United States released in 1969 (US Coast and Geodetic Survey, after Costa and Baker, 1981)

TABLE 13.4 Examples of reservoir-induced seismicity

Dam	Country	Height (m)	Volume × 10⁶m³	Year of impounding	Year of largest earthquake	Magnitude
Koyna	India	103	2780	1964	1967	6.5
Kremasta	Greece	165	4750	1965	1966	6.3
Hsinfeng-kiang	China	105	10500	1959	1961	6.1
Kariba	Zimbabwe	128	160368	1959	1963	5.8
Hoover	USA	221	36703	1936	1939	5.0
Marathon	Greece	63	41	1930	1938	5.0(?)

Source: Gough (1978).

In particular this has happened through (i) the creation of large reservoirs, (ii) the excavation of mines, and (iii) the injection of fluids into pores and cracks in crustal rocks (Gough, 1978). The generation of earthquakes by dams which are over 100 m high is well documented (Bolt *et al.*, 1975; Gupta and Rastogi, 1976; Simpson, 1976; Table 13.4). At the Koyna dam in India such an earthquake killed 177 people. However, not all dams of this elevation cause earthquakes, only those sited on rocks which may be critically stressed already. In particular the risk of such

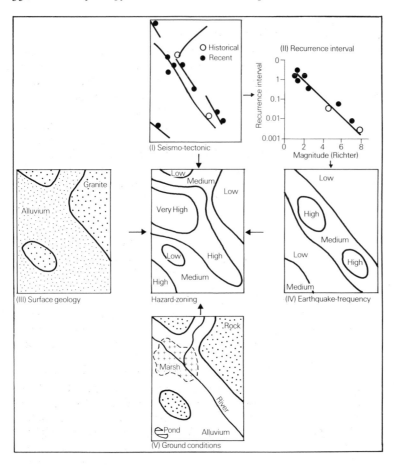

FIG. 13.14 A scheme of earthquake hazard zoning (after Cross *et al.*, 1985)

earthquakes appears to be greatest where there is hydraulic continuity between the body of impounded water and deeper groundwater.

Earthquakes induced by mining are different in that rather than additional stress being applied to the rocks there is a change in their strength as a result of excavation (Cook, 1976). The excavation causes a change in the stress field, only part of which can be stored as elastic strain; the remainder is released by closure of the excavation. Some of the energy used in this response appears as seismic energy (Gough, 1978).

Earthquakes have also been generated through the injection of liquids deep into the crust, although experience and monitoring in this context is very limited. In 1962 toxic fluid waste was forced under pressure through a borehole that penetrated fractured metamorphic rocks more than 3638 m below the Denver Basin. Soon afterwards small earthquakes began to occur (Evans, 1966). Nearly critical stress already existed in the rock mass, and the additional water pressure triggered failure.

13.8 Environmental Management and Earthquakes

The management response to earthquakes has been varied (Nichols, 1974). Management in areas prone to earthquakes needs to be considered in terms of: policy matters relating to an earthquake event; introduction of policy into land-use zonation and/or building codes; adherence to the building codes; response of the general population to either an earthquake or a public and authorative prediction that an earthquake will occur.

(a) *Policy*

Few mandatory policies exist concerning land management to provide protection against earthquakes. Where they do exist they tend to be policies regarding the design and construction of buildings rather than being based on land zoning. In the United States the first widespread and mandatory code was adopted in California in 1934 as a reaction to the 1933 Long Beach, California, earthquake. The code has been revised and updated, and similar codes now exist elsewhere in the United States.

(b) *Land-use zonation/building codes*

Since earthquake motion is greatest on loose unconsolidated sediments, it is desirable to leave all such ground as open land. Buildings should not be placed across fault lines, nor should they straddle two rocks of differing types each of which may have a different motion in an earthquake. Nowhere in the world, as far as we know, is there legislation to ensure that such sites are specifically avoided.

It is known that the resistance of a building to earthquake damage is a function of the building materials, the height of the building, its shape, the quality of construction, and its position relative to adjacent buildings. Most danger to life during an earthquake comes from falling buildings, and building collapse is related to these construction characteristics through the way the building responds to ground vibrations. Most damage occurs when the natural vibrations of the building coincide with those of the ground motion. As Degg (1986) showed for Mexico City, if two adjacent buildings are constructed with differing geometries they will react in different ways to the vibration and their relative motions may cause them to collide and destroy each other. Seed *et al.* (1972) showed that during the 1967 earthquake in Caracas, building damage was greatest for the tallest buildings (more than 14 storeys), especially those on deep soils (>150 m) with a long fundamental vibration period. Buildings of 5–9 storeys were most damaged on shallow soils (about 60 m deep). Most earthquake damage to buildings comes in response to the strong horizontal waves of energy that are transmitted along the ground surface. Building design standards take account of vertical stresses; few allow for large horizontal stresses.

In the United States the general standard of reference is the Uniform Building Code of the International Conference of Building Officials (Office of Emergency Preparedness, 1972).

(c) *Population response*

In 1977 the National Science Foundation attempted to describe the likely reaction of people in San Francisco to a prediction that an earthquake of magnitude M = 7.3 was going to occur (Fig. 13.15). The consequence they predicted was a net decrease in population and a decline in the number of retail firms, with a consequent reduction in the risk of injury or even death. There have been many predictions that a large earthquake will once more devastate San Francisco. How do the people of San Francisco conform to the prediction?

The best way to minimize earthquake damage is to bring about appropriate land use, land zoning, and building regulations that are based on a full understanding of both the site in question and the likely response of that site to an earthquake. Since in many cases site occupancy has already gone too far to permit avoidance (who would try to move San Francisco or Mexico City?), earthquakes will remain as the world's single most costly natural hazard.

Jackson and Mukerjee (1974) found that as memory of the 1906 earthquake faded the urban expansion of San Francisco has led to new development on to land, such as artificially reclaimed land, which is likely to be highly susceptible to earthquake activity. Other susceptible areas include those composed of soft sandstone, marine sands, dune sands, and river alluvium, and sites on or close to the active faults. Ground movement, although the primary effect of an earthquake, is not the only damaging one. Secondary effects such as landslides, tsunamis, floods, and fire can be even more destructive (Plate 13.2).

The possible management responses to the threat of further earthquake activity in San Francisco are many and varied. They are important in that for planning purposes it is deemed to be responsible to anticipate a major earthquake here once every 60 to 100 years (Steinbrugge, 1968), which places the event well within the lifetime of many of the present occupants of the area.

Adjustments available include those by the *homeowner*: choose a stable site; purchase an earthquake-resistant dwelling; hold protection against fire; and obtain insurance. Those available to the *municipality* include: land-use planning in

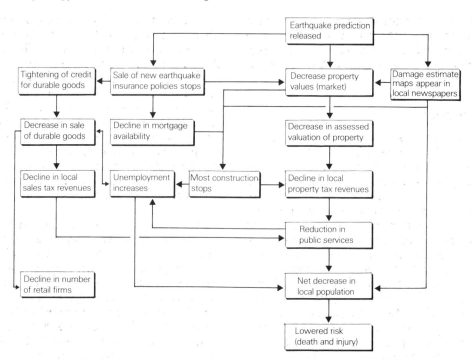

FIG. 13.15 A scenario likely to follow a prediction that an earthquake of magnitude M = 7.3 is about to occur in the USA (after National Science Foundation, 1977)

PLATE 13.2 Road disrupted by fissures formed during the Campania-Lucíania earthquake of 23 November 1980 near Bella, Potenza Province, Italy. The features are either due to landslides induced by the earthquake or by surface faulting (photo courtesy C. Vita-Finzi)

the light of an earthquake-hazard zoning map (assessment); slope stabilization against potential landslides; enforcing of building codes; and preparing disaster plans and maintaining rescue services (e.g. fire, medical).

13.9 Volcanoes

There is a very close relationship between the distribution of earthquakes (Fig. 13.1) and that of active volcanoes. Both tend to occur along active plate margins, as described in Sect. 13.3. But, whereas there are in excess of 100 000 000 earthquakes each year, there are only about 760 active volcanoes. In addition, violent earthquakes can bring about a loss of life many times greater than that of extreme volcanic eruptions. That is not to minimize the costs of volcanic activity (Table 13.5). For example, 1500 were killed by volcanic activity on Bali, Indonesia, in 1963; 30 000 were killed by Mt Palée, Martinique, in 1902, an event exceeded only by the 36 000 killed by the tsunami that followed the eruption of Krakatoa, Indonesia, in 1883. In this context volcanoes are seen as a hazard, but in some places, such as in Iceland, they provide a source of geothermal energy, and hence are a valuable resource.

Volcanoes are fed by magma chambers out of which both magma and gases may escape through weaknesses in the overlying rocks. The viscosity of the magma will determine how easily it will flow. Basaltic lava, with a lower silica content, tends to be more fluid than those magmas such as andesite and rhyolite which are richer in silica. Basalt lava can therefore flow over long distances at rates of several metres per second. The gases within the magma, when contained under pressure, may explode and cause volcanic ejecta to be projected through the air. More dangerous, however, are those cases where the gases form an asphyxiating mass.

In the USA the most significant volcanic eruption of recent years was that of Mount St Helens on 18 May 1980. Its effects included ejected material, poisonous gases, mudflows, and associated earthquakes. More than 120 houses were destroyed and about 70 people killed. Damage, including that resulting from ash-falls, was estimated to be $US2bn (Costa and Baker, 1981).

In general, volcanic activity is likely to take place at known sites, although the extent of its influence is much less predictable. For example, although Mount St Helens was being monitored even before the eruption took place, and although

TABLE 13.5 Volcanic hazards

Lava flows	Rate varies from 3 m/sec (basalt) to less than 1 m/day; fastest near source. Located in valley floors; if viscous, can flow for several hundred kilometres. Seldom costs lives. Property damage may be high. (e.g. lava out of Heimaey, Iceland, in 1973 buried 200 houses and a fish processing plant)
Dome eruption	Pressure under viscous and solidifying lava causes eruption, ejection of rocks and emission of gases (e.g. eruption of Mt Pelée, 1902–3)
Ejected material	Debris may be thrown across thousands of square kilometres, Ash can be carried by the wind. Large deposits cause buildings to collapse (e.g. causing 16 000 deaths during burial of Pompeii, AD 79). Can also destroy crops, as in 1915 when eruption of Tambora, Indonesia led to about 75 000 people starving as a result of lost crops.
Nuées ardentes	(Glowing avalanches) the fluidized cloud of hot ($>600\ °C$) ash, dust and gas flows at 100 km/hr for distances up to 10 kms (e.g. 30 000 killed, Mt Pelée, 1902 when town of St Pierre on Martinique was destroyed by a nuée ardente).
Poisonous gases	Toxic gases emitted include H_2S, CO_2, SO_2, and CO, and may be carried tens of kilometres downwind. Toxic to plants, animals, and humans (e.g. Laki eruption, Iceland in 1783 caused crop failure, 10 000 people died from starvation).
Volcanic mudflows (lahars)	These mudflows are the most damaging of all volcanic events. Flowing at great speeds they can cover 100 km (e.g. Armero, Columbia killing 20 000 in 1985, see Ch. 5).
Tsunamis	The sea wave produced by a volcanic explosion (or earthquake) can travel long distances and flood coastal settlements (e.g. Krakatoa in 1883 produced a tsunami up to 40 m high, destroying about 300 settlements, claiming 36 000 people)

a hazard/risk map of the area had already been compiled, the extent of the volcanicity was completely underestimated. One of the lessons from Mount St Helens, in terms of environmental management, is that even in a sophisticated society with a high level of geological knowledge volcanoes can cause both damage and death.

As with all hazards, if a policy of complete avoidance is possible, it is desirable. Clearly such avoidance often is not attainable, and management has to look to methods for coping in the event of volcanic activity. Costa and Baker (1981) identified the following management strategies: prepare reliable evacuation plan; construct protective works; organize sources of relief and rehabilitation; make hazard zoning/mapping to define major risk sites; define statutory prohibition on building within these sites.

Geomorphological mapping of volcanic hazards is especially useful in hazard zoning. It relies on the morphology of the volcanic landforms and the character of their deposits for identifying the present character of the volcano. The next volcanic event may not behave as the last one did, so it is important to assess how the existing geomorphology will influence the next course of events. Such an assessment requires information about the present level of activity, and whether gases, lava, or ejected matter may be expected. The character of the volcanic form may indicate what is most likely, as may the past history of events, and without special volcanological and geophysical investigations that may be all that there is to go on.

13.10 Conclusion

Current geological activity as expressed through earthquakes and volcanoes continues to have an impact on man. Indeed, as cities and areas of industrialization expand their impact is likely to increase. Through morphotectonic studies geomorphologists are able to make a considerable contribution to the study both of past earthquake and volcanic events, and to a prediction of where they are likely to recur in the future. Fuller benefits derive from an integration of the morphotectonic approach with the seismological approach, and it is in such an integration that there lies a better future for earthquake prediction.

14 Problems in Limestone Terrain

14.1 Introduction

Many rock types are associated with distinctive landforms and some present special applied geomorphological problems. The distinctiveness of volcanic terrain has been discussed in the previous chapter; granite country often includes features that arise from joint-controlled weathering, such as *tors* and deep weathering mantles, and can pose distinctive foundation problems; clay and shale landscapes, some of the most extensive on Earth and much neglected by geomorphologists, can include special problems of slope failure and drainage. But *limestone terrain* (often called *karst*) has attracted the attention of geomorphologists for many years because it covers 7–20 per cent of the total land area (Fig. 14.1), its landforms are highly distinctive, and it is often accompanied by serious management problems.

14.2 Limestone Terrain: Distinctiveness and Problems

(a) *Distinctiveness*

The distinctiveness of the terrain arises from the fact that limestone is relatively highly soluble in water and leaves relatively little weathering residue. The high solubility is reflected especially in surface and subterranean solution (both of which create distinctive features), the disappearance of surface drainage underground, and the development of surface forms that arise from subsidence or collapse following subterranean solution.

Limestone is a carbonate rock normally comprising over 50 per cent $CaCO_3$, mainly in the form of calcite; a similarly soluble rock, dolomite, comprises predominantly $MgCO_3$; and there are various mixtures of magnesium and calcium carbonate in between pure limestones and pure dolomites. For this discussion, 'limestone' will be used to refer to all the soluble carbonate rocks. Within limestones there is much lithological and structural variability that serves to differentiate the distinctiveness of limestone terrain. For example, the form of the carbonate can vary considerably even within a single deposit, from transported grains to particles formed *in situ*, from finely to coarsely crystalline, from shell fragments to cement, and from primary deposits to secondary replacements (e.g. Trudgill, 1985). Similarly, there may be a range of porosities and permeabilities, varied jointing patterns, and a full range of structural features (folds, faults, etc.). A distinction is commonly drawn between highly permeable and porous, fine-grained and relatively weak limestones, like the Chalk; and massive, coarsely crystalline, and relatively impermeable rocks such as the Carboniferous Limestone.

The topography and evolution of terrain on limestone has frequently been summarized in models that are thought to be generally applicable. Two examples are shown in Fig. 14.2. The first, which owes much to early workers like J. Cvijić, shows the evolution of a limestone bed between two relatively impermeable beds; the second summarizes a contrasting example, the evolution of a form of karst in China. Other models are reviewed in Jennings (1986), Sweeting (1972), and Trudgill (1985). The details of limestone topography are also reviewed in these texts.

The distinctiveness of limestone does, to some extent, override the effects of climate in producing distinctive landforms. But the rates of the denuding processes, and their relative importance do vary between climates. In general, rates are considered to be relatively low in arid and semi-arid areas; relatively high in seasonally wet Mediterranean areas such as the true karst areas of Yugoslavia; and highest in wet tropical regions (e.g. Jakucs, 1977).

Nevertheless, limestone solution is the key

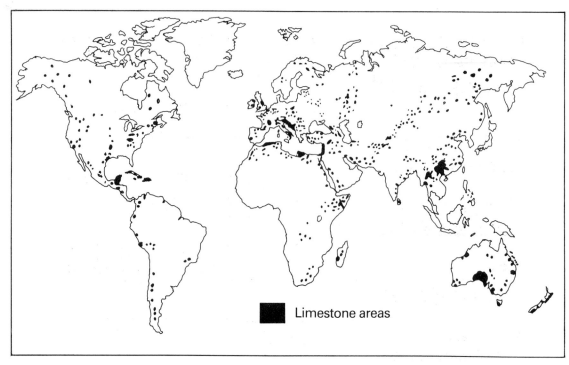

FIG. 14.1 Karst regions of the Earth (after Jakucs, 1977)

process everywhere. The chemistry of the process is well known but rather complex. Fundamentally, it depends on reactions between gases (e.g. CO_2), liquids (e.g. H_2CO_3), and solids (e.g. $CaCO_3$) in the atmosphere, in soil and superficial sediments, at the 'rockhead' (between bedrock and overlying material), and within the bedrock. For example, a weak carbonic acid is created by the solution of CO_2 in water (Equation 12.6). This process depends on several factors, including the *amount* of free CO_2 in the air, the soil, or the rock; the fact that solubility of CO_2 *falls* as temperature rises; and the length of time the water and gas are in contact (e.g. Pricknett *et al.*, 1976). The commonest reactions of the acid with the limestone are shown in Equation 12.10 and by:

$$CaMg\,(CO_3)_2 + 2H_2CO_3 \rightleftharpoons Ca^{++} + Mg^{++} + 4HCO_3^- \quad (14.1)$$

The *rate of solution* depends on several additional factors including the amount of water in contact with rock and its duration, the pH of the water, and the fact that the rate of solution increases with temperature. In addition, carbonic acid may be very significantly supplemented or its effect over-ridden by other acids, especially from vegetation or polluted atmospheres (see Chapter 12).

The amount of carbonate that can be dissolved by a given level of acidity can be predicted. It is convenient to envisage a potential equilibrium between the amount of calcium in solution and the amount of available carbon dioxide. The ability of a solution to take up calcium is referred to as its *aggressiveness*, and this will normally decline (i.e. the hardness of the solution will increase and the pH will rise) as the solution passes through the rock (Fig. 14.3; Trudgill, 1985). The key variables controlling limestone solution processes are summarized in Fig. 14.4.

(b) *Distinctive problems*

Distinctive applied problems are associated with the distinctive limestone terrain and processes. Bare limestone surfaces, or rock surfaces below soil, peat, glacial fill, or other sediment, develop distinctive features such as small pits and grooves. Such phenomena are of little practical importance, except perhaps in so far as they provide attractive landscapes and attractive stone for garden rockeries (Plate 14.1). But where the limestone is

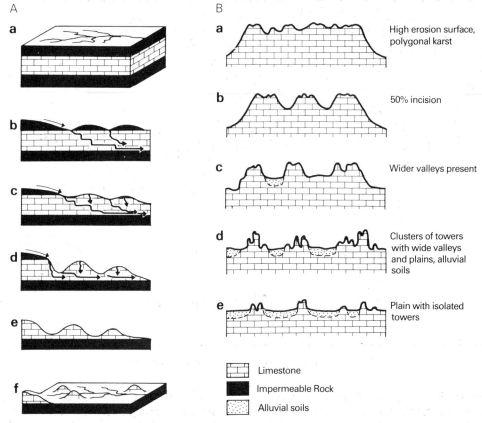

A

a

b

c

d

e

f

B

a — High erosion surface, polygonal karst

b — 50% incision

c — Wider valleys present

d — Clusters of towers with wide valleys and plains, alluvial soils

e — Plain with isolated towers

Limestone

Impermeable Rock

Alluvial soils

FIG. 14.2 Two examples of the evolution of limestone terrain: (A) dissection of a limestone block between impermeable rock layers. (B) Stages in the evolution of Chinese karst (after Trudgill, 1985)

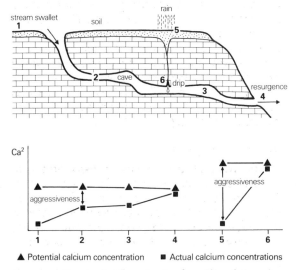

FIG. 14.3 Changes in the aggressiveness of limestone water (after Trudgill, 1985)

quarried and used in buildings, similar features may develop and present a serious problem of conservation (Chapter 12). Soils associated with limestone terrain are often distinctive, such as the residual *terra rossa* soils of the Mediterranean area, and they may present special problems to agriculture.

But undoubtedly the most serious problems of applied geomorphology in limestone terrain arise from the distinctive hydrology of such areas. This is important in the ways it affects water supply, pollution, and dam and reservoir construction. It is also fundamental to the major problem of surface subsidence or collapse which arises from solution, underground drainage, and cave formation.

(c) *Literature*

Limestone literature has its roots in the early work of Cvijić (e.g. 1895, 1918) and others, especially in

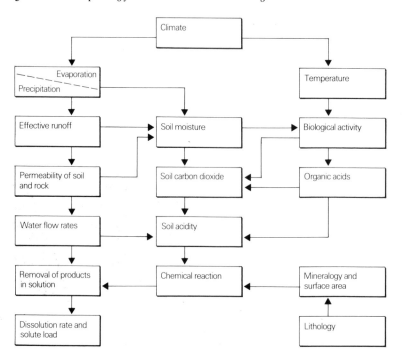

FIG. 14.4 Major variables in the control of limestone solution processes (after Trudgill, 1985)

the classic karst areas of Europe. Recent major reference works are by Sweeting (1972), Jakucs (1977), Trudgill (1985), and Jennings (1985). In addition, much recent research is included in P. Fénelon's (1968, 1975) two-volume *Phénomènes Karstique*, the special issue of the journal *Norois* on *Karstologie* (1977), and a supplement of the

Zeitschrift für Geomorphologie on *Karst Processes* (Sweeting and Pfeffer, 1976). Most of these works contribute little directly to the understanding of problems of environmental management in limestone terrain. But they all provide basic information essential to such an understanding. Specifically applied contributions are relatively

PLATE 14.1 A scenic resource and material for garden rockeries: limestone pavement in the Malham area, Yorkshire

few, and are to be found in various specialist publications (e.g. Smith, 1977; Sheedy *et al.*, 1982; Edmonds, 1983; Newton, 1984).

14.3 Problems of Limestone Hydrology

One of the most important problems of karst hydrology relates to pollution. Sink-holes and other shallow surface depressions, and abandoned limestone quarries, provide attractive loci for waste disposal, especially because streams are often not available for this purpose in limestone country. But the use of convenient hollows for waste disposal creates the possibility that the waste may find its way through the subterranean drainage system to emerge at springs some distance away as polluted flow. For example, Atkinson and Smith (1974) demonstrated the path and rate that pollution and runoff from a proposed motorway in Hampshire, UK, would follow through the Chalk, and predicted the extent and duration of pollution at the springs such flow would affect. And Aley (1972) traced spring contamination back to sinkhole dumps in Missouri, USA.

In order to tackle these problems, it is necessary to understand the nature of karst drainage systems and to monitor the paths and rates of flow. A simple model of limestone drainage, and the associated discharge hydrographs and hardness curves are shown in Fig. 14.5. The rates of movement and the aggressivity of the flows depend very much on the type of water movement. An important distinction can be made, for example, between *diffuse flow* that seeps through the rock and *free flow* that passes through open conduits. The former tends to move more slowly and to be saturated on emergence at springs, with relatively constant hardness. Free flow, on the other hand, can move very quickly through the system, and as a result hardness varies with discharge and may still be aggressive when it emerges. The proportion of diffuse and free flow at springs can vary (Fig. 14.6), and the proportion fed directly from surface runoff via swallets may be small. White (1969) classified aquifers in terms of flow types (Table 14.1).

The movement of water and pollutants through a system can be monitored using a variety of tracers (Smith, 1977) such as particulate tracers (e.g. *Lycopodium* spores), dyes, and radioactive tracers. Such monitoring can reveal the paths

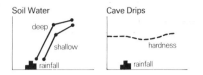

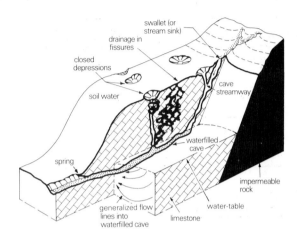

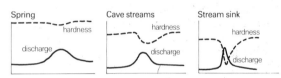

FIG. 14.5 Drainage of a limestone area through caves, depressions, and fissures, and associated changes of hardness at different sampling points (after Smith, 1977)

between surface sinks and risings (springs) and the rates and quantities of fluid movement. An example of such tracer-based patterns in the Mendip Hills, south-west England, is shown in Fig. 14.7.

A second hydrological problem in limestone terrain arises from the construction of dams and reservoirs, for there is always a dangerous possibility that there will be leakage or even complete drainage of a reservoir into the underlying limestone. The key to solving this problem, of course, is adequate appraisal of terrain prior to construction (Smith, 1977). If the site cannot be avoided and the basement is likely to leak, anticipatory remedial measures are possible. These include grouting limestone with cement to seal it, the cleaning of cavities and refilling them

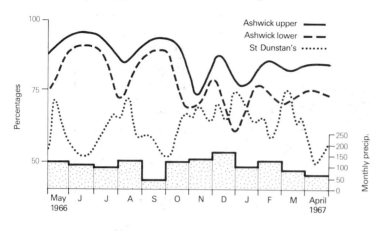

FIG. 14.6 Percolation water (diffuse flow) as a percentage of total discharge at two springs in the Mendip Hills, south-west England (after Drew, 1970)

with concrete ('dentistry'), and otherwise rendering the base of the reservoir relatively impermeable. For example, the Logan Martin Dam on the Coosa River in Alabama, USA, is constructed on faulted limestone. To make construction feasible, seepage control works costing $3 million, including nearly 85 000³ of grouting, were required (La Moreaux, 1967).

14.4 Collapse and Subsidence

Perhaps the most serious geomorphological problem in limestone terrain for engineers and planners is the possibility of surface collapse or subsidence. The limestone literature records many examples of surface failure that cause damage to property or lines of communication, and occasionally involve

TABLE 14.1 Types of carbonate aquifer systems in regions of low to moderate relief

Flow type	Hydrological control	Associated cave type
I. Diffuse flow	Gross lithology Shaley limestones; crystalline dolomites; high primary porosity.	Caves rare, small, have irregular patterns.
II. Free flow	Thick, massive soluble rocks	Integrated conduit cave systems.
A. Perched	Karst system underlain by impervious rocks near or above base level.	Cave streams perched—often have free air surface.
1. Open	Soluble rocks extend upwards to level surface.	Sinkhole inputs; heavy sediment load; short channel morphology caves.
2. Capped	Aquifer overlain by impervious rock.	Vertical shaft inputs; lateral flow under capping beds; long integrated caves.
B. Deep	Karst system extends to considerable depth below base level.	Flow is through submerged conduits.
1. Open	Soluble rocks extend to land surface.	Short tubular abandoned caves likely to be sediment choked.
2. Capped	Aquifer overlain by impervious rocks.	Long, integrated conduits under caprock. Active level of system inundated.
III. Confined flow	Structural and stratigraphic controls	
A. Artesian	Impervious beds which force flows below regional base level.	Inclined 3-D network caves.
B. Sandwich	Thin beds of soluble rock between impervious beds.	Horizontal 2-D network caves.

Source: White (1969).

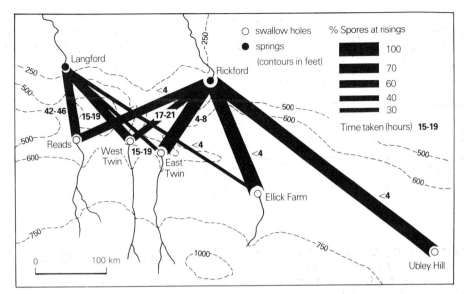

FIG. 14.7 Underground drainage and flow times for the north-central Mendip Hills, south-west England (after Drew *et al.*, 1968)

loss of life (e.g. La Moreaux, 1967; West and Dumbleton, 1972; Sperling *et al.*, 1977; Newton, 1984; Culshaw and Wallham, 1987). The problem is serious because it is difficult to predict the loci of collapse or subsidence even in areas where there is some surface manifestation that collapse has occurred in the past, and it is especially difficult where there is no surface manifestation because of a superficial cover of unconsolidated material.

The key to the geomorphological solution to this problem lies in the nature of surface depressions that usually occur on limestone surfaces. Some of these depressions arise from surface solution, others from subsidence and collapse. The main features are *swallow-holes* (where surface streams disappear underground), *dolines* or *sink-holes* (closed depressions, Plate 14.2), and *solution pipes* (a pipe-like solution cavity typically infilled with deposits). Each of these features (Fig. 14.8) reveals something about surface conditions. For example, limestone solution can lead to surface destabilization which produces dolines of various types (Fig. 14.9; Beck, 1984): *collapse dolines* arise from dissolution below the top of the rockhead; *subsidence dolines* are formed by the advance of solution along discontinuities at the rockhead beneath a superficial cover; *solution dolines* are formed by solution beneath a soil cover; and *alluvial dolines* occur when major discontinuities in the limestone are enlarged by solution beneath a

superficial deposit (e.g. Kennie and Edmonds, 1986). Clearly, different kinds of dolines present different kinds of hazard: for example, solution dolines pose little threat; whereas collapse dolines pose a hidden, possibly catastrophic hazard. It is especially important to be sure that surface forms are interpreted correctly but this is often extremely difficult. For example, surface hollows on the chalk have been variously explained as marl pits, pingos (a periglacial feature), solution pits, or dolines, and 'the forms of hollows are poor guides to their modes of origin' (Prince, 1979, p. 116).

There are several major processes that can lead to doline formation. The most obvious is the subterranean solution which advances to a point where subsidence or collapse occurs. But other triggers for surface change can include heavy rainstorms, the introduction of high water concentrations into the ground, the imposition of static or active loads (usually from construction) on to a metastable surface (Fig. 14.10), and the lowering of groundwater (e.g. Sheedy *et al.*, 1982; Edmonds, 1983).

Groundwater withdrawal is certainly a major cause of subsidence in limestone terrain (e.g. Foose, 1968). For example, in Alabama, USA, it is estimated that over 4000 doline-type sink-holes, which vary in size from a few metres to over 30 m in depth and are up to 3.2 km in diameter, have been produced since 1900 for this reason (Newton,

PLATE 14.2 A hazard for foundation engineering: collapse doline in the Swabian Jura, south-east of Reutlingen, West Germany

Types of Dolines (after Sweeting)

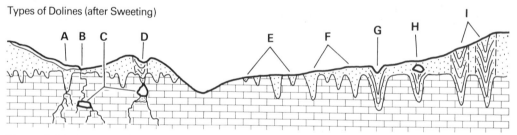

FIG. 14.8 Solution features in limestone terrain: A,B: swallow holes; E,F,I, solution pipes; D,G, dolines; G, solution-widened joints; C,D, caverns (after Kennie and Edmonds, 1986)

1984). These human-induced features are especially hazardous because they can form very quickly, in significant numbers, and in populated areas. Most of the features in Alabama are produced by vertical or lateral enlargement of cavities in unconsolidated deposits (including residual clays) overlying the limestone where water-tables have declined (Fig. 14.11); the principal process appears to be the downward migration of superficial sediments into the underlying openings, until the roof of the cavity can no longer be supported. The withdrawal of groundwater leads to four processes that foster subsidence: loss of buoyant support to the roofs of cavities; increase in the hydraulic head

and velocity of water movement so that blockage of cavities is less likely and cavities can be more easily created in unconsolidated sediments; water-level fluctuations at the base of unconsolidated deposits, which can cause repeated wetting and drying that in turn help to develop cavities in them; and induced recharge, whereby moving of water from the surface underground can help to create cavities in unconsolidated sediments above the limestone. A contrasted example is provided by accelerated sink-hole development in the Transvaal dolomites, in South Africa, as a result of lowering of water tables to permit mining shafts to be constructed: the 'result' has been the loss of over

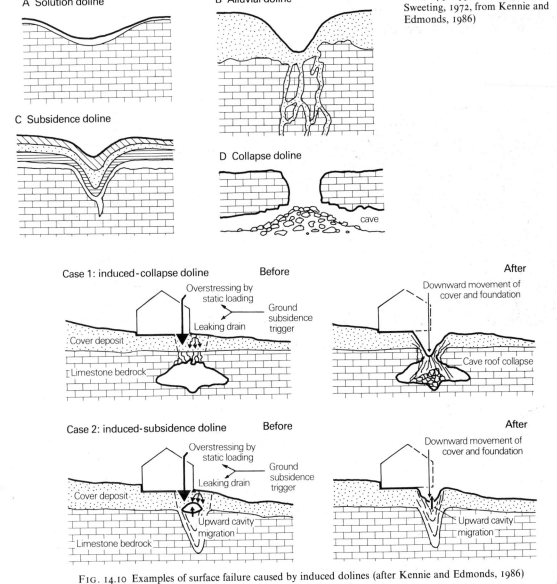

FIG. 14.9 Types of doline (after Sweeting, 1972, from Kennie and Edmonds, 1986)

FIG. 14.10 Examples of surface failure caused by induced dolines (after Kennie and Edmonds, 1986)

30 lives and massive damage to property (e.g. Brink, 1984).

Edmonds (1983) and Kennie and Edmonds (1986) have attempted to use the evidence of surface hollows on the English chalklands to predict areas liable to subsidence. They first identified the various features at the surface, using geological survey and other published data, remote sensing imagery for regional studies, and ground geophysics (e.g. seismic survey) and direct drilling at specific sites of interest. Remote sensing is a useful reconnaissance tool for regional appraisal because it not only helps the identification of specific landforms, but it also can reveal stressed vegetation zones in areas of excess or deficient moisture and fracture traces and other lineaments—features that may provide evidence of potential loci of failure. The second stage involved

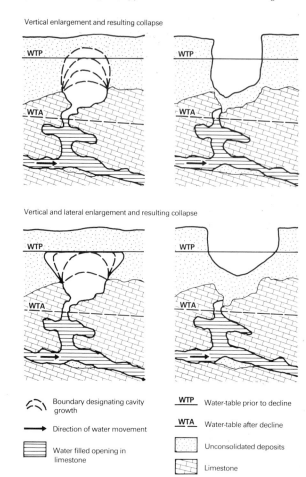

Vertical enlargement and resulting collapse

Vertical and lateral enlargement and resulting collapse

Boundary designating cavity growth

Direction of water movement

Water filled opening in limestone

WTP — Water-table prior to decline

WTA — Water-table after decline

Unconsolidated deposits

Limestone

FIG. 14.11 Development of cavities and dolines in unconsolidated sediments, Alabama, USA (redrawn from Newton, 1984)

determining regional variations in the density of solution features in the Chalk (Fig. 14.12). This analysis could be extended to include other morphometric attributes of the surface topography and drainage network that might provide indicators of surface activity. These attributes might include karst resurgence density, stream density on limestone, swallow-hole–resurgence ratio, and indices of doline geometry, such as width, depth, and length (e.g. Williams, 1969; Jennings, 1975; Ogden, 1984; Trudgill, 1985). The next stage in Edmonds's (1983) analysis was to explain the density of solution features in terms of a spatial analysis of causative factors. Such factors were found to include proximity to the feather edge of overlying Tertiary beds, the presence of an

overlying permeable deposit, and a coincidence with the relatively pure, soft, and porous Upper Chalk (Edmonds, 1983). From this analysis it is possible to identify areas where subsidence is both more common and more likely and where engineering work requires detailed, appropriate site investigation. Similar approaches to doline prediction have been adopted elsewhere (e.g. Ogden, 1984). In addition, physical *precursors* of collapse may aid prediction. These include such features as ground cracking, new areas of ponding, areas of vegetational stress, and turbidity in wells (Ogden, 1984).

Ogden (1984) synthesized approaches to doline prediction in terms of ten analytical steps: doline delineation; morphometric analysis; lithological controls appraisal; comparison of cavern occurrence with doline development; comparison of doline development with topographic factors and with depth to water-table; search for surface precursors; use of shallow geophysical appraisal techniques; assessment of human activities on doline development; and development of a collapse prediction model from all previous work.

Measures designed to prevent subsidence at specific sites are similar to those that prevent leakage from reservoirs. The main methods attempt to reinforce the surface strength, for example by grouting with cement or by creating a concrete raft. The former technique was used to reinforce the limestone site of the Huntsville, Alabama, USA, airport (La Moreaux, 1967); both were used to strengthen a section of the M40 motorway on Chalk near High Wycombe, Bucks., UK (West and Dumbleton, 1972). Other approaches include avoidance, infilling of cavities or deliberate collapse of cavities followed by filling, and minimizing collapse-generating processes (e.g. Sowers, 1984).

14.5 Environmental Management of Limestone Terrain

The hydrological, collapse, and subsidence problems of limestone terrain can be identified and analysed prior to development, and they can be avoided, ameliorated, or controlled. The most satisfactory approach usually involves prior planning on a regional scale, or site-specific investigations and actions prior to, during, or after development. An example of the integration of a geomorphologically based regional survey into a

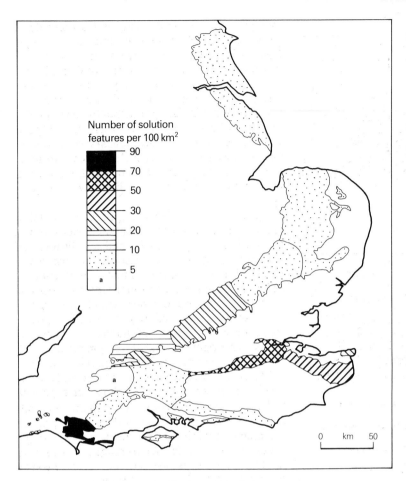

FIG. 14.12 Density of solution features on the Chalk outcrop of England (reproduced by permission of the Geological Society from Edmonds, 1983)

TABLE 14.2 Site evaluation procedure for critical areas in Bucks County carbonate belts, Pennsylvania, USA

Test element	Critical areas		
	I least hazardous	II moderate hazard	III most hazardous
Site inspection	×	×	×
Site mapping	(as verification)	×	×
Air-photograph analysis	—	×	×
Pit excavation	×	×	×
Trench excavation perpendicular to strike	—	×	×
Infiltration/percolation (two per acre in subdivisions)	×	×	×
Permeability testing for alternative systems	—	(substitute for perc. testing)	
Subdivision review			
Air-photograph analysis	×	×	×
Pit excavations			
Fixed-grid soil probes	—	—	×

Source: Sheedy *et al.* (1982).

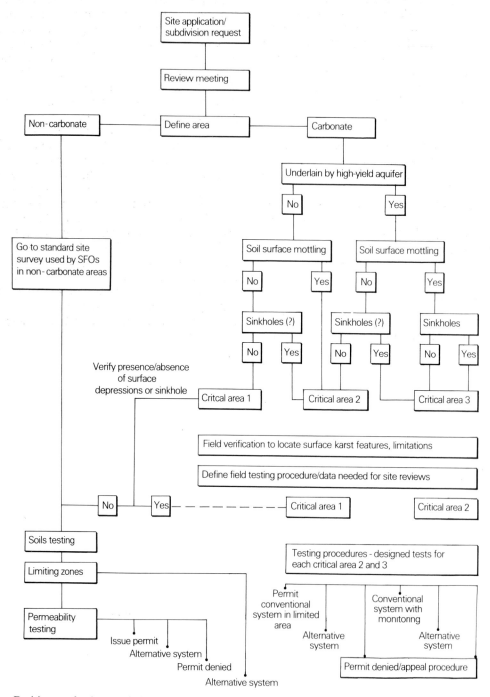

FIG. 14.13 Decision-tree for the appraisal of sewage system proposals in limestone terrain, Pennsylvania, USA (after Sheedy *et al.*, 1982)

specific management situation is provided by Sheedy *et al.* (1982) in the context of domestic, on-site sanitary waste disposal in Bucks County, Pennsylvania. Here, the disposal of waste from septic tanks into the underlying limestone can cause problems of pollution and can be accompanied by structural failures of tanks due to collapse. But the problems are not uniform throughout the region. Thus the county commissioned a geological/geomorphological/sewage-disposal survey which resulted in the classification of hazardous locations. Then, when an application is received for a sewerage system, the sewerage enforcement officer follows a *decision analysis tree* (Fig. 14.13) to determine if the site is on one of the more hazardous locations. If it is, a *site evaluation procedure* is adopted (Table 14.2) in which the developer's engineer is required to provide additional data prior to the provision of a permit.

14.6 Conclusion

Limestone provides perhaps the most distinctive suite of landforms and applied problems of any rock type. In the management of limestone terrain, the geomorphologist is likely to be mainly concerned with analysing the topography and subterranean features in order to help predict the location of potential surface failures, and with monitoring the routes of water and sediment flow through subterranean drainage systems to aid the development of water supplies and to prevent pollution. The techniques for this work are readily available. Because limestone scenery is unusual and of recreational value, and limestone is a valuable resource, there is often a conflict in limestone environments between conservation and exploitation. One recent development aimed at reducing this conflict is to plan quarry-blasting, exploitation, and restoration so that new landforms are created in quarries that are analogous in scale, form, and stability to those in the natural landcape (Gagen and Gunn, 1987). Beyond limestone terrain, limestone is widely used as a building material where, as Chapter 12 showed, its response to weathering processes, especially in polluted areas, poses quite distinct management problems.

References

ABRAHAMS, A. D., 1986, *Hillslope Processes* (Allen & Unwin, London).

ADAMS, D. P., 1975, 'Geomorphic evidence for late Holocene tilting in southern San Mateo County, California', *J. Research of the US Geol. Surv. 3*, 613–8.

ADAMS, J., 1980a, 'Paleoseismicity of the Alpine-fault seismic gap, New Zealand', *Geology, 8*, 72–6.

—— 1980b, 'Active tilting of the United States midcontinent: geodetic and geomorphic evidence', *Geology, 8*, 442–6.

AGNESI, V. and MACALUSON, T., MARTELEONE, S. and PIPITONE, G., 1984, 'Mass movements in western Sicily, Italy', in J.-C. Flageollet (ed.), 'Mouvements de terrains', *Série Documents du BRGM, 83*, 471–6.

AITCHISON, G. D. and GRANT, K., 1967, 'The PUCE programme of terrain description, evaluation and interpretation for engineering purposes', in *4th Reg. Conf. for Africa on Soil Mech. and Foundn Eng.* (Cape Town), *1*, 1–8.

ALBERTSON, M. L. and SIMONS, D. B., 1964, 'Fluid mechanics', in V. T. Chow (ed.), *Handbook of Applied Hydrology* (McGraw-Hill, New York), 7.1–7.49.

ALEXANDER, D., 1982, 'Leonardo da Vinci and fluvial geomorphology', *Am. J. Sci. 282*, 735–55.

ALEY, T., 1972, 'Groundwater contamination from sinkhole dumps', *Caves and Karst, 14*, 17–23.

ALLEN, C. R., 1975, 'Geological criteria for evaluating seismicity', *Geol. Soc. Am. Bull. 86*, 1041–57.

ALLUM, J. A. E., 1966, *Photogeology and Regional Mapping* (Pergamon Press, Oxford).

American Society of Photogrammetry, 1960, *Manual of Photographic Interpretation* (George Banta Co. Inc., Menosha, Wisc.).

—— 1975, *Manual of Remote Sensing* (Falls Church, Va.).

American Society for Testing Materials, 1971a, *Standard Test for Resistance of Concrete to Rapid Freezing and Thawing* (C666).

—— 1971b, *Specification for Facing Brick* (C216), fn. 1, 167–8.

ANDREWS, J. T., 1975, *Glacial Systems* (Duxbury Press, Mass.).

ANON, 1972, 'The preparation of maps and plans in terms of engineering geology', *Q. J. Eng. Geol. 5*, 295–382.

APPLETON, J., 1975, *The Experience of Landscape* (Wiley, London).

Arizona Dept. of Transportation, 1975, *Soil Erosion and Dust Control on Arizona Highways* (Arizona Dept. of Transportation, Phoenix), 4 vols.

ARMBRUST, D. V. and DICKERSON, J. D., 1971, 'Temporary wind erosion control: cost and effectiveness of 34 commercial materials', *J. Soil and Water Conservation, 26*, 154–7.

ARMORGIE, C. J., BALL, D. J., and LAXEN, D. P. H., 1984, 'London's air quality: the value and use of a computer dispersion model', *Clean Air, 13*, 127–34.

ARNOLDUS, H. M. J., 1980, 'An approximation of the rainfall factor in the Universal Soil loss equation', in M. de Boodt and D. Gabriels (eds.), *Assessment of Erosion* (Wiley, Chichester), 127–42.

ARNOULD, M., and FREY, P. 1977, '*Analyse détaillée des réponses à l'enquête internationale sur les glissements de terrain*' (UNESCO, Paris).

ASHURST, J. and DIMES, F. G., 1979, *Stone in Building* (Architectural Press, London).

ATKINSON, B., 1970, 'Weathering and performance', in J. W. Simpson and P. J. Horrobin (eds.), *The Weathering and Performance of Building Materials* (Medical and Technical Publishing, Aylesbury), 1–40.

ATKINSON, T. C. and SMITH, D. I., 1974, 'Rapid groundwater flow in fissures in the chalk—an example from south Hampshire', *Q. J. Eng. Geol. 7*, 197–205.

ATTEWELL, P. B. and FARMER, I. W., 1976, *Principles of Engineering Geology* (Wiley, New York).

ATWATER, B., 1978, 'Central San Mateo County, California: Land use controls arising from erosion of sea cliffs, landsliding and fault movement', in G. D. Robinson and A. M. Spieker (eds.), 'Nature to be Commanded', *US Geol. Surv. Prof. Paper, 950*, 11–20.

BABCOCK, E. A., 1971, 'Detection of active faulting using oblique infrared aerial photography in the Imperial Valley, California', *Bull. Geol. Soc. Am. 82*, 3189–96.

BABU, R., TEJWANI, K. G., AGARWAL, M. C. and BHUSHAN, L. S., 1978, 'Distribution of erosion index and isoerodent map of India', *Indian J. Soil Cons. 6*, 1–12.

BADGLEY, P. C., 1965, *Structural and Tectonic Principles* (Harper & Row, New York).

BAGNOLD, R. A., 1941, *The Physics of Blown Sand and Desert Dunes* (Methuen, London).

—— 1953, 'Forme des dunes de sable et régime des vents', in *Actions Eoliennes* (CNRS, Paris Coll. Int. 35), 23–32.

BALLING, N., 1980, 'The land uplift in Fennoscandia, gravity field anomalies and isostasy', in N.-A. Morner, *Earth Rheology, Isostasy and Eustasy, Proc. of Symposium Stockholm, 1977* (Wiley, Chichester), 297–321.

BARANOVA, Y. P. and BISKE, S. F., 1971, 'Importance of stratigraphic and geomorphological research in design of neotectonic maps (north-east of USSR)', *Inter. Geol. Rev. 13*, 507–13.

BARATA, F. E., 1969, 'Landslides in the tropical region of Rio de Janeiro', *Proc. 7th Inter. Conf. Soil Mech. and Foundn. Eng.* (Mexico), 2, 507–16.

BARNES, H. H., 1967, 'Roughness characteristics of natural channels', *US Geol. Surv. Water-Supply Paper, 1849*.

BARNES, J. W., 1981, *Basic Geological Mapping* (Geological Society of London Handbook Series, Open Univ. Press, Milton Keynes).

BARRY, R. G. and CHORLEY, R. J., 1982, *Atmosphere, Weather and Climate* (Methuen, London).

BARSCH, D. (ed.), 1979, 'The geomorphological approach to environment', *Geojournal, 3*, 329–416.

BEAUMONT, T. E., 1979, 'Remote sensing for the location and mapping of engineering construction materials in developing countries', *Q. J. Eng. Geol. 12*, 147–58.

BECK, B. F. (ed.), 1984, *Sinkholes: Their Geology, Engineering and Environmental Impact* (Balkema, Rotterdam).

BECKETT, P. H., *et al.*, 1972, 'Terrain evaluation by means of a data bank', *Geog. J. 138*, 430–56.

BELLY, P. Y., 1964, 'Sand movement by wind', *US Corps of Engineers, Coastal Research Center, Tech. Mem. 1*.

BENEDICT, J. B., 1976, 'Frost creep and gelifluction features: A review', *Quat. Res. 6*, 55–76.

BENNETT, H. H., 1939, *Soil Conservation* (McGraw-Hill, New York).

BENNETT, J. P., 1974, 'Concepts of mathematical modelling of sediment yield', *Water Resources Research, 10*, 485–92.

BERENDSEN, H. J. A. (ed.), 1986, 'Het Landschap van de Bommelerwaard', *Nederlandse Geografische Studies* (Amsterdam/Utrecht), 10.

BERG, R. L. and WRIGHT, E. A. (eds.), 1984, *Frost Action and Its Control* (Am. Soc. Civil Eng., New York).

BERGSMA, E., 1980, 'Method of reconnaissance survey of erosion hazard near Merida, Spain', in M. de Boodt and D. Gabriels (eds.), *Assessment of Erosion* (Wiley, Chichester), 55–66.

—— 1981, 'Indices of rain erosivity', *ITC Journal* (1981), 460–84.

BIANCOTTI, A., 1979, 'Rapporti fra morfologia e tetto-nica nella pianura cuneese', *Geografia Fisicia e Dinamica Quaternaria, 2*, 51–6.

BIBBY, J. S. and MACKNEY, D., 1969, 'Land use capability classification', *Soil Surv. England and Wales Tech. Mngr. 1*.

BIE, S. W. and BECKETT, P. H. T., 1970, 'The costs of soil survey', *Soils and Fertilisers, 33*, 203–16.

BIRD, E. C. F., 1979, 'Coastal processes', in K. J. Gregory and D. E. Walling (eds.), *Man and Environmental Processes* (Dawson, Folkestone), 82–101.

—— 1984, *Coasts: An Introduction to Coastal Geomorphology* (Blackwell, Oxford), 3rd edn.

—— 1985, *Coastal Changes: A Global Review* (Wiley, Chichester).

BIRKELAND, P. W., 1974, *Pedology, Weathering and Geomorphological Research* (OUP, New York).

BISAL, F. and HSIEH, J. 1966, 'Influence of moisture on erodibility of soil by wind', *Soil Science, 102*, 143–6.

BISSET, R., 1980, 'Methods of environmental impact analysis: recent trends and future prospects', *J. Environmental Management, 2*, 97–143.

BJORNSSON, H., 1980, 'Avalanche activity in Iceland, climatic conditions and terrain features', *J. Glaciology, 26*, 13–23.

BLACHUT, T. J. and MULLER, F., 1966, 'Some fundamental considerations of glacier mapping', *Can. J. Earth Sci. 3*, 747–9.

BLACK, R. F., 1976, 'Periglacial features indicative of permafrost; ice and soil wedges', *Quat. Res. 6*, 3–26.

BLAIKIE, P., 1985, *The Political Economy of Soil Erosion in Developing Countries* (Longman, London).

—— and BROOKFIELD, H., 1987, *Land Degradation and Society* (Methuen, London).

BOLT, B. A., 1978, *Earthquakes* (Freeman, San Francisco).

—— HORN, W. L., MACDONALD, G. A., and SCOTT, R. F., 1975, 'Hazards from earthquakes', in B. A. Bolt *et al.* (eds.), *Geological Hazards* (Springer-Verlag, New York), 1–62.

BOOTHROYD, J. C., 1978, 'Mesotidal inlets and estuaries', in R. A. Davis (ed.), *Coastal Sedimentary Environments* (Springer-Verlag, New York), 287–60.

BORCHERDT, R. D. (ed.), 1975, 'Studies for seismic zonation of the San Francisco Bay Region', *US Geol. Surv. Prof. Paper, 941A*.

BOULTON, G. S., 1975, 'The genesis of glacial tills: a framework for geotechnical interpretation', in *The Engineering Behaviour of Glacial Materials* (Midlands Soil Mechanics and Foundation Engineering Society), 52–9.

—— and EYLES, N., 1979, 'Sedimentation by valley glaciers: a model and genetic classification', in C. Schlüchter (ed.), *Moraines and Varves: Origins, Genesis, Classification*, Proc. INQUA symposium on genesis and lithology of Quaternary deposits (A. H. Balkema, PO Box 1675, Rotterdam, Netherlands), 11–23.

BOULTON, G. S. AND PAUL, M. A., 1976, 'The influence of genetic processes on some geotechnical properties of glacial tills', *Q. J. Eng. Geol. 9*, 159–94.

BOURNE, R., 1931, 'Regional survey and its relation to stock-taking of the agricultural resources of the British Empire', *Oxford Forestry Memoirs, 13*.

BOVIS, M. J., 1982, 'The spatial variation of soil loss and soil loss controls', in C. E. Thorn (ed.), *Space and Time in Geomorphology* (Allen & Unwin, London), 1–24.

BOYCE, R. C., 1975, 'Sediment routing and sediment delivery ratios', in *US Dept. Agriculture, Agric. Research Service, ARS-S-40*, 61–5.

BRABB, E. E., PAMPEYAN, E. H., and BONILLA, M. G., 1972, 'Landslide susceptibility in San Mateo County, California', *US Geol. Surv. Misc. Field Studies Map, MF 360* (scale 1:62 500).

BRENNAN, A. M. (Compiler) 1983, 'Permafrost: a bibliography 1978–82', *Glaciological Data, 14* (US Nat. Oceanic and Atmospheric Admin., Boulder, Colo.).

BRICE, J., ATKINS, D., and LAW, D., 1984/5, 'The London NO2 survey: progress report', *London Environmental Bull. 2*.

BRIGGS, D. J. and FRANCE, J., 1982, 'Mapping, soil erosion by wind for regional environmental planning', *J. Environmental Management, 15*, 159–68.

BRINK, A. B. A., 1984, 'A brief review of the South African sinkhole problem', in B. F. Beck (ed.), *Sinkholes* (Balkema, Rotterdam), 123–7.

—— and PARTRIDGE, T. C., 1967, 'Kyalami land system: an example of physiographic classification for the storage of terrain data', *Proc. 4th Reg. Conf. for Africa on Soil Mech. and Foundn. Eng.* (Cape Town), *1*, 9–14.

—— —— and MATTHEWS, G. B., 1970, Airphoto interpretation in terrain evaluation, *Photo Interpretation, 5*, 15–30.

—— —— WEBSTER, R., and WILLIAMS, A. A. B., 1968, 'Land classification and data storage for the engineering use of natural materials', *Proc. 4th Conf. Austr. Road Res. Board*, 1624–47.

British Standards Institution, 1975, *Methods of Testing Soils for Civil Engineering Purposes* (BS 1377).

BROOKES, A., GREGORY, K. J., and DAWSON, F. H., 1983, 'An assessment of river channelization in England and Wales', *Science of the Total Environment, 27*, 97–111.

BROOKFIELD, M. E. and AHLBRANDT, T. S. (eds.), 1983, 'Eolian sediments and processes', *Developments in Sedimentology, 38*.

BROWN, J. and YEN, Y., 1982, The second national Chinese conference on Permafrost, Lanzhou, China, 12–18 Oct. 1981, *Cold Regions Research and Engineering Laboratory, SR-82-3*.

BROWN, R. D. jun. and KOCKELMAN, W. J., 1983, 'Geologic principles for prudent land use', *US Geol. Surv. Prof. Paper, 946*.

BROWN, R. J. E., 1965, 'Factors influencing discontinuous permafrost in Canada', in T. L. Péwé (ed.), *The Periglacial Environment* (McGill-Queen's Univ. Press, Montreal), 11–54.

—— 1966, 'Influence of vegetation on permafrost', *Proc. Int. Perm. Conf.* (Nat. Acad. Sciences, Washington, DC), 20–5.

—— (ed.), 1969, *Proceedings of the 3rd Canadian Conference on Permafrost* (Nat. Research Council of Canada, Ottawa).

—— 1970, *Permafrost in Canada* (Univ. Toronto Press, Toronto).

BRUNE, G. M., 1951, 'Sediment records in the midwestern United States', *Int. Ass. Sci. Hydrology, 3*, 29–38.

BRUNSDEN, D., 1979, 'Weathering', in C. Embleton and J. Thornes (eds.), *Processes in Geomorphology* (Arnold, London), 73–129.

—— 1984, 'Mudslides', in D. Brunsden and D. B. Prior (eds.), *Slope Instability* (Wiley, Chichester), 363–418.

—— DOORNKAMP, J. C., FOOKES, P. G., JONES, D. K. C., and KELLY, J. M. N., 1975, Large-scale geomorphological mapping and highway engineering design', *Q. J. Eng. Geol. 8*, 227–53.

—— —— and JONES, D. K. C., 1978, 'Applied geomorphology: a British view', in C. Embleton, D. Brunsden, and D. K. C. Jones (eds.), *Geomorphology: Present Problems and Future Prospects* (OUP, Oxford), 251–62.

—— JONES, D. K. C., and DOORNKAMP, J. C., 1979, 'The Bahrain Surface Materials Resources Survey and its application to planning', *Geog. J. 145*, 1–35.

—— —— MARTIN, R. D. and DOORNKAMP, J. C., 1981, 'The geomorphological, character of part of the low Himalaya of Eastern Nepal', *Zeit. für Geom. Suppl. 37*, 25–42.

—— and KESEL, R. H., 1973, 'Slope development on a Mississippi river bluff in historical time', *J. Geol. 81*, 576–97.

—— and PRIOR, D. B. (eds.), 1984, *Slope Instability* (Wiley, Chichester).

—— and THORNES, J. B., 1979, 'Landscape sensitivity and change', *Trans. Inst. Brit. Geog. 4*, 463–84.

BRUUN, P. and LACKEY, J. B., 1962, 'Engineering aspects of sediment transport', in T. Fluhr and R. F. Legget (eds.), *Reviews in Engineering Geology* (Geol. Soc. Am.), *1*, 39–103.

BRYAN, K., 1925, 'The Papago Country, Arizona', *US Geol. Surv. Water-Supply Paper, 499*.

BRYAN, R. B., 1970, An improved rainfall simulator for use in erosion research', *Can. J. Earth Sci. 7*, 1552–61.

—— 1976, 'Considerations on soil erodibility indices and sheetwash', *Catena, 3*, 99–111.

BULL, W. B., 1964a, 'Alluvial fans and near-surface subsidence in western Fresno County, California', *US Geol. Surv. Prof. Paper, 437-A*.

—— 1964b, 'Geomorphology of segmented alluvial fans

in western Fresno County, California', *US Geol. Surv. Prof. Paper, 352-E,* 89–129.

—— 1977, 'The alluvial fan environment', *Prog. Phys. Geog. 1,* 222–70.

—— and MCFADDEN, L. D., 1977, 'Tectonic geomorphology north and south of the Garlock fault, California', in D. O. Doehring (ed.), *Geomorphology in Arid regions* (Publ. in Geomorphology, State Univ. New York, Binghamton), 115–38.

BURINGH, P. and VINK, A. P. A., 1965, 'The importance of geology in air photo-interpretation for soil mapping', *Revertera Number, Int. Training Centre for Aerial Survey* (Delft), *Ser. V. 33,* 16–23.

BURNETT, A. W. and SCHUMM, S. A., 1983, 'Alluvial-river response to neotectonic deformation in Louisiana and Mississippi', *Science, 222,* 48–50.

BURTON, I. R., KATES, R. W. and SNEAD, R. R., 1969, 'The human ecology of coastal flood hazard in Megapolis', *Univ. Chicago, Dept. Geography Research Paper, 115.*

—— —— and WHITE, G. F., 1979, *The Environment as Hazard* (OUP, Oxford).

BUSH, P., COOKE, R. U., BRUNSDEN, D., DOORNKAMP, J. C., and JONES, D. K. C., 1980, 'Geology and geomorphology of the Suez city region, Egypt', *J. Arid Environments, 3,* 265–81.

BUTTERFIELD, G. R., 1973, 'The susceptibility of High County soils to erosion by wind', *Proc. Calibration Control Seminar* (Mussey University, NZ).

CAMPBELL, R. H., 1980, 'Landslide map showing field classification, Point Dume Quadrangle, California', *US Geol. Surv. Misc. Field Studies Map, MF 1167* (Scale 1:24 000).

CANNON, P. J., 1973, 'The application of RADAR and infrared imagery to quantitative geomorphological problems', *2nd Annual Remote Sensing Earth Resources Conference V. Tenn. 2,* 503–19.

CANTER, L. W., 1977, *Environmental Impact Assessment* (McGraw-Hill, New York).

CANUTI, P. *et al.*, 1987, 'Slope stability mapping in Tuscany, Italy', in V. Gardiner (ed.), *Int. Geomorphology, 1986,* (Wiley, Chichester), 231–40.

CAPDEVILLE, J.-M., CASSOUDEBAT, M., GOTTIS, M. and POITIER, J.-M., 1978, 'Essai d'identification de mouvements néotectoniques à partir du traitement de données topographiques', *Sciences de la Terre et Mesures, Colloque International, Orléans,* Mai 1977, 425–34.

CARBOGNIN, L., 1985, 'Land subsidence: a worldwide environmental hazard', *Nature and Resources, 21,* 2–11.

CARLSTON, C. W., 1963, 'Drainage density and stream flow', *US Geol. Surv. Prof. Paper, 422-C.*

—— 1968, 'Slope-discharge relations for eight rivers in the United States', *US Geol. Surv. Prof. Paper, 600-D,* 45–7.

CARR, A. P., 1962, 'Cartographic record and historical accuracy', *Geography, 47,* 135–44.

CARRARA, A., 1984, 'Landslide hazard mapping: aims and methods', in J.-C. Flageollet (eds.), 'Mouvements de Terrains', *Série Documents du BRGM, 83,* 141–51.

—— CATALANO, E., SORRISO VALVO, M., REALI, C. and OSSO, I., 1978, 'Digital terrain analysis for land evaluation', *Geologia Applicata e Idrogeologia, 13,* 69–127.

—— and MERENDA, L., 1974, 'Metodologia per un consimento degli eventi franosi in Calabria', *Geologia Applicata e Idrogeologia, 9,* 237–55.

—— —— 1976, 'Landslide inventory in northern Calabria, southern Italy', *Bull. Geol. Soc. Am. 87,* 1153–62.

CASSINIS, R., 1977, 'Use of remote sensing from space platforms for regional geological evaluation and for planning ground exploration', *Geophys. Prospecting, 25,* 636–57.

CATLOW, J. and THIRLWALL, C. G., 1976, 'Environmental Impact Analysis', *Dept. of the Environment (UK) Research Report, 2* (Dept. of the Environment, UK).

CERMAK, J. E., 1971, 'Laboratory simulation of the atmospheric boundary layer', *Am. Inst. Aeronautics and Astronautics J. 9,* 1746–54.

CHAMBERLAIN, T. C., 1897, 'The method of multiple working hypotheses', *J. Geol. 5,* 837–8.

CHAMPETIER DE RIBES, G., 1979, 'Données géologiques et géotechniques et plans d'occupation des sols', *Bull. Maison Laboratoire des Ponts et Chaussées, 99.*

CHANDLER, R. J., 1970, 'Solifluction on low-angled slopes in Northamptonshire', *Q. J. Eng. Geol. 3,* 65–9.

CHAPMAN, K., 1981, 'Issues in environmental impact assessment', *Progr. Human Geog. 5,* 190–296.

CHEPIL, W. S., 1945, 'Dynamics of wind erosion: II. Initiation of soil movement', *Soil Science, 60,* 397–411.

—— 1950, 'Properties of soil which influence wind erosion: the governing principle of surface roughness', *Soil Science, 69,* 149–62.

—— 1951, 'Properties of soil which influence wind erosion: V. Mechanical stability of structure', *Soil Science, 72,* 465–78.

—— 1957, 'Width of field strips to control wind erosion', *Manhattan Tech. Bull.* (Agric. Experiment Station, Kansas State College), *92.*

—— 1959a, 'Equilibrium of soil grains at the threshold of movement by wind', *Proc. Soil Sci. Soc. Am. 23,* 422–8.

—— 1959b, 'Wind erodibility of farm fields', *J. Soil and Water Conservation, 14,* 214–19.

—— and MILNE, R. A., 1941, 'Wind erosion of soil in relation to roughness of surface', *Soil Science, 52,* 417–31.

—— SIDDOWAY, F. H., and ARMBRUST, D. V., 1963, 'Climatic index of wind erosion conditions in the Great Plains', *Proc. Soil Sci. Soc. Am. 27,* 449–52.

—— and WOODRUFF, N. P., 1957, 'Sedimentary characteristics of dust storms: II Visibility and dust concentration', *Am J. Sci. 255,* 104–14.

CHEPIL, W. S. and WOODRUFF, N. P. 1963, 'The physics of wind erosion and its control', *Advances in Agronomy, 15*, 211–302.

—— —— SIDDOWAY, F. H., and ARMBRUST, D. V., 1963, 'Mulches for wind and water erosion control', *US Dept. Agriculture, ARS,* 41–84.

CHESTER, D. K., 1980, 'The evaluation of Scottish sand and gravel resources', *Scot. Geog. Mag. 96,* 51–62.

—— 1982, 'Predicting the quality of sand and gravel deposits in areas of fluvioglacial deposition', *Dept. Geog. Univ. Liverpool Research Paper, 10.*

CHIEN, N., 1956, 'Graphic design of alluvial channels', *Proc. Am. Soc. Civil. Eng. 121,* 1267–80.

CHORLEY, R. J., 1959, 'The geomorphic significance of some Oxford soils', *Am. J. Sci. 257,* 503–15.

—— and KENNEDY, B. A., 1971, *Physical Geography: A Systems Approach* (Prentice-Hall, London).

—— SCHUMM, S. A., and SUGDEN, D. E., 1984, *Geomorphology* (Methuen, London).

CHOW, V. T. (ed.), 1964, *Handbook of Applied Hydrology* (McGraw-Hill, New York).

CHRISTIAN, C. S., 1957, 'The concept of land units and land systems', *Proc. 9th Pacific Sci. Conf. 20,* 74–81.

—— and STEWART, G. A., 1952, *Summary of General Report on Survey of Katherine-Darwin Region, 1946* (CSIRO, Australia), *Land Research Series, 1.*

—— and —— 1968, 'Methodology of integrated surveys', *Aerial Surveys and Integrated Studies, Proc. Toulouse Conf.* (UNESCO, Paris), 233–80.

CHURCH, M., 1980, 'Records of recent geomorphological events', in R. A. Gullingford, D. A. Davidson, and J. Lewin (eds.), *Timescales in Geomorphology* (Wiley, Chichester), 13–29.

CLARK, A. H., COOKE, R. U., MORTIMER, C., and SILLITOE, R. H., 1967, 'Relationships between supergene mineral alteration and geomorphology, southern Atacama Desert, Chile: an interim report', *Trans. Inst. Min. and Met. B, 76,* B89–B96.

CLARK, R. H. and PERSOAGE, N. P., 1970, 'Some implications of crustal movement in engineering planning, Symposium on Recent Crustal Movements, Ottawa, Canada, 1969', *Can. J. Earth Sci. 7,* 628–33.

CLARKE, B. L. and ASHURST, J., 1972, *Stone Preservation Experiments* (HMSO, London).

CLAYTON, K. M., 1980, 'Coastal protection along the East Anglian coast, UK', *Zeit. für Geom. Suppl. Bd. 34,* 165–72.

CLEMENTS, T., STONE, R. O., MANN, J. F., and EYMANN, J. L., 1963, 'A study of windborne and sand and dust in desert areas', *US Army Natick Labs., Earth Sciences Division, Tech. Rep. ES-8.*

Coastal Engineering Research Center, 1966, 'Shore protection, planning and design', *CERC Tech. Rep. 4* (CERC, Washington, DC).

—— 1973, *Shore Protection Manual* (US Army Corps of Engineers, Washington, DC), 3 vols.

Coastal Zone '78, 1978, *Proc. Symposium on Technical, Environmental, Socio-Economic and Regulatory Aspects of Coastal Zone Management, San Francisco, 1978* (Am. Soc. Civ. Eng., New York).

COATES, D. R. (ed.), 1974, *Glacial Geomorphology,* Publications in Geomorphology (State Univ. New York, Binghampton).

—— (ed.), 1976, *Geomorphology and Engineering* (Dowden, Hutchinson & Ross, Stroudsberg).

—— 1981, *Environmental Geology* (Wiley, New York).

—— 1983, 'Large-scale land subsidence', in R. Gardner and H. Scoging (eds.), *Mega-Geomorphology* (OUP, Oxford), 212–33.

—— 1984, 'Geomorphology and public policy', in J. E. Costa and P. J. Fleisher (eds.), *Developments and Applications of Geomorphology* (Springer-Verlag, Berlin), 97–132.

COCCOSSIS, H. N., 1985, 'Management of coastal regions: the European experience', *Nature and Resources, 21,* 20–8.

COLE, D. C. H. and LEWIS, J. G., 1960, 'Progress report on the effect of soluble salts on stability of compacted soils', *Proc. 3rd Austr.-NZ Conf. Soil Mech. Foundn. Eng.,* Sydney 22–6 Aug., 29–31.

COLE, G. W., 1984, 'A method for determining field wind erosion rates from wind-tunnel-derived functions', *Trans. Am. Soc. Agric. Eng. 27,* 110–16.

COLLIS, L., and FOX, R. A., 1985, *Aggregates: Sand, Gravel and Crushed Rock Aggregates for Construction Purposes* (Geol. Soc. London, Eng. Geol., Spec. Publ., 1).

CONWAY, B. W., FORSTER, A., NORTHMORE, K. J., and BARCLAY, W. J., 1980, 'South Wales coalfield landslip survey', *Inst. Geol. Sci. Spec. Surveys Div., Eng. Geol. Unit, Report No. EG 76/10.*

COOK, N. G. W., 1976, 'Seismicity associated with mining', *Eng. Geol. 10,* 99–122.

COOKE, R. U., 1981, 'Salt weathering in deserts', *Proc. Geol. Ass. 92,* 1–16.

—— 1982, 'The assessment of geomorphological problems in dryland urban areas', *Zeit. für Geom. NF Suppl. 44,* 119–28.

—— 1984, *Geomorphological Hazards in Los Angeles* (Allen & Unwin, London).

—— BRUNSDEN, D., DOORNKAMP, J. C., and JONES, D. K. C., 1982, *Urban Geomorphology in Drylands* (Clarendon Press, Oxford).

—— and MORTIMER, C., 1971, 'Geomorphological evidence of faulting in the southern Atacama Desert, Chile', *Revue de Géomorphologie Dynamique, 20,* 71–8.

—— and WARREN, A., 1973, *Geomorphology in Deserts* (Batsford, London; California UP, San Francisco).

CORBEL, J., 1959, Erosion en terrain calcaire: vitesse d'érosion et morphologie', *Ann. Géog. 68,* 97–120.

CORRENS, C. W. 1949, Growth and dissolution of crystals under linear pressure', in *Disc. Faraday Soc., 5, Crystal Growth* (Butterworth, London), 267–71.

CORTE, A. E., 1969, 'Geocryology and Engineering', *Reviews in Eng. Geol. 2.*

COSTA, J. E. and BAKER, V. R., 1981, *Surficial Geology, Building with the Earth* (Wiley, New York).

—— and P. J. FLEISHER, 1984, *Developments and Applications of Geomorphology* (Springer-Verlag, Berlin).

COTTON, C. A., 1948, *Landscape* (CUP, Cambridge), 2nd edn.).

CRAIG, R. F., 1978, *Soil Mechanics* (Van Nostrand Reinhold, New York).

CRAIG, R. G. and CRAFT, J. L. (eds.), 1982, *Applied Geomorphology* (Allen & Unwin, London).

CRAIK, K. H., 1972, 'Appraising the objectivity of landscape dimensions', in J. V. Krutilla (ed.), *Natural Environments* (Resources for the Future, Washington, DC), 292–346.

CROFTS, R., 1973, 'Slope categories in environmental management', *Dept. Geography, UCL, Occasional Paper*.

—— 1975, 'The landscape component approach to landcape evaluation', *Trans. Inst. Brit. Geog. 66*, 124–9.

—— and COOKE, R. U., 1974, 'Landscape evaluations: a comparison of techniques', *Dept. Geography, UCL, Occasional Papers, 25*.

CROSS, M., DEGG, M., and JOHNSON, J., 1985, 'The escalating earthquake risk in the Middle East', *Risk and Loss Management*, 16–19.

CROZIER, M. J., 1984, 'Field assessment of slope instability', in D. Brunsden and D. B. Prior (eds.), *Slope Instability* (Wiley, Chichester), 103–42.

—— 1986, *Landslides: Causes, Consequences and Environment* (Croom Helm, London).

CULLEY, E. W. *et al.*, 1984, 'Summary of sulphur dioxide, and smoke concentrations for years 1981/1982 and 1982/1983', *GLC Scientific Services Branch Report, DG/ESD/R131a*.

CULLINGFORD, R. A. *et al.* (eds.), 1980, *Timescales in Geomorphology* (Wiley, Chichester).

CULSHAW, M. G. and WALTHAM, A. C., 1987, 'Natural and artificial cavities as ground engineering hazards', *Q. J. Eng. Geol. 20*, 139–50.

CURRAN, P. J., 1985, *Principles of Remote Sensing* (Longman, Harlow).

CURTIS, C. D., 1976, 'Chemistry of rock weathering: fundamental reactions and controls', in E. Derbyshire (ed.), *Geomorphology and Climate* (Wiley, London), 25–57.

CURTIS, L. F., DOORNKAMP, J. C., and GREGORY, K. J., 1965, 'The description of relief in the field study of soils', *J. Soil. Sci. 16*, 16–30.

CUTLER, D. F. and RICHARDSON, I. B. K., 1980, *Tree Roots and Buildings* (Construction Press, London).

CVIJIĆ, J., 1895, *Karst* (Geographical Monograph, Beograd).

—— 1918, 'L'hydrographie souterraine et l'évolution morphologique du Karst', *Rev. Géog. Alpine, 6*, 375–426.

CZUDEK, T. and DEMEK, J., 1970, 'Thermokarst in Siberia and its influence on the development of lowland relief', *Quaternary Res. 1*, 103–20.

DALRYMPLE, J. B., BLONG, R. J., and CONACHER, A. J., 1968, 'A hypothetical nine unit land surface model', *Zeit. für Geom. 12*, 60–76.

DARBYSHIRE, J. and DRAPER, L., 1963, 'Forecasting wind-generated sea waves', *Engineering, 195*, 482–4.

DAVIDSON-ARNOTT, R., NICKLING, W., and FAHEY, B. D., (eds.), 1982, *Research in Glacial, Glaciofluvial and Glaciolacustrine Systems* (GeoBooks, Norwich).

DAVIES, J. L., 1964, 'A morphogenic approach to world coastlines', *Zeit. für Geom. 8*, 127–42.

—— 1973, *Geographical Variation in Coastal Development* (Hafner, New York).

DAVIES, R. A. (ed.), 1979, *Coastal Sedimentary Environments* (Springer-Verlag, New York).

—— 1978, 'Beach and nearshore zone', in R. A. Davis (ed.), *Coastal Sedimentary Environments* (Springer-Verlag, New York), 237–85.

—— and ETHINGTON, R. L. (eds.), 1976, *Beach and Nearshore Sedimentation* (Society of Economic Paleontologists and Mineralogists, *Spec. Publ. 24*).

DAY, M. J. and GOUDIE, A. S., 1977, 'Field assessment of rock hardness using the Schmidt hammer test', *Brit. Geomorphological Res. Group Tech. Bull. 18*, 19–29.

DEAN, R. G. and MAURMEYER, E. M., 1983, 'Models for beach-profile response', in P. D. Komar (ed.), *Handbook of Coastal Processes and Erosion* (CRC Press, Boca Raton, Fla.), 151–65.

DEARMAN, W. R. and FOOKES, P. G., 1974, 'Engineering geological mapping for civil engineering practice in the United Kingdom', *Q. J. Eng. Geol. 7*, 223–56.

DE BENITO, G. A., 1974, 'Sand dune stabilisation at El Aaiun, West Sahara', *Int. J. Biometeorology, 18*, 142–4.

DE BOODT, M. and GABRIELS, D., (eds.), 1980, *Assessment of Erosion* (Wiley, Chichester).

DEGG, M., 1986, *The 1985 Mexican Earthquake* (Reinsurance Offices Association, London).

—— 1987a, 'History of natural hazards', *Reactions* (Jan.), 19–21.

—— 1987b, 'Lessons from the Mexico City earthquake', *Spectrum, 205*, 5–7.

DE GRAFF, J. V. and ROMESBURG, H. C., 1980, 'Regional landslide—susceptibility assessment for wildland management: a matrix approach', in D. R. Coates and J. D. Vitek (eds.), *Thresholds in Geomorphology* (Unwin, London), 401–14.

DEMEK, J. (ed.), 1972, *Manual of Detailed Geomorphological Mapping* (Academia, Prague).

—— and EMBLETON, C. (eds.), 1978, *Guide to Medium-Scale Geomorphological Mapping* (IGU, Stuttgart).

DEPARTMENT OF THE ENVIRONMENT, 1976, *Aggregates: The Way Ahead* (HMSO, London).

DE PLOEY, J. and GABRIELS, D., 1980, 'Measuring soil loss and experimental studies', in M. J. Kirkby and

R. P. C. Morgan (eds.), *Soil Erosion* (Wiley, Chichester), 63–108.

DERBYSHIRE, E., 1975, 'The distribution of glacial soils in Great Britain', in Midlands Soil Mechanics and Foundation Engineering Society, *The Behaviour of Glacial Materials*, 7–17.

—— and LOVE, M. A., 1986, 'Glacial environments', in P. G. Fookes and P. R. Vaughan (eds.), *A Handbook of Engineering Geomorphology* (Surrey Univ. Press, Guildford).

DICKERT, T. G. and DOMENY, K. (eds.), 1974, *Environmental Impact Assessment: Guidelines and Commentary* (Univ. of California, Berkeley, Calif.).

DIETRICH, W. E., DUNNE, T., HUMPHREY, N. F. and REID, L. M., 1982, 'Construction of sediment budgets for drainage basins', in F. J. Swanson *et al.* (eds.), *Sediment Budgets and Routing in Forested Drainage Basins* (US Dept. Agriculture Forest Service, Pacific NW Forest and Range Experiment Station General Tech. Rep., PNW-141), 5–23.

DIN, S. H. S. EL, 1977, 'Effects of the Aswan High Dam on the Nile flood and on the estuarine and coastal circulation pattern along the Mediterranean Egyptian coast', *Limnology and Oceanography, 22*, 194–207.

DINGMAN, S. L. and PLATT, R. H., 1977, 'Floodplain zoning: implications of hydrologic and legal uncertainty', *Water Resources Research, 13*, 519–23.

DOLAN, R., 1973, 'Barrier islands: natural and controlled', in D. R. Coates (ed.), *Coastal Geomorphology* (Publ. in Geomorphology, State Univ. New York, Binghamton), 263–78.

—— and HAYDEN, B., 1983, 'Patterns and prediction of shoreline change', in P. D. Komar (ed.), *Handbook of Coastal Processes and Erosion* (CRC Press, Boca Raton, Fla.), 123–49.

—— —— and MAY, S., 1983, 'Erosion of the US shorelines', in P. D. Komar (ed.), *Handbook of Coastal Processes and Erosion* (CRC Press, Boca Raton, Fla.), 285–99.

—— —— —— and MAY, P., 1982, 'Erosion hazards along the mid-Atlantic coast', in R. G. Craig and J. L. Craft (eds.), *Applied Geomorphology* (Allen & Unwin, London), 165–80.

DOORNKAMP, J. C., 1971, 'Geomorphological mapping', in S. H. Ominde (ed.), *Studies in East African Geography and Development* (Heinemann, London and Nairobi), 9–28.

—— 1972, 'Trend-surface analysis of planation surfaces, with an East African case study', in R. J. Chorley (ed.), *Spatial Analysis in Geomorphology* (Methuen, London).

—— 1982a, *Applied Geography* (Nottingham Mngrs. in Applied Geography, Dept. Geography, Univ. Nottingham).

—— 1982b, 'The physical basis for planning in the Third World', *Third World Planning Review, 4*, 11–31; 111–28; 213–46.

—— 1985, *The Earth Sciences and Planning in the Third World* (Liverpool Univ. Press, Liverpool).

—— 1986, 'Geomorphological approaches to the study of neotectonics', *J. Geol. Soc. (London), 143*, 335–42.

—— BRUNSDEN, D., JONES, D. K. C., COOKE, R. U., and BUSH, P. R., 1979, 'Rapid geomorphological assessments for engineering', *Q. J. Eng. Geol. 12*, 189–204.

—— —— and —— 1980, *Geology, Geomorphology and Pedology of Bahrain* (GeoBooks, Norwich).

—— *et al.*, 1986, 'Environmental geology mapping—an international review', *Proc. Eng. Geol. Conf. Plymouth* (Eng. Geol. Group, Geol. Soc., London), 225–35.

—— and HAN, M., 1985, 'Morphotectonic research in China and its application to earthquake prediction', *Progr. Phy. Geog. 7*, 353–81.

—— and KING, C. A. M., 1971, *Numerical Analysis in Geomorphology* (Arnold, London).

DOUGLAS, I., 1967, 'Man, vegetation and the sediment yield of rivers', *Nature, 215*, 925–8.

DOUGLAS, L. A., 1980, 'The use of soils in estimating the time of last movement of faults', *Soil Science, 129*, 345–52.

DREW, D. P., 1970, 'The significance of percolation water in limestone catchments', *Groundwater, 8*, 8–11.

—— NEWSON, M. D., and SMITH, D. I., 1968, 'Mendip karst hydrology research project, phase three', *Wessex Cave Club Occ. Publ. 2.*

DRIJVER, C. A. and MARCHLAND, N., 1985, *Taming the Floods: Environmental Aspects of Floodplain Development in Africa* (Centre for Environmental Studies, Leiden).

DRURY, S. A., 1987, *Image Interpretation in Geology* (Allen & Unwin, London).

DUMAS, B., GUEREMY, P., LHENAFF, R., and RAFFY, J., 1984, 'Mouvements de terrain et risques associés: presentation d'un essai cartographique', in J. C. Flageollet (ed.), 'Mouvements de terrains', *Série Documents du BRGM, 83*, 163–171.

DUNNE, T., 1979, 'Sediment yield and land use in tropical catchments', *J. Hydr. 42*, 281–300.

—— and AUBRY, A. F., 1986, 'Evaluation of Horton's theory of sheetwash and rill erosion on the basis of field experiments', in A. D. Abrahams (ed.), *Hillslope Processes* (Allen & Unwin, London), 31–53.

—— and LEOPOLD, L. B., 1978, *Water in Environmental Planning* (Freeman, San Francisco).

DUNSTAN, L. M., 1966, 'Some aspects of planning in relation to mineral resources', *Chartered Surveyor, 99*, 67–73.

DYER, K., 1979, *Estuarine Hydrography and Sedimentation* (CUP, Cambridge).

—— 1986, *Coastal and Estuarine Sediment Dynamics* (Wiley, Chichester).

EASTAFF, D. J., BRIGGS, C. J., and MCELHINNEY, M. D., 1978, 'Middle East-geotechnical data collection', *Q. J. Eng. Geol. 11*, 51–63.

ECKHOLM, E. P., 1978, *Losing Ground: Environmental Stress and World Food Problems* (Pergamon, London).

EDMONDS, C. N., 1983, 'Towards the prediction of subsidence risk upon the Chalk outcrop', *Q. J. Eng. Geol. 16*, 261–66.

EIBY, G. A., 1980, *Earthquakes* (Heinemann, Auckland).

ELLIS, J. B., 1987, 'Sediment-water quality interactions in urban rivers', in V. Gardiner (ed.), *International Geomorphology 1986, 1* (Wiley, Chichester), 287–301.

ELLISON, W. D., 1947, 'Soil erosion studies', pts. I–VII, *Agricultural Engineering, 28*, 145–6, 197–201, 245–8, 297–300,349–51, 353, 402–5, 408, 442–4, 450.

EL-SWAIFY, S. A., MOLDENHAUER, W. C., and LO, A., 1985, *Soil Erosion and Conservation* (Soil Cons. Soc. Am., Ankeny, Iowa).

ELWELL, H. A., 1984, 'Soil loss estimation, a modelling technique', in R. F. Hadley and D. E. Walling (eds.), *Erosion and Sediment Yield* (GeoBooks, Norwich), 15–36.

EMBLETON, C., 1982, *Glaciology in the Service of Man* (Inaugural lecture, King's College, London).

EMBLETON, C., and KING, C. A. M., 1975b, *Periglacial Geomorphology* (Arnold, London).

EMBLETON, C. and THORNES, J. B. (eds.), 1979, *Process in Geomorphology* (Arnold, London).

Engineering Group Working Party, 1972, 'The preparation of maps and plans in terms of engineering geology', *Q. J. Eng. Geol.*, 5,293–381.

—— 1982, 'Land surface evaluation for engineering practice', *Q. J. Eng. Geol. 15*, 265–316.

Environment Canada, 1986, 'Wetlands in Canada: A valuable resource', *Lands Directorate Fact Sheet, 84-4.*

EVANS, D. M., 1966, 'Man-made earthquakes in Denver', *Geotimes, 10*, 11–18.

EVANS, I., 1970, 'Salt crystallization and rock weathering: a review', *Rev. Géom. 19*, 153–77.

EVERETT, D. H., 1961, 'The thermodynamics of frost damage to porous solids', *Trans. Faraday Soc. 57*, 1541–51.

EYLES, N. (ed.), 1983, *Glacial Geology: An Introduction for Engineers and Earth Scientists* (Pergamon, Oxford).

FANIRAN, A., and JEJE, L. K., 1983, *Humid Tropical Geomorphology* (Longman, London).

FAO, 1960, 'Soil erosion by wind and measures for its control on agricultural land', *Agric. Development Paper, 71.*

—— 1965, 'Soil erosion by water: some measures for its control on cultivated lands', *Agric. Development Paper, 81.*

—— 1977, 'Assessing soil erosion', *FAO Soils Bull. 34.*

—— 1978, 'Final summary report on the application of LANDSAT imagery to the soil degradation mapping at 1:5 000 000', *AGLT Bull. 1/78.*

FARRES, P. J. and CONSEN, S. M., 1985, 'An improved method of aggregate stability measurement', *Earth Surface Processes and Landforms, 10*, 321–9.

FASSINA, V., 1978, 'A survey on air pollution and deterioration of stonework in Venice', *Atmospheric Environment, 12*, 2205–11.

FEDDEMA, J. J. and MEIERDING, T. C., 1987, 'Marble weathering and air pollution in Philadelphia', *Atmospheric Environment, 21*, 143–57.

FENELON, P., 1968, 1975, *Phénomènes Karstiques* (Centre du Centre National de la Recherche Scientifique, Paris), 2 vols.

FERGUSON, H. F. (ed.), 1974, 'Geological mapping for environmental purposes', *Geol. Soc. Am., Div. Eng. Geol., Eng. Geol. Case-histories, 10.*

FERRIANS, O. J., KACHADOORIAN, R. and GREENE, G. W., 1969, 'Permafrost and related engineering problems in Alaska', *US Geol. Surv. Prof. Paper, 678.*

FINES, K. D., 1968, 'Landscape evaluation: a research project in East Sussex', *Regional Studies, 12*, 41–55.

FISCHER, D. W., 1985, 'Shoreline erosion: a management framework', *J. Shoreline Management, 1*, 37–50.

FISHER, P. S. and SKIDMORE, E. L., 1970, 'WEROS: A Fortran IV program to solve the wind-erosion equation', *US Dept. Agriculture, ARS, 41–174.*

FITZPATRICK, E. A., 1971, *Pedology: A Systematic Approach to Soil Science* (Oliver & Boyd, Edinburgh).

FLAWN, P. T., 1970, *Environmental Geology* (Harper & Row, New York).

FLEISHER, P. J., 1984, 'Maps in applied geomorphology', in J. E. Costa and P. J. Fleisher (eds.), *Developments and Applications of Geomorphology* (Springer-Verlag, Berlin), 171–202.

FLEMAL, R. C., 1976, 'Pingos and pingo scars: their characteristics, distribution and utility in reconstructing former permafrost environments', *Quat. Res. 6*, 37–53.

FLEMING, G. and AL KADHIMI, A., 1982, 'Sediment modelling and data sources: a compromise in assessment', in *Recent Developments in the Explanation and Prediction of Erosion and Sediment Yield* (Int. Ass. Sci. Hydr., Publ. 137), 251–9.

FLEMING, R. W., VARNES, D. J. and SCHUSTER, R. L., 1979, 'Landslide hazards and their reduction', *Am. Planning Assoc. J. 45*, 428–39.

FOLK, R. L. and WARD, W. C., 1957, 'Brazos River bar, a study of the significance of grain size parameters', *J. Sed. Petrol. 27*, 3–27.

FOOKES, P. G., 1980, 'An introduction to the influence of natural aggregates on the performance and durability of concrete', *Q. J. Eng. Geol. 13*, 207–29.

—— DEARMAN, W. R. and FRANKLIN, J. A., 1971, 'Some engineering aspects of rock weathering with field examples from Dartmoor and elsewhere', *Q. J. Eng. Geol. 4*, 139–85.

—— and FRENCH, W. J., 1977, 'Soluble salt damage to

surfaced roads in the Middle East', *J. Inst. Highway Engs.* *24*, 10–20.

FOOKES, P. G., GORDON, D. L. and HIGGINGBOTTOM, I. E., 1975, Glacial landforms, their deposits and engineering characteristics', in Midlands Soil Mechanics and Foundation Engineering Society, *Engineering Behaviour of Glacial Materials*, 18–51.

—— and GRAY, J. M., 1987, 'Geomorphology and civil engineering', in V. Gardiner (ed.), *Int. Geomorphology 1986*, *1* (Wiley, Chichester), 83–105.

—— and HIGGINBOTTOM, I. E., 1980, 'Some problems of construction aggregates in desert areas with particular reference to the Arabian peninsula', *Proc. Inst. Civ. Engs.* *68*, 39–67; 69–90.

—— and HORSWILL, P., 1969, 'Discussion on engineering grade zones', *Proc. Conf. In Situ Testing Soils and Rocks* (Inst. Civ. Eng., London), 53–7.

—— and SWEENEY, H., MANBY, C. N. D., and MARTIN, R. T. P., 1985, 'Geological and geotechnical engineering aspects of low-cost roads in mountainous terrain', *Eng. Geol.* *21*, 1–152.

—— and VAUGHAN, P. R., 1986, *A Handbook of Engineering Geomorphology* (Surrey Univ. Press, Blackie, Glasgow).

FOOTE, L., 1972, *Soil Erosion and Water Pollution Prevention* (Nat. Association of County Engineers Action Guide Series, USA).

FORT, M. B., WHITE, P. G., and SHRESTHA, B. L., 1984, '1:50 000 geomorphic hazards mapping in Nepal', in J.-C. Flageollet (ed.), 'Mouvements de Terrains', Série Documents du BRGM, *83*, 186–94.

FOURNIER, F., 1960, *Climat et erosion: la relation entre l'érosion du sol par l'eau et les précipitations atmosphériques* (PUF, Paris).

FOURNIQUET, J., VOGT, J., and WEKER, G., 1981, 'Seismicity and recent crustal movements in France', *Tectonophysics*, *71*, 195–216.

FOX, T., 1978, 'Modeling coastal environments', in R. A. Davis (ed.), *Coastal Sedimentary Environments* (Springer-Verlag, New York), 385–413.

FRENCH, H. M., 1974, 'Active thermokarst processes, Eastern Banks Island, Western Canadian Arctic', *Can. J. Earth Sci.* *11*, 785–94.

—— 1975, 'Man-induced thermokarst, Sachs Harbour Airstrip, Banks Island, Northwest Territories', *Can. J. Earth Sci.* *12*, 132–44.

—— 1976, *The Periglacial Environment* (Longman, London).

—— 1980, 'Periglacial geomorphology and permafrost', *Prog. Phys. Geog.* *4*, 254–61.

FREVERT, R. K., SCHWAB, G. O., EDMINSTER, T. W., and BARNES, K. K., 1955, *Soil and Water Conservation Engineering* (Wiley, New York).

FREY, R. W. and BASAN, P. B., 1978, 'Coastal salt marshes', in R. A. Davis (ed.), *Coastal Sedimentary Environments* (Springer-Verlag, New York), 101–69.

FROST, R. E., 1961, 'Aerial photography in Arctic and

Subarctic engineering', *Am. Soc. Civ. Eng.* *126*, 116–45.

FRUTIGER, H., 1980, 'History and actual state of legalization of avalanche zoning in Switzerland', *J. Glaciology*, *26*, 313–24.

FRYBERGER, S. G., 1979, 'Dune forms and wind regime', in E. D. McKee (ed.), 'A study of global sand seas', US Geol. Surv. Prof. Paper, *1052*, 137–69.

FULTON, A. R. G., JONES, D. K. C., LAZZARI, S., 1987, 'The role of geomorphology in post-disaster reconstruction: the case of Basilicata, Southern Italy', in V. Gardiner (ed.), *International Geomorphology 1986*, (Wiley, Chichester), 241–62.

GAGEN, P. J. and GUNN, J., 1987, 'A geomorphological approach to restoration blasting in limestone quarries', in B. Beck and W. L. Wilson (eds.), *Proc. 2nd Multidisciplinary Conference on Sinkholes and the Environmental Impacts of Karst* (Balkema, Rotterdam), 457–61.

GALON, R., 1962, *Instruction to the Detailed Geomorphological Map of the Polish Lowland* (Polish Acad. of Science, Geogr. Inst. of Geomorphology and Hydrography of the Polish Lowland at Torun).

GALVIN, C. J., 1968, 'Breaker-type classification on three laboratory beaches', *J. Geophys. Res.*, *73*, 3651–9.

GANDEMER, J., 1977, 'Wind environment around buildings: aerodynamic concepts', *Proc. 4th Int. Conf. on Wind Effects on Buildings and Structures*, 423–32.

GARDNER, M. E., 1968, 'Preliminary report on the engineering geology of the Boulder quadrangle, Boulder County, Colorado', *US Geol. Surv. Open-file Report*.

—— and JOHNSON, C. G., 1978, 'Engineering geologic maps for regional planning', in R. O. Utgard *et al.* (eds.), *Geology in the Urban Environment* (Burgess Publishing Co.), 256–64.

GAURI, K. L., 1978, 'The preservation of stone', *Sci. Am.* *238*, 104–110.

—— and HOLDREN, G. C., jun., 1981, 'Pollutant effects on stone monuments', *Environment Science and Technology*, *15*, 386–90.

GAVRILOV, A. V., KONDRATYEVA, K. A., PIZHANKOVA, YE. I., and DUNAYEVA, YE. N., 1983, 'Methodology for using air photos and satellite imagery in permafrost surveys', in *Proc. Permafrost 4th Int. Conf.* (Nat. Acad. Press, Washington, DC), 339–41.

GAYDOS, L. and WITMER, R. E., 1983, 'Mapping on Arctic land cover utilizing LANDSAT digital data', in *Proc. 4th Permafrost Int. Conf.* (Nat. Acad. Press, Washington, DC), 343–6.

GEOMORPHOLOGICAL SERVICES LTD., 1986, *Environmental Geology Mapping* (Dept. of the Environment, London), 2 vols.

—— 1988, *Applied Earth Science Mapping for Planning and Development*, Torbay, Devon (Dept. of the Environment, London).

GERRARD, A. J., 1981, *Soils and Landforms* (Allen & Unwin, London).

GERSON, R., AMIT, R., and GROSSMAN, S., 1985, *Dust Availability in Desert Terrains* (Inst. Earth Sciences, Hebrew Univ. of Jerusalem, Jerusalem).

GHADIRI, H. and PAYNE, D., 1980, 'A study of soil splash using cine photography', in M. De Boodt and D. Gabriels (eds.), *Assessment of Erosion* (Wiley, Chichester), 185–92.

GHETTI, A. and BATISSE, M., 1983, 'The overall protection of Venice and its lagoon', *Nature and Resources*, *19*, 7–19.

GIBBARD, P. C., 1977, 'Pleistocene history of the Vale of St Albans', *Phil. Trans. Roy. Soc.* B280, 445–83.

GILBERT, G. K., 1914, 'Transportation of debris by running water', *US Geol. Surv. Prof. Paper, 86.*

—— 1917, 'Hydraulic mining debris by running water', *US Geol. Surv. Prof. Paper, 105.*

GILEWSKA, S., 1967, 'Different methods of showing the relief on the detailed geomorphological maps', *Zeit. für Geom. 11*, 481–90.

GILLULY, J., WATERS, A. C., and WOODFORD, A. D., 1968, *Principles of Geology* (Freeman, San Franciso), 3rd edn.

GLEN, J. W., ADIE, R. J., JOHN, D. M., and HOMER, D. R. (eds.), 1980, 'Symposium on snow in motion', *J. Glaciology, 26.*

GLYMPH, L. M., 1975, 'Evolving emphases in sediment-yield predictions', in *US Dept. Agriculture, Agric. Research Service, ARS-S-40*, 1–4.

GODFREY, P. J. and GODFREY, M. M., 1973, 'Comparison of ecological and geomorphic interactions between altered and unaltered barrier island systems in North Carolina', in D. R. Coates (ed.), *Coastal Geomorphology* (Publ. in Geomorphology State Univ. New York, Binghamton), 239–58.

GOLDRICH, S. S., 1938, 'A study of rock-weathering', *J. Geol. 46*, 17–58.

GOUDIE, A. S. (ed.), 1981, *Geomorphological Techniques* (Allen & Unwin, London).

—— 1983, 'Dust storms in space and time', *Prog. Phys. Geog. 7*, 502–30.

—— and COOKE, R. U., 1984, 'Salt efflorescences and saline lakes; a distributional analysis', *Geoforum, 15*, 563–82.

———— and EVANS, I., 1970, 'Experimental investigation of rock weathering by salts', *Area, 1970*, 42–8.

—— and PYE, K. (eds.), 1983, *Chemical Sediments and Geomorphology* (Academic Press, London).

—— WARREN, A., JONES, D. K. C., and COOKE, R. U., 1987, 'The character and possible origins of the aeolian sediments of the Wahiba Sand Sea, Oman', *Geog. J. 153*, 231–56.

GOUGH, D. I., 1978, 'Induced seismicity', in *The Assessment and Mitigation of Earthquake Risk* (UNESCO, Paris), 91–117.

GRAF, W. L., 1977, 'The rate law in fluvial geomorphology', *Am. J. Sci. 277*, 178–91.

—— 1979, 'Mining and channel response', *Ann. Ass. Am. Geog. 69*, 262–75.

—— 1985, *The Colorado River: Instability and Basin Management* (Assoc. of Am. Geog., Washington, DC).

GRAHAM, J., 1984, 'Methods of stability analysis', in D. Brunsden and D. B. Prior (eds.), *Slope Instability* (Wiley, Chichester), 171–215.

GRANT, K., 1968, 'A terrain evaluation system for engineering', *Div. Soil Mech. Tech. Paper, 2* (CSIRO, Australia).

—— 1972, 'Terrain classification for engineering purposes of the Melbourne area, Victoria, *Div. Appl. Geomech. Paper, 11* (CSIRO, Australia).

—— 1974, 'The PUCE programme for terrain evaluation for engineering purposes, II, Procedures for terrain classification', *Div. Appl. Geomech., Tech. Paper, 19* (CSIRO, Australia).

—— and LODWICK, G. D., 1968, 'Storage and retrieval of information in a terrain classification system', *Proc. Conf. Aust. Road Res. Board, 4*, 1667–76.

GRAVE, N. A., 'A geocryological aspect of the problem of environmental protection', *Proc. 4th Permafrost Int. Conf.*, (Nat. Acad. Press, Washington, DC), 369–73.

—— 1984, 'Development and environmental protection in the permafrost zone of the USSR: a review', *Final Proc. 4th Permafrost Int. Conf.* (Nat. Acad. Press, Washington, DC), 226–30.

GRAY, R. E. and BRUHN, R. W., 1984, 'Coal mine subsidence: Eastern United States', *Geol. Soc. Am., Reviews in Eng. Geol. 6*, 123–49.

GREELEY, R. and IVERSON, J. D., 1985, *Wind as a Geological Process* (CUP, Cambridge).

—— WHITE, B. R., POLLACK, J. B., IVERSEN, J. D., and LEACH, R. N., 1977, 'Dust storms on Mars: considerations and simulations', *NASA Tech. Mem. 78423.*

GREEN, C. H. and PENNING-ROWSELL, E. C., 1985, 'Evaluating the intangible benefits and costs of a flood alleviation proposal', *J. Inst. Water Engs. and Scientists, 40*, 229–48.

GREEN, C. P., McGREGOR, D. F. H., and EVANS, A. H., 1982, 'Development of the Thames drainage system in Early and Middle Pleistocene times', *Geol. Mag. 119*, 281–90.

GREGORY, K. J., 1976, 'Lichens and the determination of river channel capacity', *Earth Surface Processes, 1*, 273–85.

—— 1977, 'The context of river channel changes', in K. J. Gregory (ed.), *River Channel Changes* (Wiley, Chichester), 1–12.

—— 1979a, 'River channels', in K. J. Gregory and D. E. Walling, *Man and Environmental Processes* (Westview Press, Boulder, Colo.), 123–43.

—— 1979b, 'Hydrogeomorphology: how applied should we become?' *Prog. Phys. Geog. 3(1)*, 84–101.

—— 1979c, 'Drainage network power', *Water Resources Research, 15*, 775–7.

GREGORY, K. J. AND WALLING, D., 1973, *Drainage Basin Form and Process* (Arnold, London).

—— and ——, 1974, 'Fluvial processes in instrumented watersheds', *Inst. Brit. Geog. Spec. Publ.*, 6.

GREGORY, S., 1980, 'Existing pattern of water supplies', in J. C. Doornkamp, K. J. Gregory, and A. S. Burn (eds.), 1980, *Atlas of Drought in Britain, 1975–76* (Institute of British Geographers, London), 65–6.

GRIGORYEV, A. A. and KONDRATYEV, K. J., 1980, 'Atmospheric dust observed from space', *WMO Bulletin*, 3–9.

GROVE, J. M., 1987, 'Glacier fluctuations and hazards', *Geog. J. 153*, 351–69.

GUPTA, M. K. and RASTOGI, B. K., 1976, *Dams and Earthquakes* (Elsevier, New York).

HAANTJENS, H. A., 1968, 'The relevance for engineering of principles, limitations and developments in land system surveys in New Guinea', *Proc. Conf. Aust. Road Res. Board*, 4, 1593–612.

HACK, J. T., 1957, 'Studies of longitudinal stream profiles in Virginia and Maryland', *US Geol. Surv. Prof. Paper, 294-B*.

HADLEY, R. F. and WALLING, D. E., 1984, *Erosion and Sediment Yield* (GeoBooks, Norwich).

—— et al., 1985, *Recent Developments in Erosion and Sediment Yield Studies* (UNESCO, Paris).

HAGEN, L. J., 1984, 'Soil aggregate abrasion by impacting sand and soil particles', *Trans. Am. Soc. Agric. Eng.* 27, 805–8.

HAILS, J. R. (ed.), 1977a, *Applied Geomorphology* (Elsevier, Amsterdam).

—— 1977b, 'Applied geomorphology in coastal-zone planning and management', in J. R. Hails (ed.), *Applied Geomorphology* (Elsevier, Amsterdam), 317–62.

—— and CARR, A. (eds.), 1975, *Nearshore Sediment Dynamics and Sedimentation* (Wiley, London).

HALL, A. M., THOMAS, M. F., and THORP, M. B., 1985, 'Late Quaternary alluvial placer development in the humid tropics: the case of the Birim Diamond Placer, Ghana', *J. Geol. Soc. London, 142*, 777–87.

HALL, R. P. M., CHERRY, S. M., GODDARD, J. W. F., and KENNEDY, G. R., 1980, 'Rain drop sizes and rainfall rate measured by dual-polarization radar', *Nature, 285*, 195–7.

HALLSWORTH, E. G., 1987, *Anatomy, Physiology and Psychology of Erosion* (Wiley, Chichester).

HAN MUKANG, 1984, 'Tectonic geomorphology and its application to earthquake prediction in China', *Proc. American Symposium on Tectonic Geomorphology* (SUNY, Binghamton).

—— and ZHAO JINGZHEN, 1980, 'Seismotectonic characteristics of Tangyinin graben, Henan province, and its earthquake risk, *Seismology and Geology*, 2, 47–58. (In Chinese.)

HANSEN, M. J., 1984a, 'Strategies for classification of landslides', in D. Brunsden and D. B. Prior (eds.) *Slope Instability* (Wiley, Chichester), 1–25.

—— 1984b, 'Landslide hazard analysis', in D. Brunsden and D. B. Prior (eds.), *Slope Instability* (Wiley, Chichester), 523–602.

HARDING, D. M. and PARKER, D. J., 1974, 'Flood hazard at Shrewsbury, UK', in G. F. White (ed.), *Natural Hazards, Local, National and Global* (OUP, New York), 43–52.

HARRIS, C., 1981, *Periglacial Mass-wasting: A Review of Research* (Brit. Geom. Res. Group Mngr. 4, GeoAbstracts, Norwich).

HARRIS, P. M., THURRELL, R. G., HEALING, R. A., and ARCHER, A. A., 1974, 'Aggregates in Britain', *Proc. Roy. Soc. London, A339*, 329–53.

HARRIS, R., 1987, *Satellite Remote Sensing* (Routledge & Kegan Paul, London).

HARRIS, S. A., 1986, *The Periglacial Environment* (Croom Helm, London).

HATANO, S., OKABE, F., WATANABE, Y., and FURUKAWA, T., 1974, 'Morphometrical map of large-scale landslide landforms in Hokusho District, Northwestern Kyushu, Japan', *Reports of Cooperative Research for Disaster Prevention, 32* (Japan National Research Center for Disaster Prevention).

HAYES, M. O., 1975, 'Morphology of sand accumulations in estuaries', in L. E. Cronin (ed.), *Estuarine Research* (Academic Press, New York), ii, 3–22.

HAYS, W. W. (ed.), 1981, 'Facing geologic and hydrologic hazards', *US Geol. Surv. Prof. Paper, 1240-B*.

HEARN, G. J. and FULTON, A., 1986, 'Hazard assessment techniques for planning purposes: a review', in M. G. Culshaw (ed.) *Planning and Engineering Geology* (Eng. Group. Geol. Soc. London), 351–66.

HEER, J. E. and HAGERTY, D. J., 1977, *Environmental Assessments and Statements* (Van Nostrand Reinhold, New York).

HEGINBOTTOM, J. A., 1984, 'The mapping of permafrost', *Can. Geog. 28*, 78–83.

HELM, D. C., 1984, 'Field-based computational techniques for predicting subsidence due to fluid withdrawal', *Geol. Soc. Am., Reviews in Eng. Geol. 6*, 1–22.

HELMER, R., 1981, 'Water-quality monitoring: A global approach, *Nature and Resources, 17*, 7–12.

HENDERSON, F. M., 1966, *Open Channel Flow* (Macmillan, London).

HESTNES, E., 1985, 'A contribution to the prediction of slush avalanches', *Am. Glaciology, 6*, 1–4.

HEWITT, K., 1983, 'Climatic hazards and agricultural development: some aspects of the problem in the Indo-Pakistan subcontinent', in K. Hewitt (ed.), *Interpretation of Calamity* (Allen & Unwin, London), 181–201.

HEWLETT, J. D., LULL, H. W., and REINHART, K. G., 1969, 'In defense of experimental watersheds', *Water Resources Research, 5*, 306–16.

HEY, R. D., BATHURST, R. D., and THORNE, C. R., 1982, *Gravel-bed Rivers* (Wiley, London).

HICKIN, E. J., 1974, 'The development of meanders in natural river-channels', *Am. J. Sci. 274*, 414–42.

—— 1978, 'Mean flow structure in meanders of the Squamish River, British Columbia', *Can. J. Earth Sci. 15*, 1833–49.

HIGGINBOTTOM, I. A. and FOOKES, P. G., 1971, 'Engineering aspects of periglacial features in Britain', *Q. J. Eng. Geol. 3*, 85–117.

HIGH, C. and HANNA, F. K., 1970, 'A method for the direct measurement of erosion on rock surfaces', *Brit. Geom. Res. Group Tech. Bull. 5*.

HJÜLSTROM, F., 1939, 'Transportation of detritus by moving water', in P. D. Trask (ed.), *Recent Marine Sediments* (Am. Ass. Petrol. Geol.), 5–31.

—— 1949, 'Climatic changes and river patterns', *Geografiska Annaler, 31*, 83–9.

HODGES, R. D. and ARDEN-CLARKE, C., 1986, *Soil Erosion in Britain* (Soil Association Ltd., Bristol).

HOLEMAN, J. N., 1968, 'The sediment yield of major rivers of the world', *Water Resources Research, 4*, 737–47.

—— 1975, 'Procedures used in the Soil Conservation Service to estimate sediment yield', in *Agric. Research Service, ARS-S-40* (US Dept. of Agriculture), 5–9.

HOLLIS, G. E. (ed.), 1979, *Man's Impact on the Hydrological Cycle in the UK* (GeoBooks, Norwich).

HOLMAN, R. A., 1983, 'Edge waves and the configuration of the shoreline', in P. D. Komar (ed.), *Handbook of Coastal Processes and Erosion* (CRC Press, Boca Raton, Fla.), 21–33.

HOLMES, P., 1975, 'Wave conditions in coastal areas', in J. Hails and A. Carr (eds.), *Nearshore Sediment Dynamics and Sedimentation* (Wiley, Chichester), 1–15.

HOLÝ, M., 1970, *Water Erosion in Czechoslovakia* (Min. of Forestry and Water Resources, Prague).

HOLZER, T. L. (ed.), 1984*a*, 'Man-induced land subsidence', *Geol. Soc. Am., Reviews in Eng. Geol. 6*.

—— 1984*b*, 'Ground failure induced by ground-water withdrawal from unconsolidated sediment', *Geol. Soc. Am., Reviews in Eng. Geol. 6*, 67–101.

HONEYBORNE, D. B., 1982, *The Building Limestones of France* (HMSO, London).

—— and HARRIS, P. B., 1958, 'The structure of porous building stone and its relation to weathering behaviour', *Colston Papers, 10*, 343–65.

—— and PRICE, C. A., 1977, 'Air pollution and the decay of limestones', *Building Research Estab. Note, 117/77*.

HOOKE, J. M. and KAIN, R. J. P., 1982, *Historical Change in the Physical Environment* (Butterworth, London).

HOOKE, R. Le B., 1972, 'Geomorphic evidence for late-Wisconsin and Holocene tectonic deformation, Death Valley, California', *Bull. Geol. Soc. Am. 83*, 2073–98.

HORSWILL, P. and HORTON, A., 1976, 'Cambering and valley bulging in the Gwash valley at Empingham, Rutland, *Phil. Trans. Roy. Soc. London, A283*, 427–62.

HORTON, R. E., 1933, 'The role of infiltration in the hydrologic cycle', *Trans. Am. Geophys. Union, 14*, 446–60.

—— 1945, 'Erosional development of streams and their drainage basins: hydrophysical approach to quantitative morphology', *Bull. Geol. Soc. Am. 56*, 275–370.

HORWITZ, E. L., 1978, *Our Nation's Wetlands: An Interagency Task Force Report* (Council of Environmental Quality, Washington, DC).

HOSKING, J. R. and TUBEY, L. W., 1969, 'Research on low-grade and unsound aggregates', *Road Research Lab. Report*, LR 23.

House of Commons, 1984, *4th Report from the Environment Committee, Session 1983–4, 1. Acid Rain* (HMSO, London).

HOUSEMAN, J., 1961, 'Dust haze at Bahrain', *Meteorology Mag. 91*, 50–2.

HOWARD, A. D. and REMSON, F., 1978, *Geology in Environmental Planning* (McGraw-Hill, New York).

HOYT, W. G. and LANGBEIN, W. B., *Floods* (Univ. Princeton Press, Princeton, NJ).

HSÜ, K. J. (ed.), 1982, *Mountain Building Processes* (Academic Press, London).

HUDEC, P. P., 1982/3, 'Aggregate tests—their relationship and significance', *Durability of Building Materials, 1*, 275–300.

HUDSON, N., 1971, *Soil Conservation* (Batsford, London).

—— 1980, 'Erosion prediction with insufficent data', in M. de Boodt and D. Gabriels (eds.), *Assessment of Erosion* (Wiley, Chichester), 279–84.

HUGHES, O. L., 1972, 'Surficial geology and land classification, Mackenzie Valley transportation corridor', *Proc. Canadian Northern Pipeline Research Conf.* (Nat. Research Council of Canada), 17–24.

Huntings Surveys Ltd., 1977, *Technical Report: Sand Dune Movement Study South of Umm Said, 1963–1976, for the Government of Qatar, Ministry of Public Works* (Hunting Surveys Ltd., Borehamwood).

HUNTLEY, D. A. and BOWEN, A. J., 1975, 'Comparison of the hydrodynamics of steep and shallow beaches', in J. Hails and A. Carr (eds.), *Nearshore Sediment Dynamics and Sedimentation* (Wiley, Chichester), 69–109.

HUTCHINSON, J. N., 1968, 'Mass movement', in R. W. Fairbridge (ed.), *Encyclopaedia of Geomorphology* (Reinhold, New York), 688–95.

—— 1971, 'Field and laboratory studies of a fall in Upper Chalk cliffs at Joss Bay, Isle of Thanet', *Roscoe Memorial Symposium, Cambridge Univ.* 1–22.

—— 1979, *Engineering in a Landscape* (Inaugural lecture, Imperial College, London).

—— 1980, 'The record of peat wastage in the East Anglian Fenlands at Holme Post, 1848–1978 AD', *J. Ecol. 68*, 229–49.

IBRAHIM, H. A. M. and DOORNKAMP, J. C., 1988,

'Constraints on urban development in Port Said', *Third World Planning Review*, 9, 325–43.

IDSO, S. B., 1976, 'Dust storms', *Sci. Am. 235*, 108–11, 113–14.

INGLE, J. C., 1966, *The Movement of Beach Sand* (Elsevier, New York).

INMAN, D. L. and BRUSH, B. M., 1973, 'The coastal challenge', *Science, 181*, 20–32.

Institution of Civil Engineers, 1972, *Report on Mining Subsidence* (ICE, London).

International Archives of Photogrammetry and Remote Sensing, 1986, *Mapping from Modern Imagery* (Proc. Symp. Commission IV, Int. Soc. for Photogrammetry and Remote Sensing and the Remote Sensing Society), *26*.

International Association of Engineering Geology, 1976, *The Contribution of Geology Towards the Management of the Environment* (20th IGC Sydney, Australia).

International Association of Hydrological Sciences, 1982, *Recent Developments in the Explanation and Prediction of Erosion and Sediment Yields* (Int. Ass. Hydr. Sci. Publ. *137*).

—— 1986, *Drainage Basin Sediment Delivery* (Int. Ass. Hydr. Sci. Publ. *159*).

IRELAND, R. L., POLAND, J. F., and RILEY, F. S., 1984, 'Land subsidence in the San Joaquin Valley, California as of 1980', *US Geol. Surv. Prof. Paper, 437–I*.

JACKS, G. V. and WHYTE, R. O., 1939, *The Rape of the Earth: A World Survey of Soil Erosion* (Faber, London).

JACKSON, E. L. and MUKERJEE, T., 1974, 'Human, adjustment to earthquake hazard of San Francisco, California', in G. F. White (ed.), *Natural Hazards* (OUP, Oxford), 160–6.

JACKSON, M. L., TYLER, S. A., WILLIS, A. L., BOURBEAN, G. A., and PENNINGTON, P. R., 1948, 'Weathering sequence of clay size minerals in soils and sediments', *J. Phys. and Coll. Chem. 52*, 1237–60.

JAHN, A., 1975, *Problems of the Periglacial Zone* (PWN, Warsaw).

JAEGER, C., 1972, *Rock Mechanics and Engineering* (CUP, Cambridge).

JAKUCS, L., 1977, *Morphogenetics of Karst Regions* (Hilger, Bristol).

JANSEN, J. M. L. and PAINTER, R. B., 1974, 'Predicting sediment yield from climate and topography', *J. Hydrol. 21*, 371–80.

JANSEN, P., BENDEGOM, L., BERG, J., VRIES, M., and ZANEN, A., 1979. *Principles of River Engineering: The Non-tidal Alluvial River* (Pitman, London).

JANSSON, M. B., 1982, *Land Erosion by Water in Different Climates, UNGI Rapport, 57* (Uppsala Univ., Dept. Physical Geog.).

JAYNES, S., 1985, 'Studies of Building Stone Weathering in South-east England (unpublished Ph.D. thesis, Univ. London).

—— and COOKE, R. U., 1987, 'Stone weathering in south-east England', *Atmospheric Environment, 21*, 1601–22.

JENNINGS, J. N., 1975, 'Doline morphometry as a morphogenetic tool: New Zealand examples', *New Zealand Geog. 31*, 6–28.

—— 1985, *Karst Morphology* (Blackwell, Oxford).

JENNY, H., 1941, *Factors of Soil Formation* (McGraw-Hill, New York).

JOHANNESSEN, C. L. and FEIEREISEN, J. J., 1982, 'Weathering of ocean cliffs by salt expansion in a mid-latitude coastal environment', *Shore and Beach, 1982*, 26–34.

JOHNSON, D. W., 1919, *Shore Processes and Shoreline Development* (Wiley, New York).

JOHNSON, J. W., 1956, 'Dynamics of nearshore sediment movement', *Bull. Am. Ass. Petrol. Geol. 40*, 2211–32.

JOHNSON, P. A., 1971, 'Soils in Derbyshire', *Soil Survey of England & Wales (Harpenden) Soil Survey Record, 4*.

JOHNSTON, G. H., 1966, 'Pile construction in permafrost', *Proc. 1st Permafrost Int. Conf.* (Nat. Academy Sciences—Nat. Research Council of Canada), *1287*, 371–4.

—— (ed.), 1981, *Permafrost, Engineering Design and Construction* (Wiley, Toronto).

JOLLIFFE, I. P., 1983, 'III. Coastal erosion and flood abatement: what are the options?' *Geog. J. 149*, 62–71.

—— and PATMAN, C. R., 1985, 'The coastal zone: the challenge', *J. Shoreline Management, 1*, 3–36.

JOLLY, J. P., 1982, 'A proposed method for accurately calculating sediment yields from reservoir deposition volumes', in *Recent Developments in the Explanation and Prediction of Erosion and Sediment Yield* (Int. Ass. Sci. Hydr. Publ. *137*), 153–61.

JONES, D. K. C., 1980, 'British geomorphology: an appraisal', *Zeit. für Geom., NF Suppl. 36*, 48–73.

—— 1983, 'Environments of concern', *Trans. Inst. Brit. Geog. New Series*, 429–57.

—— BRUNSDEN, D., and GOUDIE, A. S., 1983, 'A preliminary geomorphological assessment of part of the Karakoram highway', *Q. J. Eng. Geol. 16*, 331–55.

—— COOKE, R. U., and WARREN, A., 1986, 'Geomorphological investigation, for engineering purposes, of blowing sand and dust hazard', *Q. J. Eng. Geol. 19*, 251–70.

JONES, J. R. and WILLETTS, B. B., 1979, 'Errors in measuring uniform aeolian sand flow by means of an adjustable trap', *Sedimentology, 26*, 463–8.

JONES, P. F. and DERBYSHIRE, E., 1983, 'Late Pleistocene periglacial degradation of lowland Britain: implications for civil engineering', *Q. J. Eng. Geol. 16*, 197–210.

JOVANOVIC, S. and VUKCEVIC, M., 1957, 'Suspended sediment regimes on some watercourses in Yugoslavia and analysis of erosion processes', *Int. Ass. Sci. Hydr. Publ. 43*, 337–59.

JUNGERIUS, P. D. (ed.), 'Soils and Geomorphology', *Catena, Supplement, 6* (Cremlingen, W. Germany).

KACHURIN, S. P., 1962, 'Thermokarst within the territory of the USSR, *Biul. Peryglac. 11*, 49–55.

KADOMURAV, H., 1980, 'Erosion by human activities in Japan', *Geol. J.* (Japan), *4*, 133–44.

KANTEY, B. A., 1971, 'Terrain evaluation—a problem in whole engineering', *Civil Engineer in South Africa* (Nov.), 407–11.

KARCZ, I. and KAFRI, U., 1975, 'Recent crustal movements along Mediterranean coastal plain of Israel', *Nature, 257*, 296–7.

KASAHARA, K., 1981, *Earthquake Mechanics* (CUP, Cambridge).

KASPERSON, R. E., 1969, 'Environmental stress and the municipal political system: the Brockton water crisis of 1961–1966', in R. E. Kasperson and J. V. Minghi (eds.), *The Structure of Political Geography* (Univ. London Press, London), 481–96.

KATES, R. W., 1965, 'Industrial flood losses', *Univ. Chicago, Dept. of Geography Res. Paper, 98.*

KELLAWAY, G. A. and TAYLOR, J. H., 1968, 'The influence of landslipping on the development of the city of Bath, England', *Proc. 23rd Intern. Geol. Congr. 12*, 65–76.

KELLER, E. A., BONKOWSKI, M. S., KORSCH, R. J., and SHLEMON, R. J., 1982, 'Tectonic geomorphology of the San Andreas fault zone in the southern Indio Hills, Coachella Valley, California', *Bull. Geol. Soc. Am. 93*, 46–56.

—— and ROCKWELL, T. K., 1984, 'Tectonic geomorphology, Quaternary chronology and palaeoseismicity', in J. E. Costa and P. J. Fleisher (eds.), *Developments and Applications of Geomorphology* (Springer-Verlag, Berlin), 203–39.

KENNIE, T. J. M. and EDMONDS, C. N., 1986, 'The location of potential ground subsidence and collapse features in soluble carbonate rocks by remote sensing techniques', *ASTM Spec. Tech. Publ.*

KIDSON, C. and CARR, A. P., 1971, 'Marking beach materials for tracing experiments', in J. A. Steers (ed.), *Introduction to Coastline Development* (Macmillan, London), 69–93.

KIENHOLZ, H., 1978, 'Maps of geomorphology and natural hazards of Grindelwald, Switzerland, scale 1 : 10 000', *Arctic and Alpine Research, 10*, 169–84.

KIERSCH, G. A., 1964, 'Vaiont Reservoir Disaster', *Civ. Eng. 34*, 32–9.

KING, C. A. M., 1972, *Beaches and Coasts* (Arnold, London), 2nd edn.

—— (ed.), 1976, *Periglacial Processes (Benchmark Papers in Geology, 27)* (Dowden, Hutchinson & Ross, Stroudsburg, Va.).

—— and McCULLAGH, M. J., 1971, A simulation model of a complex recurved spit', *J. Geol. 79*, 22–37.

KING, L. C., 1951, *South African Scenery* (Oliver & Boyd, Edinburgh), 2nd edn.

—— 1962, *Morphology of the Earth* (Oliver & Boyd, Edinburgh).

KING, R. B., 1970, 'A parametric approach to land system classification', *Geoderma, 4*, 37–46.

—— 1987, 'Review of geomorphic description and classification in land resource surveys', in V. Gardiner (ed.), *International Geomorphology 1986 Part II*, (Wiley, Chichester), 384–403.

KIRK, R. M., 1982, 'Public policy, planning and the assessment of coastal erosion, in R. D. Bedford and A. P. Sturman (eds.), *Canterbury at the Crossroads: Issues for the Eighties* (Geog. Soc. Misc. Series, 8, Christchurch, NZ), 182–96.

KIRKBY, M. J., 1969, 'Infiltration, throughflow and overland flow: and erosion by water on hillslopes', in R. J. Chorley (ed.), *Water, Earth and Man* (Methuen, London), 215–38.

—— 1974, 'Hydrological slope models: the influence of climate', *Univ. Leeds Geog. Dept. Working Paper, 89.*

—— 1980, 'The problem'; 'and modelling water erosion processes', in M. J. Kirkby and R. P. C. Morgan (eds.), *Soil Erosion* (Wiley, Chichester), 1–16 and 183–216.

—— 1984, 'Modelling cliff development in South Wales: Savigear reviewed', *Zeit. für Geom, NF, 28*, 405–26.

—— and CHORLEY, R. J., 1967, 'Throughflow, overland flow and erosion', *Bull. Int. Ass. Sci. Hydr. 12*, 5–21.

—— and MORGAN, R. P. C. (eds.), 1980, *Soil Erosion* (Wiley, Chichester).

KLEIN, M., 1984, 'Weathering rates of limestone tombstones measured in Haifa, Israel', *Zeit. für Geom. 28*, 105–11.

KLIMASZEWSKI, M., 1956, 'The principles of the geomorphological survey of Poland', *Przeglad Geograficzny, 28* (Suppl.), 32–40.

—— 1961, 'The problems of the geomorphological and hydrographic map on the example of the Upper Silesian industrial district', *Problems of Applied Geography, Geographical Studies, 25* (Polish Acad. Sci., Institute of Geography, Warsaw), 73–81.

—— (ed.), 1963, 'Problems of geomorphological mapping', *Geographical Studies, 46* (Polish Acad. Sci., Institute of Geography, Warsaw).

KLIMENKO, L. V. and MOSKALEVA, L. A., 1979, 'Frequency of occurrence of dust storms in the USSR, *Meteorologia i Gidrologiya, 9*, 93–7.

KLINGEBIEL, A. A. and MONTGOMERY, P. H., 1961, 'Land-capability classification', *Soil Conservation Service, Agric. Handbook, 210* (US Dept. Agriculture).

KLUGMAN, M. A. and CHUNG, P., 1976, 'Slope-stability study of the Regional Municipality of Ottawa-Carleton, Ontario, Canada, *Ontario Geol. Survey, Misc. Paper, MP 68.*

KNIGHTON, D., 1984, *Fluvial Forms and Processes* (Arnold, London).

KNOTT, J. M., 1980, 'Reconnaissance assessment of

erosion and sedimentation in the Cañada de Los Alamos Basin, Los Angeles and Venture Counties, California', *US Geol. Surv. Water-Supply Paper, 2061.*

KOMAR, P. D., 1976, *Beach Processes and Sedimentation* (Prentice-Hall, Englewood Cliffs, NJ).

—— (ed.), 1983*a*, *CRC Handbook of Coastal Processes and Erosion* (CRC Press, Boca Raton, Fla.).

—— 1983*b*, 'Coastal erosion in response to the construction of jetties and breakwaters', in P. D. Komar (ed.), *Handbook of Coastal Processes and Erosion* (CRC Press, Boca Raton, Fla.), 191–204.

—— 1983*c*, 'Computer models of shoreline changes', in P. D. Komar (ed.), *Handbook of Coastal Processes and Erosion* (CRC Press, Boca Raton, Fla.), 205–16.

KUDRYAVTSEV, W. A., 1978, *General Permafrost Science: Geocryology* (Moscow State Univ., Moscow). (In Russian.)

KUGLER, H., 1975, 'Geomorphologische Erkundung und agrarische Landnutzung, *Geog. Ber. 80,* 190–204.

KUNREUTHER, H. and SHEAFFER, J. R., 1970, 'An economically meaningful and workable system for calculating flood insurance rates', *Water Resources Research,* 6, 659–67.

KUPPER, M. and PISSART, A., 1974, 'Vitesse d'érosion en Belgique de calcaires d'âge primaire, exposés a l'air libre ou soumis à l'action de l'eau courante', *Abh. Ak. Wissenschaften in Gottingen, Math.-Phys. Klasse III,* 29, 39–50.

LACEY, G., 1929–30, 'Stable channels in alluvium', *Proc. Inst. Civil Eng. 229,* 259–92.

LA CHAPELLE, E. R., 1977, 'Snow avalanches: a review of current research and applications', *J. Glaciology, 19,* 313–24.

—— 1980, 'The fundamental processes in conventional avalanche forecasting', *J. Glaciology, 26,* 75–84.

LACHENBRUCH, A. H., 1966, 'Contraction theory of ice-wedge polygons: a qualitative discussion', *Proc. 1st Permafrost Int. Conf.* (Nat. Acad. Sciences—Nat. Res. Council Canada), *1287,* 65–71.

—— 1970, 'Some estimates of the thermal effects of a heated pipeline in permafrost', *US Geol. Surv. Circular, 632.*

LAMBE, T. W. and WHITMAN, R. V., 1979, *Soil Mechanics* (Wiley, New York).

LA MOREAUX, P. E., 1967, 'Hydrology of limestone terrain', *Proc. of Symp. on Groundwater Hydrology, Am. Water Resources Assn. 4,* 72–83.

LANDRY, J., 1979, *Zones exposées à des prisques liés aux mouvements du sols et des sous-sol. Région de Lons-le-Saunier à Poligny (Jura)* (Bureau de Recherche Géologique et Minière, Orléans).

LANE, E. W., 1955, 'The importance of fluvial morphology in hydraulic engineering', *Am. Soc. Civil Eng. Proc. 81,* 1–17.

LANGBEIN, W. B., 1962, 'Hydraulics of river channels as related to navigability', *US Geol. Surv. Water-Supply Paper, 1539-W.*

—— 1964, 'Geometry of river channels', *J. of the Hydraulics Div., Am. Soc. Civil Engs. 90,* 301–12.

—— and SCHUMM, S. A., 1958, 'Yield of sediment in relation to mean annual precipitation', *Trans. Am. Geophys. Union,* 39, 1076–84.

LANGE, A. L., 1970, 'The detection of prehistoric earthquakes from fractured cave structures', *Caves and Karst, 12,* 9–14.

LARONNE, J. B. and MOSLEY, M. P. (eds.), 1982, *Erosion and Sediment Yield* (Hutchinson, Ross, Stroudsburg, Penn.).

LAWRANCE, C. J., 1972, 'Terrain evaluation in West Malaysia: Part 1: Terrain classification and survey methods', *Report LR 506* (Transport and Road Research Laboratory, Crowthorne, UK).

LAWSON, A. C., 1908, 'The California earthquake of April 18, 1906', *Report of the State Earthquake Investigation Commission, Carnegie Inst. Washington Publ. 87.*

LEARY, E., 1983, *The Building Limestones of the British Isles* (HMSO, London).

LEARY, P. C., MALIN, P. E., STRELITZ, R. A., and HENYEY, T. L., 1981, 'Possible tilt phenomena observed as water level anomalies along the Los Angeles aqueduct', *Geophys. Research Letters, 8,* 225–8.

LEE, I. K., WHITE, W., and INGLES, O. G., 1983, *Geotechnical Engineering* (Pitman, London).

LEGGET, R. F. (ed.), 1976, 'Glacier Till: An Interdisciplinary Study', *Roy. Soc. Can., Spec. Publ. 12.*

—— and MACFARLANE, I. C. (eds.), 1972, *Proceedings of the Canadian Northern Pipeline Research Conference* (Nat. Research Council of Canada, Ottawa).

LEIGHTON, F. B., 1971, 'The role of consulting geologists in urban geology, in *Environmental Planning and Geology* (US Geological Survey and other agencies, Washington, DC), 82–9.

——, 1976, 'Urban landslides: targets for land-use planning in California', *Geol. Soc. Am., Spec. Paper, 173,* 89–96.

LEITH, C. K. and MEAD, W. J., 1911, 'Origin of the iron ores of central and north-eastern Cuba', *Trans. Am. Inst. Mining Engs. 42,* 90–102.

LENSEN, G. J., 1968, 'Analysis of progressive fault displacement during downcutting at the Branch River terraces, South Island, New Zealand', *Bull. Geol. Soc. Am. 79,* 543–56.

LEOPOLD, L. B., 1962, 'The VIGIL network', *Int. Ass. Sci. Hydrology Bull. 7,* 5–9.

—— 1968, 'Hydrology for urban land planning—a guidebook on the hydrologic effects of urban land use', *US Geol. Surv. Circular, 620.*

—— 1969*a*, 'Quantitative comparison of some aesthetic factors among rivers', *US Geol. Surv. Circular, 620.*

—— 1969*b*, 'Landscape aesthetics', *Natural History* (Oct.), 35–46.

—— 1973, 'River channel change with time: an example', *Geol. Soc. Am. Bull. 84,* 1845–60.

LEOPOLD, L. B., CLARKE, F. E., HANSHAW, B. B., and BALSLEY, J. R., 1971, 'A procedure for evaluating environmental impact', *US Geol. Surv. Circular, 645.*

—— EMMETT, W. W., and MYRICK, R. M., 1966, 'Channel and hillslope processes in a semiarid area, New Mexico', *US Geol. Surv. Prof. Paper, 352-6,* 191–253.

—— and LANGBEIN, W. B., 1963, 'Association and indeterminacy in geomorphology', in C. C. Albritton (ed.), *The Fabric of Geology* (Addison-Wesley, Reading), 184–92.

—— and MADDOCK, T., jun., 1953, 'The hydraulic geometry of stream channels and some physiographic implications', *US Geol. Surv. Prof. Paper, 252.*

—— and —— 1954, *The Flood Control Controversy* (Ronald Press, New York).

—— and MILLER, J. P., 1956, 'Emphemeral streams—hydraulic factors and their relation to the drainage net', *US Geol. Surv. Prof. Paper, 282-A.*

—— and WOLMAN, M. G., 1957, 'River channel patterns, braided, meandering and straight', *US Geol. Surv. Prof. Paper, 282-B.*

—— —— and MILLER, J. P., 1964, *Fluvial Processes in Geomorphology* (Freeman, San Francisco).

LEWIS, A. J. (ed.), 1976, 'Geoscience applications of imaging RADAR systems', *RSEMS, 3.*

LIED, K. and BAKKEHI, S., 1980, 'Empirical calculations of snow-avalanche run-out distance based on topographic parameters', *J. Glaciology, 26,* 165–77.

LILLESAND, T. and KEIFER, R., 1979, *Remote Sensing and Image Interpretation* (Wiley, New York).

LINDSEY, C. G., DOESBURG, J. M., and VILLARIO, R. W., 1982, 'A review of long-term rock durability', *Proc. 5th Symp., Uranium Tailing Management, Colorado State Univ.* 101–15.

LINELL, K. A., 1973, 'Long-term effects of vegetative cover on permafrost stability in an area of discontinuous permafrost', in Nat. Academy of Sciences, *North American Contribution: Permafrost, 2nd International Conference,* 688–93.

—— and TEDROW, J. C. F., 1981, *Soil and Permafrost Surveys in the Arctic* (Clarendon Press, Oxford).

LINTON, D. L., 1968, 'The assessment of scenery as a natural resource', *Scott. Geog. Mag. 84,* 218–38.

LITTLE, A. L., 1969, 'The engineering classification of residual tropical soils', *7th Proc. Int. Conf. Soil Mech. and Foundn. Eng. 1,* 1–10.

LIVESEY, R. H., 1975, 'Corps. of Engineers methods for predicting sediment yields', *Agric. Research Service, ARS-S-40* (US Dept. Agriculture), 16–32.

LIVINGSTON, R., KANTZ, M., and DORSHEIMER, J., 1981, *Stone Deterioration Studies at Bowling Green Custom House, 1980–1981, Interim Reports,* unpublished ms.

LOCK, W. W., ANDREWS, J. T., and WEBBER, P. J., 1979, 'A manual of lichenometry', *British Geom. Res. Group Tech. Bull. 26.*

LOFGREN, B. E., 1968, 'Analysis of stresses causing land subsidence', *US Geol. Surv. Prof. Paper, 600-B,* B219–B225.

—— 1969, 'Land subsidence due to the application of water', in D. J. Varnes and G. Kiersch (eds.), *Geol. Soc. Am., Reviews in Eng. Geol. 2,* 271–303.

—— and RUBIN, M., 1975, 'Radiocarbon dates indicate rates of graben downfaulting, San Jacinto Valley, California', *J. Res. of the US Geol. Surv. 3,* 45–6.

LOMNITZ, C., 1974, *Global Tectonics and Earthquake Risk* (Elsevier, Amsterdam).

LONGWELL, C. R., FLINT, R. F., and SANDERS, J. E., 1969, *Physical Geology* (Wiley, New York).

LOUGHNAN, F. C., 1969, *Chemical Weathering of the Silicate Minerals* (Elsevier, New York).

LUCINI, P., 1973, 'The potential landslides forecasting of the "Argille Varicolori Seagliose" complex in IGM 174 SE Map, Savignano di Puglia (Compania)', *Geol. Appl. e Idrogeol. 8,* 311–16.

LUCKAT, S., 1981, 'Quantitative Inter-suchung des Ein flusses von Luftverunreinigungen bei der Zerstörung von Naturstein', *Staub-Reinhalt, 41,* 440–42.

LUNDQUIST, J., 1969, 'Earth and ice mounds: a terminological discussion', in T. L. Péwé (ed.), *The Periglacial Environment: Past and Present* (McGill, Queen's Univ. Press, Montreal), 203–15.

LUSBY, G. C., 1977, 'Determination of runoff and sediment yield by rainfall simulation', in T. J. Toy (ed.), *Erosion* (GeoAbstracts, Norwich), 19–30.

LUSTIG, L. K., 1965, 'Sediment yield of the Castaic watershed, western Los Angeles County, California: a quantitative geomorphic approach', *US Geol. Surv. Prof. Paper, 422-F.*

LYLES, L., 1981, 'The US wind erosion problem', in *Proc. Am. Soc. Agric. Engs. Conf. on Crop Production with Conservation in the 80s* (Am. Soc. Agric. Eng. Publ. 7/81), 16–24.

—— 1983, 'Erosive wind energy distributions and climatic factors for the West', *J. Soil and Water Conservation, 38,* 106–9.

—— HAGEN, J., and SKIDMORE, E. L., 1983, 'Soil conservation: principles of erosion by wind', in *Dryland Agriculture* (Am. Soc. Agr. et. al., Agronomy Mngr. 23), 177–88.

—— and TACARKO, J., 1982, 'Emergency tillage to control wind erosion: influences on winter wheat yields', *J. Soil and Water Conservation, 37,* 344–7.

MABBUTT, J. A., 1968, 'Review of concepts of land classification', in G. A. Stewart (ed.), *Land Evaluation* (Macmillan, Melbourne), 11–28.

—— 1977, *Desert Landforms* (MIT Press, Cambridge, Mass.).

McANERNEY, J. M., 1966, 'Terrain interpretation from radar imagery', *Proc. 4th Symp. on Remote Sensing of Environment* (Univ. Ann Arbor, Mich.), 731–50.

McCAIG, M., 1983, 'Predicting sediment yield in drainage basins where there are no sediment records' (unpublished ms., Geomorphological Services Ltd.).

McCann, S. B., 1982, 'Coastal landforms', *Prog. in Phys. Geog. 6*, 439–45.

McCoy, R. M., 1969, 'Drainage network analysis with K-band radar imagery', *Geog. Rev. 59*, 493–512.

MacDonald, H. C., 1969, 'Geological evaluation of radar imagery from Darien Province, Panama', *Modern Geol. 1*, 1–64.

McFarlane, M. J., 1971, 'Lateritization and landscape development in Kyagwe, Uganda, *Q. J. Geol. Soc. London, 126*, 501–39.

McGregor, D. F. M. and Green, C. P., 1983, 'Lithostratigraphic sub-divisions in the gravels of the proto-Thames between Hemel Hempstead and Watford', *Proc. Geol. Ass. 94*, 83–5.

Mackay, J. R., 1966, 'Pingos in Canada', *Proc. 1st Permafrost Int. Conf.* (Nat. Acad. Sciences—Nat. Res. Council Canada), *1287*, 71–6.

—— 1970, 'Disturbances to the tundra and forest environment of the western Arctic', *Can. Geotech. J. 7*, 420–32.

—— 1972, 'The world of underground ice', *Ann. Ass. Am. Geog. 62*, 1–22.

McKee, E. D. (ed.), 1979, 'A study of global sand seas', *US Geol. Surv. Prof. Paper, 1052*.

McLellan, A. G., 1975, 'The aggregate dilemma,' *Bull. Conservation Council of Ontario, 1975*, 12–20.

McRoberts, E. C. and Morgenstern, N. R., 1974, 'The stability of thawing slopes', *Can. Geotech. J. 11*, 447–69.

Mahr, T. and Malgot, J., 1978, 'Zoning maps for regional and urban development based on slope stability', *International Association of Engineering Geology, 3rd International Congress Sect. I, 1*, 124–37.

Mainguet, M., 1983, 'Tentative megamorphological study of the Sahara', in R. Gardner and H. Scoging (eds.), *Mega-Geomorphology* (Clarendon Press, Oxford), 113–33.

—— 1984, 'Space observations of Saharan aeolian dynamics', in F. El-Baz (ed.), *Deserts and Arid Lands* (Nijhoff, The Hague), 59–77.

Malgot, J., Baliak, F., Cabalova, D., Mahr, T., and Nemcok, A., 1973, *Mapa svahovyeh deformacii Hardlovskej kotling A. Zakladna mapa; B. Mapa igenniersko-geologickeho rajonovania.*

—— and Mahr, T., 1979, 'Engineering geological mapping of the West Carpathian landslide areas', *Int. Assoc. Eng. Geol. Bull. 19*, 116–21.

Marsh, G. P., 1864, *Man and Nature: or, Physical Geography as Modified by Human Action* (Scribners, New York).

Martin, J. C. and Serdengecti, S., 1984, 'Subsidence over oil and gas fields', *Geol. Soc. Am., Reviews in Eng. Geol. 6*, 23–34.

Mather, P. M., 1987, *Computer Processing of Remotely Sensed Imagery: An Introduction* (Wiley, Chichester).

Matula, M., 1981, 'Recommended symbols for engineering geology mapping, *Bull. Int. Assoc. Eng. Geol. 24*, 227–34.

Mausbacher, R., 1983, 'Geomorphological mapping and environmental planning in permafrost regions', in *Proc. Permafrost 4th Inst. Conf.* (Nat. Acad. Press, Washington, DC), 811–15.

Mayer, L. and Nash, D. (eds.), 1987, *Catastrophic Flooding* (Unwin Hyman, London).

Mayuga, M. N. and Allen, D. R., 1969, 'Subsidence in the Wilmington oil field Long Beach, California, USA', *Publ. Inst. Sci. Hyd. 88*, 66–79.

Mehra, S. R., Chadda, L. R., and Kapier, R. N., 1955, 'Role of detrimental salts in soil stabilization, with and without cement: 1. Effect of sodium sulphate', *Indian Concr. J. 29*, 336–7.

Mellor, M., 1968, 'Avalanches, cold regions science and engineering', *Hanover NH Pt. III Sect. A3a* (US Cold Regions Research and Engineering Laboratory).

Melton, M. A., 1958, 'Correlation structure of morphometric properties of drainage systems and their controlling agents', *J. Geol. 66*, 442–60.

Meneroud, J.-P. and Calvino, A., 1976, 'Carte ZERMOS', *Zones exposées à des risques liés aux mouvements du sol et du sous-sol à 1:25 000 region de le Mayenne Verubie (Alpes-Maritimes)* (Bureau de Recherches Geologiques et Minières, Orléans).

Meybeck, M., 1976, 'Total mineral dissolved transport by world major rivers', *Hydrol. Sci. Bull. 21*, 265–84.

—— 1982, 'Carbon nitrogen and phosphorous transport by world rivers', *Am. J. Sci. 282*, 401–50.

Meyer, L. D. and Wischmeier, W. H., 1969, Mathematical Simulation of the Process of Soil Erosion by Water, *Trans. Am. Soc. Agric. Engs. 12*, 754–8.

Middleton, H. E., 1930, 'Properties of soils which influence soil erosion', *US Dept. Agr. Tech. Bull. 178*.

Midlands Soil Mechanics and Foundation Engineering Society, 1975, *The Engineering Behaviour of Glacial Materials* (reprinted by GeoBooks, 1978).

Millington, A. C., Robinson, D. A. and Brown, T. J., 1982, 'Establishing soil loss and erosion hazard maps in a developing country: a West African example', in D. Walling (ed.), *Recent Developments in the Explanation and Prediction of Erosion and Sediment Yield (Inst. Ass. Hydr. Sciences Publ. 137)*, 283–92.

—— and Townshend, J. G. R. (eds.), 1987a, *Remote Sensing of Geomorphology* (Croom Helm, London).

—— and —— 1987b, 'The potential of satellite remote sensing for geomorphological investigations: an overview', in V. Gardiner (ed.), *International Geomorphology 1986*, II (Wiley, Chichester), 331–42.

Milne, G., 1935, Composite units for the mapping of complex soil associations, *Trans. 3rd Int. Congr. Soil Sci. 1*, 345–7.

Ministry of Natural Resources, Government of Ontario, 1977, *A Policy for Mineral Aggregate Resource Management in Ontario.*

Ministry of Town and Country Planning, 1948, *Report of the Advisory Committee on Sand and Gravel* (HMSO, London), Part 1.

Ministry of Overseas Development, 1970, *The Work of the Land Resources Division* (M.o.D., Tolworth, UK).

MITCHELL, C. W., 1973, *Terrain Evaluation* (Longman, London).

—— and HOWARD, J. A., 1978, 'Final summary report on the application of LANDSAT imagery to soil degradation mapping at 1:5 000 000', FAO *AGLT Bull. 1/78*.

MITCHELL, J. K., 1976, *Fundamentals of Soil Behaviour* (Wiley, London).

—— and BUBENZER, G. D., 1980, 'Soil loss estimation', in M. J. Kirkby and R. P. C. Morgan (eds.) *Soil Erosion* (Wiley, Chichester), 17–62.

MOGI, K., 1985, *Earthquake Prediction* (Academic Press, London).

MOLLARD, J. D., no date, *Landforms and Surface Materials of Canada: A Stereoscopic Airphoto Atlas and Glossary* (J. D. Mollard, Regina, Sask.).

—— 1972, 'Airphoto terrain classification and mapping for northern feasibility studies', *Proc. Canadian Northern Pipeline Res. Conf.* (Nat. Research Council of Canada), 105–28.

—— and PIHLAINEN, J. A., 1966, 'Airphoto interpretation applied to road selection in the Arctic', *Proc. 1st Permafrost Int. Conf.* (Nat. Acad. Sciences—Nat. Research Council), *1287*, 381–7.

MOLNAR, I., 1964, 'Soil conservation—economic and social considerations', *J. Austr. Inst. App. Sci. 30*, 247–57.

MOORE, G. W. and KENNEDY, M. P., 1975, Quaternary faults at San Diego Bay, California, *J. Research of the US Geol. Surv. 3*, 589–96.

MORALES, C. (ed.), 1979, *Saharan Dust (Scope, 14)* (Wiley, Chichester).

MORGAN, C., 1983, 'The non-independence of rainfall erosivity and soil erodibility', *Earth Surface Processes and Landforms, 8*, 323–8.

MORGAN, R. P. C., 1979, *Soil Erosion* (Longman, London).

—— 1980, 'Soil erosion and conservation in Britain', *Progr. Phys. Geog. 4*, 24–47.

—— 1985, 'Soil erosion measurement and soil conservation research in cultivated areas of the UK', *Geog. J. 151*, 11–20.

—— 1986, *Soil Erosion and Conservation* (Longman, London).

——FINNEY, H. J., LAWE, H., MERRITT, E., and NOBLE, C. A., 1986, 'Plant cover effects on hillslope runoff and erosion: evidence from two laboratory experiments', in A. D. Abrahams (ed.), *Hillslope Processes* (Allen & Unwin, London), 77–96.

MORGENSTERN, N. R. and SANGREY, D. A., 1978, 'Methods of stability analysis', in R. L. Schuster and R. J. Krizek (eds.), 'Landslides—Analysis and control', *Transport Research Board Special Report, 176* (Nat. Acad. Sciences, Washington, DC).

MORISAWA, M., 1968, *Streams, Their Dynamics and Morphology* (McGraw-Hill, New York).

—— 1985, *Rivers* (Longman, London).

—— and HACK, J. T., 1985, *Tectonic Geomorphology* (Allen & Unwin, London).

MUIR WOOD, A. M. and FLEMING, C. A., 1981, *Coastal Hydraulics* (Macmillan, London), 2nd edn.

MULCAHY, M. J., 1960, 'Laterites and lateritic soils in south-western Australia', *J. Soil Sci. 11*, 206–24.

MULLER, S. W., 1947, *Permafrost or Permanently Frozen Ground and Related Engineering Problems* (Edwards, Ann Arbor, Mich.).

MUNDAY, T. J., 1984, 'Earth science applications of radar remote sensing', in *Radar Remote Sensing over the Land Surface* (Remote Sensing Society, London).

MUNN, R. E., 1979, *Environmental Impact Assessment (Scope, 5)* (Wiley, Chichester), 2nd edn.

MUTCHLER, C. K. and YOUNG, R. A., 1975, 'Soil detachment by raindrops', in US Dept. Agriculture, *Present and Prospective Technology for Predicting Sediment Yields and Sources* (Agric. Research Service, USDA, *ARS-S-40*), 113–17.

NARASIMHAN, T. N. and GOYAL, K. P., 1984, 'Subsidence due to geothermal withdrawal', *Geol. Soc. Am., Reviews in Eng. Geol. 6*, 35–66.

National Academy of Sciences, 1966, *Proceedings of the 1st Permafrost International Conference* (Nat. Acad. Sciences, Washington, DC—Nat. Research Council of Canada, Ottawa), Publication *1287*.

—— 1973, *North American Contribution: Permafrost, 2nd International Conference, Yakutsk, USSR* (Nat. Acad. Sciences, Washington, DC).

—— 1978, *USSR Contribution: Permafrost 2nd International Conference* (Nat. Acad. Sciences, Washington, DC).

—— 1983, *Proceedings of the 4th International Permafrost Conference* (Nat. Acad. Press, Washington, DC).

National Coal Board, 1975, *Subsidence Engineer's Handbook* (NCB, London).

National Institute for Road Research, 1971 (Revised draft dated 1976), 'The production of soil engineering maps for roads and the storage of materials data', *NIRR Tech. Recomm. for Highways, 2*.

National Research Council of Canada, 1978, 1979, *Proceedings of the 3rd International Conference on Permafrost, Edmonton, Alberta* (NRCC, Ottawa), 2 vols.

—— 1981, *Permafrost* (Research Institute of Glaciology, Cryopedology and Desert Research, Lanchou, China, trans. as NRCC *Technical Translation, 2006*).

—— 1982, *Proceedings of the 4th Canadian Permafrost Conference* (NRCC, Ottawa).

National Science Foundation, 1977, 'Earthquake prediction: is it better not to know?' *Mosaic, 8*, 8–14.

National Water Council, 1978, *River Water Quality, the Next Stage: Review of Discharge Consent Conditions* (Nat. Water Council, UK).

Natural Environment Research Council, 1975, *Flood Studies Report* (5 vols.) (NERC, London).

Nature Conservancy Council, 1983, *Nature Conservation and River Engineering* (NCC, UK).

NAUMOV, Y. V., 1975, 'Die Erforschung rezenter Erdkrustenbewegungen—eine der wichtigsten Aufgaben der Geodasie', *Vermessungstechnik, 23,* 405–8.

NEEF, G. and VEEH, H. H., 1977, 'Uranium series ages and late Quaternary uplift in the New Hebrides', *Nature, 269,* 682–3.

NELSON, F. F., 1986, 'Permafrost distribution in central Canada: applications of a climate-based predictive model', *Ann. Ass. Am. Geog. 76,* 550–69.

NEMCOK, A. and RYBAR, J., 1968, 'Synoptic map of Czechoslovak landslide areas', *Geol. Ustar Ceskoslov. Akad. Ved.* (Prague).

NETTERBERG, F., 1979, 'Salt damage to roads: an interim guide to its diagnosis, prevention and repair', *IMIESA (Inst. Municipal Engineers of Southern Africa), 4.*

NEWSON, M. D., 1975, *Flooding and Flood Hazard in the United Kingdom* (Clarendon Press, Oxford).

NEWTON, J. G., 1984, 'Sinkholes resulting from groundwater withdrawals in carbonate terranes: an overview', *Geol. Soc. Am., Reviews in Eng. Geol. 6,* 195–202.

NICHOLS, T. C., 1974, 'Global summary of human response to natural hazards: earthquakes', in G. F. White (ed.), *Natural Hazards* (OUP, Oxford), 274–84.

NICKLESS, E., 1982, 'Environmental geology of the Glenrothes District, Fife Region', *Institute of Geological Sciences Rep. 82/15.*

NIKONOV, A. A., 1977, 'Contemporary tectogenic movements of the earth's crust', *Int. Geol. Rev. 19,* 1245–58.

—— SHEBALINA, T. YU., 1979, 'Lichenometry and earthquake age determination in central Asia', *Nature, 280,* 675–7.

NILSEN, T. H., WRIGHT, R. H., VLASIC, T. C., and SPANGLE, W. E., 1979, 'Relative slope stability and land-use planning in the San Francisco Bay Region, California', *US Geol. Surv. Prof. Paper, 944.*

NIXON, M., 1963, 'Flood regulation and river training in England and Wales', *Institution of Civ. Engineers, Symposium on Conservation of Water Resources,* sessions III and IV, 35–48.

NORDSTROM, K. F. and ALLEN, J. R., 1980, 'Geomorphologically compatible solutions to beach erosion', *Zeit. für Geom., Suppl. 34,* 142–54.

NOROIS, 1977, *Karstologie (Norois, 24).*

NUMMEDAL, D., 1983, 'Barrier islands', in P. D. Komar (ed.), *Handbook of Coastal Processes and Erosion* (CRC Press, Boca Raton), 77–121.

NYE, J. F., 1976, 'Water flow in glaciers: jökulhaups, tunnels and veins', *J. Glaciology, 17,* 181–207.

Office of Emergency Preparedness, 1972, *Disaster Preparedness* (Executive Office of the President, Report to Congress, Washington, DC), 3 vols.

OGDEN, A. E., 1984, 'Methods of describing and predicting the occurrence of sinkholes', in B. F. Beck (ed.), *Sinkholes* (Balkema, Rotterdam), 177–82.

OLLIER, C. D., 1977a, 'Applications of weathering studies', in J. R. Hails (ed.), *Applied Geomorphology* (Elsevier, Amsterdam), 9–50.

—— 1977b, 'Terrain classification, principles and applications', in J. R. Hails (ed.), *Applied Geomorphology* (Elsevier, Amsterdam), 277–316.

—— 1981, *Tectonics and Landforms* (Longman, London).

—— 1984, *Weathering* (Longman, London), 2nd edn.

——, WEBSTER, R., LAWRANCE, C. J., and BECKETT, P. H. T., 1967, 'The preparation of a land classification map at 1:1 000 000 of Uganda', *Actes du IIème Symp. Internat. de Photo-Interpretation* (Paris) *Sect. IV.1,* 114–22.

ONESTI, L. J., 1985, 'Meteorological conditions that initiate slushflows in the Central Brooks Range, Alaska', *Ann. Glaciology, 6,* 23–5.

OPEN UNIVERSITY, 1981, *Earth Structure,* S 237, Block 2, (Open University Press, Milton Keynes).

OSBORN, H. B., SIMANTON, J. R., and RENARD, K. G., 1976, 'Use of the universal soil loss equation in the semiarid Southwest', in *Soil Erosion Prediction and Control* (Soil Cons. Soc. Am., Ankeny, Iowa), 41–9.

OTA, Y., 1975, 'Late Quaternary vertical movement in Japan estimated from deformed shorelines', in R. P. Suggate, and M. M. Cresswell (eds.), 'Quaternary Studies', Selected Papers from 9th INQUA Congress, Christchurch, New Zealand, Dec. 1973, *Roy. Soc. NZ Bull. 13,* 231–9.

PALL, R., DICKINSON, W. T., GREEN, D., and McGIRR, R., 1982, 'Impact of soil characteristics on soil erodibility', in *Int. Ass. Hydr. Sci. Publ. 137,* 39–47.

PALMER, D. F. and HENYEY, T. L., 1971, 'San Fernando earthquake of 9 February 1971 pattern of faulting', *Science, 172,* 712–15.

Panel on Earthquake Prediction, 1976, *Predicting Earthquakes* (Nat. Acad. Sciences, Washington, DC).

Panel on the Public Policy Implications of Earthquake Prediction, 1975, *Earthquake Prediction and Public Policy* (Nat. Acad. Sciences, Washington, DC).

PANIZZA, M., 1987, 'The geomorphological hazard assessment and the analysis of geomorphological risk', in V. Gardiner (ed.), *International Geomorphology, 1986,* 1 (Wiley, Chichester), 225–9.

PARKER, D. J., 1981, 'Flood mitigation through nonstructural measures: a critical appraisal', *Int. Conf. on Flood Disaster, New Delhi, 3–5 Dec. 1981,* 1–30.

PARKER, G. G., 1963, 'Piping, a geomorphic agent in landform development of the drylands', *Int. Ass. Sci. Hydrol. Publ., 65,* 103–13.

PATERSON, W. S. B., 1975, *The Physics of Glaciers* (Pergamon, Oxford).

PATON, T. R., 1978, *The Formation of Soil Material* (Allen & Unwin, London).

PATTERSON, J. L., 1970, 'Evaluation of the streamflow data program for Arkansas', *US Geol. Surv. Prof. Paper, 700-D*, D244–D256.

PATTON, F. D., 1970, 'Significant geologic factors in rock slope stability', in P. W. J. van Rensburg (ed.), *Planning Open Pit Mines* (S. Afr. Inst. Mining and Metallurgy, Johannesburg), 143–51.

PATTON, P. C., and SCHUMM, S. A., 1975, 'Gully, erosion. Northwestern Colorado: a threshold phenomenon', *Geology, 3*, 88–90.

PELTIER, L., 1950, 'The geographic cycle in periglacial regions as it is related to climatic geomorphology', *Ann. Ass. Am. Geog. 49*, 214–36.

PENN, S., ROYCE, C. J., and EVANS, C. J., 1983, 'The periglacial modification of the Lincoln scarp', *Q. J. Eng. Geol. 16*, 309–18.

PENNING-ROWSELL, E. C., 1981a, 'Fluctuating fortunes in gauging landscape values', *Prog. Human Geog. 5*, 25–41.

—— 1981b, 'Non-structural approaches to flood control: flood plain land use regulation and flood warning schemes in England and Wales', *Int. Commission on Irrigation and Drainage Cong. 11*, 193–211.

—— and CHATTERTON, J. B., 1977, *The Benefits of Flood Alleviation: A Manual of Assessment Techniques* (Gower Publishing, Aldershot).

—— PARKER, D. J., and HARDING, D., 1986, *Floods and Drainage: British Policies for Hazard Reduction* (Allen & Unwin, London).

PERLA, R. I. and MARTINELLI, H., 1976, 'Avalanche Handbook', *Agric. Handbook, 489* (Forest Service, US Dept. Agriculture).

PERRY, R. A., 1962, 'General Report on Lands of the Alice Springs Area 1956–7', *Land Research Series, 6* (CSIRO, Australia).

PETHICK, J., 1984, *An Introduction to Coastal Geomorphology* (Arnold, London).

PETLEY, D. J., 1984, 'Ground investigation, sampling and testing for studies of slope instability', in D. Brunsden and D. B. Prior (eds.), *Slope Instability* (Wiley, Chichester), 67–101.

PETROV, M. P., 1976, *Deserts of the World* (Wiley, New York).

PETTS, G. E., 1979, 'Complex response of river channel morphology subsequent to reservoir construction', *Progr. Phys. Geog. 3*, 329–62.

—— 1983, *Rivers* (Butterworth, London).

—— 1984, *Impounded Rivers: Perspectives for Ecological Management* (Wiley, Chichester).

—— 1986, 'Water quality characteristics of regulated rivers', *Prog. Phys. Geog. 10*, 492–516.

—— and FOSTER, I., 1985, *Rivers and Landscape* (Arnold, London).

PÉWÉ, T. L. (ed.), 1981, 'Desert dust: origin, characteristics, and effect on man', *Geol. Soc. Am. Spec. Paper, 186*.

PIHLAINEN, J. A., 1962, 'Inuvik, N.W.T., engineering site information', *Nat. Research Council Canada, Techn. Paper, 135.*

PIOTROVSKI, M. V., SIMONOV, Y. G., and ARISTARKHOVA, L. B., 1972, 'Detailed geomorphological mapping in mineral prospecting', in J. Demek (ed.), *Manual of Detailed Geomorphological Mapping* (Academia, Prague), 267–77.

PITEAU, D. R., 1970, 'Geological factors significant to the stability of slopes cut in rock', in P. W. J. van Rensburg (ed.), *Planning Open Pit Mines* (S. Afr. Inst. Mining and Metallurgy, Johannesburg), 33–53.

POLAND, J. F., 1969, 'Land subsidence in the western United States', in R. A. Olson and M. M. Wallace (eds.), *Geologic Hazards and Public Problems* (Office of Emergency Preparedness, Region 7, Santa Rosa, Calif.), 77–96.

—— (ed.), 1984, *Guidebook to Studies of Land Subsidence Due to Groundwater Withdrawal* (UNESCO, Paris).

—— and DAVIES, G. H., 1969, 'Land subsidence due to withdrawal of fluids', in D. J. Varnes and G. Kiersch (eds.), *Geol. Soc. Am., Reviews in Eng. Geol. 2*, 187–269.

POOLE, A. B., FOOKES, P. G., DIBB, T. E., and HUGHES, D. W., 1983, 'Durability of rocks in breakwaters', in *Breakwaters—Design and Construction* (Telford, London), 31–52.

POPP, N., 1971, 'Hydrogeographische und geomorphologische Gesichtspunkte zum Problem der rezenten vertikalen Krustenbewegungen in Rumanien', *Zeit. für Geom. 15*, 445–59.

PORCHER, M. and GUILLOPE, P., 1979, 'Cartographie des risques ZERMOS appliquées à des plans d'occupation des sols en Normandie', *Bull. Mason Laboratoire des Ponts et Chaussées, 99.*

Port of Long Beach, 1971, *Vertical Movement of Long Beach Harbor District* (Port of Long Beach, Long Beach).

POTTS, A. S., 1970, 'Frost action in rocks: some experimental data', *Trans. Inst. Brit. Geog. 49*, 109–24.

PRENDERGAST, A. D. (ed.), 1983, *Soil Erosion: Abridged Workshop Proceedings, Florence, Oct. 1982 (EEC Agric. Series, EUR 8427).*

PRICE, C. A., 1981, 'Brethane stone preservative', *Building Research Establishment Current Paper, 1/81.*

PRICE, L. W., 1972, 'The periglacial environment', *Ass. Am. Geog. Commission on College Geography Resource Paper, 14.*

PRICKNETT, R. G., BRAY, L. G., and STENNER, R. D., 1976, 'The chemistry of cave waters', in T. D. Ford and C. H. D. Cullingford (eds.), *The Science of Speleology* (Academic Press, London), 213–66.

PRINCE, H. C., 1979, 'Marl pits or dolines of the Dorset Chalklands?' *Trans. Inst. Brit. Geog. 4*, 116–17.

PRIOR, D. B. and COLEMAN, J. M., 1979, 'Submarine

landslides—geometry and nomenclature, *Zeit. für Geom.* 23, 415–26.

PRIOR, D. B. and COLEMAN, J. M. 1980, 'Sonograph mosaics of submarine slope instabilities, Mississippi River Delta', *Marine Geology, 36*, 227–39.

—— and —— 1984, 'Submarine slope instability', in D. Brunsden and D. B. Prior (eds.), *Slope Instability* (Wiley, Chichester), 419–55.

—— and RENWICK, W. H., 1980, 'Landslide morphology and processes on some coastal slopes in Denmark and France', *Zeit. für Geom., Suppl., 34*, 63–86.

RABOT, C., 1920, 'Les catastrophes glaciares dans la vallée de Chamonix au XVIIe siècle et les variations climatiques', *La Nature, 46*, 129–34.

RANG, M. C., KLEIJN, C. E., and SCHOUTEN, C. J., 1987, 'Mapping of soil pollution by application of classical geomorphological and pedological field techniques', in V. Gardiner (ed.), *International Geomorphology, 1* (Wiley, Chichester), 1029–44.

RAO, D. P., 1975, 'Applied geomorphological mapping for erosion surveys: an example of the Oliva Basin, Calabria', *ITC Journal, 3*, 341–51.

RAPP, A., LE HOUEROU, H. N., and LUNDHOLM, B., 1976, 'Can desert encroachment be stopped?' *Swedish Natural Science Research Council Ecological Bull. 24.*

RAUSCH, D. L. and HEINEMANN, H. G., 1984, 'Measurement of reservoir sedimentation', in R. F. Hadley and D. E. Walling (eds.), *Erosion and Sediment Yield* (GeoBooks, Norwich), 179–200.

REDDING, J. and LORD, A., 1982, 'Designing for the effects of windblown sand along the new Jeddah-/Riyadh/Dammam expressway', *Arup Journal, 17*, 11–15.

REILINGER, R., BROWN, L., and POWERS, D., 1980, 'New evidence for tectonic uplift in the Diablo Plateau region, west Texas', *Geog. Research Letters, 7*, 181–4.

RIB, H. T. and LIANG, T., 1978, 'Recognition and identification', in R. L. Schuster and R. J. Krizek (eds.), 'Landslides—Analysis and Control', *Transport Research Board Spec. Rep., 176* (Nat. Acad. Science, Washington, DC).

RICHARDS, K. S., 1982, *Rivers: Form and Process in Alluvial Channels* (Methuen, London).

—— ARNETT, R. R., and ELLIS, S., 1985, *Geomorphology and Soils* (Allen & Unwin, London).

RICHTER, R. D., SAPLACO, S. R., and NOWAK, P. F., 1985, 'Watershed management problems in humid tropical uplands', *Nature and Resources, 21*, 10–21.

RICHTER, D. H. and MATSON, N. A., 1971, 'Quaternary faulting in the Eastern Alaska Range', *Bull. Geol. Soc. Am. 82*, 1529–40.

RICHTER, G., 1980, 'Soil erosion mapping in Germany and Czechoslovakia', in M. de Boodt and D. Gabriels (eds.), *Assessment of Erosion* (Wiley, Chichester), 29–54.

RICHARD, P., 1979, 'Blowing: what it costs and ways to prevent it', *Arable Farming, 6*, 79–83.

RITCHIE, A. M., 1958, 'Recognition and identification of landslides', in E. B. Eckel (ed.), *Landslides and Engineering Practice (Highway Research Board, Spec. Rep. 29)*, 48–68.

ROBERTS, D. V. and MELICKIAN, G. E., 1970, 'Geologic and other natural hazards in desert areas', *Dames and Moore Eng. Bull. 37*, 1–12.

ROBERTS, H. H., SUHAYDA, J. N., and COLEMAN, J. M., 1983, 'Sediment deformation, and transport on low-angle slopes: Mississippi River Delta', in D. R. Coates and J. D. Vitek (eds.), *Thresholds in Geomorphology* (Allen & Unwin, London), 131–67.

ROBERTSON, A. M., 1970, 'The interpretation of geological factors for use in slope theory', in P. W. J. van Rensburg (ed.), *Planning Open Pit Mines* (S. Afr. Inst. for Mining and Metallurgy, Johannesburg), 55–71.

ROBINSON, A. H. W., 1975, 'Cyclical changes in shoreline development at the entrance of Teignmouth Harbour, Devon, England', in J. Hails and A. Carr (eds.), *Nearshore Sediment Dynamics and Sedimentation* (Wiley, Chichester), 181–200.

ROBINSON, D. G., LAURIE, I. C., WAGER, J. F., and TRAILL, A. L. (eds.), 1976, *Landscape Evaluation* (Manchester Univ. Press, Manchester).

ROBINSON, D. N., 1969, 'Soil erosion by wind in Lincolnshire, March, 1968', *East Midland Geographer, 4*, 351–62.

ROBINSON, G. D. and SPIEKER, A. M. (eds.), 'Nature to be Commanded . . .', *US Geol. Surv. Prof. Paper, 950.*

RODDA, J. C., 1967, 'The significance of characteristics of basin rainfall and morphometry in a study of floods in the United Kingdom', *Int. Ass. Sci. Hydr., Leningrad Symposium*, 835–45.

ROGERS, G. F., MALDE, H. E., and TURNER, R. M., 1984, *Bibliography of Repeat Photography for Evaluating Landscape Change* (Univ. of Utah Press, Salt Lake City).

ROGERS, T. H. and NASON, R. D., 1971, 'Active displacement on the Calaveras fault zone at Hollister, California', *Bull. Seismological Soc. Am. 61*, 399–416.

ROOSE, E. J., 1976, 'The use of the universal soil loss equation to predict erosion in West Africa', in Soil Cons. Soc. Am., *Soil Erosion: Prediction and Control* (Soil Cons. Soc. Am., Ankeny, Iowa), 60–74.

—— 1980, 'Approach to the definition of rain erosivity and soil erodibility in West Africa', in M. de Boodt and D. Gabriels (eds.), *Assessment of Erosion* (Wiley, Chichester), 153–64.

ROSENFELD, C. L., 1980, 'Observations on the Mt. St Helens eruptions', *Am. Sci. 68*, 494–509.

—— 1984, 'Remote sensing techniques for geomorphologists', in J. E. Costa and P. J. Fleisher (eds.), *Developments and Applications of Geomorphology* (Springer-Verlag, Berlin), 1–37.

ROSS, K., 1980, 'Characterisation of porous bodies by UV fluorescence microscopy', *Building Research Establishment Note, 98/80.*

Ross, K., 1984, 'The assessment of the durability of limestone and other porous materials: a review', *Building Research Establishment Note, 86/84.*

Rossi-Manaresi, R. and Vannucci, S., 1978, 'Note on conservation treatment of stone objects', *Colloque International sur Alteration et Protection des Monuments en Pierre, Paris, 5–9 Juin 1978.*

—— 1972, 'On the treatment of stone sculptures in the past', *Proc. Meeting of the Joint Committee for the Conservation of Stone, Bolgna, 1–3 Oct. 1971*, 81–104.

Roth, E. S., 1965, 'Temperature and water content as factors in desert weathering', *J. Geol. 73*, 454–68.

Rouse, W. C., 1984, 'Flow slides', in D. Brunsden and D. B. Prior (eds.), *Slope Instability* (Wiley, Chichester), 491–522.

Rowe, P. B., Countryman, C. M., and Storey, H. C., 1954, 'Hydrologic analysis used to determine effects of fire on peak discharge and erosion rates in southern California watersheds', *US Dept. Agriculture Forest Service, California Forest and Range Experiment Station Report* (Berkeley, Calif.).

Rubey, W. W., 1952, 'Geology and mineral resources of the Hardin and Drussels quadrangles (in Illinois)', *US Geol. Surv. Prof. Paper, 218*, 175.

Rupke, J. and De Jong, M. G. G., 1983, 'Slope collapse destroying ice-marginal topography in the Walgau (Vorarlberg, Austria)', *Materialien zur Physiogeographie, 5*, 33–41.

Russell, J. R., 1981, 'Sedimentation in the proposed Kotmale Reservoir, Sri Lanka', *SE Asian Regional Symposium on Problems of Soil Erosion and Sedimentation, Bangkok, Thailand, 27–29 Jan. 1981*, 405–18.

Ruxton, B. P., 1968, 'Measure of the degree of chemical weathering of rocks', *J. Geol. 76*, 518–27.

Sabins, F. F., 1978, *Remote Sensing* (Freeman, San Francisco).

Sarntheim, M., 1978, 'Sand deserts during glacial maximum and climatic optimum', *Nature, 272*, 43–6.

Saunders, M. K. and Fookes, P. G., 1970, 'A review of the relationship of rock weathering and climate and its significance to foundation engineering', *Eng. Geol. 4*, 289–325.

Savigear, R. A. G., 1952, 'Some observations on slope development in South Wales', *Trans. Inst. Brit. Geog. 18*, 31–52.

—— 1965, 'A technique of morphological mapping', *Ann. Assoc. Am. Geog. 53*, 514–38.

Sayward, J. M., 1984, 'Salt action on concrete', *US Army Corps. of Engineers, Cold Regions Research and Engineering Laboratory Special Report, 84–25.*

Schaffer, R. J., 1932, *The Weathering of Building Stones* (HMSO, London).

—— 1959, 'Testing building stone', *Architectural Engineering Monumental Stone, 16* and *17*.

Scheidegger, A., 1965, 'The algebra of stream-order numbers', *US Geol. Surv. Prof. Paper, 525B, B187–9.*

Schick, A., 1974, 'Alluvial fans and desert roads: a problem in applied geomorphology', *Report of the Commission on Present-Day Geomorphological Processes* (Göttingen), 418–25.

Schmitz, G. 1980, 'A rural development project for environmental management in Lesotho', *ITC Journal, 2*, 349–63.

Schumm, S. A., 1963, 'Sinuosity of alluvial rivers on the Great Plains', *Bull. Geol. Soc. Am. 74*, 1089–100.

—— 1967, 'Meander wavelength of alluvial rivers', *Science, 157*, 1549–50.

—— 1969, 'River metamorphosis', *J. Hydraulics Div., Proc. Am. Soc. Civil Eng. 95*, 255–73.

—— 1971, 'Fluvial geomorphology: The historical perspective', in H. W. Shen (ed.), *Fluvial Geomorphology in River Mechanics* (Water Resources Publ., Fort Collins, Colo.), 4.1–4.30.

—— 1973, 'Geomorphic thresholds and complex response of drainage systems', in M. E. Morisawa (ed.), *Fluvial Geomorphology* (State Univ. of New York, Binghamton), 299–310.

—— 1977, *The Fluvial System* (Wiley, New York).

—— 1985, 'Explanation and extrapolation in geomorphology: seven reasons for geologic uncertainty', *Trans. Japanese Geophys. Union, 6*, 1–18.

—— and Beathard, R. M., 1976, 'Geomorphic thresholds: an approach to river management', in *Rivers* (Vol. 1), *Third Symposium of the Waterways, Harbors and Coastal Engineers Division of the American Society of Civil Engineers*, 707–24.

—— Harvey, M. D., and Watson, C. C., 1984, *Incised Channels: Morphology, Dynamics and Control* (Water Resources Publ., Colo.).

—— and Lichty, R. W., 1965, 'Time, space and causality in geomorphology', *Am. J. Sci. 263*, 110–19.

Schuster, R. L., 1978, 'Introduction', in R. L. Schuster and R. J. Krizek (eds.), *Landslides: Analysis and Control, Transport Research Board Special Report, 176* (Nat. Acad. Sciences, Washington, DC), 1–10.

—— and Krizek, R. J. (eds.), 1978, *Landslides: Analysis and Control, Transport Research Board Special Report, 176* (Nat. Acad. Sciences, Washington, DC).

Scott, R. F. and Schoustra, J. J., 1968, *Soil Mechanics and Engineering* (McGraw-Hill, New York).

Scott, K. M. and Williams, R. D., 1978, 'Erosion and sediment yields in the Transverse Ranges, Southern California', *US Geol. Surv. Prof. Paper, 1030.*

Scott, R. M., Healy, P. A., and Humphreys, G. S., 1985, 'Land units of Chimbu Province Papua New Guinea, *CSIRO Division of Water and Land Resources Natural Resources Series, 5.*

Seed, H. B., Whitman, R. V., Dezfulian, H., Dorby, R., and Idriss, J. M., 'Soil conditions and building damage in 1967 Caracas earthquake', *J. Soil Mech. Div. Amer. Soc. Civil Eng. 98*, 787–806.

Seevers, P. M., Lewis, D. T., and Drew, J. V., 1975, 'Use of ERTS-1 imagery to interpret the wind erosion hazard in Nebraska's Sandhills', *J. Soil and Water Conservation, 30*, 181–3.

SELBY, M. J., 1985, *Earth's Changing Surface* (OUP, Oxford).

SEWELL, W. R. D. and COPPOCK, J. T., 1976, 'Achievements and prospects', in J. T. Coppock and W. R. D. Sewell (eds.), *Spatial Dimensions of Public Policy* (OUP, Oxford), 257–62.

SHARMA, T., UPRETI, B. N., and VASHI, N. M., 1980, 'Kali Gandaki gravel deposits of central west Nepal—their neotectonic significance', *Tectonophysics, 62*, 127–39.

SHARP, A. D., TRUDGILL, S. T., COOKE, R. U., PRICE, C. A., CRABTREE, R. W., PICKLES, A. M., and SMITH, D. I., 1982, 'Weathering of the balustrade on St Paul's cathedral, London', *Earth Surface Processes and Landforms, 7*, 387–9.

SHARPE, C. F. S., 1938, *Landslides and Related Phenomena* (Pageant, NJ).

SHEAFFER, J. R., 1960, 'Floodproofing: an element in a flood damage reduction program', *Univ. Chicago, Dept. of Geog. Research Paper, 65*.

SHEEDY, K. A., LEIS, W. M., THOMAS, A., and BEERS, W. F., 1982, 'Land use in carbonate terrain: problems and case study solutions', in R. G. Craig and J. L. Craft (eds.), *Applied Geomorphology* (Allen & Unwin, London), 202–13.

SHEPARD, F. P., 1976, 'Coastal classification and changing coastlines', *Geoscience and Man, 14*, 53–64.

SHERLOCK, R. L., 1922, *Man as a Geological Agent* (Witherby, London).

SHREVE, R. L., 1966, 'Statistical law of stream numbers', *J. Geol. 74*, 17–37.

SIDLE, R. C., PEARCE, A. J., and O'LOUGHLIN, C. L., 1985, *Hillslope Stability and Land Use* (Am. Geophys. Union, Washington, DC).

SIGOV, V. A. and ROMASHOVA, V. H., 1969, Problems of quantitative assessment of the most recent tectonic movements of the Urals, *Geology and Mineral Resources of the Urals Part 1* (Sverdlovsk) 126–7. (In Russian.)

SILVESTER, R., 1974, *Coastal Engineering* (Elsevier, Amsterdam).

SIMMONDS, A., 1977, 'Benefits from beaches', *Geog. Mag.* (Oct.), 1–4.

SIMONS, D. B., REESE, A. J., RUH-MINGLI, and WARD, T. J., 1976, A simple method for estimating sediment yield, in *Soil Erosion: Prediction and Control* (Soil Cons. Soc., Ankeny, Iowa), 234–41.

—— and RICHARDSON, E. V., 1961, 'Forms of bed roughness in alluvial channels', *Proc. Am. Soc. Civil Eng. J. Hydraulics Div. 87* (HY3 pt. I), 87–105.

SIMPSON, D. W., 1976, 'Seismicity changes associated with reservoir loading', *Eng. Geol. 10*, 123–50.

SINGER, M. J., BLACKARD, J., GILLOGLEY, E., and ARULANANDAM, K., 1978, *Engineering and Pedological Properties of Soils as They Affect Soil Erodibility*

(California Water Resources Center, Davis, Contributions, *166*).

SIREYJOL, P., 'Communication sur la construction du port de Cotonou (Dahomey)', *Proc. 9th Conf. Coastal Eng.* (Am. Soc. Civ. Eng.).

SKEMPTON, A. W., 1948, 'The rate of softening of stiff, fissured clays', *Proc. 2nd. Intern. Conf. Soil Mech. Foundn. Eng., Rotterdam, 2*, 50–3.

—— 1964, 'Long-term stability of clay slopes' (Fourth Rankin Lecture), *Géotechnique, 14*, 77–102.

—— and WEEKS, A. G., 1976, 'The Quaternary history of the lower Greensand escarpment in Weald Clay vale near Sevenoaks, Kent', *Phil. Trans. Roy. Soc. London, A283*, 493–526.

SKIDMORE, E. L., FISHER, P. S., and WOODRUFF, N. P., 1970, 'Wind erosion equation: computer solution and application', *Proc. Soil Sci. Am., 34*, 931–5.

—— and WOODRUFF, N. P., 1968, 'Wind erosion forces in the United States and their use in predicting soil loss', *US Dept. Agriculture Handbook, 346*.

SKOULIKIDIS, D. and PAPAKONSTANTINOU-ZIOTIS, P., 1981, 'Mechanism of sulphation by atmospheric SO_2 of the limestones and marbles of the ancient monuments and statues', *Br. Corrosion J. 16*, 63–9.

SLAYMAKER, O., 1980, 'Geomorphic field experiments: inventory and prospect', *Zeit für Geom., Suppl.-Bd. 35*, 183–94.

—— (ed.) 1981, 'High mountains', *Zeit. für Geom., Suppl. 35*.

—— 1982, 'The nature of field experiments in geomorphology', *Studia Geomorphologia Carpatho-Balcanica, 15*, 11–17.

SLOSSEN, J. E., 1969, 'The role of engineering geology in urban planning', *Colorado Geol. Surv. Spec. Paper, 1*, 8–15.

SMALLEY, J. P. and CAPPA, J., 1971, 'Geologic parameters of slope stability in the Upper Santa Ynez drainage system', *US Dept. Agriculture Forest Service*, unpublished report.

SMITH, D. D. and WISCHMEIER, W. H., 1962, 'Rainfall erosion', *Advances in Agronomy, 14*, 109–48.

SMITH, D. I., 1977, 'Applied geomorphology and hydrology of Karst regions', in J. R. Hails (ed.), *Applied Geomorphology* (Elsevier, Amsterdam), 85–118.

—— and PENNING-ROWSELL, E. C., 1982, 'An evaluation of house-raising as a flood mitigation strategy for Lismore, NSW', *Australian Nat. University Centre for Resource and Environmental Studies Working Paper, 7*.

SMITH, K. and TOBIN, G., 1979, *Human Adjustment to the Flood Hazard* (Longman, London).

Soil Conservation Society of America, 1977, *Soil Erosion: Prediction and Control* (Soil. Cons. Soc. Am., Ankeny, Iowa).

Soil Science Society of America, 1977, *Minerals in Soil Environments* (Soil Science Soc. Am., Madison, Wisc.).

Soil Survey Staff, 1951, *Soil Survey Manual* (US Dept. Agriculture Handbook, *18*).

Soons, J. M. and Selby, M. J. (eds.), 1982, *Landforms of New Zealand* (Longman Paul, Auckland).

Sorriso-Valvo, M., 1984, 'Deep-seated gravitational slope deformations in Calabria (Italy)', in J. C. Flageollet (ed.), 'Mouvements de terrains', *Série Documents du BRGM 83*, 81–90.

Sowers, G. F., 1984, 'Correction and protection in limestone terrain', in B. F. Beck (ed.), *Sinkholes* (Balkema, Rotterdam), 373–8.

Spangle, W. and Associates, 1974, *Application of Earth-Science Information in Urban Land Use Planning* (unpublished report, NTIS, Springfield, Va).

Speight, J. G., 1968, 'Parametric description of land form', in G. A. Stewart (ed.), *Land Evaluation* (Macmillan, Melbourne), 239–60.

—— 1974, 'A parametric approach to landform regions', *Inst. Brit. Geog., Spec. Publ. 7*, 213–30.

—— 1976, 'Numerical classification of landform elements from air photo data', *Zeit für Geom. Suppl. 25*, 154–68.

Sperling, C. H. B. and Cooke, R. U., 1985, 'Laboratory simulation of rock weathering by salt crystallization and hydration processes in hot, arid environments', *Earth Surface Processes and Landforms, 10*, 541–55.

—— Goudie, A. S., Stoddart, D. R., and Poole, G. G., 1977, 'Dolines of the Dorset Chalklands and other areas in southern Britain', *Trans. Inst. Brit. Geog. 2*, 205–23.

Steinbrugge, K. V., 1968, *Earthquake Hazard in the San Francisco Bay Area: A Continuing Problem in Public Policy* (Univ. California, Berkeley, Institute of Government Studies).

Stephens, J. C., Allen, L. H., and Chen, E., 1984, 'Organic soil subsidence', *Geol. Soc. Am., Reviews in Eng. Geol. 6*, 107–22.

—— and Speir, W. H., 1969, 'Subsidence of organic soils in the USA', *Pub. Int. Ass. Sci. Hyd. 89*, 523–34.

Stevens, M. A., Simons, D. B, and Schumm, S. A., 1975, 'Man-induced changes of Middle Mississippi River', *Proc. Am. Soc. Eng., J. Waterways Harbors, Coast. Eng. Div. 101*, 119–33.

Stevenson, P. C. and Sloane, D. J., 1980, 'The evolution of a risk-zoning system for landslide areas in Tasmania, Australia', *Proc. 3rd Australian and New Zealand Geomechanics Conf., Wellington*.

Stewart, G. A. and Perry, R. A., 1953, 'Survey of Townsville-Bowen Region (1950)', *Land Research Series, 2* (CSIRO 1120, Australia).

Stocking, M. A. and Elwell, H. A., 1973, 'Soil erosion hazard in Rhodesia', *Rhodesian Agric. J. 70*, 93–101.

Stoddart, D. R., 1971, 'Coral reefs and islands and catastrophic storms', in J. A. Steers (ed.), *Applied Coastal Geomorphology* (Macmillan, London), 155–97.

Strahler, A. N., 1964, 'Quantitative geomorphology of drainage basins and channel networks', in V. T. Chow (ed.), *Handbook of Applied Hydrology* (McGraw-Hill, New York), Sect. 4-II.

Strakhov, N. M., 1967, *Principles of Lithogenesis*, Vol. 1 (Consultants Bureau, New York).

Strand, R. I., 1975, 'Bureau of Reclamation procedures for predicting sediment yield', in *US Dept. Agriculture, Agric. Research Service, ARS-5-40*, 10–15.

Sunamura, T., 1983, 'Processes of sea cliff and platform erosion', in P. D. Komar (ed.) *Handbook of Coastal Processes and Erosion* (CRC Press, Boca Raton, Fla.), 233–65.

Sundborg, A., 1983, 'Sedimentation problems in river basins', *Nature and Resources, 19*, 10–21.

Sutcliffe, J. V., 1978, 'Methods of flood estimation: a guide to the flood studies report', *Inst. Hydrology Rep. 49*.

Swanson, F. J., Janda, R. J., Dunne, T., and Swanston, D. N., 1982, *Sediment Budgets and Routing in Forested Drainage Basins* (US Dept. Agriculture Forest Service, Pacific Northwest Forest and Range Experiment Station, *General Tech. Report, PNW-141*).

Sweeting, M. M., 1972, *Karst Landforms* (Macmillan, London).

—— and Pfeffer, K.-H., 1976, 'Karst Processes', *Zeit. für Geom., Suppl. 26*.

Swiss Reinsurance Company, 1978, *Atlas on Seismicity and Volcanism* (Zurich).

Taber, S., 1929, 'Frost heaving', *J. Geol. 37*, 428–61.

—— 1930, 'The mechanics of frost heaving', *J. Geol. 38*, 303–17.

Tank, R. W., 1973, *Focus on Environmental Geology* (OUP, Oxford).

Taylor, F. A. and Brabb, E. E., 1972, 'Map showing distribution and cost by counties of structurally damaging landslides in the San Francisco Bay Region, California, winter of 1968–69', *US Geol. Surv. Misc. Field Studies Map, MF 327* (scale 1:1 000 000).

Taylor, F. W., Isacks, B. L., Jouannic, C., Bloom, A. L. Dubois, J., 1980, 'Co-seismic and Quaternary vertical tectonic movements, Santo and Malekula Islands, New Hebrides Island Arc.', *J. Geophys. Res. 85*, 5367–81.

Temple, P. H., 1964, 'Evidence of lake level changes from the northern shoreline of Lake Victoria, Uganda', in R. W. Steel and R. M. Prothero (eds.) *Geographers and the Tropics: Liverpool Essays* (University of Liverpool Press, Liverpool), 31–56.

Terzaghi, K., 1950, 'Mechanisms of landslides', in *Application of Geology to Engineering Practice* (Berkey Volume, Geol. Soc. Am.), 83–125.

—— 1962, 'Stability of steep slopes on hard unweathered rock', *Géotechnique, 12*, 251–70.

—— and Peck, R. B., 1948, *Soil Mechanics in*

Engineering Practice (Wiley, New York), 2nd edn., 1967.

TERZAGHI, R. D., 1965, 'Sources of error in joint surveys', *Géotechnique, 15*, 287–304.

THOMAS, H. P. and FERRELL, J. E., 1983, 'Thermokarst features associated with buried sections of the Trans-Alaska pipeline', in *Proc. 4th Permafrost Int. Conf.* (Nat. Acad. Press, Washington, DC), 1245–50.

THOMSON, S., 1980, 'A brief review of foundation construction in the western Canadian Arctic', *Q. J. Eng. Geol. 13*, 67–76.

THORN, R. B. and ROBERTS, A. G., 1981, *Sea Defence and Coast Protection Works* (Telford, London).

THORNBURY, W. D., 1954, *Principles of Geomorphology* (Wiley, New York).

THORNES, J. B., 1979a, *River Channels* (Macmillan, London).

—— 1979b, 'Research and application in British geomorphology', *Geoforum, 10*, 253–9.

—— 1985, 'The ecology of erosion', *Geography, 70*, 222–35.

THROWER, N. J. W. and COOKE, R. U., 1968, 'Scales for determining slope from topographic maps', *Prof. Geog. 20*, 181–6.

THURM, H., BANKWITZ, P., HERRMANN, H., and BANKWITZ, E., 1971, 'Neue Aspekte rezenter Erdkrusten bewegungen in Gebiet der Deutschen Demokratischen Republik', *Petermanns Geographische Mitteilungen, 115*, 123–59.

THURRELL, R. G., 1981, 'The identification of bulk mineral resources: the contribution of the Institute of Geological Sciences', *Quarry Management and Products, London, 8*, 181–93.

TOWNSHEND, J. R. G. (ed.), 1981, *Terrain Analysis and Remote Sensing* (Allen & Unwin, London).

TOY, T. J. (ed.), 1977, *Erosion: Research Techniques, Erodibility and Sediment Delivery* (GeoAbstracts, Norwich).

TRICART, J., 1956, 'Étude expérimentale du problème de la gelivation', *Biul. Peryglacjalny, 4*, 285–318.

—— 1959, 'Presentation d'une feuille de la carte geomorphologique du delta du Senegal au 1:50,000', *Rev. de Géomorph. Dynamique, 11*, 106–16.

—— 1961, 'Notice explicative de la carte geomorphologique du delta du Senegal', *Bureau de Recherches Géologiques et Minières, 8*, 1–137.

—— 1965, *Principes et Méthodes de la Géomorphologie* (Masson, Paris).

—— 1966, 'Géomorphologie et aménagement rural (example du Venezuela)', *Coopération Technique, 44–5*, 69–81.

—— 1970, *Geomorphology of Cold Environments* (Macmillan, New York).

TRIMBLE, S. W., 1974, *Man-induced Soil Erosion on the Southern Piedmont, 1700–1970* (Soil Cons. Soc. Am., Ankeny, Iowa).

—— 1977, 'The fallacy of stream equilibrium in contemporary denudation studies', *Am. J. Sci. 277*, 876–87.

—— 1981, 'Changes in sediment storage in the Coon Creek Basin, Driftless Area, Wisconsin, 1853–1975', *Science, 214*, 181–3.

—— 1983, 'A sediment budget for Coon Creek Basin in the Driftless Area, Wisconsin, 1853–1977', *Am. J. Sci. 283*, 454–74.

—— and LUND, S. W., 1982, 'Soil conservation and the reduction of erosion and sedimentation in the Coon Creek Basin, Wisconsin', *US Geol. Surv. Prof. Paper, 1234*.

TRUDGILL, S. T., 1976, 'Rock weathering and climate: quantitative and experimental aspects', in E. Derbyshire (ed.), *Geomorphology and Climate* (Wiley, London), 59–99.

—— 1982, *Weathering and Erosion* (Butterworth, London).

—— 1985, *Limestone Geomorphology* (Longman, London).

—— CRABTREE, R. W., VILES, H., and COOKE, R. U., 1989, 'Remeasurement of weathering rates, St Paul's Cathedral, London, *Earth Surface Processes and Landforms, 14*, 175–96.

—— HIGH, C., and HANNA, F. K., 1981, 'Improvements to the micro-erosion meter (MEM)', *British Geomorphological Research Group, Tech. Bull. 29*, 3–17.

TSURIELL, D. E. (ed.), 1974, 'Sand binding and desert plants', *Int. J. Biometeorology, 18*, 85–181.

TSYTOVICH, N. A., 1975, *The Mechanics of Frozen Ground* (McGraw-Hill, New York). (Trans. from the Russian.)

TUFNELL, L., 1984, *Glacier Hazards* (Longman, London).

UGOLINI, F. E., 1966, 'Soil investigations in the Lower Wright Valley, Antarctica', *Proc. 1st Permafrost Int. Conf.* (Nat. Acad. Sciencies, Washington, DC) 55–61.

UNESCO, 1969, 'Land subsidence', *Publications of the Institute of Scientific Hydrology, 88* (2 vols.).

—— 1970, *International Legend for Hydrogeological Maps* (UNESCO, Paris), 1976, *Engineering Geology Maps, a Guide to their Preparation* (UNESCO, Paris).

UN, 1977, 'World map of desertification at a scale of 1:25 000 000', *UN Conference on Desertification, Nairobi, 74/2*.

UNSTEAD, J. F., 1933, 'A system of regional geography', *Geography, 18*, 175–87.

US Dept. Agriculture, 1957, *Yearbook of Agriculture— Soil* (US Dept. Agriculture, Washington, DC).

—— 1962, 'Agricultural land resources—capabilities, uses, conservation needs', *Agr. Inf. Bull. 263*.

—— 1963, *Proceedings of the Federal Inter-Agency Sedimentation Conference, 1963 (US Dept. Agriculture, Misc. Publ. 970)*.

—— 1968, 'Snow avalanches: a handbook of forecasting and control measures', *Agricultural Handbook, 194*.

US Dept. Agriculture, 1970, 'First aid for flooded homes and farms', *Agricultural Handbook, 38*.

—— 1975, *Present and Prospective Technology for Predicting Sediment Yields and Sources* (US Dept. Agriculture, Agric. Research Service, *ACS-S-40*).

—— 1976, *Proceedings of the 3rd Federal Inter-Agency Sedimentation Conference, 1976* (Water Resources Council, Washington, DC).

US Dept. Housing and Urban Development, 1969, *Proc. National Conference on Sediment Control, Washington, DC, 14–16 Sept. 1969* (USDHUD and others, Washington Environmental Planning Paper).

—— 1974, *National Flood Insurance Program* (USD-HUD, Washington, DC).

—— 1974, *National Flood Insurance Program* (USD-HUD, Washington, DC).

US Geological Survey, 1979, 'Slope map, Connecticut Valley Urban Area, central New England', *US Geol. Surv. Miscellaneous Investigations Series, Map I-1074-F*.

—— 1982, 'Goals and tasks of the landslide part of a ground-failure hazard reduction program', *US Geological Survey Circular, 880*.

USOROH, E. J., 1977, *Coastal Development in the Lagos Areas* (Ph.D. thesis, Univ. of Ibadan).

VALENTIN, H., 1954, 'Der Landverlust in Holderness, Ostengland, von 1852 bis 1952', *Die Erde, 6*, 296–315.

VALLEJO, G. DE, 1977, 'Engineering geology for urban planning and development with an example from Tenerife (Canary Islands)', *Bull. Int. Assoc. Eng. Geol. 15*, 37–43.

VANICEK, P. and NAGY, D., 1980, 'The map of contemporary vertical coastal movement in Canada', *Eos, 61*, 145–7.

VAN LOPIK, J. R. and KOLB, C. R., 1959, 'A technique for preparing desert terrain analogues', *US Army Corps of Engineers Waterways Experiment Station, Vicksburg, Miss. Tech. Rep., 3–506*.

VANN, G, 1965, 'Location and evaluation of sand and gravel deposits by geophysical methods and drilling', in Inst. Mining and Metallurgy (eds.), *Opencast Mining, Quarrying and Alluvial Mining, 3–19*.

VANONI, V. A. (ed.), 1975, *Sedimentation Engineering* (Am. Soc. Civ. Eng., New York).

VARNES, D. J., 1958, 'Landslide types and processes', in E. B. Eckel (ed.), 'Landslides and Engineering Practice', *Highway Research Board, Washington Spec. Rep. 29; NAS-NRC Publ. 544*, 20–47.

—— 1978, 'Slope movements and types and processes', in *Landslides: Analysis and Control, Spec. Rep. 176*, (Transportation Res., Board Nat. Acad. Sci., Washington), 11–33.

—— 1984, *Landslide Hazard Zonation: A review of principles and practice* (UNESCO, Paris).

VEATCH, J. D., 1933, 'Agricultural land classification and land types of Michigan', *Michigan Agric. Exp. Stat. Spec. Bull. 231*.

VEENENBOS, J., 1956, 'Methods of soil and land classification surveys', *Sols Afr. 4*, 122–35.

VERSTAPPEN, H. TH., 1966, 'The role of landform classification in integrated surveys', *Actes du le Symposium International de Photo-Interpretation* (Paris), VI.35–VI. 39.

—— 1970, 'Introduction to the ITC system of geomorphological survey', *Koninklijk Nederlands Aardrijkkundig Genootschap, Geografisch Niewe Reeks, 4.1*, 85–91.

—— 1977a, 'Geomorphology and terrain analysis of Saba and St Eustatius (Neth. Antilles)', *ITC Journal, 4*, 675–82.

—— 1977b, *Remote Sensing in Geomorphology* (Elsevier, Amsterdam).

—— 1983, *Applied Geomorphology: Geomorphological Surveys for Environmental Development* (Elsevier, Amsterdam).

—— and VAN ZUIDAM, R. A., 1968, 'ITC system of geomorphological survey', *ITC Textbook of Photo-Interpretation* (Delft), Ch. VII, 2.

Victoria Water Resources Council, 1978, *Flood Plain Management in Victoria* (VWRC, Melbourne).

VISHER, G. S., 1969, 'Grain size distribution and depositional processes', *J. Sed. Petrol. 39*, 1074–106.

VITA-FINZI, C., 1973, *Recent Earth History* (Macmillan, London).

—— 1986, *Recent Earth Movements* (Academic Press, London).

VOIGHT, B. (ed.), 1978, *Rockslides and Avalanches I: Natural Phenomena* (Elsevier, Amsterdam).

WALLACE, R. E., 1977, 'Profiles and ages of young fault scarps, north-central Nevada', *Bull. Geol. Soc. Am. 88*, 1267–81.

WALLING, D., 1980, 'Water in the catchment ecosystem', in A. M. Gower (ed.), *Water Quality in Catchment Ecosystems* (Wiley, Chichester), 1–47.

WALLING, D. E. (ed.), 1982, 'Recent developments in the explanation and prediction of erosion and sediment yield', *Int. Ass. Hydr. Sci. Publ. 137*.

—— and KLEO, A. H. A., 1979, 'Sediment yield of rivers in areas of low precipitation: a global view', in *The Hydrology of Areas of Low Precipitation (Int. Ass. Sci., Hydr. Publ. 128)*, 479–93.

WANG, Y., REN, M. E., and ZHU, D., 1986, 'Sediment supply to the continental shelf by the major rivers of China', *J. Geol. Soc. London, 143*, 935–44.

WARD, P. R. B., 1984, 'Measurement of sediment yields', in R. F. Hadley and D. E. Walling (eds.), *Erosion and Sediment Yield* (GeoBooks, Norwich), 37–71.

WARD, R., 1978, *Floods: A Geographical Perspective* (Macmillan, London).

WARREN, A. and MAIZELS, J. K., 1977, 'Ecological change and desertification', in UNCOD, *Desertification: Its Causes and Consequences* (Pergamon, Oxford) 169–260.

WASHBURN, A. L., 1967, 'Instrumental observations of

mass wasting in the Mesters Vig District, NE Greenland, *Medd. am Gronland, 166.*

WASHBURN, A. L. 1979, *Geocryology: A Survey of Periglacial Processes and Environments* (Arnold, London).

WATSON, A., 1985, 'The control of windblown sand and moving dunes: a review of the methods of sand control in deserts, with observations from Saudi Arabia', *Q. J. Eng. Geol. 18*, 237–52.

WATSON, R. A., 1969, 'Explanation and prediction in geology', *J. Geol. 77*, 488–94.

—— and WRIGHT, H. E., jun. 1969, 'The Saidmarreh Landslide, Iran', *Geol. Soc. Am. Spec. Paper, 123*, 115–39.

WATT COMMITTEE ON ENERGY, 1984, 'Acid Rain', *Report 14.*

WAYLAND, E. J., 1929, 'Rift valleys and Lake Victoria', *International Geological Congress*, Session IV, Vol. II, Sect. VI (Pretoria), 323–53.

WEBB, R. H. and WILSHIRE, H. G., 1983, *Environmental Effects of Off-Road Vehicles* (Springer-Verlag, New York).

WEBSTER, R. and BECKETT, P. H. T., 1970, 'Terrain classification and evaluation using air photography: A review of recent work at Oxford', *Photogrammetria, 26*, 51–7.

WEEKS, A. G., 1969, 'The stability of natural slopes in south-east England as affected by periglacial activity', *Q. J. Eng. Geol. 2*, 49–61.

WEINERT, H. H., 1965, 'Climatic factors affecting the weathering of igneous rocks", *Agric. Met. 2*, 27–42.

WEISE, O. R., 1983, *The Periglacial Geomorphology and Climate in Glacier Free Cold Regions* (Gebruder Borntraeger, Berlin). (In German.)

WELLMAN, H. W., 1972, 'Rate of horizontal fault displacement in New Zealand', *Nature, 9237*, 275–7.

Wessex Water Authority, 1979, *Land Drainage Survey Report, Water Act 1973, Sect. 24/5* (WWA, Bristol).

WESSON, R. L., HELLEY, E. J., LAJOIE, K. R., and WENTWORTH, C. M., 1975, 'Faults and future earthquakes', in R. D. Borcherdt (ed.), 'Studies for Seismic Zonation of the San Francisco Bay Region', *US Geol. Surv. Prof. Paper, 941–A*, A5–A30.

WEST, G. and DUMBLETON, M. J., 1972, 'Some observations on swallow holes and mines in the Chalk', *Q. J. Eng. Geol. 5*, 171–7.

WESTERN, S., 1978, *Soil Survey Contracts and Quality Control* (OUP, Oxford).

WHALLEY, W. B., 1976, *Properties of Materials and Geomorphological Explanation* (OUP, Oxford).

WHITE, G. F., 1964, 'Choice of adjustment to floods', *Univ. Chicago, Dept. of Geog. Research Paper, 93.*

—— 1974*a*, 'Flood damage prevention policies', *Nature and Resources, 11*, 2–7.

—— 1975*b*, *Flood Hazards in the United States: A Research Assessment* (Inst. of Behavioural Science, Univ. of Colorado).

WHITE, W. B., 1969, 'Conceptual models of carbonate aquifers', *Groundwater, 7*, 15–21.

WHITNEY, M. I., 1978, 'The role of vorticity in developing lineation in wind erosion', *Bull. Geol. Soc. Am. 89*, 1–18.

WILLIAMS, J. R., 1985, 'The physical components of the EPIC model', in S. A. El-Swaify and W. C. Moldenhauer (eds.), *Soil Erosion and Conservation* (Soil Cons. Soc. Am., Ankeny, Iowa), 272–84.

—— JONES, C. A. and DYKE, P. T., 1984, 'A modeling approach to determining the relationship between erosion and soil productivity', *Trans. Am. Soc. Agric. Engs. 27*, 129–44.

WILLIAMS, P. J., 1979, *Pipelines and Permafrost: Physical Geography and Development in the Circumpolar North* (Longman, London).

WILLIAMS, P. W., 1969, 'The geomorphic effects of groundwater', in R. J. Chorley (ed.), *Water, Earth and Man* (Methuen, London), 269–94.

—— 1978, 'Karst research in China', *Trans. Brit. Cave Res. Ass. 5*, 29–46.

WILLIAMSON, A. N., 1974, 'Mississippi river flood maps from ERTS-1 digital data', *Water Resources Bull. 10*, 1050–9.

WILKS, D. H. and PRICE, W. A., 1975, 'Trends in the application of research to solve coastal engineering problems', in J. Hails and A. Carr (eds.), *Nearshore Sediment Dynamics and Sedimentation* (Wiley, London), 111–21.

WILSHIRE, H. G., 1980, 'Human causes of accelerated wind erosion in California's deserts', in D. R. Coates and J. D. Vitek (eds.), *Thresholds in Geomorphology* (Allen & Unwin, London), 415–33.

WILSON, G. and HENRY, G., 1942, 'The settlement of London due to underdrainage of the London Clay', *J. Inst. Civ. Eng. 19*, 100–27.

WILSON, L., 1973, 'Variations in mean annual sediment yield as a function of mean annual precipitation', *Am. J. Sci. 273*, 335–49.

WILSON, M. E. and WOOD, S. H., 1980, 'Tectonic tilt rates derived from lake-level measurements, Salton Sea, California', *Science, 207*, 183–6.

WILSON, S. J. and COOKE, R. U., 1980, 'Wind erosion', in M. J. Kirkby and R. P. C. Morgan (eds.), *Soil Erosion* (Wiley, Chichester), 217–51.

WIMAN, S., 1963, 'A preliminary study of experimental frost weathering', *Geog. Ann. 45*, 113–21.

WINKLER, E. M., 1966, 'Corrosion rates of carbonate rocks for construction', *Eng. Geol. 4*, 52–8.

—— 1970, 'The importance of air pollution in the corrosion of stone and metals', *Eng. Geol. 4*, 327–34.

—— 1973, *Stone: Properties, Durability in Man's Environment* (Springer-Verlag, New York).

—— and WILHELM, E. J., 1970, 'Salt burst by hydration pressures in architectural stone in urban atmosphere', *Geol. Soc. Am. Bull., 81*, 567–72.

WISCHMEIER, W. H., 1977, 'Use and misuse of the universal soil loss equation', in *Soil Erosion: Prediction and Control* (Soil Cons. Soc. Am., Ankeny, Iowa), 371–8.

WISCHMEIER, W. H. and SMITH, D. D., 1978, 'Predicting rainfall erosion losses—a guide to conservation planning', *US Dept. Agriculture Handbook, 537*.

WOHLRAB, B., 1969, 'Effects of mining subsidences on the ground water and remedial measures', *Pub. Int. Ass. Sci. Hydr. 89*, 502–12.

WOLMAN, M. G., 1967, 'A cycle of erosion and sedimentation in urban river channels', *Geog. Annaler, 49A*, 385–96.

—— 1971, 'Evaluating alternative techniques of flood-plain mapping', *Water Resources Research, 7*, 1383–92.

—— 1977, 'Changing needs and opportunities in the sediment field', *Water Resources Research, 13*, 50–4.

—— and GERSON, R., 1978, 'Relative scales of time and effectiveness of climate in watershed geomorphology', *Earth Surface Processes, 3*, 189–208.

WOODRUFF, N. P., LYLES, L., SIDDOWAY, F. H., and FRYNEAR, D. W., 1972, 'How to control wind erosion', *US Dept. Agriculture Information Bull. 354*.

—— and SIDDOWAY, F. H., 1965, 'A wind erosion equation', *Proc. Soil Sci. Soc. Am. 29*, 602–8.

WOOLDRIDGE, S. W., 1932, 'The cycle of erosion and the representation of relief', *Scot. Geog. Mag. 48*, 30–6.

—— and LINTON, D. L., 1955, *Structure, Surface and Drainage in South-east England* (Philip, London).

WORSTER, D., 1979, *Dust Bowl* (OUP, New York).

WRIGHT, L. D., 1978, 'River deltas', in R. A. David (ed.), *Coastal Sedimentary Environments* (Springer-Verlag, New York), 5–68.

WYSS, M. (ed.), 1975, *Earthquake Prediction and Rock Mechanics* (Birkhauser, Basel).

YAMAMOTO, S., 1977, 'Recent trend of land subsidence in Japan', *Science Reports of the Tokyo Kyioku Daigaku, C, 13*, 1–7.

YATES, A. B. and STANLEY, D. R., 1966, 'Domestic water supply and sewage disposal in the Canadian north', *Proc. 1st Permafrost Int. Conf.* (Nat. Acad. Sciences—Nat. Research Council of Canada), *1287*, 413–19.

YOCHELSON, E. L., 1980, 'The scientific ideas of G. K. Gilbert', *US Geol. Surv. Spec. Paper, 183*.

YOUNG, A., 1972, *Slopes* (Oliver & Boyd, Edinburgh).

—— 1973, 'Rural land evaluation', in J. A. Dawson and J. C. Doornkamp (eds.), *Evaluating the Human Environment* (Arnold, London), 5–33.

ZACHAR, D., 1982, *Soil Erosion* (Elsevier, Amsterdam).

ZARUBA, Q. and MENCL, V., 1969, *Landslides and Their Control* (Elsevier, Amsterdam).

ZENKOVICH, V. P., 1967, *Processes of Coastal Development* (Oliver & Boyd, Edinburgh).

ZHUKOVSKIY, YU. S., 1980, 'On the interaction between tectonic presence and river network of left bank inflows of the Lower Zena Basin', *Izvestiya Vsesoyuznogo Geograficheskogo Obshchestva, 112*, 345–7. (In Russian.)

ZINGG, A. W., 1940, 'Degree and length of slope as it effects soil loss in runoff', *Agric. Eng., 21*, 59–64.

ZUIDAM, R. VAN, 1985–6, *Aerial Photo Interpretation in Terrain Analysis and Geomorphological Mapping* (Smits, The Hague).

INDEX